Neurodegenerative Diseases: From Molecular Basis to Therapy

Neurodegenerative Diseases: From Molecular Basis to Therapy

Editors

Marcello Ciaccio
Luisa Agnello

MDPI • Basel • Beijing • Wuhan • Barcelona • Belgrade • Manchester • Tokyo • Cluj • Tianjin

Editors
Marcello Ciaccio
Department of Biomedicine,
Neurosciences and Advanced
Diagnostics
University of Palermo
Palermo
Italy

Luisa Agnello
Department of Biomedicine,
Neurosciences and Advanced
Diagnostics
University of Palermo
Palermo
Italy

Editorial Office
MDPI
St. Alban-Anlage 66
4052 Basel, Switzerland

This is a reprint of articles from the Special Issue published online in the open access journal *International Journal of Molecular Sciences* (ISSN 1422-0067) (available at: www.mdpi.com/journal/ijms/special_issues/Neurodegenerative_Therapy).

For citation purposes, cite each article independently as indicated on the article page online and as indicated below:

LastName, A.A.; LastName, B.B.; LastName, C.C. Article Title. *Journal Name* **Year**, *Volume Number*, Page Range.

ISBN 978-3-0365-7075-4 (Hbk)
ISBN 978-3-0365-7074-7 (PDF)

Contents

About the Editors

Marcello Ciaccio

Marcello Ciaccio is Full Professor of Clinical Biochemistry and Laboratory Medicine and President of the School of Medicine at the University of Palermo. He is Director of the Institute of Clinical Biochemistry, Clinical Molecular Medicine and Laboratory Medicine, Department of Biomedicine, Neurosciences and Advanced Diagnostics at the University of Palermo. He is also Director of the Department of Laboratory Medicine - A.O.U.P. Palermo "P. Giaccone". He is member of the Commission for National Scientific Qualification (ASN) 2021-2023 for the Scientific Sector 05 / E3-Clinical Biochemistry and Clinical Molecular Biology. He is past and elective President of the Italian Society of Clinical Biochemistry and Clinical Molecular Biology (SIBioC). He is part of the Editorial Board of several peer-reviewed journals and he serves as a Reviewer for many prestigious journals. He is author of more than 500 scientific publications on national and international Journals.

Luisa Agnello

Luisa Agnello is Researcher of Clinical Biochemistry and Laboratory Medicine at the Institute of Clinical Biochemistry, Clinical Molecular Medicine and Laboratory Medicine, Department of Biomedicine, Neurosciences and Advanced Diagnostics at the University of Palermo. She was speaker at several national and international Conference. She coordinated several research projects and is Author of more than 100 articles on peer-reviewed journals.

Preface to "Neurodegenerative Diseases: From Molecular Basis to Therapy"

Neurodegenerative diseases (NDs) are a heterogeneous group of complex diseases characterized by neuronal loss and progressive degeneration of different areas of the nervous system. NDs represent a significant health problem worldwide, with an increasing incidence rate. Although the exact pathogenesis of NDs remains unclear, a complex interaction among genetic, epigenetic, and environmental factors has been proposed. To date, no effective therapeutics have been developed to slow, halt, or prevent any NDs. Thus, information on the molecular mechanisms underlying the pathogenesis of NDs is strongly sought after. This Special Issue collected papers discussing advances in the field of NDs.

Marcello Ciaccio and Luisa Agnello

Editors

International Journal of *Molecular Sciences*

Editorial

Neurodegenerative Diseases: From Molecular Basis to Therapy

Luisa Agnello [1] and Marcello Ciaccio [1,2,*]

1 Department of Biomedicine, Neurosciences and Advanced Diagnostics, Institute of Clinical Biochemistry, Clinical Molecular Medicine and Clinical Laboratory Medicine, University of Palermo, 90127 Palermo, Italy
2 Department of Laboratory Medicine, University Hospital "P. Giaccone", 90127 Palermo, Italy
* Correspondence: marcello.ciaccio@unipa.it

Neurodegenerative diseases (NDs) are a heterogeneous group of complex diseases characterized by neuronal loss and progressive degeneration of different areas of the nervous system [1,2].

NDs represent a significant health problem worldwide, with an increasing incidence rate. Although the exact pathogenesis of NDs remains unclear, a complex interaction among genetic, epigenetic, and environmental factors has been proposed. To date, no effective therapeutics have been developed to slow, halt, or prevent any NDs. Thus, information on the molecular mechanisms underlying the pathogenesis of NDs is strongly sought after.

This Special Issue titled "Neurodegenerative Diseases: From Molecular Basis to Therapy" collected papers discussing advances in the field of NDs, including multiple sclerosis (MS) and Parkinson's disease (PD), focusing on the underlying pathobiological mechanisms.

PD represents the second most common neurodegenerative disorder worldwide after Alzheimer's Disease. In recent decades, significant advances have been achieved in the field of PD research. In an interesting review, Fonseca Cabral et al. described the emerging role of genetic variants, epigenomic modifications, and microbiome in the onset and progression of PD [3]. The exploration of these molecular aspects in an integrative and multidisciplinary manner is pivotal to the rise of personalized medicine.

In an experimental PD model, Fieblinger et al. showed that the inhibition of caspase-3 prevented the loss of dendritic spines and long-term depression in spiny projection neurons without interfering with the ongoing degeneration of nigrostriatal dopamine neurons [4]. Thus, these findings support a possible role of caspase-3 in PD pathogenesis.

Recently, Rab1 emerged as a promising candidate for therapeutic strategies [5]. It belongs to the family of GTPases, which are involved in regulating membrane traffic. Interestingly, Rab1 has been linked to alfa-synuclein toxicity in PD. In a cellular model of PD and human samples, it was overexpressed in surviving nigral neurons, highlighting its protective role in PD. Rab1 expression/function could be modulated by small molecules interacting with selective regions of this protein. Thus, there is an ongoing search for therapeutic approaches modifying Rab1-dependent α-syn toxicity.

Another emerging therapeutic target is the dopamine D3 receptor (D3R). In PD experimental models, it has been shown that D3R-signalling promotes disease progression by favouring neuroinflammation and by stimulating the pathogenic CD4+ T cell response. Moreover, microglial activation is suppressed in D3R-deficient mice. Noteworthily, molecules able to block D3R attenuated the cerebral inflammation and, consequently, slowed the progression of PD. According to such evidence, Broome et al. evaluated the effect of the anxiolytic drug buspirone, a potent D3R antagonist in an experimental PD model [6]. The authors showed that buspirone had a neuroprotective effect and improved mitochondrial function and antioxidant activities. These findings encourage further investigation of buspirone in PD.

Research has also focused on natural compounds as an alternative treatment for PD. Recently, chicoric acid (CA), a polyphenolic acid extracted from chicory and echinacea, which has antiviral, antioxidative, and anti-inflammatory activities, has been tested as a

Citation: Agnello, L.; Ciaccio, M. Neurodegenerative Diseases: From Molecular Basis to Therapy. *Int. J. Mol. Sci.* **2022**, *23*, 12854. https://doi.org/10.3390/ijms232112854

Received: 11 October 2022
Accepted: 22 October 2022
Published: 25 October 2022

candidate in experimental studies. Wang et al. showed that CA prevented dopaminergic neuronal lesions, motor deficits, and glial activation in PD mice, along with increments in striatal brain-derived neurotrophic factor, dopamine, and 5-hydroxyindoleacetic acid. Furthermore, it regulated immunological response by reducing the levels of proinflammatory cytokines [7]. The authors hypothesised that the neuroprotective effect could be related to the manipulation of CA on the brain–spleen and brain–gut axes in PD. Further studies are warranted to confirm the beneficial effect of CA and to elucidate the underlying mechanisms.

Hepp-Rehfeldt et al. investigated the potential beneficial effect of luteolin, a flavonoid present in many fruits and vegetables, on the central nervous system [8]. Specifically, the authors evaluated the antioxidant and anti-inflammatory activities of the glycosylated form of luteolin, known as luteolin-7-O-glucoside (Lut7), in an in vitro human neurodegenerative model (SH-SY5Y cells induced with 6-OHDA) in both undifferentiated and differentiated cells. They found that Lut7 has several neuroprotective effects, especially, a high antioxidant capacity.

In recent decades, the neurotoxic effect of mercury has emerged in several studies, but little evidence is available on the negative effect of inorganic mercury (IHg) species, which have been detected in both contaminated food and cells of central nervous system origin. IHg represents one of the main targets of mercury associated with the neurological symptomatology of mercurial poisoning. Bittencourt et al. explored the effects of long-term exposure to IHg in adult rats' cerebellum by evaluating the proteomic profile associated with the motor dysfunction outcome, including molecular, biochemical, and morphological approaches [9]. The global proteomic profile revealed several molecules involved in different biological processes, such as synaptic signalling, energy metabolism, and nervous system development, with all being associated with increased cytotoxicity and apoptosis and leading to poor motor coordination and balance. Thus, these findings reveal new molecular mechanisms involved in mercury toxicity.

Another important research area includes multiple sclerosis. The aetiology of MS is still not fully understood, although the involvement of several mechanisms has been proposed. Among these proposed mechanisms, excitotoxicity has emerged as the mechanism behind the excess glutamate, the main excitatory neurotransmitter. Glutamate-induced neuronal degeneration is mainly mediated by N-methyl-D-aspartate (NMDA) receptors, and it is characterised by the formation of reactive oxygen species (ROS) and the activation of both caspase-dependent and caspase-independent cell death. Dąbrowska-Bouta et al. evaluated the effect of memantine, the uncompetitive NMDA receptor antagonist, on the modulation of neurological deficits and oxidative stress in experimental models of MS [10]. The authors showed that the pharmacological inhibition of ionotropic NMDA glutamate receptors has several beneficial effects, including an improvement in the physical activity of rats, a reduction in neurological deficits such as paralysis of the tail and hind limbs, and the modulation of oxidative stress. These findings provide evidence of a new possible therapeutic strategy for MS treatment.

In an exciting review, Van Schaik et al. described the multifaceted roles of fibronectin in MS pathogenesis and discuss promising therapeutic targets and agents to overcome fibronectin-mediated inhibition of remyelination [11].

Finally, in recent decades, miRNA emerged as promising tools in several clinical conditions, including MS. Pietrasik et al. described the current state of knowledge on miRNA panel expression in the different forms of MS [12].

Funding: This research received no external funding.

Conflicts of Interest: The authors declare no conflict of interest.

References

1. Agnello, L.; Gambino, C.M.; Sasso, B.L.; Bivona, G.; Milano, S.; Ciaccio, A.M.; Piccoli, T.; La Bella, V.; Ciaccio, M. Neurogranin as a Novel Biomarker in Alzheimer's Disease. *Lab. Med.* **2021**, *52*, 188–196. [CrossRef] [PubMed]
2. Cheng, Y.-J.; Lin, C.-H.; Lane, H.-Y. From Menopause to Neurodegeneration—Molecular Basis and Potential Therapy. *Int. J. Mol. Sci.* **2021**, *22*, 8654. [CrossRef] [PubMed]
3. Cabral, G.F.; Schaan, A.P.; Cavalcante, G.C.; Sena-Dos-Santos, C.; de Souza, T.P.; Port'S, N.M.S.; Pinheiro, J.A.D.S.; Ribeiro-Dos-Santos, Â.; Vidal, A.F. Nuclear and Mitochondrial Genome, Epigenome and Gut Microbiome: Emerging Molecular Biomarkers for Parkinson's Disease. *Int. J. Mol. Sci.* **2021**, *22*, 9839. [CrossRef] [PubMed]
4. Fieblinger, T.; Li, C.; Espa, E.; Cenci, M.A. Non-Apoptotic Caspase-3 Activation Mediates Early Synaptic Dysfunction of Indirect Pathway Neurons in the Parkinsonian Striatum. *Int. J. Mol. Sci.* **2022**, *23*, 5470. [CrossRef] [PubMed]
5. Martínez-Menárguez, J.; Martínez-Alonso, E.; Cara-Esteban, M.; Tomás, M. Focus on the Small GTPase Rab1: A Key Player in the Pathogenesis of Parkinson's Disease. *Int. J. Mol. Sci.* **2021**, *22*, 12087. [CrossRef] [PubMed]
6. Broome, S.T.; Castorina, A. The Anxiolytic Drug Buspirone Prevents Rotenone-Induced Toxicity in a Mouse Model of Parkinson's Disease. *Int. J. Mol. Sci.* **2022**, *23*, 1845. [CrossRef] [PubMed]
7. Wang, N.; Li, R.; Feng, B.; Cheng, Y.; Guo, Y.; Qian, H. Chicoric Acid Prevents Neuroinflammation and Neurodegeneration in a Mouse Parkinson's Disease Model: Immune Response and Transcriptome Profile of the Spleen and Colon. *Int. J. Mol. Sci.* **2022**, *23*, 2031. [CrossRef] [PubMed]
8. Rehfeldt, S.C.H.; Silva, J.; Alves, C.; Pinteus, S.; Pedrosa, R.; Laufer, S.; Goettert, M.I. Neuroprotective Effect of Luteolin-7-O-Glucoside against 6-OHDA-Induced Damage in Undifferentiated and RA-Differentiated SH-SY5Y Cells. *Int. J. Mol. Sci.* **2022**, *23*, 2914. [CrossRef] [PubMed]
9. Bittencourt, L.O.; Chemelo, V.S.; Aragão, W.A.B.; Puty, B.; Dionizio, A.; Teixeira, F.B.; Fernandes, M.S.; Silva, M.C.F.; Fernandes, L.M.P.; de Oliveira, E.H.C.; et al. From Molecules to Behavior in Long-Term Inorganic Mercury Intoxication: Unraveling Proteomic Features in Cerebellar Neurodegeneration of Rats. *Int. J. Mol. Sci.* **2021**, *23*, 111. [CrossRef] [PubMed]
10. Dąbrowska-Bouta, B.; Strużyńska, L.; Sidoryk-Węgrzynowicz, M.; Sulkowski, G. Memantine Modulates Oxidative Stress in the Rat Brain following Experimental Autoimmune Encephalomyelitis. *Int. J. Mol. Sci.* **2021**, *22*, 11330. [CrossRef] [PubMed]
11. van Schaik, P.E.M.; Zuhorn, I.S.; Baron, W. Targeting Fibronectin to Overcome Remyelination Failure in Multiple Sclerosis: The Need for Brain- and Lesion-Targeted Drug Delivery. *Int. J. Mol. Sci.* **2022**, *23*, 8418. [CrossRef] [PubMed]
12. Pietrasik, S.; Dziedzic, A.; Miller, E.; Starosta, M.; Saluk-Bijak, J. Circulating miRNAs as Potential Biomarkers Distinguishing Relapsing–Remitting from Secondary Progressive Multiple Sclerosis. A Review. *Int. J. Mol. Sci.* **2021**, *22*, 11887. [CrossRef] [PubMed]

International Journal of *Molecular Sciences*

MDPI

Review

Targeting Fibronectin to Overcome Remyelination Failure in Multiple Sclerosis: The Need for Brain- and Lesion-Targeted Drug Delivery

Pauline E. M. van Schaik [1], Inge S. Zuhorn [2,*] and Wia Baron [1,*]

[1] Section Molecular Neurobiology, Department of Biomedical Sciences of Cells & Systems, University of Groningen, University Medical Center Groningen, Antonius Deusinglaan 1, 9713 AV Groningen, The Netherlands; p.e.m.van.schaik@umcg.nl

[2] Department of Biomedical Engineering, University of Groningen, University Medical Center Groningen, Antonius Deusinglaan 1, 9713 AV Groningen, The Netherlands

* Correspondence: i.zuhorn@umcg.nl (I.S.Z.); w.baron@umcg.nl (W.B.); Tel.: +31-50-3616178 (I.S.Z.); +31-503611652 (W.B.); Fax: +31-503616190 (W.B.)

Abstract: Multiple sclerosis (MS) is a neuroinflammatory and neurodegenerative disease with unknown etiology that can be characterized by the presence of demyelinated lesions. Prevailing treatment protocols in MS rely on the modulation of the inflammatory process but do not impact disease progression. Remyelination is an essential factor for both axonal survival and functional neurological recovery but is often insufficient. The extracellular matrix protein fibronectin contributes to the inhibitory environment created in MS lesions and likely plays a causative role in remyelination failure. The presence of the blood–brain barrier (BBB) hinders the delivery of remyelination therapeutics to lesions. Therefore, therapeutic interventions to normalize the pathogenic MS lesion environment need to be able to cross the BBB. In this review, we outline the multifaceted roles of fibronectin in MS pathogenesis and discuss promising therapeutic targets and agents to overcome fibronectin-mediated inhibition of remyelination. In addition, to pave the way for clinical use, we reflect on opportunities to deliver MS therapeutics to lesions through the utilization of nanomedicine and discuss strategies to deliver fibronectin-directed therapeutics across the BBB. The use of well-designed nanocarriers with appropriate surface functionalization to cross the BBB and target the lesion sites is recommended.

Keywords: blood–brain barrier; extracellular matrix; fibronectin; liposomes; multiple sclerosis; nanomedicine; oligodendrocytes; PLGA; remyelination; therapeutic targets

Citation: van Schaik, P.E.M.; Zuhorn, I.S.; Baron, W. Targeting Fibronectin to Overcome Remyelination Failure in Multiple Sclerosis: The Need for Brain- and Lesion-Targeted Drug Delivery. *Int. J. Mol. Sci.* **2022**, *23*, 8418. https://doi.org/10.3390/ijms23158418

Academic Editors: Marcello Ciaccio and Luisa Agnello

Received: 30 April 2022
Accepted: 23 July 2022
Published: 29 July 2022

1. Introduction

Multiple sclerosis (MS) is a neuroinflammatory and neurodegenerative disease that is characterized by the presence of demyelinated lesions. MS is generally diagnosed between the third and fifth decade of life, with women being two to three times more likely to be affected than men [1]. Initially, MS can present itself as a clinically isolated syndrome (CIS) [2] when patients typically face symptoms indicatory of a demyelinating insult to the optic nerve, spinal cord, brainstem, or cerebral hemisphere [1]. A second neurological event converts the CIS to clinically definitive MS. Initially, patients experience spontaneous recovery due to endogenous remyelination of the lesioned area. Most patients with this relapsing–remitting (RRMS) disease course will enter a secondary progressive disease (SPMS) phase in which their disability progressively worsens. A small proportion of patients progressively deteriorate without showing relapses and remissions at first and are classified as primary progressive MS (PPMS) patients [3,4].

The exact cause of MS is yet unknown, though several risk modifiers were identified. Genetic association studies revealed several MS risk gene variants, many of which are involved with immune functioning [5]. Environmental factors include viral infections [6,7], vitamin D deficiency, cigarette smoking, and diet [8]. Current therapies are either symptomatic in nature or rely on immune-modulating strategies, thereby delaying the time and severity of new lesion formation. However, these therapies do not prevent disease progression and often fail in progressive MS patients. Progressive axonal loss is key to the continuous and irreversible neurological decline in progressive MS [9]. In addition to ensuring saltatory conduction, oligodendrocytes secrete via myelin metabolic and trophic factors that maintain the integrity and survival of axons [10,11]. Therefore, next to primary axon damage, a major cause of axonal loss in chronic stages of MS is secondary axon degeneration because of remyelination failure [12]. Hence, to halt disease progression, the development of treatments that preserve axons, i.e., via the promotion of remyelination, is an essential therapeutic goal.

Oligodendrocytes (OLGs) are responsible for myelinating neuronal axons in the CNS and mature from oligodendrocyte progenitor cells (OPCs). Remyelination in MS lesions is often insufficient despite the presence of OPCs and/or surviving mature OLGs in most lesions [13,14]. Therefore, therapeutic interventions must overcome the pathogenic MS lesion environment. Perturbed remodeling of the extracellular matrix (ECM) in MS lesions likely plays a causative role in remyelination failure [15,16]. In this review, we focus on the ECM protein fibronectin (Fn) that, in its aggregated form, persists in MS lesions and impairs OPC differentiation and remyelination [17,18]. We outline the beneficial roles of Fn in the neurovascular unit and the detrimental roles of Fn in MS pathology and discuss therapeutic strategies and agents to prevent Fn aggregation and/or to overcome Fn-mediated inhibition of remyelination. In addition, to translate these therapeutic strategies for myelin regeneration to the clinic and consider the beneficial role of Fn in the neurovascular unit, we discuss the need and strategies for the brain- and lesion-targeted delivery of nanomedicine.

2. Multiple Sclerosis: An Unmet Need for a Remyelination-Based Therapy to Stop Disease Progression

MS has long been perceived as an autoimmune disease mediated by autoreactive T and B cells, which is an 'outside-in hypothesis' that has been substantiated by successful disease-modifying immunomodulatory therapies [19]. Demyelinating plaques arise due to an autoimmune response against myelin, mediated by CD8+ and CD4+ T cells, called autoimmune encephalomyelitis [20]. Oligoclonal bands detected in the cerebrospinal fluid (CSF) simultaneously indicate the presence of immunoglobulin-producing B cells [21]. Especially in the early stages of RRMS, immunomodulatory therapies show high efficacy, indicating that inflammation appears to play a crucial role in disease development [3,22]. Alternatively, intrinsic neuronal or glial disturbances may initiate a cascade of inflammation, which is coined the 'inside-out hypothesis' [3,9]. This is corroborated by the finding that brain atrophy manifests early and is a more important determinant of disease progression than lesion load. Additionally, brain atrophy in PPMS may exceed that observed in RRMS [23]. Furthermore, due to the low efficacy of immunomodulatory therapies in the progressive phase of MS, alternative disease mechanisms appear to play a role in disease progression.

While remyelination is known to occur in the early stages of the disease, this regenerative capacity of the CNS diminishes with age, ultimately leading to the accumulation of permanently demyelinated lesions and aggravated clinical disability [24]. In RRMS, new lesions form during relapses, which in the white matter are characterized by inflammation and blood–brain barrier (BBB) damage. MS lesions display profound heterogeneity, leading to the development of several lesion classification systems, each of which focuses on different aspects of the lesion stage and activity. Kuhlmann et al., proposed an updated classification system of demyelinated white matter lesions based on the presence and distribution of macrophages/microglia, resulting in the classification of active, mixed active/inactive, or inactive lesions [25]. Typical actively demyelinating lesions as observed in RRMS are not commonly seen in PPMS, though it was recently shown that a large majority of PPMS patients had mixed active/inactive lesions [26]. This indicates that ongoing demyelination and inflammation may be ubiquitously present. In addition, chronic diffuse inflammation of (normal-appearing) white matter and cortical demyelination are frequently observed in progressive MS [27–29]. Cortical lesions are classified based on location and are less associated with the infiltration of immune cells, as is generally seen in actively demyelinating lesions of the white matter [30]. The amount of intracortical and leukocortical lesions shows a strong correlation with clinical impairment [26]. For example, a substantial proportion of (progressive) MS patients develops a form of cognitive impairment, which strongly correlates with cortical demyelination [31,32]. Particularly, reductions in information processing speed, working memory, and executive functioning are reported [33,34], which reflect deficits in frontal lobe functioning [35]. Furthermore, a direct link between white matter lesion volume and cognitive impairment in early-stage MS was recently reported [36]. Due to the functional relationship between demyelinating lesions and clinical symptoms, remyelination can ameliorate clinical symptoms, including cognitive and physical functioning [37]. Indeed, post-mortem inspections of brain lesions demonstrated that MS patients with a higher load of remyelinated lesions had lower clinical disability scores [26,38].

Remyelination of lesioned areas occurs, causing the appearance of so-called 'shadow plaques', or partly remyelinated areas at the border of lesions [25]. Remyelinating areas are more pronounced in active than in inactive lesions and are hardly present in mixed active/inactive lesions [14]. Nevertheless, only a small proportion of lesions fully remyelinate, which is a process that negatively correlates with the amount of TMEM119+ and iNOS+ (indicative of an inflammatory phenotype) myeloid cells present in the lesion [14] and diminishes with age and disease chronicity [39]. Thus, demyelination and subsequent remyelination are two antagonistic processes in which inflammatory and neurodegenerative processes concurrently contribute to the development and maintenance of lesions. Next to mitigating the excessive inflammatory response promoting demyelination, strategies aimed at enabling remyelination may contribute to alleviating the disease burden [3,40].

3. Remyelination Failure in MS: Perturbed ECM Remodeling in White Matter Lesions

Recent evidence has demonstrated that pre-existing surviving mature OLGs retain their myelinating capacity after a demyelinating insult [41]. Moreover, remyelination in shadow plaques can mainly be attributed to pre-existing mature OLGs rather than newly differentiating OPCs [42]. Nevertheless, OLGs surviving a demyelinating insult in zebrafish were shown to make few and not-well-targeted myelin sheaths, which is a finding that corresponds to observations in remyelinated MS lesions [43]. In contrast, OPC-based remyelination, i.e., the generation of newly formed OLGs, is more efficient, as evident in experimental models of demyelination. OPCs are present in the adult brain throughout the lifespan and preserve the capacity to differentiate into mature OLGs [13]. The absence of pro-oligodendrogenic factors and the presence of anti-oligodendrogenic factors in MS lesions were postulated as complicit elements in preventing OPC maturation [44]. Indeed, most chronically demyelinated lesions contain OPCs that apparently fail to differentiate and mature [45–48], while in a subset of mixed active/inactive lesions, few OPCs are present, which is likely due to cellular expression of chemorepellent factors [47]. Both cell-intrinsic factors, such as the maturation and differentiation stages of OLG lineage cells [49], and cell-extrinsic factors, such as the composition of the ECM, determine the inhibitory milieu at the injury site [50]. Moreover, other inhibitory factors present in the lesioned area, such as infiltrating lymphocytes and inflammatory mediators, oxidative stress, and irreversible damage to OPCs and OLGs, may all be conducive to the non-permissive milieu [51,52].

General aging has been hypothesized to play an important role in the reduced regenerative capacity of the brain [39,53,54], as age is also the best predictor of disease progression [55,56]. In rats, remyelination following toxin-induced demyelination slowed down with age [57] due to an impairment in OPC recruitment and differentiation [53]. Inefficient epigenetic downregulation of OLG genes that inhibit OPC maturation may underlie this [58], as well as a failure of OPCs to respond to pro-differentiation signals with aging [54,59]. Whether the intrinsic aging of OPCs or the aging environment induces these deficits has been elusive, though recent data hint at a significant role of environmental cues, namely, a gradual stiffening of the extracellular microenvironment and general brain tissue stiffening may contribute to OPC malfunctioning over time [60]. Thus, OPCs increasingly lose their capacity to proliferate, migrate, and differentiate with age, likely due to a non-permissive stiffened aging environment. In situations where demyelination is exacerbated, as in MS, this poses an increased risk as OLGs surviving demyelinating insults and OPCs also exhibit diminished myelinating capacities and face additional disease-specific alterations in stiffness [60].

The composition of the ECM is a major determinant of tissue stiffness. The ECM restrains the movement of cells by forming a physical scaffold and is simultaneously important for maintaining healthy brain homeostasis by directing cell differentiation, growth, and migration [61–63]. It is composed of an interactive network of fibrous-forming proteins, such as collagens, elastin, Fn, laminins, glycoproteins, proteoglycans, and glycosaminoglycans. Cells receive and integrate signals from the ECM via specified surface receptors with an affinity for one of the ECM constituents [63]. Concurrently, signaling molecules, such as growth factors, cytokines, and chemokines, can be stored temporarily within the matrix and released when needed, meaning the cells and the extracellular milieu form a bi-directional synergy [63,64]. In turn, cytokine release (i.e., TFG-β, TNF-α, and IFN-γ) during inflammation can affect ECM synthesis and turnover, thereby causing changes in the ECM composition [16,65]. In particular, due to their repeating glycosaminoglycan (GAG) chains, proteoglycans have the ability to bind cytokines and growth factors [66]. For example, heparan sulfate chains are known to bind basic fibroblast growth factor (bFGF) [67]. Furthermore, fibronectin domains were found to bind growth factors, in particular vascular endothelial growth factor (VEGF) [68] and hepatic growth factor [69]. Subsequent proteolytic cleavage of ECM proteins can result in a directed release of these factors in the extracellular milieu, thereby contributing to local cell differentiation and proliferation [66].

The interaction between ECM and OLG lineage cells decides whether remyelination can occur based on ECM rigidity and activation of intracellular signaling pathways. A softer matrix inhibits cell differentiation and myelination, while gliosis is stimulated by a stiff matrix [70,71]. OPCs are mechanosensitive [72] and in vitro data suggest that a relatively stiff matrix favors OPC proliferation and differentiation, while a soft matrix is beneficial for myelination [73]. Distinct ECM proteins differentially affect OPC differentiation. While Fn induces OPC proliferation and impedes OPC differentiation [17,74], laminin stimulates the expression of mature OLG markers and myelin components, including myelin basic protein (MBP) and proteolipid protein (PLP) [75–77]. Hence, changes in the ECM protein composition modulate reparative processes by allowing for altered cell behavior, be it beneficial or detrimental [15]. In acute demyelinating conditions, a transient change in ECM components occurs, which consists of increased tissue stiffness that reverses upon remyelination. In chronic demyelinating conditions, these changes are not reversed and are accompanied by enhanced ECM deposition [78]. This demonstrates that an adequate response to acute changes in the lesioned area is stalled in areas of chronic demyelination. However, the transient deposition of ECM proteins is a natural response to CNS injury, particularly the deposition of chondroitin sulfate proteoglycans (CSPGs) and Fn at the lesion [50,79]. Astrocytes are the main source of Fn and CSPGs in the CNS [80], which form a glial scar at the lesioned area through a process of reactive astrogliosis [81]. CSPGs and Fn deposited in demyelinated lesions aid OPC recruitment but impair OPC differentiation and myelination, indicating that timely Fn and CSPG removal is required for efficient remyelination to occur [17,82–84].

Hence, the composition of the ECM and perturbed remodeling during inflammatory and demyelinating insults typically hampers the establishment of a remyelination-permissive milieu. Theoretically, this implies that ECM-mediated inhibition of OPC differentiation in MS lesions may be therapeutically targeted by degradation enzymes that remove the inhibitory ECM proteins. Indeed, enzymatic digestion of CSPGs with simultaneous supplementation of growth factors aids OPC differentiation and migration after injury [85,86]. However, as CSPGs are also components of the interstitial ECM in the healthy adult brain and Fn is also a component of the BBB BM, targeted delivery of ECM-degradation enzymes to MS lesions is a prerequisite to avoid unwanted side effects. Alternatively, as shown for CSPGs, blocking the transient deposition of ECM proteins upon injury and interfering with ECM-mediated signaling appear to be feasible approaches to prevent or overcome impaired OPC differentiation [87–91]. Nevertheless, the complete absence of an ECM-remodeling response does not necessarily benefit healthy regeneration. Before discussing brain- and lesion-targeted delivery approaches for remyelination therapeutic agents, we first present an overview of the beneficial roles of Fn at the BBB and detrimental roles of Fn in MS pathology and provide potential therapeutic agents to overcome Fn-mediated remyelination failure. For CSPG-targeting approaches to promote remyelination in MS, we refer to excellent recent reviews [92,93].

4. Fibronectin: Multifaceted Roles in the CNS and in the Pathogenesis of MS

The Fn gene (*FN1*) contains three repeating domains (I, II, and III), each of which can be found in other molecules, indicating that Fn evolved through exon shuffling [94–96]. Though Fn is transcribed from a single gene, 20 different human splice variants are known, suggesting that the protein has a multitude of functions depending on the splicing of the pre-mRNA [95]. Broadly speaking, Fn is present in plasma and body fluids in its soluble dimeric form, while an insoluble variant with a cellular origin can be found in the ECM of tissues. Cellular Fn exclusively expresses extradomain A (EDA, EIIIA in rodents) and/or EDB (EIIIB in rodents) and shows higher heterogeneity than plasma Fn due to its role in the ECM modeling of different tissues [96]. Fn is a ligand for integrins of the $\beta1$, $\beta3$, $\beta5$, and $\beta6$ families [95–98]. Integrins are cell surface heterodimers composed of an α subunit noncovalently linked to a β subunit. Of these, about 20 heterodimers are known, each of which binds to specific ligands [99]. They form a physical link between the ECM and the cytoskeleton of cells, thereby allowing for the transduction of extracellular signals [100,101] and the control of cell behavior.

4.1. Fibronectin and Its Role in BBB Functioning

In the healthy adult brain, the presence of Fn is restricted to the BBB. The BBB consists of a layer of tightly connected endothelial cells that line the brain capillaries and play an important role in maintaining brain homeostasis (Figure 1). The polarized brain endothelial cell monolayer differentially harbors lipids and proteins at its luminal (blood-side) and abluminal (brain-side) membrane [102]. For BBB endothelial cells to maintain their tight barrier function, close contact with astrocytes and pericytes is necessary. Perivascular astrocytic end feet make up the glia limitans, which fully cover the BBB endothelial cells and part of the pericytes. Microglial processes can enter through interspersed slits in the glia limitans, thereby allowing for direct contact with the endothelial basal lamina. This interface is important for controlling water and ion exchange between blood and the brain [103]. The basal lamina, also known as the basement membrane (BM), is a sheet-like ECM structure that provides cells with an adhesive substrate to grow and migrate on and allows for the modulation and transmission of intracellular signals and mechanosensitive physical cues [99,104]. The BBB BM is composed of two parts: an endothelial BM that lines the vascular wall, and one that forms the parenchymal BM of the glia limitans, which is produced by astrocytes (Figure 1) [105–107]. The two parts combine to form a protein network that contributes to the maintenance and barrier tightness of the endothelial BBB [61]. In healthy conditions, the parenchymal BM protects the brain parenchyma from leukocyte infiltration, while in inflammatory disease states, its degradation by matrix metalloproteinases (MMPs) disrupts this function [108]. Pericytes, which are believed to derive from migrating mesenchymal cells, neural crest cells, or macrophages [109–112], are embedded in the BM of the BBB, where direct exchange of cellular signals is mediated by GAP junctions. Hence, the BBB is not solely formed by tightly connected endothelial cells but also consists of endothelial cells, astrocytes, pericytes, and an ECM, which is referred to as the neurovascular unit (NVU) [113].

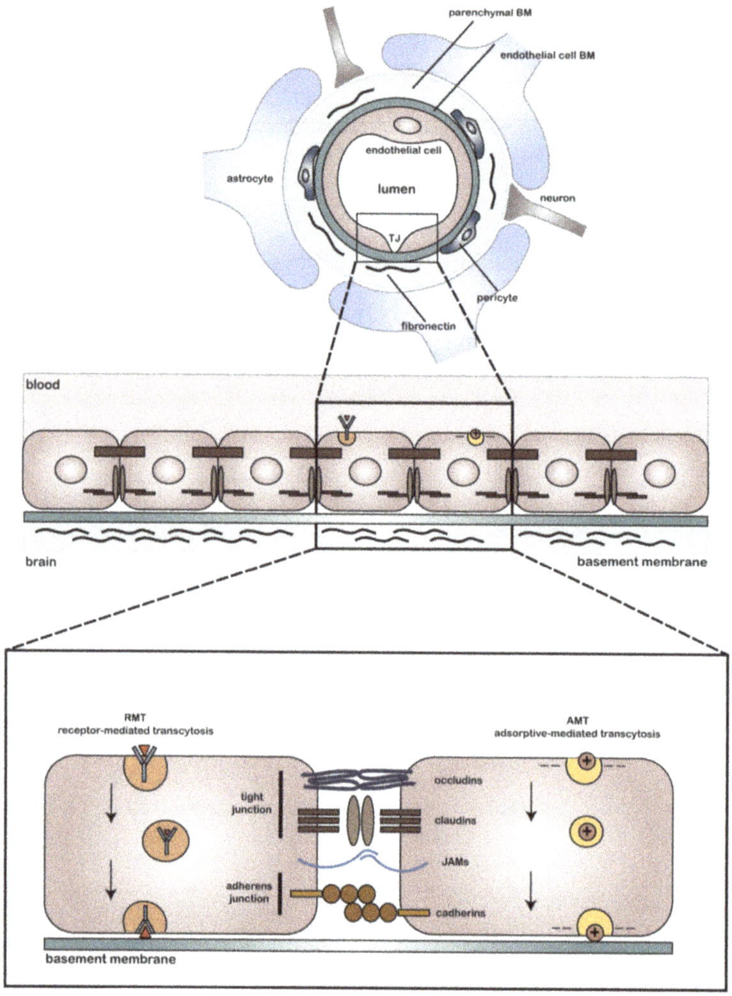

Figure 1. Schematic representation of the blood–brain barrier and active transport mechanisms. The blood–brain barrier (BBB) is formed by a monolayer of specialized endothelial cells, which form together with pericytes, astrocytes, and the basement membrane (BM) to create the neurovascular unit. The BM is a thin sheet of supporting extracellular matrix, including fibronectin, and is composed of endothelial BM and the astrocyte-derived parenchymal BM. Endothelial cells are tightly connected via tight and adherens junctions, which prevent the paracellular passage of molecules. Active transport mechanisms across the BBB include receptor-mediated transcytosis (RMT) or adsorptive-mediated transcytosis (AMT). RMT involves ligand–receptor binding, followed by endocytosis of the receptor complex, intracellular trafficking, and exocytosis at the basal membrane [114]. Conversely, cationic molecules can interact with the negatively charged membrane, thereby inducing transcellular transport of the positively charged molecule. See the text for more details. JAM—junction adhesion molecule; TJ—tight junction.

The main components of the BM are Fn, laminin, and collagen type IV [115], which are tethered to nidogens and proteoglycans [116,117]. As the composition of the BM plays a pivotal role in providing endothelial cells with a supportive growth substrate, changes in the BM protein composition induce functional changes in endothelial cell phenotypes [118]. The importance of the supporting cell types for BBB integrity is highlighted by the lower endothelial barrier integrity and an altered cytoplasmic anchoring of tight junctions (TJs) in monocultures [119] compared with co-cultures [120,121]. Indeed, endothelial cells grown on an ECM produced by astrocytes and pericytes show enhanced barrier impermeability compared with cells grown on non-astrocyte/pericyte-derived ECM [122]. A pericyte-derived ECM most rapidly increased the barrier resistance of endothelial cells, induced stronger expression of the TJ proteins occludin and claudin-5, and contained the largest relative amount of Fn, indicating that Fn is essential for proper barrier induction [61]. The Fn-binding integrins $\alpha 5\beta 1$ and $\alpha v\beta 3$ induce proliferation and aid the survival of BBB endothelial cells via the MAP kinase signaling pathway [123], while on laminin endothelial cells enter growth arrest on laminin via integrin $\alpha 2\beta 1$ [124]. Alongside regulating interactions with ECM components, integrins organize the proper functioning of BBB-specific molecules. For example, VE-cadherins form the molecular basis of adherens junctions that regulate the permeability of the BBB endothelium. Like integrin receptors, VE-cadherins are able to transduce extracellular signals and coordinate cell attachment and migration [106]. Fn treatment of endothelial cell layers can disrupt VE-cadherin-mediated cell–cell interaction through interaction with integrin $\alpha v\beta 3$ [125]. Direct activation of integrin $\alpha v\beta 3$ similarly disrupts the endothelial monolayer integrity by mislocalizing VE-cadherins [126]. These results demonstrate a direct link between endothelial–ECM interactions and functional properties of BBB endothelial cells, as reviewed in [127].

The altered composition of ECM proteins surrounding the BBB in disease states plays a causative role in disturbing barrier integrity in several neurological diseases. In MS, the BBB becomes disrupted, meaning the barrier's tightness is reduced, followed by an increase in leakiness [106]. In addition, monocytes and T cells cross the BBB endothelium, after which they gain access to the vascular and astroglial BM and brain parenchyma [128]. MMPs are known to degrade the ECM surrounding the blood vessels, thereby allowing for leukocyte infiltration into the CNS [129,130]. Furthermore, adhesion molecules, such as ICAM-1, act as receptors for leukocytes and are upregulated in MS lesions [131], thereby facilitating their migration across the BBB [132]. In active and mixed active/inactive lesions, mononuclear cells accumulate between the astroglial and vascular BM, which leads to a widened perivascular space and the appearance of so-called perivascular cuffs [107]. The ECM of these perivascular cuffs has a differential composition, with fiber-like networks containing several laminin isoforms, Fn, collagen IV, and heparan sulfate proteoglycans. This diverging composition may aid the migration of leukocytes [107]. Astrocytes are likely a source of the altered ECM in the BM, including extracellular Fn [133], and the expression of a splice variant of Fn that binds integrin $\alpha 4\beta 1$ may facilitate leukocyte migration into the CNS [134] (Figure 2). Blocking α4-integrins indeed inhibits leukocyte infiltration in EAE and alleviates clinical symptoms [135,136]. Natalizumab, which is an antibody against α4-integrins used in the treatment of MS and Crohn's disease, similarly blocks peripheral leukocyte trafficking to the CNS and was shown to be highly effective in preventing relapses in RRMS [137,138]. Nevertheless, natalizumab treatment carries the substantial risk of developing progressive multifocal leukoencephalopathy (PML) with long-term treatment [139].

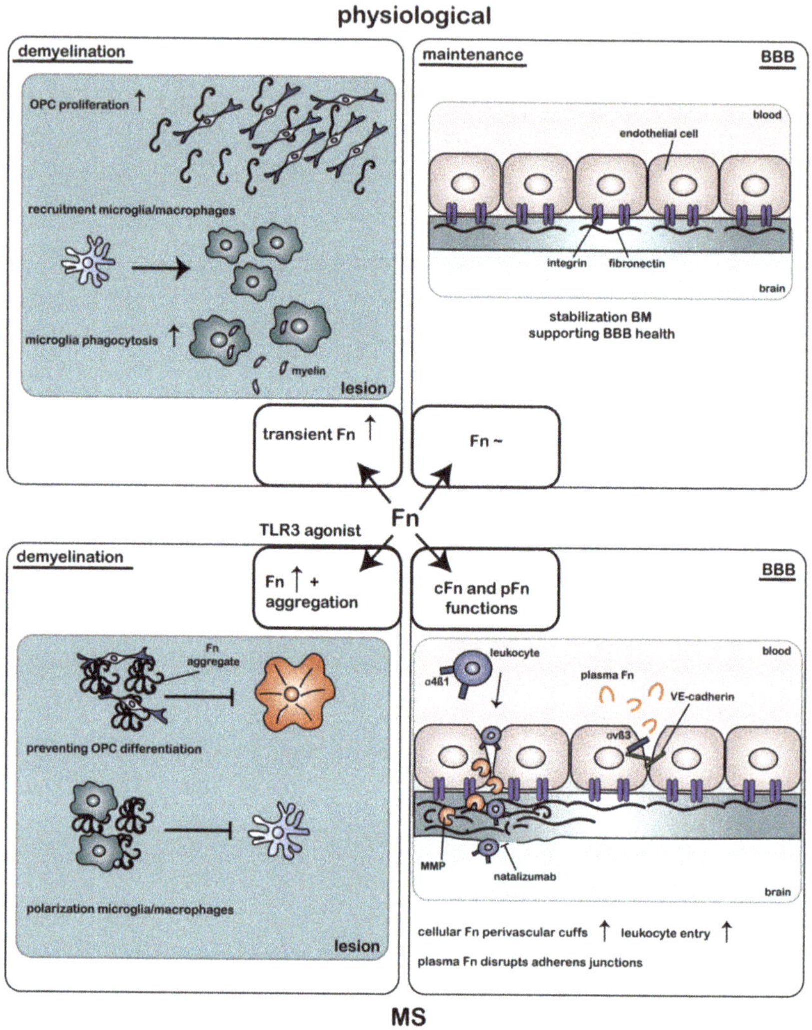

Figure 2. Role of fibronectin in the physiological adult brain, upon demyelination, and in multiple sclerosis. In the non-injured adult brain, fibronectin (Fn) expression is limited to the basement membrane (BM) of the blood–brain barrier (BBB), stabilizing the BM, thereby supporting BBB maintenance. Upon demyelination in the non-MS brain, Fn expression in the parenchyma is transiently upregulated, which aids (1) oligodendrocyte progenitor cell (OPC) proliferation in the lesioned area, (2) activation and recruitment of pro-inflammatory microglia and macrophages, and (3) myelin debris removal via phagocytosis. Conversely, upon demyelination in the MS brain, Fn expression persists and aggregates under the influence of Toll-like receptor 3 (TLR3) activation. Fn aggregates (1) impede remyelination by preventing OPC differentiation and (2) inhibit the switch from a pro-inflammatory to anti-inflammatory phenotype in microglia and macrophages. In addition, in an MS BM, Fn accumulates in perivascular cuffs near the BBB, which aids leukocyte transmigration across the BBB via integrin α4β1–Fn interaction. Natalizumab, which is a clinically approved MS medication, stops leukocyte entry into the CNS by blocking α4-integrins. Furthermore, several matrix metalloproteinases (MMPs) are upregulated in MS, which degrade Fn and other BM constituents, contributing to BBB destabilization and leukocyte entry into the brain parenchyma. Furthermore,

in MS, plasma Fn (pFn) may interact with integrin αvβ3 expressed on endothelial cells, thereby destabilizing VE-cadherins and increasing the BBB permeability. Upward arrow indicates 'enhanced'. cFn—cellular fibronectin; pFn—plasma fibronectin; VE-cadherin—vascular endothelial-cadherin.

4.2. Fibronectin as a Vasculogenic Regulator in MS

During angiogenesis new blood vessels are being formed, which requires signaling from growth factors, e.g., vascular endothelial growth factor (VEGF) and bFGF [140,141], and the ECM through integrins [123,142,143]. Fn was found to promote cell survival, proliferation, and migration of endothelial cells [123,144–146], while laminin induces differentiation and stabilization [123,147,148]. Thus, high expression levels of Fn are required during the formation of new blood vessels to support endothelial cell proliferation, while the presence of laminin is essential for their maintenance. Alterations in vasculature properties contribute to disease development and maintenance in experimental models of MS. For example, in EAE, VEGF is upregulated in the spinal cord during relapses and is correlated to demyelination and cell infiltrate levels [149,150]. An increase in vascular density (neo-angiogenesis) may be a regenerative response to hypoxic conditions [151,152], but can eventually lead to the formation of abnormal, leaky blood vessels (pathological angiogenesis). Angiogenesis under hypoxic conditions is (at least partly) regulated by Fn–integrin α5β1 interactions [153]. In EAE, an increase in the number of blood vessels in the white matter of the spinal cord is observed during the pre-symptomatic phase of the disease, which is accompanied by an elevation in Fn and α5β1 levels. This indeed reinforces the notion that α5-integrins mediate endothelial cell proliferation, thereby bolstering the formation of new blood vessels in EAE [154]. However, it is unclear whether this temporal increase in angiogenesis is mainly detrimental or beneficial. In MS, an increase in blood vessels is reported at lesions [155], indicating angiogenic remodeling. In addition, MS lesions have increased vessel expression of Fn, which correlates with the degree of inflammation [156]. Indeed, while plasma Fn levels are low under homeostatic conditions, they sharply increase during episodes of heightened inflammation, such as those observed during MS relapses and acquired vascular damage [157,158]. Whether this increase in plasma Fn or enhanced expression of cellular Fn by endothelial cells or astrocytes and deposition in the BMs contributes to angiogenesis remains to be determined.

4.3. Fibronectin (Aggregates) as a Remyelination Inhibitor in MS

In the adult brain, Fn expression in the parenchyma is very low. However, transient Fn expression by resident cells is a common response to tissue injury [159]. Several studies similarly reported a transient upregulation of Fn in toxin-induced demyelinated CNS lesions, which declined at the onset of remyelination of the lesioned area [17,84,160,161]. The source of this interstitial Fn could partly be attributed to nearby cellular Fn-producing astrocytes [17,74]. In addition, the reduction in BBB integrity in MS results in blood proteins, including plasma Fn, gaining access to the brain parenchyma [162] and implicates BBB breakdown as one of the primal factors in disease onset [162–164]. The appearance of Fn in demyelinated lesions has a bifold effect on OPCs (Figure 2). Fn promotes proliferation and migration of OPCs via integrins αvβ3 and αvβ1, respectively [74,165–167], while Fn hinders the maturation of OPCs into fully differentiated OLGs and the formation of new myelin membranes [168–170]. The latter may be beneficial by allowing first for myelin debris removal and OPC recruitment to the lesion. Degradation-resistant Fn aggregates are observed in inflammation-mediated demyelination, including at the relapse phase in EAE and in chronic demyelinated MS lesions. The persistent presence of Fn (aggregates) in MS lesions impedes myelin biogenesis in a β1-dependent manner [169], which is an effect that is dominant over laminin-2-mediated positive signals for myelin biogenesis [171]. Therefore, the presence of Fn (aggregates) in lesions explains why spontaneous regeneration in MS does not occur (Figure 2). This hypothesis was confirmed by studies showing that intralesional injection of Fn aggregates into toxin-induced demyelinated lesions inhibits

OPC differentiation and remyelination [17,18]. Moreover, remyelinated lesions contain hardly any Fn aggregates [17].

Microglia are innate immune cells specific to the brain, where they exist as distinct subtypes that are involved in several regulatory functions. These include the removal of remyelination-inhibiting myelin debris [172–174] after injury in conjunction with infiltrated peripheral-derived macrophages [175,176] and the generation of pro-regenerative signaling factors [177]. In fact, while a pro-inflammatory response is initially required to remove myelin debris [178], a switch to a more regenerative phenotype of microglia is required for remyelination to proceed [179]. This illustrates how the presence of an inflammatory response in MS is not unequivocally deleterious [180]. Microglia and macrophages are located in MS lesions and express the appropriate receptors for binding Fn [181–183]. Therefore, Fn aggregates in MS lesions may not only influence the behavior of OPCs, but also that of microglia and macrophages (Figure 2). Indeed, microglia and macrophages that are grown on aggregated Fn adopt an activated phenotype consisting of an amoeboid morphology and the expression of both pro- and anti-inflammatory markers [184]. If the pro-inflammatory phenotype of microglia and macrophages in the lesion is sustained, this may impair the subsequent remyelination process [185]. Of relevance, immunohistochemical analysis demonstrated that microglia and macrophages in MS lesions still display pro-inflammatory markers [14,186,187]. Remarkably, and in contrast to Fn's effect on OPCs, soluble dimeric (plasma) Fn and aggregated Fn differentially affect microglia and macrophages. The enhanced expression of Fn during an inflammatory insult increases the expression of integrins $\alpha4\beta1$ and $\alpha5\beta1$, activates microglia [181], and increases the expression of MMP9 [188]. The $\beta1$-integrin-induced proliferation of microglia is regulated by cAMP-dependent PKA signaling, which plays a negative regulatory role in $\beta1$-integrin translocation [189]. By doing so, Fn plays a role in the activation and recruitment of microglia to the inflamed area (Figure 2). Simultaneously, soluble Fn containing the EIIIA domain was demonstrated to stimulate inflammatory processes through TLR4 activation [190,191], thereby stimulating microglia phagocytosis [192], migration, and proliferation [193]. As such, the appearance of soluble Fn plays a beneficial role in regenerating the lesioned area. Nevertheless, aggregated Fn does not bind TLR4, and its effect on microglial activation is $\beta1$-independent [184]. Thus, aggregated Fn differentially affects the lesion environment from soluble Fn and does not aid lesion regeneration. This shows that for successful remyelination, an upregulation of $\alpha5$-integrins and soluble Fn may initially be important, as these signal to microglia to remove myelin debris present at the lesioned site.

Taken together, in the adult brain, Fn plays a significant regulatory role in (1) maintaining BBB integrity, (2) increasing angiogenesis upon injury, and (3) initial OPC and microglia recruitment to demyelinated lesions. However, its aggregation in MS lesions results in a gain of function, as its aggregated form is resistant to degradation and is an impeding factor in OPC maturation and the sustained presence of pro-inflammatory microglia within MS lesions (Figure 2). Given that Fn signaling to OPCs is dominant over laminin-mediated signals [171], we next discuss therapeutic strategies to specifically overcome Fn-aggregate-mediated inhibition of OPC differentiation to overcome remyelination failure.

5. Promoting Remyelination in MS: Therapeutic Strategies to Overcome Fn-Mediated Inhibition of Remyelination Failure

Fn aggregates impair OPC differentiation and remyelination, either directly, or indirectly via Fn-mediated microglia and macrophage dysfunction. Accordingly, a therapeutic benefit will be achieved by counteracting the negative signals of Fn aggregates (Table 1). One strategy to stimulate remyelination in chronically demyelinated lesions is by utilizing factors that can aid OPC differentiation in the presence of aggregated Fn, i.e., blocking or bypassing signals from Fn aggregates to cellular receptors on OPCs and microglia. For instance, exposure to ganglioside GD1a stimulates OPC differentiation, maturation, and myelination in cuprizone-induced demyelinated lesions that contain externally injected Fn aggregates [18]. Although the underlying mechanism remains to be determined, the effect of GD1a is evoked by a PKA-mediated signaling pathway and is mimicked by increasing cAMP levels [18]. Theoretically, agents that increase or prolong cAMP levels, such as PDE inhibitors that are beneficial in experimental models of MS and/or are currently used in clinical trials [194–197], may overcome Fn-mediated inhibition of OPC differentiation, and thus, benefit remyelination. On the other hand, agents that modulate intracellular signaling pathways may induce unwanted side effects in other cell types, such as microglia and neurons in healthy and injured tissue. GD1a's effect on OPC maturation and differentiation is, however, (cell-type) specific and only effective in an Fn-containing environment [18], making it a promising therapeutic agent for the treatment of chronically demyelinated MS lesions.

Another strategy to overcome remyelination failure is to prevent Fn aggregate formation, by allowing for timely Fn degradation. The formation of Fn aggregates is associated with an inflammatory process and encompasses disrupted Fn fibrillogenesis [198]. The assembly of fibrillar Fn into a network of high molecular weight fibrils is mediated at the astrocyte surface by integrin α5β1. Soluble Fn dimers bind the receptor, assemble into high-molecular-weight Fn, and self-associate using non-covalent bonds [199–201]. As stated before, *FN1* has many splice variants, of which some have specific relevance to fibrillogenesis [202]. These include variants containing cellular Fn-specific EIIIA and/or EIIIB domains. The relative abundance of either domain changes the conformational shape of Fn, thereby altering its function [203–207]. A 'double inflammatory hit' mechanism induces Fn aggregation and involves an initial exposure of astrocytes to pro-inflammatory cytokines that are associated with a demyelinating event, resulting in altered Fn splicing and a relative upregulation of the EIIIA-containing Fn [198]. A subsequent hit with a TLR3 agonist interferes with Fn cell-surface binding, thereby increasing Fn aggregation [198]. The degradation of myelin can result in the release of endogenous TLR3 agonists, such as stathmin [208], which is upregulated in myelin obtained from MS lesions [209]. Thus, timely treatment with factors that interfere with Fn splicing and/or TLR3 signaling in astrocytes may preclude Fn aggregation [198].

A possible reason for the absence of Fn aggregates in experimental toxin-induced demyelination models is not only the absence of a combination of BBB disturbances and a prolonged inflammatory component in toxin-mediated demyelination but also the efficient clearance of Fn before it has the chance to aggregate. Due to the efficient clearance of myelin debris and Fn in toxin-induced demyelination models, the likelihood of encountering a TLR3 agonist is small, hence Fn aggregates are unlikely to form during Fn fibrillogenesis. Simultaneously, the upregulation of MMPs, which are endogenous proteinases able to digest ECM components, during the earliest phase of toxin-induced demyelination may aid in the timely removal of Fn [210,211]. In chronic MS lesions, a lack of MMP7 activity may underlie the impairment in Fn clearance [211], thereby increasing the possibility of TLR3-mediated Fn aggregation due to prolonged inflammation and inefficient myelin debris clearance [198]. In addition, MMP7 is pivotal in cleaving aggregated Fn [211]. MMP7 and MMP3 are increasingly expressed in actively demyelinating MS lesions [212–214], implying their natural upregulation after a demyelinating insult, while MMP7 is absent in inactive MS lesions [211]. This data hints at the possibility of utilizing MMPs, particularly MMP7, as a relevant therapeutic target for MS to clear Fn (aggregates). Thus, a locally induced upregulation of MMP7 could prepare the lesioned area for subsequent remyelination, mainly by aiding the removal of aggregated Fn. Notably, MMP7 is a powerful enzyme, which in addition to Fn aggregates, also cleaves other ECM proteins, as well as cell surface receptors and growth factors [215]. This emphasizes the importance of local, targeted, and controlled MMP7 delivery to Fn aggregates.

Additionally, Fn aggregation may be influenced by heat shock proteins (HSPs). HSPs are intracellular chaperones that aid in the folding of denatured proteins [216]. Their extracellular presence increases in response to injury and under stress [217,218]. Proteomic analysis indicates that Fn aggregates serve as a scaffold for HSP70, which in turn induce both pro- and anti-inflammatory phenotypes in microglia and macrophages [184]. Furthermore, HSP70 increases the expression of ECM proteins, such as collagen I and Fn via transforming growth factor type β1 (TGF-β1) [219]. An exacerbated increase in HSP70 expression in response to heat shock and LPS stimulation was found in immune cells from MS patients compared with healthy subjects [220]. Furthermore, HSP47 and HSP90β are associated with Fn aggregates [184]. While HSP47 is involved in fibrillar collagen deposition [221,222], HSP90β contributes to the unfolding of Fn dimers to facilitate Fn fibrillogenesis [223]. Notably, the presence of antibodies against HSP90β is elevated in CSF of MS patients and implicated in OPC death [224], indicating a role for HSP90β in MS pathogenesis. These data point to the idea that, in MS lesions, HSP dysfunction may contribute to Fn aggregation and that the accumulation of HSPs in Fn aggregates impairs the functioning of glial cells.

Table 1. Potential therapeutic strategies to overcome the fibronectin-aggregate-mediated inhibition of remyelination failure.

Strategy	Method	Mechanism of Action	Reference
Prevent Fn expression	Prevent TG2 expression or activity	Mediates Fn expression and deposition	[161,225]
Prevent Fn aggregation	Modulate Fn splicing	Induces conformational changes in Fn to increase cell surface binding	[198]
Prevent Fn aggregation	Prevent TLR3 signaling (astrocytes)	Prevents the release of Fn fibrils from the cell surface	[198]
Prevent Fn aggregation	Modulate HSP90β activity	Contributes to the unfolding of Fn to facilitate Fn fibrillogenesis	[223]
Degrade Fn aggregates	Increase MMP7 expression and activity	Cleaves Fn, including Fn aggregates	[211]
Bypass Fn aggregates	Treat with ganglioside GD1a	Overcomes the Fn-mediated inhibition of OPC maturation via a PKA-mediated signaling pathway	[18]
Bypass Fn aggregates	Treat with PDE inhibitors	Prolongs cAMP levels, thereby potentially activating PKA, and enhances CNS remyelination	[194–197]

cAMP—cyclic adenosine monophosphate; CNS—central nervous system; Fn—fibronectin; HSP90β—heat shock protein 90 beta; MMP7—matrix metalloproteinase 7; PDE—phosphodiesterase; PKA—protein kinase A; TG2—tissue transglutaminase 2; TLR3—Toll-like receptor 3.

In conclusion, promising options to overcome the impairment of remyelination by Fn aggregates are (1) bypassing its signals to OLG lineage cells by GD1a, (2) preventing its aggregation by interfering with TLR3 signaling and/or HSP function, and (3) facilitating its clearance via the lesional delivery of MMPs, such as MMP7 (Table 1). Via stereotactic intralesional injection, the pharmacological effect of GD1a was documented [18], while the efficacy of the other potential therapeutic strategies still requires testing in relevant experimental models. Notably, as in the absence of astrocytic or plasma Fn, remyelination still occurs [74] and the transient increase in Fn upon toxin-induced demyelination is redundant for remyelination, making the degradation of its aggregates or even premature degradation of Fn feasible approaches to overcome remyelination failure in MS. For the latter, downregulation of tissue transglutaminase 2 (TG2) is an attractive option, as astrocytic TG2 mediates Fn expression and deposition [161,225].

Nevertheless, the beneficial effects of Fn should be considered when designing therapeutic strategies. In particular, Fn's involvement in BBB maintenance complicates targeting Fn for myelin regeneration in MS, as altering the functioning or presence of this protein may adversely affect the BBB. For example, MMP7 may negatively affect BBB functioning by promoting the breakdown of the BBB when administered peripherally. Indeed, in patients suffering from traumatic brain injury serum levels of MMP7 correlated with dynamic contrast-enhanced magnetic resonance imaging (DCE-MRI), which measures the BBB permeability and breakdown [226]. Furthermore, elevated MMP9 serum levels were observed in RRMS patients, with higher serum levels predicting new gadolinium-positive lesions, i.e., active lesions with a disrupted BBB [227]. Targeting Fn aggregates directly by increasing MMP activity may therefore not be the most suitable tactic for preventing the remyelination block observed in MS, especially when these changes are effectuated on a systemic rather than local level. Moreover, despite BBB alterations in the early stages of MS, PPMS is characterized by lesions with a different inflammatory profile, where the BBB remains largely intact and remyelination is marginal [228]. This highlights the necessity for a 'two-step approach', i.e., brain-targeted and locally delivered therapeutics. Nevertheless, this is complicated when treating brain diseases [229,230], mainly due to the presence of the BBB, for which solutions will be discussed next.

6. MS Therapeutics: Drug Delivery Vehicles for Delivery to the Brain

Plasma proteins and other compounds that are neurotoxic at high concentrations damage neurons and other brain-resident cells when allowed free access to the brain. For example, the blood plasma components thrombin and plasmin can induce apoptosis or lead to seizures [231,232], while high levels of the excitatory neurotransmitter glutamate are toxic to neurons [233]. Therefore, passive diffusion of hydrophilic and hydrophobic compounds across the BBB is restricted by the presence of TJs and efflux transporters, respectively, while active transport through substrate-specific transporters and receptor-mediated transcytosis allows for the regulated uptake and excretion of compounds [234] in order to maintain brain homeostasis (Figure 1).

The encapsulation of drugs in a carrier system that is able to cross the BBB seems a promising strategy for obtaining brain penetration of medicinal compounds while providing several additional advantages [235]. Techniques are being used to develop nanocarriers with a high drug-loading profile [236] and those that aid controlled and sustained release of the drug of interest [237–240]. By doing so, the need for frequent dosing is reduced. Nanovehicles can improve the bio-availability of hydrophobic and hydrophilic compounds by providing protection against chemical and biological degradation and improving target-site delivery [241–243]. Additionally, nanoparticle design can be optimized for the development of precision medicine, where the personal characteristics of patients in conjunction with specified nanomedicine engineering allow for patient-specific disease treatments [244].

However, systemically delivered nanocarriers still face several hurdles, including the acidic environment of the gastrointestinal tract (when administered orally), clearance by the liver and spleen, clearance by immune cells, and physical barriers that prevent easy access to target sites [245]. Overall, for in vivo efficacy, nanocarriers need to show high stability, low toxicity, prevent clearance by the reticuloendothelial system, and efficiently accumulate at the target site [229,246,247]. Hence, the physicochemical properties of the nanocarrier (i.e., size, shape, charge, and type of material) and the properties of the biological barriers that hinder their transport to the target site need to be taken into account [245,248], as well as how these properties influence the interaction between the nanocarriers and cell barriers. Furthermore, a myriad of nanocarrier modifications are invented, which are aimed at the sustained release of the drug, target specificity, and circumvention of intra- and extracellular clearance. For example, tuning the charge of lipid nanoparticles (LNPs) resulted in tissue-specific gene delivery by LNPs [249]. Furthermore, analyte-responsive hydrogels bind or release drugs of interest in a controlled manner, i.e., glucose oxidase-containing hydrogels can interact with glucose in the environment, swell, and subsequently release insulin [250–252]. Moreover, reversible PEGylation enhances the stability and circulation time of nanocarriers in vivo without preventing target cell uptake and drug release [253].

Commonly studied nanocarriers are lipid- or polymer-based and include liposomes, polymersomes, micelles, dendrimers, nanogels, nano-emulsions, and exosomes [229,254,255].

Unfortunately for brain-targeted therapeutics, the brain endothelium (BBB) appears more difficult for nanoparticles to penetrate than lung, liver, and kidney endothelial barriers [256]. Several studies utilized nanocarrier formulations with either the aim to improve the delivery of established drugs that reduce the number and severity of relapses in MS or to test experimental therapeutic agents in experimental models of MS aimed at alleviating disease progression (Table 2) [257–279].

6.1. Lipid-Based Nanoparticles as Drug Delivery Vehicle for RRMS Treatment

Liposomes are small biocompatible vesicles that consist of one or multiple lipid bilayers. Commonly used lipids include cholesterol, phosphatidylcholine, and soy lecithin [280]. Liposomes can entrap hydrophilic and lipophilic compounds in their aqueous core and lipid bilayer(s), respectively [229,281], thereby making liposomes versatile nanocarriers for a multitude of therapeutics. The addition of PEG moieties to the outer surface of liposomes endows them with stealth properties, limiting an immune response and reducing their plasma clearance [282]. The biodistribution and brain penetration of liposomes were tested in relevant experimental models for MS. The temporary BBB disturbance observed in RRMS is equally observed in experimental MS models that simulated its inflammatory profile, such as EAE. This allowed researchers to test liposomal treatments for MS that are not specifically targeted to the BBB [257]. For example, in EAE, ^{99m}Tc-DTPA-labeled liposomes accumulated more in the brain and spinal cord compared with healthy controls, which is an effect that may rely on BBB disruption and/or infiltrating macrophages that transport liposomes to the CNS (Figure 3) [258]. Notably, DTPA is a low-molecular-weight molecule that can cross the BBB during injury or inflammation-induced endothelial permeability. Compared with free DTPA, DTPA conjugated to liposomes accumulated more in lesioned areas, which is an effect that could be attributed to monocytes that phagocytosed liposomes before significant BBB damage had occurred (Figure 3) [257]. In addition to employing infiltrating liposome-laden monocytes as a means of transporting anti-inflammatory drugs to target inflammatory lesions in MS [259], Schweingruber et al., provided evidence that glucocorticoids, which are approved MS therapeutics, entrapped in liposomes induced macrophages to adopt an anti-inflammatory phenotype rather than inducing T cell apoptosis [262]. Similarly, glucocorticoids entrapped in inorganic–organic hybrid nanoparticles (IOH-NP) were shown to exclusively modulate macrophage functioning [283]. Conversely, using a long-circulating prednisolone liposome formulation, Schmidt et al., demonstrated that glucocorticoid administration via liposomes restored BBB integrity, reduced inflammation caused by T cells, diminished macrophage infiltration into the CNS, and slowed down disease progression in EAE [261]. Active targeting of glucocorticoid liposomes to the brain was also achieved by labeling liposomes with glutathione, which is a BBB-targeting ligand [264].

Another approved MS medication is dimethyl fumarate (DMF), which is an orally administered anti-inflammatory drug with neuroprotective properties [284,285]. Due to its low brain permeability and oral administration, it requires high and frequent dosing. Accordingly, the drug may benefit from incorporation into a nanocarrier system. Indeed, the encapsulation of DMF into solid lipid nanoparticles increased the half-life and bioavailability of the orally administered drug in rats [265]. In the cuprizone model, DMF-loaded lipid-based nanoparticles administered orally once a day improved remyelination to the same extent as a three-times a day free oral DMF treatment [266]. Hence, incorporation into nanoparticles reduces the need for excessive dosing, particularly for a drug with a fast clearance rate. Other than the encapsulation of approved MS medication in liposomes, the encapsulation of autoantigens [267] and MBP-peptides [275,276,278,286] in liposomes was successfully tested as a potential disease-modifying medication. Thus, there is compelling evidence that the encapsulation of approved or experimental therapeutics in liposomes allows for the safe administration of drugs in MS patients that cannot be administered in free-form due to their instability in circulation and/or can have beneficial effects on top of the curative response of the drug itself, such as less frequent dosing.

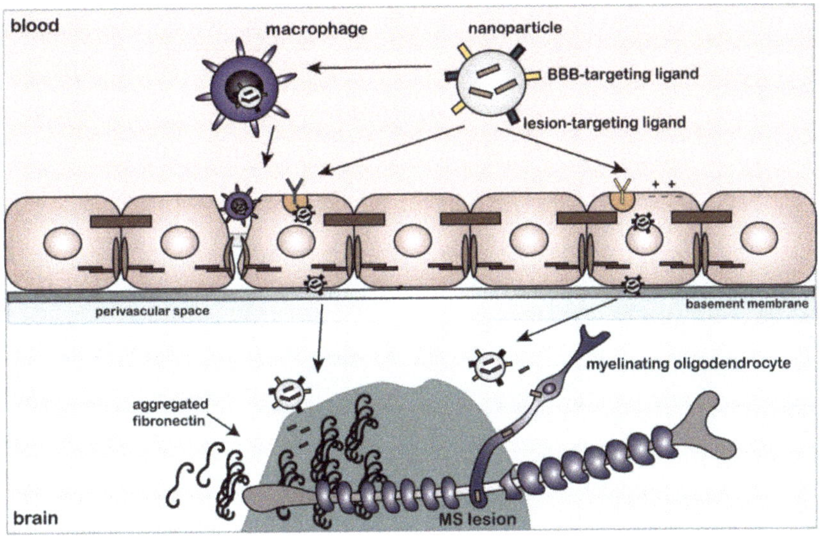

Figure 3. Strategies to deliver MS therapeutics that overcome the fibronectin-mediated inhibition of remyelination failure in the brain. Therapeutic compounds encapsulated in nanoparticles that contain a blood–brain barrier (BBB)-targeting ligand or carry a positive charge enter the brain via transcytosis through receptor-mediated transcytosis (RMT) or adsorptive-mediated transcytosis (AMT), respectively. Alternatively, in relapsing–remitting MS, surveilling monocytes may phagocytose nanoparticles and transport these across the compromised BBB during relapses. Furthermore, the identification of lesion-specific BBB alterations (i.e., upregulation of receptors at the BBB near lesions) would aid the targeting of lesion-directed medication. In the brain, lesion targeting of therapeutic-containing nanoparticles may be achieved by cell-specific ligands targeting receptors that are present on, e.g., oligodendrocyte lineage cells, or ligands targeting the altered, and therefore specific, environment in MS lesions. For example, targeting specific splice variants of fibronectin that are abundant in fibronectin aggregates can aid the cell- and lesion-specific delivery of the therapeutic compound. This 'two-step approach' utilizes ligands that facilitate transcytosis across brain endothelium (BBB) and ligands that direct the delivery of therapeutics to MS lesions.

Table 2. Nanoparticles used for treatment in experimental MS models and in MS patients.

Treatment	Administration Means	Administration Time Point	Outcome Measure	Reference
		Experimental MS Models		
^{99m}Tc-DTPA-loaded liposomes in EAE	Intravenous	At induction of disease	Biodistribution of liposomes	[258]
MOG_{40-55}-loaded liposomes in EAE	Intraperitoneal	At induction of disease	Preventive and preclinical treatment effects on EAE development	[267]
MBP-loaded liposomes in EAE	Subcutaneous	At disease onset for 6 days	Effect of different MBP isoforms on EAE progression	[275]
Prednisolone-loaded liposomes in EAE	Intravenous	At peak of disease	Effect on EAE progression, BBB permeability, and drug biodistribution	[261]
(Methyl)prednisolone-loaded liposomes in EAE	Intravenous	At peak of disease	Effect on EAE progression and macrophage functioning	[262]

Table 2. *Cont.*

Treatment	Administration Means	Administration Time Point	Outcome Measure	Reference
Methylprednisolone-loaded liposomes in EAE	Intravenous	Prophylactic, at disease onset, and disease peak	Brain-targeted effect on EAE symptoms	[264]
MOG-loaded PLGA particles in EAE	Intravenous/ subcutaneous	Prophylactic	Effect on EAE development	[268]
MOG-anti-Fas-PD-L1-Fc-CD47-Fc-TGFβ-loaded PLGA particles in EAE	Intravenous	At disease onset and disease peak	Modulation of auto-reactive T cells in EAE and disease progression	[269]
MOG-IL10-loaded PLGA particles in EAE	Subcutaneous	Prophylactic, at disease onset, and disease peak	Effect of 'inverse vaccination' on EAE progression	[270]
PLP-coupled PLGA particles in EAE	Intravenous	At disease onset	Treatment of EAE and nanoparticle uptake in vitro by antigen-presenting cells	[272]
PHCCC-loaded PLGA particles in EAE	Subcutaneous	From induction of disease, every 3 or 5 days	Effect on DC activation and EAE disease progression	[273]
miR-219a-5p liposomes, PLGA particles, and extracellular vesicles in EAE	Intranasal	2 and 8 days post-induction of disease (before symptom onset)	Effect on remyelination in EAE	[274]
Curcumin-loaded HPPS in EAE	Intravenous	8, 10, 12, and 14 days post-induction of disease	Restriction of immune cell infiltration of the brain in EAE by modulation of monocytes	[259]
PLP-coupled PLGA particles in relapsing–remitting EAE	Intravenous	At disease onset, disease peak, and disease remission	Prevention and treatment of relapsing EAE disease	[271]
(Methyl)prednisolone-loaded liposomes in chronic relapsing EAE	Intravenous	At first peak of disease	Effect on disease progression, their effect on relapse risk, and macrophage CNS infiltration	[263]
Dimethyl-fumarate-loaded solid lipid nanoparticles in cuprizone	Oral	Daily cuprizone and nanoparticles for 30 days	Effect on remyelination	[266]
LIF-loaded PLGA particles in focal demyelination	Intralesional	8 days post-lysolecithin lesioning	Effect on OPC differentiation in vitro and remyelination in vivo	[279]
		MS Patients		
MBP-loaded liposomes	Subcutaneous	Once a week for 6 weeks	Safety profile of CD206-targeted liposomal MBP treatment in RRMS and SPMS patients	[286]
MBP-loaded liposomes	Subcutaneous	Once a week for 6 weeks	Serum cytokine analysis and Th1/Th2 ratio in RRMS and SPMS patients	[278]

BBB—blood–brain barrier; CNS—central nervous system; DTPA—diethylenetriaminepentacetate; EAE—experimental autoimmune encephalomyelitis; HPPS—high-density lipoprotein-mimicking peptide-phospholipid scaffold; LIF—leukemia inhibitory factor; MOG—myelin oligodendrocyte glycoprotein; MS—multiple sclerosis; OPC—oligodendrocyte progenitor cell; PHCC—N-phenyl-7-(hydroxyimino) cyclopropa[b]chromen-1a-carboxamide; PLGA—poly(lactic-co-glycolic acid); RRMS—relapsing–remitting MS; SPMS—secondary progressive MS.

6.2. Polymer-Based Nanoparticles as a Drug Delivery Vehicle for RRMS Treatment

Alongside lipid-based nanoparticles, polymeric nanoparticles were demonstrated to be an effective and safe drug delivery system, though for clinical applications, the toxicity profile of polymeric particles needs to be critically assessed. Poly(lactic-co-glycolic acid) (PLGA) is a synthetic polymer that shows potential for clinical applications due to its biocompatibility, biodegradability, and low immunogenicity profile. Despite its promise, few PLGA-based medications are currently approved for clinical use [287]. Variations in product design may underlie the low success rate of PLGA nanospheres on the market. Since minor alterations in the production process may alter the pharmacodynamics of the desired product, a strong emphasis needs to be placed on the optimization of particle properties [288]. PLGA particles consist of polymerized lactic acid and glycolic acid subunits. The proportion of lactic acid to glycolic acid determines the hydrophobicity of the particle, with higher proportions of lactic acid conferring a higher degree of hydrophobicity. Simultaneously, lactic-acid-rich PLGA particles degrade slower than particles relatively high in glycolic acid, though particles with equal amounts of lactic acid and glycolic acids degrade the fastest [289,290]. The therapeutic effectiveness of PLGA particles depends on their physicochemical properties, drug loading efficiency, and drug release behavior, which altogether determine successful drug delivery. Biodistribution and (immune) clearance of PLGA nanoparticles are largely determined by their size and surface charge (zeta potential) [291]. Properly designed PLGA particles demonstrated improved delivery of therapeutics in vivo. For example, the incorporation of the chemotherapeutic docetaxel in pegylated PLGA particles showed minimal liver accumulation in rats, enhanced accumulation in tumors in mice, and induced tumor shrinkage in humans at a lower dose than when free docetaxel was administered [292].

PLGA particles are tested as vehicles to induce immune tolerance in experimental models of MS. Inhibition of the inflammatory phenotype of autoreactive T cells and a delay in disease onset were achieved by injecting PLGA particles containing myelin oligodendrocyte glycoprotein (MOG) prior to the induction of EAE [268]. Similarly, an 'inverse vaccination' treatment with MOG-PLGA, PLP-PLGA, and IL10-PLGA particles inhibited EAE development and equally ameliorated EAE progression when administered post-EAE induction [270,271]. In vitro data suggest that disease-relevant peptide-conjugated PLGA nanoparticles diminished inflammatory signaling in macrophages and dendritic cells, i.e., antigen-presenting cell types with a known role in nanoparticle clearance from blood circulation. These in turn reduced T cell proliferation and induced T cell apoptosis [272]. Other compounds (indirectly) affecting T cell polarization also benefit from encapsulation in nanoparticles. For example, (hydroxyimino)cyclopropa[b]chromen-1a-carboxamide (PHCCC) affected glutamate metabolism in dendritic cells, which indirectly affected T cell polarization through cytokine secretion. The incorporation of PHCCC in PLGA particles resulted in a controlled release of PHCCC, thereby allowing for a reduction in dosing frequency from daily to once every three days in mice [273].

6.3. Drug Delivery Vehicles for Treatment of Progressive MS

To date, most approved MS drugs modulate the peripheral immune system with the purpose of reducing the inflammatory response associated with relapses [293]. Hence, most previously described studies targeting MS rely on the presence of a disrupted BBB or a significant inflammatory response characterized by infiltrating macrophages. As stated earlier, treatment aimed at restoring remyelination in progressive MS when inflammation has largely subsided requires BBB-penetrating capabilities of drugs or nanocarriers that ideally recognize chronic lesions (i.e., with a low inflammatory profile) and/or target cells within the lesion. Few nanoparticle studies have specifically targeted OLGs in MS or assessed nanoparticle accumulation in the brain when the BBB is largely intact to determine whether particles could cross the BBB. A recently published study by Osorio-Querejeta et al., compared liposomes, PLGA particles, and extracellular vesicles (exosomes) for the delivery of miR-219a-5p, which is a microRNA capable of inducing OPC differentiation and myelination [274]. Liposomes and PLGA particles were more efficiently taken up by OPCs in vitro, though exosomes were more effective at inducing OPC differentiation, as assessed by the expression of myelin-related genes. Additionally, in an in vitro BBB model, exosomes crossed the BBB more easily than liposomes or PLGA particles. Intranasal delivery of miR-219a-5p-containing exosomes after EAE induction enhanced remyelination and attenuated clinical disability scores compared with treatment with control exosomes. This study demonstrated the effectiveness of remyelination-inducing therapy, albeit in conditions where inflammation was present. As inflammation-induced alterations in the vasculature are inherent to EAE, it is likely that particles could reach the brain through a breached BBB. Indeed, the binding of cationic liposomes to endoneural vessels in the spinal cord occurred throughout the disease course of EAE, which was not observed for control animals. This binding effect was correlated to changes in animal's vasculature and inflammatory profile [294]. For this reason, testing remyelination therapy in non- or low-inflammatory conditions is necessary to understand the applicability of these therapies to progressive MS.

Leukaemia inhibitory factor (LIF) is a known pro-myelinating factor shown to improve remyelination [295–297]. A single treatment with LIF-containing PLGA particles induced differentiation of OPCs into mature OLGs by activating pSTAT-3 signaling in vitro. In vivo, these particles increased myelin thickness, as well as the percentage of remyelinated axons after a focal demyelinating insult [279]. OPC-specific targeting was achieved by decorating nanoparticles using anti-NG2 antibodies, thereby avoiding off-target effects. As LIF is rapidly degraded in vivo, the utilization of PLGA particles improved the stability of the drug [279]. However, as particles were injected directly into the demyelinated lesions, systemic stability and the BBB-traversing capability of the particles were not assessed. Hence, an effective drug delivery system for the treatment of progressive MS that is systemically administered and targeted to the brain needs to be developed, for which considerations and potential strategies are reflected upon in the next section.

7. Progressive MS Treatment: Considerations for Designing a Brain-Targeted Drug Delivery System

The ideal drug delivery system for the treatment of progressive MS shows no toxicity, has high specificity for the target site (i.e., the brain and specifically demyelinated lesions), only releases the drug when it has arrived at the lesioned area, and is biodegradable and/or biocompatible. To achieve such a highly specific delivery of the therapeutic compound, several complicating factors must be addressed. Although endocytosis followed by intracellular disintegration of the nanoparticle and subsequent drug release is required for the drug to reach its intracellular targets, first, transcytosis across the endothelial cells of the BBB is needed to get from the blood to the brain. Several strategies for blood-to-brain transport are developed, as well as alternatives to circumvent the BBB, which are discussed next in more detail.

7.1. Receptor-Mediated Transcytosis (RMT)

A commonly used brain-targeting approach uses the conjugation of BBB endothelial cell-recognizing ligands, targeting peptides, or antibodies to nanoparticle formulations, which allows them to cross the BBB. Examples include nanoparticles decorated with ligands for the transferrin (Tf), insulin, lipoprotein (LRP), lactoferrin (LfR), and diphtheria toxin receptors; the GM1-binding G23 peptide; and glutathione [264,298–308]. Many of these brain-targeting peptides and ligands cross the BBB through ATP-dependent receptor-mediated transcytosis (RMT) [309]. RMT is the process by which ligand binding to membrane-bound receptors induces internalization of the ligand–cargo complex through endocytosis, followed by intracellular vesicular trafficking and exocytosis of the cargo at the opposite side of the endothelial cell monolayer (Figure 1) [114]. The use of specific coatings and nanoparticle materials may hence prove to be important for organ-targeted drug design [256]. For example, a poloxamer-188 coating caused the adsorption of blood apolipoproteins to the surface of PLGA particles, which then induced BBB transcytosis through an interaction with the LDL receptor (LRP1) [310]. Poloxamer-188-coated PLGA particles demonstrated efficient BBB transcytosis in vitro and successful delivery of an anti-viral HIV drug to macrophages and microglia [311,312].

7.2. Adsorptive-Mediated Transport (AMT)

Another strategy to target the brain is the utilization of positively charged moieties that mimic the transport of polycationic proteins, such as protamine, across the BBB [313]. The use of cationic polymers promotes BBB transport through adsorptive-mediated transport (AMT) [313,314]. AMT relies on electrostatic interactions between the negatively charged endothelial cell membrane and the positively charged molecule (Figure 1). Based on the same electrostatic interactions, the cationic polymer poly(β-amino ester) (PbAE), when mixed with siRNA, self-assembled into 100 nm sized nanoparticles and released the siRNA content when exposed to the reducing environment of the cytosol [315]. In an iPSC-derived human BBB transwell model co-cultured with glioblastoma cells, the PbAE particles delivered siRNA to glioblastoma cells after transcytosis by the in vitro BBB. Successive in vivo experiments demonstrated that these particles reached and delivered siRNA to orthotopically implanted patient-derived glioblastoma cells in mice after intravenous administration [316].

Some cell-penetrating peptides (CPPs), such as HIV-1 trans-activating protein (TAT), also employ AMT [317]. The decoration of nanoparticles with CPPs strongly increased transcytosis and improved targeted brain delivery and controlled release of nanoparticle contents [318–320]. Liposomes decorated with a combination of the cyclic Arg-Gly-Asp (cRGD) peptide, which binds to integrin $\alpha v\beta 3$ at the BBB [321] and induces clathrin-mediated endocytosis in cells [322], with a histidine-rich pH-sensitive cell-penetrating peptide (TH) that evades lysosomal degradation [323] resulted in their efficient transcytosis across the BBB endothelium. After binding to integrin $\alpha v\beta 3$, nanoparticles were internalized by glioma cells due to the positive charge of the CPP, which was achieved via histidine protonation in the acidic microenvironment around the tumor cells [318].

7.3. Focused Ultrasound

Besides ligand-based nanoparticle modifications aimed at transporting drugs across tightly connected BBB endothelial cells, temporarily opening the BBB through focused ultrasound (FUS) is also considered a viable method of brain-targeted drug delivery [324]. The technique, which was first described approximately twenty years ago [325,326], involves the local application of pulsed sonication. In combination with gas-filled microbubbles, reversible openings in the BBB can be achieved, through which therapeutics gain access to the brain [324]. Since then, FUS has demonstrated promising benefits for the treatment of neurodegenerative diseases [327–330]. FUS appears to rely on both paracellular and transcellular transport mechanisms. Thus, FUS induces the temporary disintegration of TJ complexes, thereby allowing for paracellular entry into the brain [331], and was also shown to increase endocytosis [332,333]. With the appropriate ultrasound settings, it has promise for the selective delivery of medication into the brain, though its safety and application in humans still need to be properly assessed. For application in MS, it is disadvantageous that individual lesions need to be targeted, which is complex when using FUS.

7.4. Intranasal Drug Delivery

A way to circumvent the BBB is by administering drugs intranasally. A portion of intranasally administered particles is expected to reach the brain via trigeminal neurons and olfactory nerves without entering the systemic circulation. The CSF can be reached directly via a route that involves the nasal epithelium and the perineuronal and subarachnoid space [334,335]. Direct nose-to-brain delivery through the olfactory bulb may involve paracellular, transcellular, and neuronal transport [335]. Thus, intranasal dispensation of nanoparticles, e.g., PLGA particles [336–338], and liposomes [339–341] may offer a means of fast and efficient delivery of brain therapeutics. However, as the nasal cavity is small, only a limited amount of the drug can be administered at each dose. Furthermore, mucociliary clearance and enzymatic degradation in the nasal cavity reduce brain uptake of the administered drug [342], causing less than 1% of the drug administered to reach the brain [343]. For this reason, enzyme inhibitors, mucoadhesives, and absorption enhancers are incorporated in intranasal formulations, which themselves can be irritating to nasal mucosa [343,344]. In addition, the efficiency of intranasal drug delivery across olfactory cells in vitro differed between PLGA and lipid carriers (with lipid carriers having a higher transcytotic ability) [320]. Furthermore, though intranasal drug delivery is promising, it appears to not evade systemic circulation completely [345]. Thus, BBB targeting through RMT or AMT may still be relevant to nanoparticles following nasal administration.

8. MS Therapeutics: Considerations for Intracellular Delivery of Therapeutic Agents

An important consideration for nanocarrier design for the treatment of MS are properties that allow systemically administered nanoparticles to release their content only after crossing the BBB and accumulation at the target site, i.e., an MS lesion (Figure 3). Upon endocytosis of nanoparticles by target cells, the therapeutic payload needs to escape from the endolysosomal system to circumvent the acidic and enzymatic environment of the lysosome that may destabilize or inactivate therapeutics [346–348]. Where hydrophobic drugs can passively cross the endosomal membrane, intracellular delivery of hydrophilic drugs, e.g., DNA, RNA, and proteins, requires permeabilization of the endosomal membrane [349–351]. At least for mesoporous silica nanoparticles, repeated administration resulted in reduced intracellular delivery of hydrophilic cargo [346], which signifies that the development of nanocarriers with a high drug-loading capacity is important. PEGylation, which is commonly used to confer stealth properties on nanoparticles to promote their blood circulation time, has an inhibiting effect on endo/lysosomal escape. The addition of PEG lipids onto gene-carrying liposomes inhibited the endosomal release of the genetic material and thus gene delivery [352–354]. To overcome this limitation, exchangeable [354], cleavable [355,356], and pH-sensitive PEG chains are used [357]. Finally, small extracellular vesicles (sEVs) were shown to enter cells via endocytosis and fuse with endosomal and/or lysosomal membranes in order to release their cargo in the cytosol [358]. As it was demonstrated that sEVs could efficiently cross the BBB and deliver a pro-myelinating drug in vivo [274], their applicability for brain-targeted medicinal delivery seems promising. Overall, nanoparticle design aims for nanoparticle stability during systemic circulation and the release of therapeutics, i.e., nanoparticle destabilization, at the target site.

9. Active Targeting to MS Lesions: Considerations for Controlled Drug Delivery to Overcome Fibronectin-Mediated Inhibition of Remyelination

While the previous sections largely focused on drug vehicles' requirement to cross the BBB, drug delivery to demyelinated lesions also requires lesion-targeting approaches. To this end, conjugation of nanoparticles with ligands, such as antibodies or peptides that interact with binding sites on lesion-resident cells or environmental factors that are only present in MS lesions, ensures controlled drug delivery. In particular, peptides are favorable for being small, easy to synthesize, and less immunogenic than antibodies. In the following, we reflect on how to control and functionally deliver medication to MS lesions to overcome the Fn-mediated inhibition of remyelination failure. The need for nanoparticles targeting Fn aggregate-bearing lesions is essential for drugs that may interfere with Fn expression and/or degradation, as these may also affect Fn in the BBB BM without lesion-specific targeting. Contrariwise, drugs that interfere with Fn aggregation or bypass the negative effect of Fn aggregates on OPC maturation and have minimal off-target effects are unlikely to interfere with Fn functioning outside of lesions. Additionally, similar delivery approaches can be considered for other remyelination-directed medications. For an overview of peptide-based targeted drug delivery to cells and ECM components in MS lesions, we refer to a recent comprehensive review [359].

9.1. Active Targeting to Fn Aggregates

Active targeting of Fn aggregates is relevant for both functional cellular delivery of therapeutic agents that bypass Fn-mediated inhibition of remyelination (e.g., GD1a), and treatment approaches aimed to clear Fn aggregates at MS lesions (e.g., MMP7), with the latter requiring extracellular release of nanoparticle content or local stimulation of MMP production. As mentioned before, Fn is not abundantly present in healthy adult brains, allowing for Fn-targeting approaches in demyelinated areas. In addition, the targeting of Fn was utilized for drug delivery to tumor tissues [360–363]. The EDA and EDB splice variants of cellular Fn are mainly expressed during fetal development but become re-expressed at locations of tumor growth [360]. Using phage display, a peptide recognizing the EDB splice variant of Fn was identified that showed specific targeting to human prostate tumor xenografts implanted in mice [364]. Moreover, imaging of xenografted tumor tissue in mice was achieved using an EDB Fn-splice-variant-binding high-affinity peptide (named an aptide) [365]. Conjugation of these aptides to liposomes or PLGA particles [366–368] and an aptide–docetaxel conjugate [369] improved chemotherapy delivery to the tumor and inhibited tumor growth (reviewed in [359]). In addition, liposomes decorated with an aptide recognizing the EDB domain of Fn showed improved drug delivery to MCF7/ADR orthotopic tumors in vivo and delayed tumor growth [363]. One of the features of Fn aggregates in demyelinated MS lesions is their relative abundance in the EDA over the EDB splice variant [74,198]. This hints at the possibility of using an EDA-recognizing ligand to target Fn aggregates in MS. Advantageously, this method of interaction with plasma Fn can be circumvented, as EDA is unique to cellular Fn. Future studies need to uncover whether such Fn-targeted nanoparticles will indeed reach MS lesions. A complicating factor is that in healthy brain tissue, EDA Fn in the basement membrane is limited to larger blood vessels, while in actively demyelinating MS lesions, EDA Fn is abundantly present in perivascular networks [107]. It would be undesirable if nanoparticles that targeted lesional Fn aggregates remain within the perivascular space rather than penetrate the lesioned parenchyma. However, this may not be a problem in chronic demyelinated MS lesions, as in chronic MS lesions, EDA Fn is less abundant in the basement membrane, while parenchymal Fn aggregates are more abundant [17] and remyelination failure is more prominent [14,17]. Alternatively, other ECM components in MS lesions can be utilized as bait for peptide-targeted drug accumulation in MS lesions [359]. For example, a CSPG-targeting peptide was used for functionalizing nanoparticles to target traumatic brain injury [370].

9.2. Active Targeting of Cells

Alternatively, functional delivery of ganglioside GD1a to OLG lineage cells, which has minimal off-target activity in healthy tissue, and specificity to overcome Fn-induced remyelination failure can be employed [18]. The use of a nanocarrier conjugated with OLG lineage cell-targeting peptides or antibodies may facilitate specific delivery to the cells of interest. To this end, anti-NG2-antibody-coated nanoparticles were successfully used to functionally deliver LIF to OPCs [279], while targeting OPCs with anti-PDGFRα antibodies offers an alternative approach [371]. Given that GD1a is hardly degraded in OLG lineage cells and the addition of GD1a to OPCs can overcome Fn-mediated inhibition of myelin membrane formation in vitro [18], GD1a incorporation in anti-NG2 or anti-PDGFRα antibody-coated nanoparticles may be a feasible approach. As an alternative, nanocarriers decorated with antibodies against GPR17 may be considered. GPR17 surface expression is restricted to immature OLGs and is absent on mature myelinating OLGs [372]. Strategies to prevent Fn synthesis (e.g., with the enzyme TG2) and/or to interfere with aggregation (e.g., with TLR3 antagonists) may benefit from functional delivery of therapeutic cargo to astrocytes present in MS lesions. Astrocyte-targeting peptides were identified [373–375]; however, given astrocyte heterogeneity per se and functional remodeling of astrocytes in response to demyelination and inflammation, the knowledge on selective targeting of astrocyte subpopulations, and therefore, the suitability of these peptides to specifically target astrocytes in MS lesions, is still limited. Nevertheless, to specifically target cells in MS lesions, exploiting a peptide that binds to a lesion-specific environmental factor, e.g., Fn or other ECM components that are upregulated in MS lesions, likely represents an effective approach to accomplishing lesion-specific cellular drug delivery method for lesion-targeted functional drug delivery [359].

10. Concluding Remarks and Future Perspectives

MS is a heterogeneous disease involving inflammation, as well as neurodegenerative processes that simultaneously contribute to disease pathology. The underlying cause of MS is still unknown and likely to differ between patients. This heterogeneity in disease pathology asks for a multitude of treatment approaches as each stage of the disease and likely each lesion type has its hallmarks. Therefore, no single MS treatment is appropriate for all MS patients, nor will it be effective for all the different disease stages. This means that MS treatment is likely to benefit from personalized medicine. During later stages of the disease, impaired remyelination results in secondary neurodegeneration that overshadows the initial flares of demyelinating insults, meaning that a sustained and gradual decline in neurological functioning appears. Current therapeutic approaches mainly focus on mitigating the inflammatory components of the disease but do not halt underlying degenerative processes affecting disease progression. Several factors that are addressed in this review, including changes in ECM composition and stiffness, contribute to remyelination failure. Moreover, BBB malfunctioning appears to underlie disease initiation, while a relatively intact BBB concomitantly impedes disease treatment in progressive MS. We focused on a particular ECM aberration that is typical of chronically demyelinated lesions, namely, the occurrence of remyelination-impairing Fn aggregates, and discussed distinct means to specifically negate the effect of Fn aggregates. Given the role of Fn in BBB maintenance and angiogenesis, the release of pro-remyelination drugs that interfere with Fn signaling at the BBB should be prevented. Therefore, to overcome Fn-mediated inhibition of remyelination in chronic demyelinated lesions, we stress the importance of employing nanocarrier systems with a 'two-step approach': decoration of drug-containing nanoparticles with ligands that facilitate transcytosis across the BBB and delivery to Fn aggregates, OPCs, or astrocytes for lesion-specific drug release. Ideally, these nanocarriers possess a high drug-loading capacity, shield the drug from environmental influences, allow for controlled release of the drug, decrease the need for frequent dosing, and reduce undesirable side effects. Lipid (liposomal)- and polymer (PLGA)-based nanomedicines are being used for (experimental) MS nanomedication. Whether one or the other would be most suitable for remyelination-based treatment of MS lesions depends on the type of therapeutic to be delivered. For instance, hydrophilic and amphiphilic drugs have a low encapsulation efficiency into PLGA particles in contrast to hydrophobic compounds [376]. In turn, liposomes can encapsulate hydrophilic compounds in their aqueous core, preventing rapid clearance and enabling sustained release [377], and amphiphilic molecules in their lipid bilayer. For example, to bypass Fn aggregate-mediated inhibition of remyelination, the encapsulation of ganglioside GD1a, which is an amphiphilic compound, into liposomes would be preferred. Notably, lesion-specific alterations of the BBB and its BM were reported in MS [378], which may provide a means to specifically target nanomedicine toward lesions. Moreover, a study with an induced pluripotent stem cell-derived in vitro MS BBB model implied the existence of intrinsic differences in BBB functioning between RRMS patients and healthy controls [379]. Thus, to improve the targeting of nanomedicine to MS lesions, future advances in MS treatment require the identification of targets at the MS BBB and knowledge of the (patient-specific) MS lesion microenvironment to identify stimuli that can be used to induce lesion-specific drug release. Alternatively, therapeutic agents, such as GD1a, that treat lesions while leaving healthy tissue unaffected may obviate the need for lesion-specific targeting.

Besides functional recovery, no optimal measure of remyelination in patients exists yet [380]. Therefore, in parallel, clinical biomarkers for remyelination should be developed and optimized to assess the effectiveness of (remyelination) therapy. MRI is currently not a reliable measure of remyelination [381]. Instead, advanced MRI techniques, such as diffusion tensor imaging, which measures the diffusion of water along axons, is particularly useful for the imaging of white matter tracts in the brain [381,382]. Other potential biomarkers or diagnostic tools include the plasma levels of neurofilament light chains [383], position emission tomography tracers for myelin [384], and multifocal visually evoked potential [385]. As of yet, none of these techniques have the sensitivity required to assess the remyelination of individual MS lesions. Nonetheless, they may still provide benefits for the global assessment of remyelination therapy effectivity. An interesting development is the use of magnetic resonance elastography (MRE), which measures the mechanical properties of tissue [386,387]. Brain viscoelasticity is reduced in MS [388,389] and appears to correlate with demyelination and ECM degradation in the cuprizone model [390] and inflammation in EAE [391]. Advantageous to MRE is that it allows for in vivo, localized imaging of inflammatory lesions. Interestingly, reduced brain viscoelasticity correlates with the upregulated expression of cellular Fn during inflammation in EAE [392]. Nevertheless, even though MRE correlates with de- and remyelination in experimental models [390], it is not a direct measure of remyelination, nor has it been established as a direct measure of de- and remyelination in MS. Moreover, it was recently observed that MRE measures of demyelinated white matter and normal-appearing white matter did not differ [393]. Thus, the search for a reliable biomarker or diagnostic tool for the detection of remyelination in patients is still ongoing and will require further optimization of existing techniques or the development of new techniques in the future.

Author Contributions: All authors contributed to the conception of the manuscript. P.E.M.v.S. performed the literature search and drafted the manuscript, while I.S.Z. and W.B. critically revised the manuscript. All authors have read and agreed to the published version of the manuscript.

Funding: This work was financially supported by the Dutch MS Research Foundation (Stichting MS Research, 16-944 MS).

Conflicts of Interest: The authors declare no conflict of interest.

References

1. Oh, J.; Vidal-Jordana, A.; Montalban, X. Multiple Sclerosis. *Curr. Opin. Neurol.* **2018**, *31*, 752–759. [CrossRef] [PubMed]
2. Montalban, X.; Tintoré, M.; Swanton, J.; Barkhof, F.; Fazekas, F.; Filippi, M.; Frederiksen, J.; Kappos, L.; Palace, J.; Polman, C.; et al. MRI Criteria for MS in Patients with Clinically Isolated Syndromes. *Neurology* **2010**, *74*, 427–434. [CrossRef] [PubMed]
3. Mahad, D.H.; Trapp, B.D.; Lassmann, H. Pathological Mechanisms in Progressive Multiple Sclerosis. *Lancet Neurol.* **2015**, *14*, 183–193. [CrossRef]
4. Lublin, F.D.; Reingold, S.C. Defining the Clinical Course of Multiple Sclerosis: Results of an International Survey. *Neurology* **1996**, *46*, 907–911. [CrossRef] [PubMed]
5. Gourraud, P.-A.; Harbo, H.F.; Hauser, S.L.; Baranzini, S.E. The Genetics of Multiple Sclerosis: An up-to-Date Review. *Immunol. Rev.* **2012**, *248*, 87–103. [CrossRef]
6. Ascherio, A.; Munger, K.L. Environmental Risk Factors for Multiple Sclerosis. Part I: The Role of Infection. *Ann. Neurol.* **2007**, *61*, 288–299. [CrossRef] [PubMed]
7. Bjornevik, K.; Cortese, M.; Healy, B.C.; Kuhle, J.; Mina, M.J.; Leng, Y.; Elledge, S.J.; Niebuhr, D.W.; Scher, A.I.; Munger, K.L.; et al. Longitudinal Analysis Reveals High Prevalence of Epstein-Barr Virus Associated with Multiple Sclerosis. *Science* **2022**, *375*, 296–301. [CrossRef] [PubMed]
8. Amato, M.P.; Derfuss, T.; Hemmer, B.; Liblau, R.; Montalban, X.; Soelberg Sørensen, P.; Miller, D.H.; Alfredsson, L.; Aloisi, F.; Ascherio, A.; et al. Environmental Modifiable Risk Factors for Multiple Sclerosis: Report from the 2016 ECTRIMS Focused Workshop. *Mult. Scler. J.* **2018**, *24*, 590–603. [CrossRef]
9. Trapp, B.D.; Nave, K.-A. Multiple Sclerosis: An Immune or Neurodegenerative Disorder? *Annu. Rev. Neurosci.* **2008**, *31*, 247–269. [CrossRef] [PubMed]
10. Fünfschilling, U.; Supplie, L.M.; Mahad, D.; Boretius, S.; Saab, A.S.; Edgar, J.; Brinkmann, B.G.; Kassmann, C.M.; Tzvetanova, I.D.; Möbius, W.; et al. Glycolytic Oligodendrocytes Maintain Myelin and Long-Term Axonal Integrity. *Nature* **2012**, *485*, 517–521. [CrossRef]
11. Lee, Y.; Morrison, B.M.; Li, Y.; Lengacher, S.; Farah, M.H.; Hoffman, P.N.; Liu, Y.; Tsingalia, A.; Jin, L.; Zhang, P.W.; et al. Oligodendroglia Metabolically Support Axons and Contribute to Neurodegeneration. *Nature* **2012**, *487*, 443–448. [CrossRef] [PubMed]

12. Irvine, K.A.; Blakemore, W.F. Remyelination Protects Axons from Demyelination-Associated Axon Degeneration. *Brain* **2008**, *131*, 1464–1477. [CrossRef] [PubMed]
13. Franklin, R.J.M.; Frisén, J.; Lyons, D.A. Revisiting Remyelination: Towards a Consensus on the Regeneration of CNS Myelin. *Semin. Cell Dev. Biol.* **2020**, *116*, 3–9. [CrossRef] [PubMed]
14. Heß, K.; Starost, L.; Kieran, N.W.; Thomas, C.; Vincenten, M.C.J.; Antel, J.; Martino, G.; Huitinga, I.; Healy, L.; Kuhlmann, T. Lesion Stage-Dependent Causes for Impaired Remyelination in MS. *Acta Neuropathol.* **2020**, *140*, 359–375. [CrossRef] [PubMed]
15. de Jong, J.M.; Wang, P.; Oomkens, M.; Baron, W. Remodeling of the Interstitial Extracellular Matrix in White Matter Multiple Sclerosis Lesions: Implications for Remyelination (Failure). *J. Neurosci. Res.* **2020**, *98*, 1370–1397. [CrossRef] [PubMed]
16. Ghorbani, S.; Yong, V.W. The Extracellular Matrix as Modifier of Neuroinflammation and Remyelination in Multiple Sclerosis. *Brain* **2021**, *144*, 1958–1973. [CrossRef]
17. Stoffels, J.M.J.; de Jonge, J.C.; Stancic, M.; Nomden, A.; van Strien, M.E.; Ma, D.; Šišková, Z.; Maier, O.; ffrench-Constant, C.; Franklin, R.J.M.; et al. Fibronectin Aggregation in Multiple Sclerosis Lesions Impairs Remyelination. *Brain* **2013**, *136*, 116–131. [CrossRef]
18. Qin, J.; Sikkema, A.H.; van der Bij, K.; de Jonge, J.C.; Klappe, K.; Nies, V.; Jonker, J.W.; Kok, J.W.; Hoekstra, D.; Baron, W. GD1a Overcomes Inhibition of Myelination by Fibronectin via Activation of Protein Kinase A: Implications for Multiple Sclerosis. *J. Neurosci.* **2017**, *37*, 9925–9938. [CrossRef]
19. Hohlfeld, R.; Wekerle, H. Autoimmune Concepts of Multiple Sclerosis as a Basis for Selective Immunotherapy: From Pipe Dreams to (Therapeutic) Pipelines. *Proc. Natl. Acad. Sci. USA* **2004**, *101*, 14599–14606. [CrossRef]
20. Lucchinetti, C.; Brück, W.; Parisi, J.; Scheithauer, B.; Rodriguez, M.; Lassmann, H. Heterogeneity of Multiple Sclerosis Lesions: Implications for the Pathogenesis of Demyelination. *Ann. Neurol.* **2000**, *47*, 707–717. [CrossRef]
21. Qin, Y.; Duquette, P.; Zhang, Y.; Talbot, P.; Poole, R.; Antel, J. Clonal Expansion and Somatic Hypermutation of V(H) Genes of B Cells from Cerebrospinal Fluid in Multiple Sclerosis. *J. Clin. Investig.* **1998**, *102*, 1045–1050. [CrossRef] [PubMed]
22. Coles, A.J. Alemtuzumab Therapy for Multiple Sclerosis. *Neurotherapeutics* **2013**, *10*, 29–33. [CrossRef] [PubMed]
23. Bermel, R.A.; Bakshi, R. The Measurement and Clinical Relevance of Brain Atrophy in Multiple Sclerosis. *Lancet Neurol.* **2006**, *5*, 158–170. [PubMed]
24. Franklin, R.J.M.; ffrench-Constant, C. Regenerating CNS Myelin—From Mechanisms to Experimental Medicines. *Nat. Rev. Neurosci.* **2017**, *18*, 753–769. [CrossRef]
25. Kuhlmann, T.; Ludwin, S.; Prat, A.; Antel, J.; Brück, W.; Lassmann, H. An Updated Histological Classification System for Multiple Sclerosis Lesions. *Acta Neuropathol.* **2017**, *133*, 13–24. [CrossRef]
26. Luchetti, S.; Fransen, N.L.; van Eden, C.G.; Ramaglia, V.; Mason, M.; Huitinga, I. Progressive Multiple Sclerosis Patients Show Substantial Lesion Activity that Correlates with Clinical Disease Severity and Sex: A Retrospective Autopsy Cohort Analysis. *Acta Neuropathol.* **2018**, *135*, 511–528. [CrossRef] [PubMed]
27. Kutzelnigg, A.; Lucchinetti, C.F.; Stadelmann, C.; Brück, W.; Rauschka, H.; Bergmann, M.; Schmidbauer, M.; Parisi, J.E.; Lassmann, H. Cortical Demyelination and Diffuse White Matter Injury in Multiple Sclerosis. *Brain* **2005**, *128*, 2705–2712. [CrossRef]
28. Ferguson, B.; Matyszak, M.K.; Esiri, M.M.; Perry, V.H. Axonal Damage in Acute Multiple Sclerosis Lesions. *Brain* **1997**, *120*, 393–399. [CrossRef] [PubMed]
29. Trapp, B.D.; Peterson, J.; Ransohoff, R.M.; Rudick, R.; Mörk, S.; Bö, L. Axonal Transection in the Lesions of Multiple Sclerosis. *N. Engl. J. Med.* **1998**, *338*, 278–285. [CrossRef]
30. Peterson, J.W.; Bö, L.; Mörk, S.; Chang, A.; Trapp, B.D. Transected Neurites, Apoptotic Neurons, and Reduced Inflammation in Cortical Multiple Sclerosis Lesions. *Ann. Neurol.* **2001**, *50*, 389–400. [CrossRef]
31. Calabrese, M.; Agosta, F.; Rinaldi, F.; Mattisi, I.; Grossi, P.; Favaretto, A.; Atzori, M.; Bernardi, V.; Barachino, L.; Rinaldi, L.; et al. Cortical Lesions and Atrophy Associated with Cognitive Impairment in Relapsing-Remitting Multiple Sclerosis. *Arch. Neurol.* **2009**, *66*, 1144–1150. [CrossRef] [PubMed]
32. Nielsen, A.S.; Kinkel, R.P.; Madigan, N.; Tinelli, E.; Benner, T.; Mainero, C. Contribution of Cortical Lesion Subtypes at 7T MRI to Physical and Cognitive Performance in MS. *Neurology* **2013**, *81*, 641–649. [CrossRef] [PubMed]
33. Chiaravalloti, N.D.; DeLuca, J. Cognitive Impairment in Multiple Sclerosis. *Lancet. Neurol.* **2008**, *7*, 1139–1151. [CrossRef]
34. Kouvatsou, Z.; Masoura, E.; Kimiskidis, V. Working Memory Deficits in Multiple Sclerosis: An Overview of the Findings. *Front. Psychol.* **2022**, *13*, 866885. [CrossRef]
35. Santiago, O.; Guàrdia, J.; Casado, V.; Carmona, O.; Arbizu, T. Specificity of Frontal Dysfunctions in Relapsing–Remitting Multiple Sclerosis. *Arch. Clin. Neuropsychol.* **2007**, *22*, 623–629. [CrossRef]
36. Engl, C.; Tiemann, L.; Grahl, S.; Bussas, M.; Schmidt, P.; Pongratz, V.; Berthele, A.; Beer, A.; Gaser, C.; Kirschke, J.S.; et al. Cognitive Impairment in Early MS: Contribution of White Matter Lesions, Deep Grey Matter Atrophy, and Cortical Atrophy. *J. Neurol.* **2020**, *267*, 2307. [CrossRef] [PubMed]
37. Cunniffe, N.; Coles, A. Promoting Remyelination in Multiple Sclerosis. *J. Neurol.* **2021**, *268*, 30–44. [CrossRef] [PubMed]
38. Bramow, S.; Frischer, J.M.; Lassmann, H.; Koch-Henriksen, N.; Lucchinetti, C.F.; Sørensen, P.S.; Laursen, H. Demyelination versus Remyelination in Progressive Multiple Sclerosis. *Brain* **2010**, *133*, 2983–2998. [CrossRef]
39. Goldschmidt, T.; Antel, J.; König, F.B.; Brück, W.; Kuhlmann, T. Remyelination Capacity of the MS Brain Decreases with Disease Chronicity. *Neurology* **2009**, *72*, 1914–1921. [CrossRef]

40. Rodgers, J.M.; Robinson, A.P.; Miller, S.D. Strategies for Protecting Oligodendrocytes and Enhancing Remyelination in Multiple Sclerosis. *Discov. Med.* **2013**, *16*, 53–63.
41. Duncan, I.D.; Radcliff, A.B.; Heidari, M.; Kidd, G.; August, B.K.; Wierenga, L.A. The Adult Oligodendrocyte Can Participate in Remyelination. *Proc. Natl. Acad. Sci. USA* **2018**, *115*, e11807–e11816. [CrossRef] [PubMed]
42. Yeung, M.S.Y.; Djelloul, M.; Steiner, E.; Bernard, S.; Salehpour, M.; Possnert, G.; Brundin, L.; Frisén, J. Dynamics of Oligodendrocyte Generation in Multiple Sclerosis. *Nature* **2019**, *566*, 538–542. [CrossRef] [PubMed]
43. Neely, S.A.; Williamson, J.M.; Klingseisen, A.; Zoupi, L.; Early, J.J.; Williams, A.; Lyons, D.A. New Oligodendrocytes Exhibit More Abundant and Accurate Myelin Regeneration than Those That Survive Demyelination. *Nat. Neurosci.* **2022**, *25*, 415–420. [CrossRef]
44. Rivera, F.J.; Steffenhagen, C.; Kremer, D.; Kandasamy, M.; Sandner, B.; Couillard-Despres, S.; Weidner, N.; Küry, P.; Aigner, L. Deciphering the Oligodendrogenic Program of Neural Progenitors: Cell Intrinsic and Extrinsic Regulators. *Stem Cells Dev.* **2010**, *19*, 595–606. [CrossRef] [PubMed]
45. Wolswijk, G. Chronic Stage Multiple Sclerosis Lesions Contain a Relatively Quiescent Population of Oligodendrocyte Precursor Cells. *J. Neurosci.* **1998**, *18*, 601–609. [CrossRef]
46. Kuhlmann, T.; Miron, V.; Cuo, Q.; Wegner, C.; Antel, J.; Brück, W. Differentiation Block of Oligodendroglial Progenitor Cells as a Cause for Remyelination Failure in Chronic Multiple Sclerosis. *Brain* **2008**, *131*, 1749–1758. [CrossRef] [PubMed]
47. Boyd, A.; Zhang, H.; Williams, A. Insufficient OPC Migration into Demyelinated Lesions Is a Cause of Poor Remyelination in MS and Mouse Models. *Acta Neuropathol.* **2013**, *125*, 841–859. [CrossRef]
48. Plemel, J.R.; Liu, W.Q.; Yong, V.W. Remyelination Therapies: A New Direction and Challenge in Multiple Sclerosis. *Nat. Rev. Drug Discov.* **2017**, *16*, 617–634. [CrossRef]
49. Galloway, D.A.; Gowing, E.; Setayeshgar, S.; Kothary, R. Inhibitory Milieu at the Multiple Sclerosis Lesion Site and the Challenges for Remyelination. *Glia* **2020**, *68*, 859–877. [CrossRef]
50. Quraishe, S.; Forbes, L.H.; Andrews, M.R. The Extracellular Environment of the CNS: Influence on Plasticity, Sprouting, and Axonal Regeneration after Spinal Cord Injury. *Neural Plast.* **2018**, *2018*, 2952386. [CrossRef]
51. Câmara, J.; ffrench-Constant, C. Lessons from Oligodendrocyte Biology on Promoting Repair in Multiple Sclerosis. *J. Neurol.* **2007**, *254*, I15–I22. [CrossRef]
52. McTigue, D.M.; Tripathi, R.B. The Life, Death and Replacement of Oligodendrocytes in the Adult Central Nervous System. *J Neurochem.* **2008**, *107*, 1–19. [CrossRef]
53. Sim, F.J.; Zhao, C.; Penderis, J.; Franklin, R.J.M. The Age-Related Decrease in CNS Remyelination Efficiency Is Attributable to an Impairment of Both Oligodendrocyte Progenitor Recruitment and Differentiation. *J. Neurosci.* **2002**, *22*, 2451–2459. [CrossRef] [PubMed]
54. Neumann, B.; Segel, M.; Chalut, K.J.; Franklin, R.J.M. Remyelination and Ageing: Reversing the Ravages of Time. *Mult. Scler. J.* **2019**, *25*, 1835–1841. [CrossRef]
55. Alroughani, R.A.; Akhtar, S.; Ahmed, S.F.; Al-Hashel, J.Y. Clinical Predictors of Disease Progression in Multiple Sclerosis Patients with Relapsing Onset in a Nation-Wide Cohort. *Int. J. Neurosci.* **2015**, *125*, 831–837. [CrossRef] [PubMed]
56. Confavreux, C.; Vukusic, S. Age at Disability Milestones in Multiple Sclerosis. *Brain* **2006**, *129*, 595–605. [CrossRef]
57. Shields, S.A.; Gilson, J.M.; Blakemore, W.; Franklin, R.J. Remyelination Occurs as Extensively but More Slowly in Old Rats Compared to Young Rats Following Fliotoxin-Induced CNS Demyelination. *Glia* **1999**, *28*, 77–83. [CrossRef]
58. Shen, S.; Sandoval, J.; Swiss, V.A.; Li, J.; Dupree, J.; Franklin, R.J.M.; Casaccia-Bonnefil, P. Age-Dependent Epigenetic Control of Differentiation Inhibitors Is Critical for Remyelination Efficiency. *Nat. Neurosci.* **2008**, *11*, 1024–1034. [CrossRef] [PubMed]
59. Neumann, B.; Baror, R.; Zhao, C.; Segel, M.; Dietmann, S.; Rawji, K.S.; Foerster, S.; McClain, C.R.; Chalut, K.; van Wijngaarden, P.; et al. Metformin Restores CNS Remyelination Capacity by Rejuvenating Aged Stem Cells. *Cell Stem Cell* **2019**, *25*, 473–485. [CrossRef] [PubMed]
60. Segel, M.; Neumann, B.; Hill, M.F.E.; Weber, I.P.; Viscomi, C.; Zhao, C.; Young, A.; Agley, C.C.; Thompson, A.J.; Gonzalez, G.A.; et al. Niche Stiffness Underlies the Ageing of Central Nervous System Progenitor Cells. *Nature* **2019**, *573*, 130–134. [CrossRef] [PubMed]
61. Zobel, K.; Hansen, U.; Galla, H.J. Blood-Brain Barrier Properties in Vitro Depend on Composition and Assembly of Endogenous Extracellular Matrices. *Cell Tissue Res.* **2016**, *365*, 233–245. [CrossRef] [PubMed]
62. Davis, G.E.; Senger, D.R. Endothelial Extracellular Matrix: Biosynthesis, Remodeling, and Functions during Vascular Morphogenesis and Neovessel Stabilization. *Circ. Res.* **2005**, *97*, 1093–1107. [CrossRef] [PubMed]
63. Theocharis, A.D.; Skandalis, S.S.; Gialeli, C.; Karamanos, N.K. Extracellular Matrix Structure. *Adv. Drug Deliv. Rev.* **2016**, *97*, 4–27. [CrossRef] [PubMed]
64. Rozario, T.; DeSimone, D.W. The Extracellular Matrix in Development and Morphogenesis: A Dynamic View. *Dev. Biol.* **2010**, *341*, 126–140. [CrossRef] [PubMed]
65. Sorokin, L. The Impact of the Extracellular Matrix on Inflammation. *Nat. Rev. Immunol.* **2010**, *10*, 712–723. [CrossRef] [PubMed]
66. Boyd, D.F.; Thomas, P.G. Towards Integrating Extracellular Matrix and Immunological Pathways. *Cytokine* **2017**, *98*, 79–86. [CrossRef] [PubMed]
67. Schonherr, E.; Hausser, H.J. Extracellular Matrix and Cytokines: A Functional Unit. *Dev. Immunol.* **2000**, *7*, 89–101. [CrossRef]

68. Wijelath, E.S.; Rahman, S.; Namekata, M.; Murray, J.; Nishimura, T.; Mostafavi-Pour, Z.; Patel, Y.; Suda, Y.; Humphries, M.J.; Sobel, M. Heparin-II Domain of Fibronectin Is a Vascular Endothelial Growth Factor-Binding Domain: Enhancement of VEGF Biological Activity by a Singular Growth Factor/Matrix Protein Synergism. *Circ. Res.* **2006**, *99*, 853–860. [CrossRef]
69. Rahman, S.; Patel, Y.; Murray, J.; Patel, K.V.; Sumathipala, R.; Sobel, M.; Wijelath, E.S. Novel Hepatocyte Growth Factor (HGF) Binding Domains on Fibronectin and Vitronectin Coordinate a Distinct and Amplified Met-Integrin Induced Signalling Pathway in Endothelial Cells. *BMC Cell Biol.* **2005**, *6*, 8. [CrossRef] [PubMed]
70. Bauer, N.G.; ffrench-Constant, C. Physical Forces in Myelination and Repair: A Question of Balance? *J. Biol.* **2009**, *8*, 78. [CrossRef] [PubMed]
71. Saha, K.; Keung, A.J.; Irwin, E.F.; Li, Y.; Little, L.; Schaffer, D.V.; Healy, K.E. Substrate Modulus Directs Neural Stem Cell Behavior. *Biophys. J.* **2008**, *95*, 4426–4438. [CrossRef] [PubMed]
72. Jagielska, A.; Norman, A.L.; Whyte, G.; Vliet, K.J.V.; Guck, J.; Franklin, R.J.M. Mechanical Environment Modulates Biological Properties of Oligodendrocyte Progenitor Cells. *Stem Cells Dev.* **2012**, *21*, 2905–2914. [CrossRef] [PubMed]
73. Urbanski, M.M.; Kingsbury, L.; Moussouros, D.; Kassim, I.; Mehjabeen, S.; Paknejad, N.; Melendez-Vasquez, C.V. Myelinating Glia Differentiation Is Regulated by Extracellular Matrix Elasticity. *Sci. Rep.* **2016**, *6*, 33751. [CrossRef] [PubMed]
74. Stoffels, J.M.J.; Hoekstra, D.; Franklin, R.J.M.; Baron, W.; Zhao, C. The EIIIA Domain from Astrocyte-Derived Fibronectin Mediates Proliferation of Oligodendrocyte Progenitor Cells Following CNS Demyelination. *Glia* **2015**, *63*, 242–256. [CrossRef] [PubMed]
75. Lourenço, T.; Paes De Faria, J.; Bippes, C.A.; Maia, J.; Lopes-Da-Silva, J.A.; Relvas, J.B.; Graõs, M. Modulation of Oligodendrocyte Differentiation and Maturation by Combined Biochemical and Mechanical Cues. *Sci. Rep.* **2016**, *6*, 21563. [CrossRef]
76. Buttery, P.C.; ffrench-Constant, C. Laminin-2/Integrin Interactions Enhance Myelin Membrane Formation by Oligodendrocytes. *Mol. Cell. Neurosci.* **1999**, *14*, 199–212. [CrossRef] [PubMed]
77. Colognato, H.; Galvin, J.; Wang, Z.; Relucio, J.; Nguyen, T.; Harrison, D.; Yurchenco, P.D.; ffrench-Constant, C. Identification of Dystroglycan as a Second Laminin Receptor in Oligodendrocytes, with a Role in Myelination. *Development* **2007**, *134*, 1723–1736. [CrossRef] [PubMed]
78. Urbanski, M.M.; Brendel, M.B.; Melendez-Vasquez, C.V. Acute and Chronic Demyelinated CNS Lesions Exhibit Opposite Elastic Properties. *Sci. Rep.* **2019**, *9*, 999. [CrossRef] [PubMed]
79. Yi, J.-H.; Katagiri, Y.; Susarla, B.; Figge, D.; Symes, A.J.; Geller, H.M. Alterations in Sulfated Chondroitin Glycosaminoglycans Following Controlled Cortical Impact Injury in Mice. *J. Comp. Neurol.* **2012**, *520*, 3295–3313. [CrossRef]
80. Carulli, D.; Rhodes, K.E.; Brown, D.J.; Bonnert, T.P.; Pollack, S.J.; Oliver, K.; Strata, P.; Fawcett, J.W. Composition of Perineuronal Nets in the Adult Rat Cerebellum and the Cellular Origin of Their Components. *J. Comp. Neurol.* **2006**, *494*, 559–577. [CrossRef]
81. Hara, M.; Kobayakawa, K.; Ohkawa, Y.; Kumamaru, H.; Yokota, K.; Saito, T.; Kijima, K.; Yoshizaki, S.; Harimaya, K.; Nakashima, Y.; et al. Interaction of Reactive Astrocytes with Type I Collagen Induces Astrocytic Scar Formation through the Integrin-N-Cadherin Pathway after Spinal Cord Injury. *Nat. Med.* **2017**, *23*, 818–828. [CrossRef] [PubMed]
82. Lau, L.W.; Keough, M.B.; Haylock-Jacobs, S.; Cua, R.; Döring, A.; Sloka, S.; Stirling, D.P.; Rivest, S.; Yong, V.W. Chondroitin Sulfate Proteoglycans in Demyelinated Lesions Impair Remyelination. *Ann. Neurol.* **2012**, *72*, 419–432. [CrossRef] [PubMed]
83. Pendleton, J.C.; Shamblott, M.J.; Gary, D.S.; Belegu, V.; Hurtado, A.; Malone, M.L.; McDonald, J.W. Chondroitin Sulfate Proteoglycans Inhibit Oligodendrocyte Myelination through PTPσ. *Exp. Neurol.* **2013**, *247*, 113–121. [CrossRef]
84. Hibbits, N.; Yoshino, J.; Le, T.Q.; Armstrong, R.C. Astrogliosis during Acute and Chronic Cuprizone Demyelination and Implications for Remyelination. *ASN Neuro* **2012**, *4*, 393–408. [CrossRef]
85. Karimi-Abdolrezaee, S.; Schut, D.; Wang, J.; Fehlings, M.G. Chondroitinase and Growth Factors Enhance Activation and Oligodendrocyte Differentiation of Endogenous Neural Precursor Cells after Spinal Cord Injury. *PLoS ONE* **2012**, *7*, e37589. [CrossRef] [PubMed]
86. Siebert, J.R.; Stelzner, D.J.; Osterhout, D.J. Chondroitinase Treatment Following Spinal Contusion Injury Increases Migration of Oligodendrocyte Progenitor Cells. *Exp. Neurol.* **2011**, *231*, 19–29. [CrossRef]
87. Kumari, S.; Mak, M.; Poh, Y.; Tohme, M.; Watson, N.; Melo, M.; Janssen, E.; Dustin, M.; Geha, R.; Irvine, D.J. Cytoskeletal Tension Actively Sustains the Migratory T-Cell Synaptic Contact. *EMBO J.* **2020**, *39*, e102783. [CrossRef]
88. Keough, M.B.; Rogers, J.A.; Zhang, P.; Jensen, S.K.; Stephenson, E.L.; Chen, T.; Hurlbert, M.G.; Lau, L.W.; Rawji, K.S.; Plemel, J.R.; et al. An Inhibitor of Chondroitin Sulfate Proteoglycan Synthesis Promotes Central Nervous System Remyelination. *Nat. Commun.* **2016**, *7*, 11312. [CrossRef]
89. Stephenson, E.L.; Zhang, P.; Ghorbani, S.; Wang, A.; Gu, J.; Keough, M.B.; Rawji, K.S.; Silva, C.; Yong, V.W.; Ling, C.C. Targeting the Chondroitin Sulfate Proteoglycans: Evaluating Fluorinated Glucosamines and Xylosides in Screens Pertinent to Multiple Sclerosis. *ACS Cent. Sci.* **2019**, *5*, 1223–1234. [CrossRef]
90. Pu, A.; Mishra, M.K.; Dong, Y.; Ghorbanigazar, S.; Stephenson, E.L.; Rawji, K.S.; Silva, C.; Kitagawa, H.; Sawcer, S.; Yong, V.W. The Glycosyltransferase EXTL2 Promotes Proteoglycan Deposition and Injurious Neuroinflammation Following Demyelination. *J. Neuroinflamm.* **2020**, *17*, 220. [CrossRef]
91. Luo, F.; Tran, A.P.; Xin, L.; Sanapala, C.; Lang, B.T.; Silver, J.; Yang, Y. Modulation of Proteoglycan Receptor PTPσ Enhances MMP-2 Activity to Promote Recovery from Multiple Sclerosis. *Nat. Commun.* **2018**, *9*, 4126. [CrossRef]
92. Pu, A.; Stephenson, E.L.; Yong, V.W. The Extracellular Matrix: Focus on Oligodendrocyte Biology and Targeting CSPGs for Remyelination Therapies. *Glia* **2018**, *66*, 1809–1825. [CrossRef] [PubMed]

93. Lau, L.W.; Cua, R.; Keough, M.B.; Haylock-Jacobs, S.; Yong, V.W. Pathophysiology of the Brain Extracellular Matrix: A New Target for Remyelination. *Nat. Rev. Neurosci.* **2013**, *14*, 722–729. [CrossRef]
94. Patel, R.S.; Odermatt, E.; Schwarzbauer, J.E.; Hynes, R.O. Organization of the Fibronectin Gene Provides Evidence for Exon Shuffling during Evolution. *EMBO J.* **1987**, *6*, 2565–2572. [CrossRef] [PubMed]
95. ffrench-Constant, C. Alternative Splicing of Fibronectin-Many Different Proteins but Few Different Functions. *Exp. Cell Res.* **1995**, *221*, 261–271. [CrossRef]
96. Pankov, R.; Yamada, K.M. Fibronectin at a Glance. *J. Cell Sci.* **2002**, *115*, 3861–3863. [CrossRef] [PubMed]
97. Plow, E.F.; Haas, T.A.; Zhang, L.; Loftus, J.; Smith, J.W. Ligand Binding to Integrins. *J. Biol. Chem.* **2000**, *275*, 21785–21788. [CrossRef]
98. Murphy, P.A.; Begum, S.; Hynes, R.O. Tumor Angiogenesis in the Absence of Fibronectin or Its Cognate Integrin Receptors. *PLoS ONE* **2015**, *10*, e0120872. [CrossRef] [PubMed]
99. Hynes, R.O. Integrins: Versatility, Modulation, and Signaling in Cell Adhesion. *Cell* **1992**, *69*, 11–25. [CrossRef]
100. Sastry, S.K.; Horwitz, A.F. Integrin Cytoplasmic Domains: Mediators of Cytoskeletal Linkages and Extra- and Intracellular Initiated Transmembrane Signaling. *Curr. Opin. Cell Biol.* **1993**, *5*, 819–831. [CrossRef]
101. Diamond, M.S.; Springer, T.A. The Dynamic Regulation of Integrin Adhesiveness. *Curr. Biol.* **1994**, *4*, 506–517. [CrossRef]
102. Wolburg, H.; Noell, S.; Mack, A.; Wolburg-Buchholz, K.; Fallier-Becker, P. Brain Endothelial Cells and the Glio-Vascular Complex. *Cell Tissue Res.* **2009**, *335*, 75–96. [CrossRef] [PubMed]
103. Mathiisen, T.M.; Lehre, K.P.; Danbolt, N.C.; Ottersen, O.P. The Perivascular Astroglial Sheath Provides a Complete Covering of the Brain Microvessels: An Electron Microscopic 3D Reconstruction. *Glia* **2010**, *58*, 1094–1103. [CrossRef] [PubMed]
104. Sekiguchi, R.; Yamada, K.M. Basement Membranes in Development and Disease. *Curr. Top. Dev. Biol.* **2018**, *130*, 143–191.
105. Owens, T.; Bechmann, I.; Engelhardt, B. Perivascular Spaces and the Two Steps to Neuroinflammation. *J. Neuropathol. Exp. Neurol.* **2008**, *67*, 1113–1121. [CrossRef]
106. Baeten, K.M.; Akassoglou, K. Extracellular Matrix and Matrix Receptors in Blood-Brain Barrier Formation and Stroke. *Dev. Neurobiol.* **2011**, *71*, 1018–1039. [CrossRef] [PubMed]
107. van Horssen, J.; Bö, L.; Vos, C.M.P.; Virtanen, I.; de Vries, H.E. Basement Membrane Proteins in Multiple Sclerosis-Associated Inflammatory Cuffs: Potential Role in Influx and Transport of Leukocytes. *J. Neuropathol. Exp. Neurol.* **2005**, *64*, 722–729. [CrossRef] [PubMed]
108. Kouwenhoven, M.; Özenci, V.; Gomes, A.; Yarilin, D.; Giedraitis, V.; Press, R.; Link, H. Multiple Sclerosis: Elevated Expression of Matrix Metalloproteinases in Blood Monocytes. *J. Autoimmun.* **2001**, *16*, 463–470. [CrossRef]
109. Korn, J.; Christ, B.; Kurz, H. Neuroectodermal Origin of Brain Pericytes and Vascular Smooth Muscle Cells. *J. Comp. Neurol.* **2002**, *442*, 78–88. [CrossRef] [PubMed]
110. Etchevers, H.C.; Vincent, C.; Le Douarin, N.M.; Couly, G.F. The Cephalic Neural Crest Provides Pericytes and Smooth Muscle Cells to All Blood Vessels of the Face and Forebrain. *Development* **2001**, *128*, 1059–1068. [CrossRef]
111. Yamanishi, E.; Takahashi, M.; Saga, Y.; Osumi, N. Penetration and Differentiation of Cephalic Neural Crest-Derived Cells in the Developing Mouse Telencephalon. *Dev. Growth Differ.* **2012**, *54*, 785–800. [CrossRef]
112. Yamamoto, S.; Muramatsu, M.; Azuma, E.; Ikutani, M.; Nagai, Y.; Sagara, H.; Koo, B.N.; Kita, S.; O'Donnell, E.; Osawa, T.; et al. A Subset of Cerebrovascular Pericytes Originates from Mature Macrophages in the Very Early Phase of Vascular Development in CNS. *Sci. Rep.* **2017**, *7*, 3855. [CrossRef] [PubMed]
113. Muoio, V.; Persson, P.B.; Sendeski, M.M. The Neurovascular Unit-Concept Review. *Acta Physiol.* **2014**, *210*, 790–798. [CrossRef]
114. Preston, J.E.; Joan Abbott, N.; Begley, D.J. Transcytosis of Macromolecules at the Blood-Brain Barrier. *Adv. Pharmacol.* **2014**, *71*, 1147–1163. [CrossRef]
115. Tilling, T.; Engelbertz, C.; Decker, S.; Korte, D.; Hüwel, S.; Galla, H.J. Expression and Adhesive Properties of Basement Membrane Proteins in Cerebral Capillary Endothelial Cell Cultures. *Cell Tissue Res.* **2002**, *310*, 19–29. [CrossRef] [PubMed]
116. Yurchenco, P.; Patton, B. Developmental and Pathogenic Mechanisms of Basement Membrane Assembly. *Curr. Pharm. Des.* **2009**, *15*, 1277–1294. [CrossRef] [PubMed]
117. Yurchenco, P.D. Basement Membranes: Cell Scaffoldings and Signaling Platforms. *Cold Spring Harb. Perspect. Biol.* **2011**, *3*, a004911. [CrossRef] [PubMed]
118. Tilling, T.; Korte, D.; Hoheisel, D.; Galla, H.-J. Basement Membrane Proteins Influence Brain Capillary Endothelial Barrier Function In Vitro. *J. Neurochem.* **2002**, *71*, 1151–1157. [CrossRef]
119. Wolburg, H.; Neuhaus, J.; Kniesel, U.; Krauß, B.; Schmid, E.M.; Öcalan, M.; Farrell, C.; Risau, W. Modulation of Tight Junction Structure in Blood-Brain Barrier Endothelial Cells: Effects of Tissue Culture, Second Messengers and Cocultured Astrocytes. *J. Cell Sci.* **1994**, *107*, 1347–1357. [CrossRef] [PubMed]
120. Canfield, S.G.; Stebbins, M.J.; Faubion, M.G.; Gastfriend, B.D.; Palecek, S.P.; Shusta, E.V. An Isogenic Neurovascular Unit Model Comprised of Human Induced Pluripotent Stem Cell-Derived Brain Microvascular Endothelial Cells, Pericytes, Astrocytes, and Neurons. *Fluids Barriers CNS* **2019**, *16*, 25. [CrossRef]
121. Nakagawa, S.; Deli, M.A.; Kawaguchi, H.; Shimizudani, T.; Shimono, T.; Kittel, Á.; Tanaka, K.; Niwa, M. A New Blood-Brain Barrier Model Using Primary Rat Brain Endothelial Cells, Pericytes and Astrocytes. *Neurochem. Int.* **2009**, *54*, 253–263. [CrossRef]
122. Hartmann, C.; Zozulya, A.; Wegener, J.; Galla, H.J. The Impact of Glia-Derived Extracellular Matrices on the Barrier Function of Cerebral Endothelial Cells: An in Vitro Study. *Exp. Cell Res.* **2007**, *313*, 1318–1325. [CrossRef]

123. Wang, J.; Milner, R. Fibronectin Promotes Brain Capillary Endothelial Cell Survival and Proliferation through αvβ1 and αvβ3 Integrins via MAP Kinase Signalling. *J. Neurochem.* **2006**, *96*, 148–159. [CrossRef]
124. Mettouchi, A.; Klein, S.; Guo, W.; Lopez-Lago, M.; Lemichez, E.; Westwick, J.K.; Giancotti, F.G. Integrin-Specific Activation of Rac Controls Progression through the G1 Phase of the Cell Cycle. *Mol. Cell* **2001**, *8*, 115–127. [CrossRef]
125. Wang, Y.; Jin, G.; Miao, H.; Li, J.Y.S.; Usami, S.; Chien, S. Integrins Regulate VE-Cadherin and Catenins: Dependence of This Regulation on Src, but Not on Ras. *Proc. Natl. Acad. Sci. USA* **2006**, *103*, 1774–1779. [CrossRef]
126. Alghisi, G.C.; Ponsonnet, L.; Rüegg, C. The Integrin Antagonist Cilengitide Activates AVβ3, Disrupts VE-Cadherin Localization at Cell Junctions and Enhances Permeability in Endothelial Cells. *PLoS ONE* **2009**, *4*, e4449. [CrossRef] [PubMed]
127. Pulous, F.E.; Petrich, B.G. Integrin-Dependent Regulation of the Endothelial Barrier. *Tissue Barriers* **2019**, *7*, 1685844. [CrossRef]
128. Yadav, R.; Larbi, K.Y.; Young, R.E.; Nourshargh, S. Migration of Leukocytes through the Vessel Wall and Beyond. *Thromb. Haemost.* **2003**, *90*, 598–606. [CrossRef]
129. Rosenberg, G.A.; Kornfeld, M.; Estrada, E.; Kelley, R.O.; Liotta, L.A.; Stetler-Stevenson, W.G. TIMP-2 Reduces Proteolytic Opening of Blood-Brain Barrier by Type IV Collagenase. *Brain Res.* **1992**, *576*, 203–207. [CrossRef]
130. Brundula, V.; Rewcastle, N.B.; Metz, L.M.; Bernard, C.C.; Yong, V.W. Targeting Leukocyte MMPs and TransmigrationMinocycline as a Potential Therapy for Multiple Sclerosis. *Brain* **2002**, *125*, 1297–1308. [CrossRef]
131. Sobel, R.A.; Mitchell, M.E.; Fondren, G. Intercellular Adhesion Molecule-1 (ICAM-1) in Cellular Immune Reactions in the Human Central Nervous System. *Am. J. Pathol.* **1990**, *136*, 1309–1316. [PubMed]
132. Greenwood, J.; Etienne-Manneville, S.; Adamson, P.; Couraud, P.-O. Lymphocyte Migration into the Central Nervous System: Implication of ICAM-1 Signalling at the Blood-Brain Barrier. *Vascul. Pharmacol.* **2002**, *38*, 315–322. [CrossRef]
133. van der Laan, L.J.W.; Groot, C.J.A.; De Elices, M.J.; Dijkstra, C.D. Extracellular Matrix Proteins Expressed by Human Adult Astrocytes in Vivo and in Vitro: An Astrocyte Surface Protein Containing the CS1 Domain Contributes to Binding of Lymphoblasts. *J. Neurosci. Res.* **1997**, *50*, 539–548. [CrossRef]
134. Guan, J.L.; Hynes, R.O. Lymphoid Cells Recognize an Alternatively Spliced Segment of Fibronectin via the Integrin Receptor α4β1. *Cell* **1990**, *60*, 53–61. [CrossRef]
135. van der Laan, L.J.W.; van der Goes, A.; Wauben, M.H.M.; Ruuls, S.R.; Döpp, A.; de Groot, C.J.A.; Kuijpers, T.W.; Elices, M.J.; Dijkstra, C.D. Beneficial Effect of Modified Peptide Inhibitor of α4 Integrins on Experimental Allergic Encephalomyelitis in Lewis Rats. *J. Neurosci. Res.* **2002**, *67*, 191–199. [CrossRef]
136. Kent, S.J.; Karlik, S.J.; Cannon, C.; Hines, D.K.; Yednock, T.A.; Fritz, L.C.; Horner, H.C. A Monoclonal Antibody to α4 Integrin Suppresses and Reverses Active Experimental Allergic Encephalomyelitis. *J. Neuroimmunol.* **1995**, *58*, 1–10. [CrossRef]
137. Yu, Y.; Schürpf, T.; Springer, T.A. How Natalizumab Binds and Antagonizes A4 Integrins. *J. Biol. Chem.* **2013**, *288*, 32314–32325. [CrossRef]
138. Shirani, A.; Stüve, O. Natalizumab: Perspectives from the Bench to Bedside. *Cold Spring Harb. Perspect. Med.* **2018**, *8*, a029066. [CrossRef]
139. Clerico, M.; Artusi, C.A.; Di Liberto, A.; Rolla, S.; Bardina, V.; Barbero, P.; De Mercanti, S.F.; Durelli, L. Natalizumab in Multiple Sclerosis: Long-Term Management. *Int. J. Mol. Sci.* **2017**, *18*, 940. [CrossRef] [PubMed]
140. Melincovici, C.S.; Boşca, A.B.; Şuşman, S.; Mărginean, M.; Mihu, C.; Istrate, M.; Moldovan, I.M.; Roman, A.L.; Mihu, C.M. Vascular Endothelial Growth Factor (VEGF)—Key Factor in Normal and Pathological Angiogenesis. *Rom. J. Morphol. Embryol.* **2018**, *59*, 455–467.
141. Hao, W.; Han, J.; Chu, Y.; Huang, L.; Zhuang, Y.; Sun, J.; Li, X.; Zhao, Y.; Chen, Y.; Dai, J. Collagen/Heparin Bi-Affinity Multilayer Modified Collagen Scaffolds for Controlled bFGF Release to Improve Angiogenesis In Vivo. *Macromol. Biosci.* **2018**, *18*, e1800086. [CrossRef]
142. Bloch, W.; Forsberg, E.; Lentini, S.; Brakebusch, C.; Martin, K.; Krell, H.W.; Weidle, U.H.; Addicks, K.; Fässler, R. β1 Integrin Is Essential for Teratoma Growth and Angiogenesis. *J. Cell Biol.* **1997**, *139*, 265–278. [CrossRef]
143. Giancotti, F.G.; Ruoslahti, E. Integrin Signaling. *Science* **1999**, *285*, 1028–1032. [CrossRef] [PubMed]
144. Ingber, D.E.; Folkman, J. Mechanochemical Switching between Growth and Differentiation during Fibroblast Growth Factor-Stimulated Angiogenesis in Vitro: Role of Extracellular Matrix. *J. Cell Biol.* **1989**, *109*, 317–330. [CrossRef]
145. McIntosh, L.C.; Muckersie, L.; Forrester, J.V. Retinal Capillary Endothelial Cells Prefer Different Substrates for Growth and Migration. *Tissue Cell* **1988**, *20*, 193–209. [CrossRef]
146. Kirkpatrick, C.J.; Kampe, M.; Rixen, H.; Fischer, E.G.; Ruchatz, D.; Mittermayer, C. In Vitro Studies on the Expansion of Endothelial Cell Monolayers on Components of the Basement Membrane. *Virchows Arch. B Cell Pathol. Incl. Mol. Pathol.* **1990**, *58*, 207–213. [CrossRef] [PubMed]
147. Grant, D.S.; Kleinman, H.K. Regulation of Capillary Formation by Laminin and Other Components of the Extracellular Matrix. *EXS* **1997**, *79*, 317–333.
148. Grant, D.S.; Tashiro, K.I.; Segui-Real, B.; Yamada, Y.; Martin, G.R.; Kleinman, H.K. Two Different Laminin Domains Mediate the Differentiation of Human Endothelial Cells into Capillary-like Structures in Vitro. *Cell* **1989**, *58*, 933–943. [CrossRef]
149. Roscoe, W.A.; Welsh, M.E.; Carter, D.E.; Karlik, S.J. VEGF and Angiogenesis in Acute and Chronic MOG(35-55) Peptide Induced EAE. *J. Neuroimmunol.* **2009**, *209*, 6–15. [CrossRef]

150. Seabrook, T.J.; Littlewood-Evans, A.; Brinkmann, V.; Pöllinger, B.; Schnell, C.; Hiestand, P.C. Angiogenesis Is Present in Experimental Autoimmune Encephalomyelitis and Pro-Angiogenic Factors Are Increased in Multiple Sclerosis Lesions. *J. Neuroinflamm.* **2010**, *7*, 95. [CrossRef]
151. Lassmann, H. Hypoxia-like Tissue Injury as a Component of Multiple Sclerosis Lesions. *J. Neurol. Sci.* **2003**, *206*, 187–191. [CrossRef]
152. Trapp, B.D.; Stys, P.K. Virtual Hypoxia and Chronic Necrosis of Demyelinated Axons in Multiple Sclerosis. *Lancet Neurol.* **2009**, *8*, 280–291. [CrossRef]
153. Li, L.; Welser-Alves, J.; van der Flier, A.; Boroujerdi, A.; Hynes, R.O.; Milner, R. An Angiogenic Role for the α5β1 Integrin in Promoting Endothelial Cell Proliferation during Cerebral Hypoxia. *Exp. Neurol.* **2012**, *237*, 46–54. [CrossRef] [PubMed]
154. Boroujerdi, A.; Welser-Alves, J.V.; Milner, R. Extensive Vascular Remodeling in the Spinal Cord of Pre-Symptomatic Experimental Autoimmune Encephalomyelitis Mice; Increased Vessel Expression of Fibronectin and the α5β1 Integrin. *Exp. Neurol.* **2013**, *250*, 43–51. [CrossRef]
155. Holley, J.E.; Newcombe, J.; Whatmore, J.L.; Gutowski, N.J. Increased Blood Vessel Density and Endothelial Cell Proliferation in Multiple Sclerosis Cerebral White Matter. *Neurosci. Lett.* **2010**, *470*, 65–70. [CrossRef] [PubMed]
156. Sobel, R.A.; Mitchell, M.E. Fibronectin in Multiple Sclerosis Lesions. *Am. J. Pathol.* **1989**, *135*, 161–168. [PubMed]
157. Peters, J.H.; Loredo, G.A.; Chen, G.; Maunder, R.; Hahn, T.J.; Willits, N.H.; Hynes, R.O. Plasma Levels of Fibronectin Bearing the Alternatively Spliced EIIIB Segment Are Increased after Major Trauma. *J. Lab. Clin. Med.* **2003**, *141*, 401–410. [CrossRef]
158. Castellanos, M.; Leira, R.; Serena, J.; Blanco, M.; Pedraza, S.; Castillo, J.; Dávalos, A. Plasma Cellular-Fibronectin Concentration Predicts Hemorrhagic Transformation after Thrombolytic Therapy in Acute Ischemic Stroke. *Stroke* **2004**, *35*, 1671–1676. [CrossRef] [PubMed]
159. Stoffels, J.M.J.; Zhao, C.; Baron, W. Fibronectin in Tissue Regeneration: Timely Disassembly of the Scaffold Is Necessary to Complete the Build. *Cell. Mol. Life Sci.* **2013**, *70*, 4243–4253. [CrossRef] [PubMed]
160. Zhao, C.; Fancy, S.P.J.; Franklin, R.J.M.; ffrench-Constant, C. Up-Regulation of Oligodendrocyte Precursor Cell αv Integrin and Its Extracellular Ligands during Central Nervous System Remyelination. *J. Neurosci. Res.* **2009**, *87*, 3447–3455. [CrossRef]
161. Espitia Pinzon, N.; Sanz-Morello, B.; Brevé, J.J.P.; Bol, J.G.J.M.; Drukarch, B.; Bauer, J.; Baron, W.; van Dam, A.M. Astrocyte-Derived Tissue Transglutaminase Affects Fibronectin Deposition, but Not Aggregation, during Cuprizone-Induced Demyelination. *Sci. Rep.* **2017**, *7*, 40995. [CrossRef]
162. Paul, J.; Strickland, S.; Melchor, J.P. Fibrin Deposition Accelerates Neurovascular Damage and Neuroinflammation in Mouse Models of Alzheimer's Disease. *J. Exp. Med.* **2007**, *204*, 1999–2008. [CrossRef] [PubMed]
163. Kermode, A.G.; Thompson, A.J.; Tofts, P.; Macmanus, D.G.; Kendall, B.E.; Kingsley, D.P.E.; Moseley, I.F.; Rudge, P.; Mcdonald, W.I. Breakdown of the Blood-Brain Barrier Precedes Symptoms and Other Mri Signs of New Lesions in Multiple Sclerosis: Pathogenetic and Clinical Implications. *Brain* **1990**, *113*, 1477–1489. [CrossRef] [PubMed]
164. Huber, J.D.; Egleton, R.D.; Davis, T.P. Molecular Physiology and Pathophysiology of Tight Junctions in the Blood -Brain Barrier. *Trends Neurosci.* **2001**, *24*, 719–725. [CrossRef]
165. Baron, W.; Colognato, H.; ffrench-Constant, C. Integrin-Growth Factor Interactions as Regulators of Oligodendroglial Development and Function. *Glia* **2005**, *49*, 467–479. [CrossRef] [PubMed]
166. Baron, W.; Shattil, S.J.; ffrench-Constant, C. The Oligodendrocyte Precursor Mitogen PDGF Stimulates Proliferation by Activation of αvβ3 Integrins. *EMBO J.* **2002**, *21*, 1957–1966. [CrossRef]
167. Milner, R.; Edwards, G.; Streuli, C.; ffrench-Constant, C. A Role in Migration for the αvβ1 Integrin Expressed on Oligodendrocyte Precursors. *J. Neurosci.* **1996**, *16*, 7240–7252. [CrossRef] [PubMed]
168. Maier, O.; van der Heide, T.; van Dam, A.M.; Baron, W.; de Vries, H.; Hoekstra, D. Alteration of the Extracellular Matrix Interferes with Raft Association of Neurofascin in Oligodendrocytes. Potential Significance for Multiple Sclerosis? *Mol. Cell. Neurosci.* **2005**, *28*, 390–401. [CrossRef] [PubMed]
169. Šišková, Z.; Baron, W.; de Vries, H.; Hoekstra, D. Fibronectin Impedes "Myelin" Sheet-Directed Flow in Oligodendrocytes: A Role for a Beta 1 Integrin-Mediated PKC Signaling Pathway in Vesicular Trafficking. *Mol. Cell. Neurosci.* **2006**, *33*, 150–159. [CrossRef]
170. Šišková, Z.; Yong, V.W.; Nomden, A.; van Strien, M.; Hoekstra, D.; Baron, W. Fibronectin Attenuates Process Outgrowth in Oligodendrocytes by Mislocalizing MMP-9 Activity. *Mol. Cell. Neurosci.* **2009**, *42*, 234–242. [CrossRef]
171. Baron, W.; Bijlard, M.; Nomden, A.; de Jonge, J.C.; Teunissen, C.E.; Hoekstra, D. Sulfatide-Mediated Control of Extracellular Matrix-Dependent Oligodendrocyte Maturation. *Glia* **2014**, *62*, 927–942. [CrossRef] [PubMed]
172. Natrajan, M.S.; Komori, M.; Kosa, P.; Johnson, K.R.; Wu, T.; Franklin, R.J.M.; Bielekova, B. Pioglitazone Regulates Myelin Phagocytosis and Multiple Sclerosis Monocytes. *Ann. Clin. Transl. Neurol.* **2015**, *2*, 1071–1084. [CrossRef] [PubMed]
173. Ruckh, J.M.; Zhao, J.W.; Shadrach, J.L.; van Wijngaarden, P.; Rao, T.N.; Wagers, A.J.; Franklin, R.J.M. Rejuvenation of Regeneration in the Aging Central Nervous System. *Cell Stem Cell* **2012**, *10*, 96–103. [CrossRef] [PubMed]
174. Kotter, M.R.; Li, W.W.; Zhao, C.; Franklin, R.J.M. Myelin Impairs CNS Remyelination by Inhibiting Oligodendrocyte Precursor Cell Differentiation. *J. Neurosci.* **2006**, *26*, 328–332. [CrossRef] [PubMed]
175. Neumann, H.; Kotter, M.R.; Franklin, R.J.M. Debris Clearance by Microglia: An Essential Link between Degeneration and Regeneration. *Brain* **2009**, *132*, 288–295. [CrossRef] [PubMed]
176. Ousman, S.S.; David, S. Lysophosphatidylcholine Induces Rapid Recruitment and Activation of Macrophages in the Adult Mouse Spinal Cord. *Glia* **2000**, *30*, 92–104. [CrossRef]

177. Cavone, L.; McCann, T.; Drake, L.K.; Aguzzi, E.A.; Oprişoreanu, A.M.; Pedersen, E.; Sandi, S.; Selvarajah, J.; Tsarouchas, T.M.; Wehner, D.; et al. A Unique Macrophage Subpopulation Signals Directly to Progenitor Cells to Promote Regenerative Neurogenesis in the Zebrafish Spinal Cord. *Dev. Cell* **2021**, *56*, 1617–1630. [CrossRef]
178. Foote, A.K.; Blakemore, W.F. Inflammation Stimulates Remyelination in Areas of Chronic Demyelination. *Brain* **2005**, *128*, 528–539. [CrossRef]
179. Lloyd, A.F.; Davies, C.L.; Holloway, R.K.; Labrak, Y.; Ireland, G.; Carradori, D.; Dillenburg, A.; Borger, E.; Soong, D.; Richardson, J.C.; et al. Central Nervous System Regeneration Is Driven by Microglia Necroptosis and Repopulation. *Nat. Neurosci.* **2019**, *22*, 1046–1052. [CrossRef]
180. McMurran, C.E.; Jones, C.A.; Fitzgerald, D.C.; Franklin, R.J.M. CNS Remyelination and the Innate Immune System. *Front. Cell Dev. Biol.* **2016**, *4*, 39. [CrossRef]
181. Milner, R.; Campbell, I.L. The Extracellular Matrix and Cytokines Regulate Microglial Integrin Expression and Activation. *J. Immunol.* **2003**, *170*, 3850–3858. [CrossRef] [PubMed]
182. Milner, R.; Relvas, J.B.; Fawcett, J.; ffrench-Constant, C. Developmental Regulation of αv Integrins Produces Functional Changes in Astrocyte Behavior. *Mol. Cell. Neurosci.* **2001**, *18*, 108–118. [CrossRef] [PubMed]
183. Tawil, N.J.; Wilson, P.; Carbonetto, S. Expression and Distribution of Functional Integrins in Rat CNS Glia. *J. Neurosci. Res.* **1994**, *39*, 436–447. [CrossRef]
184. Sikkema, A.H.; Stoffels, J.M.J.; Wang, P.; Basedow, F.J.; Bulsink, R.; Bajramovic, J.J.; Baron, W. Fibronectin Aggregates Promote Features of a Classically and Alternatively Activated Phenotype in Macrophages. *J. Neuroinflamm.* **2018**, *15*, 218. [CrossRef] [PubMed]
185. Miron, V.E.; Boyd, A.; Zhao, J.W.; Yuen, T.J.; Ruckh, J.M.; Shadrach, J.L.; van Wijngaarden, P.; Wagers, A.J.; Williams, A.; Franklin, R.J.M.; et al. M2 Microglia and Macrophages Drive Oligodendrocyte Differentiation during CNS Remyelination. *Nat. Neurosci.* **2013**, *16*, 1211–1218. [CrossRef] [PubMed]
186. Peferoen, L.A.N.; Vogel, D.Y.S.; Ummenthum, K.; Breur, M.; Heijnen, P.D.A.M.; Gerritsen, W.H.; Peferoen-Baert, R.M.B.; van der Valk, P.; Dijkstra, C.D.; Amor, S. Activation Status of Human Microglia Is Dependent on Lesion Formation Stage and Remyelination in Multiple Sclerosis. *J. Neuropathol. Exp. Neurol.* **2015**, *74*, 48–63. [CrossRef]
187. Vogel, D.Y.S.; Vereyken, E.J.F.; Glim, J.E.; Heijnen, P.D.A.M.; Moeton, M.; van der Valk, P.; Amor, S.; Teunissen, C.E.; van Horssen, J.; Dijkstra, C.D. Macrophages in Inflammatory Multiple Sclerosis Lesions Have an Intermediate Activation Status. *J. Neuroinflamm.* **2013**, *10*, 35. [CrossRef]
188. Milner, R.; Crocker, S.J.; Hung, S.; Wang, X.; Frausto, R.F.; del Zoppo, G.J. Fibronectin- and Vitronectin-Induced Microglial Activation and Matrix Metalloproteinase-9 Expression Is Mediated by Integrins α5β1 and αvβ5. *J. Immunol.* **2007**, *178*, 8158–8167. [CrossRef]
189. Nasu-Tada, K.; Koizumi, S.; Inoue, K. Involvement of β1 Integrin in Microglial Chemotaxis and Proliferation on Fibronectin: Different Regulations by ADP through PKA. *Glia* **2005**, *52*, 98–107. [CrossRef]
190. Goos, M.; Lange, P.; Hanisch, U.K.; Prinz, M.; Scheffel, J.; Bergmann, R.; Ebert, S.; Nau, R. Fibronectin Is Elevated in the Cerebrospinal Fluid of Patients Suffering from Bacterial Meningitis and Enhances Inflammation Caused by Bacterial Products in Primary Mouse Microglial Cell Cultures. *J. Neurochem.* **2007**, *102*, 2049–2060. [CrossRef] [PubMed]
191. Okamura, Y.; Watari, M.; Jerud, E.S.; Young, D.W.; Ishizaka, S.T.; Rose, J.; Chow, J.C.; Strauss, J.F. The Extra Domain A of Fibronectin Activates Toll-like Receptor 4. *J. Biol. Chem.* **2001**, *276*, 10229–10233. [CrossRef] [PubMed]
192. Ribes, S.; Ebert, S.; Regen, T.; Czesnik, D.; Scheffel, J.; Zeug, A.; Bunkowski, S.; Eiffert, H.; Hanisch, U.K.; Hammerschmidt, S.; et al. Fibronectin Stimulates Escherichia Coli Phagocytosis by Microglial Cells. *Glia* **2010**, *58*, 367–376. [CrossRef] [PubMed]
193. Summers, L.; Kielty, C.; Pinteaux, E. Adhesion to Fibronectin Regulates Interleukin-1 Beta Expression in Microglial Cells. *Mol. Cell. Neurosci.* **2009**, *41*, 148–155. [CrossRef] [PubMed]
194. Medina-Rodríguez, E.M.; Bribián, A.; Boyd, A.; Palomo, V.; Pastor, J.; Lagares, A.; Gil, C.; Martínez, A.; Williams, A.; De Castro, F. Promoting in Vivo Remyelination with Small Molecules: A Neuroreparative Pharmacological Treatment for Multiple Sclerosis. *Sci. Rep.* **2017**, *7*, 43545. [CrossRef] [PubMed]
195. Syed, Y.A.; Baer, A.; Hofer, M.P.; González, G.A.; Rundle, J.; Myrta, S.; Huang, J.K.; Zhao, C.; Rossner, M.J.; Trotter, M.W.B.; et al. Inhibition of Phosphodiesterase-4 Promotes Oligodendrocyte Precursor Cell Differentiation and Enhances CNS Remyelination. *EMBO Mol. Med.* **2013**, *5*, 1918–1934. [CrossRef] [PubMed]
196. de Santana Nunes, A.K.; Rapôso, C.; de Oliveira, W.H.; Thomé, R.; Verinaud, L.; Tovar-Moll, F.; Peixoto, C.A. Phosphodiesterase-5 Inhibition Promotes Remyelination by MCP-1/CCR-2 and MMP-9 Regulation in a Cuprizone-Induced Demyelination Model. *Exp. Neurol.* **2016**, *275*, 143–153. [CrossRef] [PubMed]
197. Bielekova, B.; Richert, N.; Howard, T.; Packer, A.N.; Blevins, G.; Ohayon, J.; McFarland, H.F.; Stürzebecher, C.S.; Martin, R. Treatment with the Phosphodiesterase Type-4 Inhibitor Rolipram Fails to Inhibit Blood–Brain Barrier Disruption in Multiple Sclerosis. *Mult. Scler.* **2009**, *15*, 1206–1214. [CrossRef] [PubMed]
198. Werkman, I.; Sikkema, A.H.; Versluijs, J.B.; Qin, J.; de Boer, P.; Baron, W. TLR3 Agonists Induce Fibronectin Aggregation by Activated Astrocytes: A Role of pro-Inflammatory Cytokines and Fibronectin Splice Variants. *Sci. Rep.* **2020**, *10*, 532. [CrossRef] [PubMed]
199. Wu, C.; Keivenst, V.M.; O'Toole, T.E.; McDonald, J.A.; Ginsberg, M.H. Integrin Activation and Cytoskeletal Interaction Are Essential for the Assembly of a Fibronectin Matrix. *Cell* **1995**, *83*, 715–724. [CrossRef]

200. Wierzbicka-Patynowski, I.; Schwarzbauer, J.E. The Ins and Outs of Fibronectin Matrix Assembly. *J. Cell Sci.* **2003**, *116*, 3269–3276. [CrossRef]
201. Ohashi, T.; Erickson, H.P. Revisiting the Mystery of Fibronectin Multimers: The Fibronectin Matrix Is Composed of Fibronectin Dimers Cross-Linked by Non-Covalent Bonds. *Matrix Biol.* **2009**, *28*, 170–175. [CrossRef] [PubMed]
202. Schwarzbauer, J.E.; DeSimone, D.W. Fibronectins, Their Fibrillogenesis, and in Vivo Functions. *Cold Spring Harb. Perspect. Biol.* **2011**, *3*, a005041. [CrossRef] [PubMed]
203. Bazigou, E.; Xie, S.; Chen, C.; Weston, A.; Miura, N.; Sorokin, L.; Adams, R.; Muro, A.F.; Sheppard, D.; Makinen, T. Integrin-α9 Is Required for Fibronectin Matrix Assembly during Lymphatic Valve Morphogenesis. *Dev. Cell* **2009**, *17*, 175–186. [CrossRef] [PubMed]
204. Astrof, S.; Crowley, D.; George, E.L.; Fukuda, T.; Sekiguchi, K.; Hanahan, D.; Hynes, R.O. Direct Test of Potential Roles of EIIIA and EIIIB Alternatively Spliced Segments of Fibronectin in Physiological and Tumor Angiogenesis. *Mol. Cell. Biol.* **2004**, *24*, 8662–8670. [CrossRef] [PubMed]
205. Tan, M.H.; Sun, Z.; Opitz, S.L.; Schmidt, T.E.; Peters, J.H.; George, E.L. Deletion of the Alternatively Spliced Fibronectin EIIIA Domain in Mice Reduces Atherosclerosis. *Blood* **2004**, *104*, 11–18. [CrossRef] [PubMed]
206. Chauhan, A.K.; Moretti, F.A.; Iaconcig, A.; Baralle, F.E.; Muro, A.F. Impaired Motor Coordination in Mice Lacking the EDA Exon of the Fibronectin Gene. *Behav. Brain Res.* **2005**, *161*, 31–38. [CrossRef]
207. Muro, A.F.; Chauhan, A.K.; Gajovic, S.; Iaconcig, A.; Porro, F.; Stanta, G.; Baralle, F.E. Regulated Splicing of the Fibronectin EDA Exon Is Essential for Proper Skin Wound Healing and Normal Lifespan. *J. Cell Biol.* **2003**, *162*, 149–160. [CrossRef]
208. Bsibsi, M.; Bajramovic, J.J.; Vogt, M.H.J.; van Duijvenvoorden, E.; Baghat, A.; Persoon-Deen, C.; Tielen, F.; Verbeek, R.; Huitinga, I.; Ryffel, B.; et al. The Microtubule Regulator Stathmin Is an Endogenous Protein Agonist for TLR3. *J. Immunol.* **2010**, *184*, 6929–6937. [CrossRef]
209. Liu, A.; Stadelmann, C.; Moscarello, M.; Bruck, W.; Sobel, A.; Mastronardi, F.G.; Casaccia-Bonnefil, P. Expression of Stathmin, a Developmentally Controlled Cytoskeleton- Regulating Molecule, in Demyelinating Disorders. *J. Neurosci.* **2005**, *25*, 737–747. [CrossRef]
210. Škuljec, J.; Gudi, V.; Ulrich, R.; Frichert, K.; Yildiz, Ö.; Pul, R.; Voss, E.V.; Wissel, K.; Baumgärtner, W.; Stangel, M. Matrix Metalloproteinases and Their Tissue Inhibitors in Cuprizone-Induced Demyelination and Remyelination of Brain White and Gray Matter. *J. Neuropathol. Exp. Neurol.* **2011**, *70*, 758–769. [CrossRef]
211. Wang, P.; Gorter, R.P.; de Jonge, J.C.; Nazmuddin, M.; Zhao, C.; Amor, S.; Hoekstra, D.; Baron, W. MMP7 Cleaves Remyelination-Impairing Fibronectin Aggregates and Its Expression Is Reduced in Chronic Multiple Sclerosis Lesions. *Glia* **2018**, *66*, 1625–1643. [CrossRef] [PubMed]
212. Anthony, D.C.; Ferguson, B.; Matyzak, M.K.; Miller, K.M.; Esiri, M.M.; Perry, V.H. Differential Matrix Metalloproteinase Expression in Cases of Multiple Sclerosis and Stroke. *Neuropathol. Appl. Neurobiol.* **1997**, *23*, 406–415. [CrossRef]
213. Cossins, J.A.; Clements, J.M.; Ford, J.; Miller, K.M.; Pigott, R.; Vos, W.; van der Valk, P.; De Groot, C.J.A. Enhanced Expression of MMP-7 and MMP-9 in Demyelinating Multiple Sclerosis Lesions. *Acta Neuropathol.* **1997**, *94*, 590–598. [CrossRef]
214. Lindberg, R.L.P.; De Groot, C.J.A.; Montagne, L.; Freitag, P.; van der Valk, P.; Kappos, L.; Leppert, D. The Expression Profile of Matrix Metalloproteinases (MMPs) and Their Inhibitors (TIMPs) in Lesions and Normal Appearing White Matter of Multiple Sclerosis. *Brain* **2001**, *124*, 1743–1753. [CrossRef]
215. Gorter, R.P.; Baron, W. Matrix Metalloproteinases Shape the Oligodendrocyte (Niche) during Development and upon Demyelination. *Neurosci. Lett.* **2020**, *729*, 134980. [CrossRef]
216. Young, J.C.; Agashe, V.R.; Siegers, K.; Hartl, F.U. Pathways of Chaperone-Mediated Protein Folding in the Cytosol. *Nat. Rev. Mol. Cell Biol.* **2004**, *5*, 781–791. [CrossRef]
217. Young, J.C. Mechanisms of the Hsp70 Chaperone System. *Biochem. Cell Biol.* **2010**, *88*, 291–300. [CrossRef] [PubMed]
218. Calderwood, S.K.; Gong, J.; Murshid, A. Extracellular HSPs: The Complicated Roles of Extracellular HSPs in Immunity. *Front. Immunol.* **2016**, *7*, 159. [CrossRef] [PubMed]
219. González-Ramos, M.; Calleros, L.; López-Ongil, S.; Raoch, V.; Griera, M.; Rodríguez-Puyol, M.; De Frutos, S.; Rodríguez-Puyol, D. HSP70 Increases Extracellular Matrix Production by Human Vascular Smooth Muscle through TGF-B1 up-Regulation. *Int. J. Biochem. Cell Biol.* **2013**, *45*, 232–242. [CrossRef]
220. Cwiklinska, H.; Mycko, M.P.; Szymanska, B.; Matysiak, M.; Selmaj, K.W. Aberrant Stress-Induced Hsp70 Expression in Immune Cells in Multiple Sclerosis. *J. Neurosci. Res.* **2010**, *88*, 3102–3110. [CrossRef] [PubMed]
221. Khan, E.S.; Sankaran, S.; Llontop, L.; Del Campo, A. Exogenous Supply of Hsp47 Triggers Fibrillar Collagen Deposition in Skin Cell Cultures in Vitro. *BMC Mol. Cell Biol.* **2020**, *21*, 22. [CrossRef]
222. Ito, S.; Nagata, K. Biology of Hsp47 (Serpin H1), a Collagen-Specific Molecular Chaperone. *Semin. Cell Dev. Biol.* **2017**, *62*, 142–151. [CrossRef] [PubMed]
223. Hunter, M.C.; O'Hagan, K.L.; Kenyon, A.; Dhanani, K.C.H.; Prinsloo, E.; Edkins, A.L. Hsp90 Binds Directly to Fibronectin (FN) and Inhibition Reduces the Extracellular Fibronectin Matrix in Breast Cancer Cells. *PLoS ONE* **2014**, *9*, e86842. [CrossRef] [PubMed]
224. Cid, C.; Alvarez-Cermeño, J.C.; Camafeita, E.; Salinas, M.; Alcázar, A. Antibodies Reactive to Heat Shock Protein 90 Induce Oligodendrocyte Precursor Cell Death in Culture. Implications for Demyelination in Multiple Sclerosis. *FASEB J.* **2004**, *18*, 409–411. [CrossRef] [PubMed]

225. Espitia Pinzón, N.; Brevé, J.J.P.; Bol, J.G.J.M.; Drukarch, B.; Baron, W.; van Dam, A.M. Tissue Transglutaminase in Astrocytes Is Enhanced by Inflammatory Mediators and Is Involved in the Formation of Fibronectin Fibril-like Structures. *J. Neuroinflamm.* **2017**, *14*, 260. [CrossRef]
226. Nichols, P.; Urriola, J.; Miller, S.; Bjorkman, T.; Mahady, K.; Vegh, V.; Nasrallah, F.; Winter, C. Blood-Brain Barrier Dysfunction Significantly Correlates with Serum Matrix Metalloproteinase-7 (MMP-7) Following Traumatic Brain Injury. *Neuroimage Clin.* **2021**, *31*, 102741. [CrossRef] [PubMed]
227. Waubant, E.; Goodkin, D.E.; Gee, L.; Bacchetti, P.; Sloan, R.; Stewart, T.; Andersson, P.B.; Stabler, G.; Miller, K. Serum MMP-9 and TIMP-1 Levels Are Related to MRI Activity in Relapsing Multiple Sclerosis. *Neurology* **1999**, *53*, 1397–1401. [CrossRef] [PubMed]
228. Lassmann, H.; van Horssen, J.; Mahad, D. Progressive Multiple Sclerosis: Pathology and Pathogenesis. *Nat. Rev. Neurol.* **2012**, *8*, 647–656. [CrossRef] [PubMed]
229. Alexander, A.; Agrawal, M.; Uddin, A.; Siddique, S.; Shehata, A.M.; Shaker, M.A.; Rahman, S.A.U.; Abdul, M.I.M.; Shaker, M.A. Recent Expansions of Novel Strategies towards the Drug Targeting into the Brain. *Int. J. Nanomed.* **2019**, *14*, 5895–5909. [CrossRef] [PubMed]
230. Nagpal, K.; Singh, S.K.; Mishra, D.N. Drug Targeting to Brain: A Systematic Approach to Study the Factors, Parameters and Approaches for Prediction of Permeability of Drugs across BBB. *Expert Opin. Drug Deliv.* **2013**, *10*, 927–955. [CrossRef]
231. Stein, E.S.; Itsekson-Hayosh, Z.; Aronovich, A.; Reisner, Y.; Bushi, D.; Pick, C.G.; Tanne, D.; Chapman, J.; Vlachos, A.; Maggio, N. Thrombin Induces Ischemic LTP (ILTP): Implications for Synaptic Plasticity in the Acute Phase of Ischemic Stroke. *Sci. Rep.* **2015**, *5*, 7912. [CrossRef]
232. Gingrich, M.B.; Traynelis, S.F. Serine Proteases and Brain Damage-Is There a Link? *Trends Neurosci.* **2000**, *23*, 399–407. [CrossRef]
233. Chong, W.; Kim, S.N.; Han, S.K.; Lee, S.Y.; Ryu, P.D. Low Non-NMDA Receptor Current Density as Possible Protection Mechanism from Neurotoxicity of Circulating Glutamate on Subfornical Organ Neurons in Rats. *Korean J. Physiol. Pharmacol.* **2015**, *19*, 177–181. [CrossRef] [PubMed]
234. Abbott, N.J.; Patabendige, A.A.K.; Dolman, D.E.M.; Yusof, S.R.; Begley, D.J. Structure and Function of the Blood-Brain Barrier. *Neurobiol. Dis.* **2010**, *37*, 13–25. [CrossRef] [PubMed]
235. Lu, C.T.; Zhao, Y.Z.; Wong, H.L.; Cai, J.; Peng, L.; Tian, X.Q. Current Approaches to Enhance CNS Delivery of Drugs across the Brain Barriers. *Int. J. Nanomed.* **2014**, *9*, 2241–2257. [CrossRef] [PubMed]
236. Liu, Y.; Yang, G.; Jin, S.; Xu, L.; Zhao, C.X. Development of High-Drug-Loading Nanoparticles. *Chempluschem* **2020**, *85*, 2143–2157. [CrossRef]
237. Musumeci, T.; Ventura, C.A.; Giannone, I.; Ruozi, B.; Montenegro, L.; Pignatello, R.; Puglisi, G. PLA/PLGA Nanoparticles for Sustained Release of Docetaxel. *Int. J. Pharm.* **2006**, *325*, 172–179. [CrossRef] [PubMed]
238. Mukherjee, B.; Santra, K.; Pattnaik, G.; Ghosh, S. Preparation, Characterization and in-Vitro Evaluation of Sustained Release Protein-Loaded Nanoparticles Based on Biodegradable Polymers. *Int. J. Nanomed.* **2008**, *3*, 487. [CrossRef]
239. Kim, D.H.; Martin, D.C. Sustained Release of Dexamethasone from Hydrophilic Matrices Using PLGA Nanoparticles for Neural Drug Delivery. *Biomaterials* **2006**, *27*, 3031–3037. [CrossRef] [PubMed]
240. Jacobson, G.B.; Shinde, R.; Contag, C.H.; Zare, R.N. Sustained Release of Drugs Dispersed in Polymer Nanoparticles. *Angew. Chemie* **2008**, *120*, 7998–8000. [CrossRef]
241. Su, C.W.; Chiang, C.S.; Li, W.M.; Hu, S.H.; Chen, S.Y. Multifunctional Nanocarriers for Simultaneous Encapsulation of Hydrophobic and Hydrophilic Drugs in Cancer Treatment. *Nanomedicine* **2014**, *9*, 1499–1515. [CrossRef] [PubMed]
242. Zogg, H.; Singh, R.; Ro, S. Current Advances in RNA Therapeutics for Human Diseases. *Int. J. Mol. Sci.* **2022**, *23*, 2736. [CrossRef] [PubMed]
243. Cheon, J.; Chan, W.; Zuhorn, I. The Future of Nanotechnology: Cross-Disciplined Progress to Improve Health and Medicine. *Acc. Chem. Res.* **2019**, *52*, 2405. [CrossRef] [PubMed]
244. Mitchell, M.J.; Billingsley, M.M.; Haley, R.M.; Wechsler, M.E.; Peppas, N.A.; Langer, R. Engineering Precision Nanoparticles for Drug Delivery. *Nat. Rev. Drug Discov.* **2021**, *20*, 101–124. [CrossRef] [PubMed]
245. Renukuntla, J.; Vadlapudi, A.D.; Patel, A.; Boddu, S.H.S.; Mitra, A.K. Approaches for Enhancing Oral Bioavailability of Peptides and Proteins. *Int. J. Pharm.* **2013**, *447*, 75–93. [CrossRef] [PubMed]
246. Odiba, A.; Ottah, V.; Ottah, C.; Anunobi, O.; Ukegbu, C.; Edeke, A.; Uroko, R.; Omeje, K. Therapeutic Nanomedicine Surmounts the Limitations of Pharmacotherapy. *Open Med.* **2017**, *12*, 271–287. [CrossRef]
247. Agrahari, V.; Agrahari, V.; Mitra, A.K. Nanocarrier Fabrication and Macromolecule Drug Delivery: Challenges and Opportunities. *Ther. Deliv.* **2016**, *7*, 257–278. [CrossRef]
248. Mahato, R.I.; Narang, A.S.; Thoma, L.; Miller, D.D. Emerging Trends in Oral Delivery of Peptide and Protein Drugs. *Crit. Rev. Ther. Drug Carrier Syst.* **2003**, *20*, 153–214. [CrossRef] [PubMed]
249. Cheng, Q.; Wei, T.; Farbiak, L.; Johnson, L.T.; Dilliard, S.A.; Siegwart, D.J. Selective ORgan Targeting (SORT) Nanoparticles for Tissue Specific MRNA Delivery and CRISPR/Cas Gene Editing. *Nat. Nanotechnol.* **2020**, *15*, 313. [CrossRef]
250. Podual, K.; Doyle, F.J.; Peppas, N.A. Glucose-Sensitivity of Glucose Oxidase-Containing Cationic Copolymer Hydrogels Having Poly(Ethylene Glycol) Grafts. *J. Control. Release* **2000**, *67*, 9–17. [CrossRef]
251. Marek, S.R.; Peppas, N.A. Insulin Release Dynamics from Poly(Diethylaminoethyl Methacrylate) Hydrogel Systems. *AIChE J.* **2013**, *59*, 3578–3585. [CrossRef] [PubMed]

252. Culver, H.R.; Clegg, J.R.; Peppas, N.A. Analyte-Responsive Hydrogels: Intelligent Materials for Biosensing and Drug Delivery. *Acc. Chem. Res.* **2017**, *50*, 170. [CrossRef]
253. Cheng, X.; Lee, R.J. The Role of Helper Lipids in Lipid Nanoparticles (LNPs) Designed for Oligonucleotide Delivery. *Adv. Drug Deliv. Rev.* **2016**, *99*, 129–137. [CrossRef]
254. Li, X.; Tsibouklis, J.; Weng, T.; Zhang, B.; Yin, G.; Feng, G.; Cui, Y.; Savina, I.N.; Mikhalovska, L.I.; Sandeman, S.R.; et al. Nano Carriers for Drug Transport across the Blood–Brain Barrier. *J. Drug Target.* **2017**, *25*, 17–28. [CrossRef] [PubMed]
255. Joshi, B.S.; Ortiz, D.; Zuhorn, I.S. Converting Extracellular Vesicles into Nanomedicine: Loading and Unloading of Cargo. *Mater. Today Nano* **2021**, *16*, 100148. [CrossRef]
256. Aliyandi, A.; Satchell, S.; Unger, R.E.; Bartosch, B.; Parent, R.; Zuhorn, I.S.; Salvati, A. Effect of Endothelial Cell Heterogeneity on Nanoparticle Uptake. *Int. J. Pharm.* **2020**, *587*, 119699. [CrossRef] [PubMed]
257. Ozbakir, B.; Crielaard, B.J.; Metselaar, J.M.; Storm, G.; Lammers, T. Liposomal Corticosteroids for the Treatment of Inflammatory Disorders and Cancer. *J. Control. Release* **2014**, *190*, 624–636. [CrossRef] [PubMed]
258. Rousseau, V.; Denizot, B.; Le Jeune, J.J.; Jallet, P. Early Detection of Liposome Brain Localization in Rat Experimental Allergic Encephalomyelitis. *Exp. Brain Res.* **1999**, *125*, 255–264. [CrossRef]
259. Lu, L.; Qi, S.; Chen, Y.; Luo, H.; Huang, S.; Yu, X.; Luo, Q.; Zhang, Z. Targeted Immunomodulation of Inflammatory Monocytes across the Blood-Brain Barrier by Curcumin-Loaded Nanoparticles Delays the Progression of Experimental Autoimmune Encephalomyelitis. *Biomaterials* **2020**, *245*, 119987. [CrossRef]
260. Doshi, A.; Chataway, J. Multiple Sclerosis, a Treatable Disease. *Clin. Med.* **2017**, *17*, 530–536. [CrossRef] [PubMed]
261. Schmidt, J.; Metselaar, J.M.; Wauben, M.H.M.; Toyka, K.V.; Storm, G.; Gold, R. Drug Targeting by Long-Circulating Liposomal Glucocorticosteroids Increases Therapeutic Efficacy in a Model of Multiple Sclerosis. *Brain* **2003**, *126*, 1895–1904. [CrossRef] [PubMed]
262. Schweingruber, N.; Haine, A.; Tiede, K.; Karabinskaya, A.; van den Brandt, J.; Wüst, S.; Metselaar, J.M.; Gold, R.; Tuckermann, J.P.; Reichardt, H.M.; et al. Liposomal Encapsulation of Glucocorticoids Alters Their Mode of Action in the Treatment of Experimental Autoimmune Encephalomyelitis. *J. Immunol.* **2011**, *187*, 4310–4318. [CrossRef] [PubMed]
263. Linker, R.A.; Weller, C.; Lühder, F.; Mohr, A.; Schmidt, J.; Knauth, M.; Metselaar, J.M.; Gold, R. Liposomal Glucocorticosteroids in Treatment of Chronic Autoimmune Demyelination: Long-Term Protective Effects and Enhanced Efficacy of Methylprednisolone Formulations. *Exp. Neurol.* **2008**, *211*, 397–406. [CrossRef]
264. Gaillard, P.J.; Appeldoorn, C.C.M.; Rip, J.; Dorland, R.; van der Pol, S.M.A.; Kooij, G.; de Vries, H.E.; Reijerkerk, A. Enhanced Brain Delivery of Liposomal Methylprednisolone Improved Therapeutic Efficacy in a Model of Neuroinflammation. *J. Control. Release* **2012**, *164*, 364–369. [CrossRef]
265. Ojha, S.; Kumar, B. Preparation and Statistical Modeling of Solid Lipid Nanoparticles of Dimethyl Fumarate for Better Management of Multiple Sclerosis. *Adv. Pharm. Bull.* **2018**, *8*, 225–233. [CrossRef]
266. Kumar, P.; Sharma, G.; Gupta, V.; Kaur, R.; Thakur, K.; Malik, R.; Kumar, A.; Kaushal, N.; Raza, K. Preclinical Explorative Assessment of Dimethyl Fumarate-Based Biocompatible Nanolipoidal Carriers for the Management of Multiple Sclerosis. *ACS Chem. Neurosci.* **2018**, *9*, 1152–1158. [CrossRef] [PubMed]
267. Pujol-Autonell, I.; Mansilla, M.J.; Rodriguez-Fernandez, S.; Cano-Sarabia, M.; Navarro-Barriuso, J.; Ampudia, R.M.; Rius, A.; Garcia-Jimeno, S.; Perna-Barrull, D.; Caceres, E.M.; et al. Liposome-Based Immunotherapy against Autoimmune Diseases: Therapeutic Effect on Multiple Sclerosis. *Nanomedicine* **2017**, *12*, 1231–1242. [CrossRef]
268. Gholamzad, M.; Ebtekar, M.; Shafiee Ardestani, M. Intravenous Injection of Myelin Oligodendrocyte Glycoprotein-Coated PLGA Microparticles Have Tolerogenic Effects in Experimental Autoimmune Encephalomyelitis. *Iran. J. Allergy Asthma Immunol.* **2017**, *16*, 271–281. [PubMed]
269. Pei, W.; Wan, X.; Shahzad, K.A.; Zhang, L.; Song, S.; Jin, X.; Wang, L.; Zhao, C.; Shen, C. Direct Modulation of Myelin-Autoreactive CD4+ and CD8+ T Cells in EAE Mice by a Tolerogenic Nanoparticle Co-Carrying Myelin Peptide-Loaded Major Histocompatibility Complexes, CD47 and Multiple Regulatory Molecules. *Int. J. Nanomed.* **2018**, *13*, 3731–3750. [CrossRef]
270. Cappellano, G.; Woldetsadik, A.D.; Orilieri, E.; Shivakumar, Y.; Rizzi, M.; Carniato, F.; Gigliotti, C.L.; Boggio, E.; Clemente, N.; Comi, C.; et al. Subcutaneous Inverse Vaccination with PLGA Particles Loaded with a MOG Peptide and IL-10 Decreases the Severity of Experimental Autoimmune Encephalomyelitis. *Vaccine* **2014**, *32*, 5681–5689. [CrossRef]
271. Hunter, Z.; McCarthy, D.P.; Yap, W.T.; Harp, C.T.; Getts, D.R.; Shea, L.D.; Miller, S.D. A Biodegradable Nanoparticle Platform for the Induction of Antigen-Specific Immune Tolerance for Treatment of Autoimmune Disease. *ACS Nano* **2014**, *8*, 2148–2160. [CrossRef]
272. Kuo, R.; Saito, E.; Miller, S.D.; Shea, L.D. Peptide-Conjugated Nanoparticles Reduce Positive Co-Stimulatory Expression and T Cell Activity to Induce Tolerance. *Mol. Ther.* **2017**, *25*, 1676–1685. [CrossRef]
273. Gammon, J.M.; Tostanoski, L.H.; Adapa, A.R.; Chiu, Y.C.; Jewell, C.M. Controlled Delivery of a Metabolic Modulator Promotes Regulatory T Cells and Restrains Autoimmunity. *J. Control. Release* **2015**, *210*, 169–178. [CrossRef]
274. Osorio-Querejeta, I.; Carregal-Romero, S.; Ayerdi-Izquierdo, A.; Mäger, I.; Nash, L.A.; Wood, M.; Egimendia, A.; Betanzos, M.; Alberro, A.; Iparraguirre, L.; et al. MiR-219a-5p Enriched Extracellular Vesicles Induce OPC Differentiation and EAE Improvement More Efficiently than Liposomes and Polymeric Nanoparticles. *Pharmaceutics* **2020**, *12*, 186. [CrossRef] [PubMed]

275. Belogurov, A.A.; Stepanov, A.V.; Smirnov, I.V.; Melamed, D.; Bacon, A.; Mamedov, A.E.; Boitsov, V.M.; Sashchenko, L.P.; Ponomarenko, N.A.; Sharanova, S.N.; et al. Liposome-Encapsulated Peptides Protect against Experimental Allergic Encephalitis. *FASEB J.* **2013**, *27*, 222–231. [CrossRef] [PubMed]
276. Belogurov, A.A.; Zargarova, T.A.; Turobov, V.I.; Novikova, N.I.; Favorova, O.O.; Ponomarenko, N.A.; Gabibov, A.G. Suppression of Ongoing Experimental Allergic Encephalomyelitis in Da Rats by Novel Peptide Drug, Structural Part of Human Myelin Basic Protein 4662. *Autoimmunity* **2009**, *42*, 362–364. [CrossRef]
277. Ivanova, V.V.; Khaiboullina, S.F.; Gomzikova, M.O.; Martynova, E.V.; Ferreira, A.M.; Garanina, E.E.; Sakhapov, D.I.; Lomakin, Y.A.; Khaibullin, T.I.; Granatov, E.V.; et al. Divergent Immunomodulation Capacity of Individual Myelin Peptides-Components of Liposomal Therapeutic against Multiple Sclerosis. *Front. Immunol.* **2017**, *1335*, 8. [CrossRef]
278. Lomakin, Y.; Belogurov, A.; Glagoleva, I.; Stepanov, A.; Zakharov, K.; Okunola, J.; Smirnov, I.; Genkin, D.; Gabibov, A. Administration of Myelin Basic Protein Peptides Encapsulated in Mannosylated Liposomes Normalizes Level of Serum TNF- α and IL-2 and Chemoattractants CCL2 and CCL4 in Multiple Sclerosis Patients. *Mediators Inflamm.* **2016**, *2016*, 2847232. [CrossRef]
279. Rittchen, S.; Boyd, A.; Burns, A.; Park, J.; Fahmy, T.M.; Metcalfe, S.; Williams, A. Myelin Repair Invivo Is Increased by Targeting Oligodendrocyte Precursor Cells with Nanoparticles Encapsulating Leukaemia Inhibitory Factor (LIF). *Biomaterials* **2015**, *56*, 78–85. [CrossRef]
280. Li, M.; Du, C.; Guo, N.; Teng, Y.; Meng, X.; Sun, H.; Li, S.; Yu, P.; Galons, H. Composition Design and Medical Application of Liposomes. *Eur. J. Med. Chem.* **2019**, *164*, 640–653. [CrossRef] [PubMed]
281. Akbarzadeh, A.; Rezaei-Sadabady, R.; Davaran, S.; Joo, S.W.; Zarghami, N.; Hanifehpour, Y.; Samiei, M.; Kouhi, M.; Nejati-Koshki, K. Liposome: Classification, Preparation, and Applications. *Nanoscale Res. Lett.* **2013**, *8*, 102. [CrossRef] [PubMed]
282. Milla, P.; Dosio, F.; Cattel, L. PEGylation of Proteins and Liposomes: A Powerful and Flexible Strategy to Improve the Drug Delivery. *Curr. Drug Metab.* **2011**, *13*, 105–119. [CrossRef] [PubMed]
283. Montes-Cobos, E.; Ring, S.; Fischer, H.J.; Heck, J.; Strauß, J.; Schwaninger, M.; Reichardt, S.D.; Feldmann, C.; Lühder, F.; Reichardt, H.M. Targeted Delivery of Glucocorticoids to Macrophages in a Mouse Model of Multiple Sclerosis Using Inorganic-Organic Hybrid Nanoparticles. *J. Control. Release* **2017**, *245*, 157–169. [CrossRef] [PubMed]
284. Papadopoulou, A.; D'Souza, M.; Kappos, L.; Yaldizli, O. Dimethyl Fumarate for Multiple Sclerosis. *Expert Opin. Investig. Drugs* **2010**, *19*, 1603–1612. [CrossRef] [PubMed]
285. Bomprezzi, R. Dimethyl Fumarate in the Treatment of Relapsing-Remitting Multiple Sclerosis: An Overview. *Ther. Adv. Neurol. Disord.* **2015**, *8*, 20–30. [CrossRef]
286. Belogurov, A.; Zakharov, K.; Lomakin, Y.; Surkov, K.; Avtushenko, S.; Kruglyakov, P.; Smirnov, I.; Makshakov, G.; Lockshin, C.; Gregoriadis, G.; et al. CD206-Targeted Liposomal Myelin Basic Protein Peptides in Patients with Multiple Sclerosis Resistant to First-Line Disease-Modifying Therapies: A First-in-Human, Proof-of-Concept Dose-Escalation Study. *Neurotherapeutics* **2016**, *13*, 895–904. [CrossRef]
287. Park, K.; Skidmore, S.; Hadar, J.; Garner, J.; Park, H.; Otte, A.; Soh, B.K.; Yoon, G.; Yu, D.; Yun, Y.; et al. Injectable, Long-Acting PLGA Formulations: Analyzing PLGA and Understanding Microparticle Formation. *J. Control. Release* **2019**, *304*, 125–134. [CrossRef]
288. Hua, Y.; Su, Y.; Zhang, H.; Liu, N.; Wang, Z.; Gao, X.; Gao, J.; Zheng, A. Poly(Lactic-Co-Glycolic Acid) Microsphere Production Based on Quality by Design: A Review. *Drug Deliv.* **2021**, *28*, 1342. [CrossRef] [PubMed]
289. Pearson, R.M.; Podojil, J.R.; Shea, L.D.; King, N.J.C.; Miller, S.D.; Getts, D.R. Overcoming Challenges in Treating Autoimmuntity: Development of Tolerogenic Immune-Modifying Nanoparticles. *Nanomed. Nanotechnol. Biol. Med.* **2019**, *18*, 282–291. [CrossRef]
290. Makadia, H.K.; Siegel, S.J. Poly Lactic-Co-Glycolic Acid (PLGA) as Biodegradable Controlled Drug Delivery Carrier. *Polymers* **2011**, *3*, 1377–1397. [CrossRef] [PubMed]
291. Chiu, H.I.; Samad, N.A.; Fang, L.; Lim, V. Cytotoxicity of Targeted PLGA Nanoparticles: A Systematic Review. *RSC Adv.* **2021**, *11*, 9433–9449. [CrossRef]
292. Hrkach, J.; Von Hoff, D.; Ali, M.M.; Andrianova, E.; Auer, J.; Campbell, T.; De Witt, D.; Figa, M.; Figueiredo, M.; Horhota, A.; et al. Preclinical Development and Clinical Translation of a PSMA-Targeted Docetaxel Nanoparticle with a Differentiated Pharmacological Profile. *Sci. Transl. Med.* **2012**, *4*, 128ra39. [CrossRef]
293. Dargahi, N.; Katsara, M.; Tselios, T.; Androutsou, M.E.; De Courten, M.; Matsoukas, J.; Apostolopoulos, V. Multiple Sclerosis: Immunopathology and Treatment Update. *Brain Sci.* **2017**, *7*, 78. [CrossRef]
294. Cavaletti, G.; Cassetti, A.; Canta, A.; Galbiati, S.; Gilardini, A.; Oggioni, N.; Rodriguez-Menendez, V.; Fasano, A.; Liuzzi, G.M.; Fattler, U.; et al. Cationic Liposomes Target Sites of Acute Neuroinflammation in Experimental Autoimmune Encephalomyelitis. *Mol. Pharm.* **2009**, *6*, 1363–1370. [CrossRef]
295. Deverman, B.E.; Patterson, P.H. Exogenous Leukemia Inhibitory Factor Stimulates Oligodendrocyte Progenitor Cell Proliferation and Enhances Hippocampal Remyelination. *J. Neurosci.* **2012**, *32*, 2100–2109. [CrossRef] [PubMed]
296. Butzkueven, H.; Zhang, J.G.; Soilu-Hanninen, M.; Hochrein, H.; Chionh, F.; Shipham, K.A.; Emery, B.; Turnley, A.M.; Petratos, S.; Ernst, M.; et al. LIF Receptor Signaling Limits Immune-Mediated Demyelination by Enhancing Oligodendrocyte Survival. *Nat. Med.* **2002**, *8*, 613–619. [CrossRef] [PubMed]
297. Slaets, H.; Hendriks, J.J.A.; van den Haute, C.; Coun, F.; Baekelandt, V.; Stinissen, P.; Hellings, N. CNS-Targeted Lif Expression Improves Therapeutic Efficacy and Limits Autoimmune-Mediated Demyelination in a Model of Multiple Sclerosis. *Mol. Ther.* **2010**, *18*, 684–691. [CrossRef]

298. Wang, P.; Xue, Y.; Shang, X.; Liu, Y. Diphtheria Toxin Mutant CRM197-Mediated Transcytosis across Blood-Brain Barrier in Vitro. *Cell. Mol. Neurobiol.* **2010**, *30*, 717–725. [CrossRef] [PubMed]
299. Lu, Q.; Cai, X.; Zhang, X.; Li, S.; Song, Y.; Du, D.; Dutta, P.; Lin, Y. Synthetic Polymer Nanoparticles Functionalized with Different Ligands for Receptor-Mediated Transcytosis across the Blood-Brain Barrier. *ACS Appl. Bio Mater.* **2018**, *1*, 1687–1694. [CrossRef] [PubMed]
300. Pardridge, W.M.; Kang, Y.S.; Buciak, J.L.; Yang, J. Human Insulin Receptor Monoclonal Antibody Undergoes High Affinity Binding to Human Brain Capillaries in Vitro and Rapid Transcytosis Through the Blood–Brain Barrier in Vivo in the Primate. *Pharm. Res.* **1995**, *12*, 807–816. [CrossRef] [PubMed]
301. Stojanov, K.; Georgieva, J.V.; Brinkhuis, R.P.; van Hest, J.C.; Rutjes, F.P.; Dierckx, R.A.J.O.; De Vries, E.F.J.; Zuhorn, I.S. In Vivo Biodistribution of Prion- and GM1-Targeted Polymersomes Following Intravenous Administration in Mice. *Mol. Pharm.* **2012**, *9*, 1620–1627. [CrossRef]
302. de Jong, E.; Williams, D.S.; Abdelmohsen, L.K.E.A.; van Hest, J.C.M.; Zuhorn, I.S. A Filter-Free Blood-Brain Barrier Model to Quantitatively Study Transendothelial Delivery of Nanoparticles by Fluorescence Spectroscopy. *J. Control.* **2018**, *289*, 14–22. [CrossRef]
303. Dos Santos Rodrigues, B.; Lakkadwala, S.; Kanekiyo, T.; Singh, J. Development and Screening of Brain-Targeted Lipid-Based Nanoparticles with Enhanced Cell Penetration and Gene Delivery Properties. *Int. J. Nanomed.* **2019**, *14*, 6497–6517. [CrossRef]
304. Georgieva, J.V.; Hoekstra, D.; Zuhorn, I.S. Smuggling Drugs into the Brain: An Overview of Ligands Targeting Transcytosis for Drug Delivery across the Blood–Brain Barrier. *Pharmaceutics* **2014**, *6*, 557–583. [CrossRef] [PubMed]
305. Su, C.H.; Tsai, C.Y.; Tomanek, B.; Chen, W.Y.; Cheng, F.Y. Evaluation of Blood-Brain Barrier-Stealth Nanocomposites for in Situ Glioblastoma Theranostics Applications. *Nanoscale* **2016**, *8*, 7866–7870. [CrossRef] [PubMed]
306. Haroon, M.M.; Saba, K.; Harshavardhan Boddedda, V.; Kumar, J.M.; Patel, A.B.; Gopal, V. Delivery of BACE1 SiRNA Mediated by TARBP-BTP Fusion Protein Reduces b-Amyloid Deposits in a Transgenic Mouse Model of Alzheimer's Disease. *J. Biosci.* **2019**, *44*, 1. [CrossRef]
307. Haroon, M.M.; Dar, G.H.; Jeyalakshmi, D.; Venkatraman, U.; Saba, K.; Rangaraj, N.; Patel, A.B.; Gopal, V. A Designed Recombinant Fusion Protein for Targeted Delivery of SiRNA to the Mouse Brain. *J. Control. Release* **2016**, *228*, 120–131. [CrossRef]
308. Zhang, H.; van Os, W.L.; Tian, X.; Zu, G.; Ribovski, L.; Bron, R.; Bussmann, J.; Kros, A.; Liu, Y.; Zuhorn, I.S. Development of Curcumin-Loaded Zein Nanoparticles for Transport across the Blood–Brain Barrier and Inhibition of Glioblastoma Cell Growth. *Biomater. Sci.* **2021**, *26*, 7092–7103. [CrossRef]
309. Simionescu, M.; Gafencu, A.; Antohe, F. Transcytosis of Plasma Macromolecules in Endothelial Cells: A Cell Biological Survey. *Microsc. Res. Tech.* **2002**, *57*, 269–288. [CrossRef]
310. Kreuter, J. Drug Delivery to the Central Nervous System by Polymeric Nanoparticles: What Do We Know? *Adv. Drug Deliv. Rev.* **2014**, *71*, 2–14. [CrossRef]
311. Gong, Y.; Chowdhury, P.; Nagesh, P.K.B.; Rahman, M.A.; Zhi, K.; Yallapu, M.M.; Kumar, S. Novel Elvitegravir Nanoformulation for Drug Delivery across the Blood-Brain Barrier to Achieve HIV-1 Suppression in the CNS Macrophages. *Sci. Rep.* **2020**, *10*, 3835. [CrossRef]
312. Gong, Y.; Zhi, K.; Nagesh, P.K.B.; Sinha, N.; Chowdhury, P.; Chen, H.; Gorantla, S.; Yallapu, M.M.; Kumar, S. An Elvitegravir Nanoformulation Crosses the Blood–Brain Barrier and Suppresses HIV-1 Replication in Microglia. *Viruses* **2020**, *12*, 564. [CrossRef] [PubMed]
313. Pardridge, W.M.; Buciak, J.L.; Kang, Y.S.; Boado, R.J. Protamine-Mediated Transport of Albumin into Brain and Other Organs of the Rat. Binding and Endocytosis of Protamine-Albumin Complex by Microvascular Endothelium. *J. Clin. Investig.* **1993**, *92*, 2224–2229. [CrossRef] [PubMed]
314. Hervé, F.; Ghinea, N.; Scherrmann, J.M. CNS Delivery via Adsorptive Transcytosis. *AAPS J.* **2008**, *10*, 455–472. [CrossRef]
315. Kozielski, K.L.; Tzeng, S.Y.; Hurtado De Mendoza, B.A.; Green, J.J. Bioreducible Cationic Polymer-Based Nanoparticles for Efficient and Environmentally Triggered Cytoplasmic SiRNA Delivery to Primary Human Brain Cancer Cells. *ACS Nano* **2014**, *8*, 3232–3241. [CrossRef]
316. Karlsson, J.; Rui, Y.; Kozielski, K.L.; Placone, A.L.; Choi, O.; Tzeng, S.Y.; Kim, J.; Keyes, J.J.; Bogorad, M.I.; Gabrielson, K.; et al. Engineered Nanoparticles for Systemic SiRNA Delivery to Malignant Brain Tumours. *Nanoscale* **2019**, *11*, 20045–20057. [CrossRef] [PubMed]
317. Chen, Y.; Liu, L. Modern Methods for Delivery of Drugs across the Blood-Brain Barrier. *Adv. Drug Deliv. Rev.* **2012**, *64*, 640–665. [CrossRef] [PubMed]
318. Shi, K.; Long, Y.; Xu, C.; Wang, Y.; Qiu, Y.; Yu, Q.; Liu, Y.; Zhang, Q.; Gao, H.; Zhang, Z.; et al. Liposomes Combined an Integrin $\alpha v\beta 3$-Specific Vector with PH-Responsible Cell-Penetrating Property for Highly Effective Antiglioma Therapy through the Blood-Brain Barrier. *ACS Appl. Mater. Interfaces* **2015**, *7*, 21442–21454. [CrossRef]
319. Shi, K.; Li, J.; Cao, Z.; Yang, P.; Qiu, Y.; Yang, B.; Wang, Y.; Long, Y.; Liu, Y.; Zhang, Q.; et al. A PH-Responsive Cell-Penetrating Peptide-Modified Liposomes with Active Recognizing of Integrin $\alpha v\beta 3$ for the Treatment of Melanoma. *J. Control. Release* **2015**, *217*, 138–150. [CrossRef] [PubMed]
320. Gartziandia, O.; Egusquiaguirre, S.P.; Bianco, J.; Pedraz, J.L.; Igartua, M.; Hernandez, R.M.; Préat, V.; Beloqui, A. Nanoparticle Transport across in Vitro Olfactory Cell Monolayers. *Int. J. Pharm.* **2016**, *499*, 81–89. [CrossRef]

321. Liu, Y.; Ran, R.; Chen, J.; Kuang, Q.; Tang, J.; Mei, L.; Zhang, Q.; Gao, H.; Zhang, Z.; He, Q. Paclitaxel Loaded Liposomes Decorated with a Multifunctional Tandem Peptide for Glioma Targeting. *Biomaterials* **2014**, *35*, 4835–4847. [CrossRef] [PubMed]
322. Liu, Y.; Ji, M.; Wong, M.K.; Il Joo, K.; Wang, P. Enhanced Therapeutic Efficacy of IRGD-Conjugated Crosslinked Multilayer Liposomes for Drug Delivery. *Biomed Res. Int.* **2013**, *2013*, 378380. [CrossRef] [PubMed]
323. Zhang, Q.; Tang, J.; Fu, L.; Ran, R.; Liu, Y.; Yuan, M.; He, Q. A PH-Responsive α-Helical Cell Penetrating Peptide-Mediated Liposomal Delivery System. *Biomaterials* **2013**, *34*, 7980–7993. [CrossRef]
324. Burgess, A.; Shah, K.; Hough, O.; Hynynen, K. Focused Ultrasound-Mediated Drug Delivery through the Blood-Brain Barrier. *Expert Rev. Neurother.* **2015**, *15*, 477. [CrossRef] [PubMed]
325. Hynynen, K.; McDannold, N.; Vykhodtseva, N.; Jolesz, F.A. Noninvasive MR Imaging-Guided Focal Opening of the Blood-Brain Barrier in Rabbits. *Radiology* **2001**, *220*, 640–646. [CrossRef] [PubMed]
326. Mesiwala, A.H.; Farrell, L.; Wenzel, H.J.; Silbergeld, D.L.; Crum, L.A.; Winn, H.R.; Mourad, H.R. High-Intensity Focused Ultrasound Selectively Disrupts the Blood-Brain Barrier in Vivo. *Ultrasound Med. Biol.* **2002**, *28*, 389–400. [CrossRef]
327. Leinenga, G.; Götz, J. Scanning Ultrasound Removes Amyloid-β and Restores Memory in an Alzheimer's Disease Mouse Model. *Sci. Transl. Med.* **2015**, *7*, 278ra33. [CrossRef] [PubMed]
328. Mooney, S.J.; Shah, K.; Yeung, S.; Burgess, A.; Aubert, I.; Hynynen, K. Focused Ultrasound-Induced Neurogenesis Requires an Increase in Blood-Brain Barrier Permeability. *PLoS ONE* **2016**, *11*, e0159892. [CrossRef] [PubMed]
329. Shin, J.; Kong, C.; Lee, J.; Choi, B.Y.; Sim, J.; Koh, C.S.; Park, M.; Na, Y.C.; Suh, S.W.; Chang, W.S.; et al. Focused Ultrasound-Induced Blood-Brain Barrier Opening Improves Adult Hippocampal Neurogenesis and Cognitive Function in a Cholinergic Degeneration Dementia Rat Model. *Alzheimers Res. Ther.* **2019**, *11*, 110. [CrossRef] [PubMed]
330. Choi, J.J.; Wang, S.; Brown, T.R.; Small, S.A.; Duff, K.E.K.; Konofagou, E.E. Noninvasive and Transient Blood-Brain Barrier Opening in the Hippocampus of Alzheimer's Double Transgenic Mice Using Focused Ultrasound. *Ultrason. Imaging* **2008**, *30*, 189–200. [CrossRef] [PubMed]
331. Sheikov, N.; McDannold, N.; Sharma, S.; Hynynen, K. Effect of Focused Ultrasound Applied with an Ultrasound Contrast Agent on the Tight Junctional Integrity of the Brain Microvascular Endothelium. *Ultrasound Med. Biol.* **2008**, *34*, 1093–1104. [CrossRef] [PubMed]
332. Pandit, R.; Koh, W.K.; Sullivan, R.K.P.; Palliyaguru, T.; Parton, R.G.; Götz, J. Role for Caveolin-Mediated Transcytosis in Facilitating Transport of Large Cargoes into the Brain via Ultrasound. *J. Control. Release* **2020**, *327*, 667–675. [CrossRef] [PubMed]
333. Meijering, B.D.M.; Juffermans, L.J.M.; van Wamel, A.; Henning, R.H.; Zuhorn, I.S.; Emmer, M.; Versteilen, A.M.G.; Paulus, W.J.; van Gilst, W.H.; Kooiman, K.; et al. Ultrasound and Microbubble-Targeted Delivery of Macromolecules Is Regulated by Induction of Endocytosis and Pore Formation. *Circ. Res.* **2009**, *104*, 679–687. [CrossRef] [PubMed]
334. Crowe, T.P.; Greenlee, M.H.W.; Kanthasamy, A.G.; Hsu, W.H. Mechanism of Intranasal Drug Delivery Directly to the Brain. *Life Sci.* **2018**, *195*, 44–52. [CrossRef]
335. Erdő, F.; Bors, L.A.; Farkas, D.; Bajza, Á.; Gizurarson, S. Evaluation of Intranasal Delivery Route of Drug Administration for Brain Targeting. *Brain Res. Bull.* **2018**, *143*, 155–170. [CrossRef]
336. Chu, L.; Wanga, A.; Ni, L.; Yan, X.; Song, Y.; Zhao, M.; Sun, K.; Mu, H.; Liu, S.; Wu, Z.; et al. Nose-to-Brain Delivery of Temozolomide-Loaded Plga Nanoparticles Functionalized with Anti-Epha3 for Glioblastoma Targeting. *Drug Deliv.* **2018**, *25*, 1634–1641. [CrossRef]
337. Nigam, K.; Kaur, A.; Tyagi, A.; Nematullah, M.; Khan, F.; Gabrani, R.; Dang, S. Nose-to-Brain Delivery of Lamotrigine-Loaded PLGA Nanoparticles. *Drug Deliv. Transl. Res.* **2019**, *9*, 879–890. [CrossRef]
338. Musumeci, T.; Serapide, M.F.; Pellitteri, R.; Dalpiaz, A.; Ferraro, L.; Dal Magro, R.; Bonaccorso, A.; Carbone, C.; Veiga, F.; Sancini, G.; et al. Oxcarbazepine Free or Loaded PLGA Nanoparticles as Effective Intranasal Approach to Control Epileptic Seizures in Rodents. *Eur. J. Pharm. Biopharm.* **2018**, *133*, 309–320. [CrossRef]
339. Zheng, X.; Shao, X.; Zhang, C.; Tan, Y.; Liu, Q.; Wan, X.; Zhang, Q.; Xu, S.; Jiang, X. Intranasal H102 Peptide-Loaded Liposomes for Brain Delivery to Treat Alzheimer's Disease. *Pharm. Res.* **2015**, *32*, 3837–3849. [CrossRef]
340. Narayan, R.; Singh, M.; Ranjan, O.P.; Nayak, Y.; Garg, S.; Shavi, G.V.; Nayak, U.Y. Development of Risperidone Liposomes for Brain Targeting through Intranasal Route. *Life Sci.* **2016**, *163*, 38–45. [CrossRef] [PubMed]
341. Wei, H.; Liu, T.; Jiang, N.; Zhou, K.; Yang, K.; Ning, W.; Yu, Y. A Novel Delivery System of Cyclovirobuxine D for Brain Targeting: Angiopep-Conjugated Polysorbate 80-Coated Liposomes via Intranasal Administration. *J. Biomed. Nanotechnol.* **2018**, *14*, 1252–1262. [CrossRef] [PubMed]
342. Agrawal, M.; Saraf, S.; Saraf, S.; Antimisiaris, S.G.; Chougule, M.B.; Shoyele, S.A.; Alexander, A. Nose-to-Brain Drug Delivery: An Update on Clinical Challenges and Progress towards Approval of Anti-Alzheimer Drugs. *J. Control. Release* **2018**, *281*, 139–177. [CrossRef]
343. Dhuria, S.V.; Hanson, L.R.; Frey, W.H. Intranasal Delivery to the Central Nervous System: Mechanisms and Experimental Considerations. *J. Pharm. Sci.* **2010**, *99*, 1654–1673. [CrossRef] [PubMed]
344. Owens, D.R.; Zinman, B.; Bolli, G. Alternative Routes of Insulin Delivery. *Diabet. Med.* **2003**, *20*, 886–898. [CrossRef] [PubMed]
345. Chow, H.-H.S.; Chen, Z.; Matsuura, G.T. Direct Transport of Cocaine from the Nasal Cavity to the Brain Following Intranasal Cocaine Administration in Rats. *J. Pharm. Sci.* **1999**, *88*, 754–758. [CrossRef] [PubMed]

346. Li, W.Q.; Sun, L.P.; Xia, Y.; Hao, S.; Cheng, G.; Wang, Z.; Wan, Y.; Zhu, C.; He, H.; Zheng, S.Y. Preoccupation of Empty Carriers Decreases Endo-/Lysosome Escape and Reduces the Protein Delivery Efficiency of Mesoporous Silica Nanoparticles. *ACS Appl. Mater. Interfaces* **2018**, *10*, 5340–5347. [CrossRef]
347. Choudhury, H.; Gorain, B.; Pandey, M.; Khurana, R.K.; Kesharwani, P. Strategizing Biodegradable Polymeric Nanoparticles to Cross the Biological Barriers for Cancer Targeting. *Int. J. Pharm.* **2019**, *565*, 509–522. [CrossRef] [PubMed]
348. Selby, L.I.; Cortez-Jugo, C.M.; Such, G.K.; Johnston, A.P.R. Nanoescapology: Progress toward Understanding the Endosomal Escape of Polymeric Nanoparticles. *Wiley Interdiscip. Rev. Nanomed. Nanobiotechnol.* **2017**, *9*, e1452. [CrossRef] [PubMed]
349. Degors, I.M.S.; Wang, C.; Rehman, Z.U.; Zuhorn, I.S. Carriers Break Barriers in Drug Delivery: Endocytosis and Endosomal Escape of Gene Delivery Vectors. *Acc. Chem. Res.* **2019**, *52*, 1750–1760. [CrossRef] [PubMed]
350. Rehman, Z.U.; Zuhorn, I.S.; Hoekstra, D. How Cationic Lipids Transfer Nucleic Acids into Cells and across Cellular Membranes: Recent Advances. *J. Control. Release* **2013**, *166*, 46–56. [CrossRef]
351. Rehman, Z.U.; Hoekstra, D.; Zuhorn, I.S. Mechanism of Polyplex- and Lipoplex-Mediated Delivery of Nucleic Acids: Real-Time Visualization of Transient Membrane Destabilization without Endosomal Lysis. *ACS Nano* **2013**, *7*, 3767–3777. [CrossRef] [PubMed]
352. Song, L.Y.; Ahkong, Q.F.; Rong, Q.; Wang, Z.; Ansell, S.; Hope, M.J.; Mui, B. Characterization of the Inhibitory Effect of PEG-Lipid Conjugates on the Intracellular Delivery of Plasmid and Antisense DNA Mediated by Cationic Lipid Liposomes. *Biochim. Biophys. Acta* **2002**, *1558*, 1–13. [CrossRef]
353. Remaut, K.; Lucas, B.; Braeckmans, K.; Demeester, J.; De Smedt, S.C. Pegylation of Liposomes Favours the Endosomal Degradation of the Delivered Phosphodiester Oligonucleotides. *J. Control. Release* **2007**, *117*, 256–266. [CrossRef]
354. Shi, F.; Wasungu, L.; Nomden, A.; Stuart, M.C.A.; Polushkin, E.; Engberts, J.B.F.N.; Hoekstra, D. Interference of Poly(Ethylene Glycol)-Lipid Analogues with Cationic-Lipid-Mediated Delivery of Oligonucleotides; Role of Lipid Exchangeability and Non-Lamellar Transitions. *Biochem. J.* **2002**, *366*, 333–341. [CrossRef] [PubMed]
355. Kirpotin, D.; Hong, K.; Mullah, N.; Papahadjopoulos, D.; Zalipsky, S. Liposomes with Detachable Polymer Coating: Destabilization and Fusion of Dioleoylphosphatidylethanolamine Vesicles Triggered by Cleavage of Surface-Grafted Poly(Ethylene Glycol). *FEBS Lett.* **1996**, *388*, 115–118. [CrossRef]
356. Kuai, R.; Yuan, W.; Qin, Y.; Chen, H.; Tang, J.; Yuan, M.; Zhang, Z.; He, Q. Efficient Delivery of Payload into Tumor Cells in a Controlled Manner by TAT and Thiolytic Cleavable PEG Co-Modified Liposomes. *Mol. Pharm.* **2010**, *7*, 1816–1826. [CrossRef] [PubMed]
357. Choi, J.S.; MacKay, J.A.; Szoka, F.C. Low-PH-Sensitive PEG-Stabilized Plasmid-Lipid Nanoparticles: Preparation and Characterization. *Bioconjug. Chem.* **2003**, *14*, 420–429. [CrossRef]
358. Joshi, B.S.; de Beer, M.A.; Giepmans, B.N.G.; Zuhorn, I.S. Endocytosis Extracellular Vesicles and Releaseof Their Cargo from Endosomes. *ACS Nano* **2020**, *14*, 4444. [CrossRef] [PubMed]
359. Rayatpour, A.; Javan, M. Targeting the Brain Lesions Using Peptides: A Review Focused on the Possibility of Targeted Drug Delivery to Multiple Sclerosis Lesions. *Pharmacol. Res.* **2021**, *167*, 105441. [CrossRef] [PubMed]
360. Kumra, H.; Reinhardt, D.P. Fibronectin-Targeted Drug Delivery in Cancer. *Adv. Drug Deliv. Rev.* **2016**, *97*, 101–110. [CrossRef] [PubMed]
361. Han, Z.; Lu, Z.R. Targeting Fibronectin for Cancer Imaging and Therapy. *J. Mater. Chem. B* **2017**, *5*, 639–654. [CrossRef]
362. Yang, C.H.; Wang, Y.; Sims, M.; Cai, C.; Pfeffer, L.M. MicroRNA-1 Suppresses Glioblastoma in Preclinical Models by Targeting Fibronectin. *Cancer Lett.* **2019**, *465*, 59–67. [CrossRef] [PubMed]
363. Saw, P.E.; Park, J.; Jon, S.; Farokhzad, O.C. A Drug-Delivery Strategy for Overcoming Drug Resistance in Breast Cancer through Targeting of Oncofetal Fibronectin. *Nanomed. Nanotechnol. Biol. Med.* **2017**, *13*, 713–722. [CrossRef] [PubMed]
364. Han, Z.; Zhou, Z.; Shi, X.; Wang, J.; Wu, X.; Sun, D.; Chen, Y.; Zhu, H.; Magi-Galluzzi, C.; Lu, Z.-R. EDB Fibronectin Specific Peptide for Prostate Cancer Targeting. *Bioconjug. Chem.* **2015**, *26*, 830–838. [CrossRef] [PubMed]
365. Kim, S.; Kim, D.; Jung, H.H.; Lee, I.-H.; Kim, J.I.L.; Suh, J.-Y.; Jon, S. Bio-Inspired Design and Potential Biomedical Applications of a Novel Class of High-Affinity Peptides. *Angew. Chem. Int. Ed.* **2012**, *51*, 1890–1894. [CrossRef] [PubMed]
366. Saw, P.E.; Kim, S.; Lee, I.; Park, J.; Yu, M.; Lee, J.; Kim, J.-I.; Jon, S. Aptide-Conjugated Liposome Targeting Tumor-Associated Fibronectin for Glioma Therapy. *J. Mater. Chem. B* **2013**, *1*, 4723–4726. [CrossRef]
367. Saw, P.E.; Park, J.; Lee, E.; Ahn, S.; Lee, J.; Kim, H.; Kim, J.; Choi, M.; Farokhzad, O.C.; Jon, S. Effect of PEG Pairing on the Efficiency of Cancer-Targeting Liposomes. *Theranostics* **2015**, *5*, 746–754. [CrossRef]
368. Gu, G.; Hu, Q.; Feng, X.; Gao, X.; Menglin, J.; Kang, T.; Jiang, D.; Song, Q.; Chen, H.; Chen, J. PEG-PLA Nanoparticles Modified with APTEDB Peptide for Enhanced Anti-Angiogenic and Anti-Glioma Therapy. *Biomaterials* **2014**, *35*, 8215–8226. [CrossRef] [PubMed]
369. Kim, H.; Lee, Y.; Lee, I.H.; Kim, S.; Kim, D.; Saw, P.E.; Lee, J.; Choi, M.; Kim, Y.C.; Jon, S. Synthesis and Therapeutic Evaluation of an Aptide–Docetaxel Conjugate Targeting Tumor-Associated Fibronectin. *J. Control. Release* **2014**, *178*, 118–124. [CrossRef] [PubMed]
370. Mann, A.P.; Scodeller, P.; Hussain, S.; Joo, J.; Kwon, E.; Braun, G.B.; Mölder, T.; She, Z.G.; Kotamraju, V.R.; Ranscht, B.; et al. A Peptide for Targeted, Systemic Delivery of Imaging and Therapeutic Compounds into Acute Brain Injuries. *Nat. Commun.* **2016**, *7*, 11980. [CrossRef] [PubMed]

371. Labrak, Y.; Heurtault, B.; Frisch, B.; Saulnier, P.; Lepeltier, E.; Miron, V.E.; Muccioli, G.G.; des Rieux, A. Impact of Anti-PDGFRα Antibody Surface Functionalization on LNC Uptake by Oligodendrocyte Progenitor Cells. *Int. J. Pharm.* **2022**, *618*, 121623. [CrossRef] [PubMed]
372. Chen, Y.; Wu, H.; Wang, S.; Koito, H.; Li, J.; Ye, F.; Hoang, J.; Escobar, S.S.; Gow, A.; Arnett, H.A.; et al. The Oligodendrocyte-Specific G Protein-Coupled Receptor GPR17 Is a Cell-Intrinsic Timer of Myelination. *Nat. Neurosci.* **2009**, *12*, 1398–1406. [CrossRef] [PubMed]
373. Terashima, T.; Ogawa, N.; Nakae, Y.; Sato, T.; Katagi, M.; Okano, J.; Maegawa, H.; Kojima, H. Gene Therapy for Neuropathic Pain through siRNA-IRF5 Gene Delivery with Homing Peptides to Microglia. *Mol. Ther. Nucleic Acids* **2018**, *11*, 203–215. [CrossRef] [PubMed]
374. Zhang, F.; Lin, Y.A.; Kannan, S.; Kannan, R.M. Targeting Specific Cells in the Brain with Nanomedicines for CNS Therapies. *J. Control. Release* **2016**, *240*, 212–226. [CrossRef] [PubMed]
375. Valori, C.F.; Possenti, A.; Brambilla, L.; Rossi, D. Challenges and Opportunities of Targeting Astrocytes to Halt Neurodegenerative Disorders. *Cells* **2021**, *10*, 2019. [CrossRef]
376. Ramazani, F.; Chen, W.; van Nostrum, C.F.; Storm, G.; Kiessling, F.; Lammers, T.; Hennink, W.E.; Kok, R.J. Strategies for Encapsulation of Small Hydrophilic and Amphiphilic Drugs in PLGA Microspheres: State-of-the-Art and Challenges. *Int. J. Pharm.* **2016**, *499*, 358–367. [CrossRef] [PubMed]
377. Li, Q.; Li, X.; Zhao, C. Strategies to Obtain Encapsulation and Controlled Release of Small Hydrophilic Molecules. *Front. Bioeng. Biotechnol.* **2020**, *8*, 437. [CrossRef] [PubMed]
378. van Horssen, J.; Brink, B.P.; de Vries, H.E.; van der Valk, P.; Bø, L. The Blood-Brain Barrier in Cortical Multiple Sclerosis Lesions. *J. Neuropathol. Exp. Neurol.* **2007**, *66*, 321–328. [CrossRef] [PubMed]
379. Nishihara, H.; Perriot, S.; Gastfriend, B.D.; Steinfort, M.; Cibien, C.; Soldati, S.; Matsuo, K.; Guimbal, S.; Mathias, A.; Palecek, S.P.; et al. Intrinsic Blood-Brain Barrier Dysfunction Contributes to Multiple Sclerosis Pathogenesis. *Brain* **2022**, *27*, awac019. [CrossRef] [PubMed]
380. Mallik, S.; Samson, R.S.; Wheeler-Kingshott, C.A.M.; Miller, D.H. Imaging Outcomes for Trials of Remyelination in Multiple Sclerosis. *J. Neurol. Neurosurg. Psychiatry* **2014**, *85*, 1396–1404. [CrossRef]
381. Ben-Shalom, I.; Karni, A.; Kolb, H. The Role of Molecular Imaging as a Marker of Remyelination and Repair in Multiple Sclerosis. *Int. J. Mol. Sci.* **2021**, *23*, 474. [CrossRef] [PubMed]
382. Ranzenberger, L.R.; Snyder, T. Diffusion Tensor Imaging. In *StatPearls*; StatPearls Publishing: Treasure Island, FL, USA, 2022.
383. Kuhle, J.; Kropshofer, H.; Haering, D.A.; Kundu, U.; Meinert, R.; Barro, C.; Dahlke, F.; Tomic, D.; Leppert, D.; Kappos, L. Blood Neurofilament Light Chain as a Biomarker of MS Disease Activity and Treatment Response. *Neurology* **2019**, *92*, e1007–e1015. [CrossRef] [PubMed]
384. De Paula Faria, D. Myelin Positron Emission Tomography (PET) Imaging in Multiple Sclerosis. *Neural Regen. Res.* **2020**, *15*, 1842. [CrossRef]
385. Klistorner, A.; Graham, S.L. Role of Multifocal Visually Evoked Potential as a Biomarker of Demyelination, Spontaneous Remyelination, and Myelin Repair in Multiple Sclerosis. *Front. Neurosci.* **2021**, *15*, 1410. [CrossRef] [PubMed]
386. Muthupillai, R.; Lomas, D.J.; Rossman, P.J.; Greenleaf, J.F.; Manduca, A.; Ehman, R.L. Magnetic Resonance Elastography by Direct Visualization of Propagating Acoustic Strain Waves. *Science* **1995**, *269*, 1854–1857. [CrossRef]
387. Hirsch, S.; Braun, J.; Sack, I. Magnetic Resonance Elastography-Physical Background and Medical Applications. *Magn. Reson. Elastography* **2017**, *14*, 263–281. [CrossRef]
388. Wuerfel, J.; Paul, F.; Beierbach, B.; Hamhaber, U.; Klatt, D.; Papazoglou, S.; Zipp, F.; Martus, P.; Braun, J.; Sack, I. MR-Elastography Reveals Degradation of Tissue Integrity in Multiple Sclerosis. *Neuroimage* **2010**, *49*, 2520–2525. [CrossRef] [PubMed]
389. Streitberger, K.J.; Sack, I.; Krefting, D.; Pfüller, C.; Braun, J.; Paul, F.; Wuerfel, J. Brain Viscoelasticity Alteration in Chronic-Progressive Multiple Sclerosis. *PLoS ONE* **2012**, *7*, e29888. [CrossRef]
390. Schregel, K.; Née Tysiak, E.W.; Garteiser, P.; Gemeinhardt, I.; Prozorovski, T.; Aktas, O.; Merz, H.; Petersen, D.; Wuerfel, J.; Sinkus, R. Demyelination Reduces Brain Prenchymal Stiffness Quantified in Vivo by Magnetic Resonance Elastography. *Proc. Natl. Acad. Sci. USA* **2012**, *109*, 6650–6655. [CrossRef] [PubMed]
391. Riek, K.; Millward, J.M.; Hamann, I.; Mueller, S.; Pfueller, C.F.; Paul, F.; Braun, J.; Infante-Duarte, C.; Sack, I. Magnetic Resonance Elastography Reveals Altered Brain Viscoelasticity in Experimental Autoimmune Encephalomyelitis. *NeuroImage. Clin.* **2012**, *1*, 81–90. [CrossRef]
392. Wang, S.; Millward, J.M.; Hanke-Vela, L.; Malla, B.; Pilch, K.; Gil-Infante, A.; Waiczies, S.; Mueller, S.; Boehm-Sturm, P.; Guo, J.; et al. MR Elastography-Based Assessment of Matrix Remodeling at Lesion Sites Associated with Clinical Severity in a Model of Multiple Sclerosis. *Front. Neurol.* **2019**, *10*, 1382. [CrossRef] [PubMed]
393. Herthum, H.; Hetzer, S.; Scheel, M.; Shahryari, M.; Braun, J.; Paul, F.; Sack, I. In Vivo Stiffness of Multiple Sclerosis Lesions Is Similar to That of Normal-Appearing White Matter. *Acta Biomater.* **2022**, *138*, 410–421. [CrossRef] [PubMed]

International Journal of *Molecular Sciences*

MDPI

Article

Non-Apoptotic Caspase-3 Activation Mediates Early Synaptic Dysfunction of Indirect Pathway Neurons in the Parkinsonian Striatum

Tim Fieblinger [1,2,*], Chang Li [1], Elena Espa [1] and M. Angela Cenci [1,*]

1 Basal Ganglia Pathophysiology Unit, Department of Experimental Medical Science, Lund University, 223 62 Lund, Sweden; chang.li@med.lu.se (C.L.); elena.espa@med.lu.se (E.E.)

2 University Medical Center Hamburg-Eppendorf, Institute for Synaptic Physiology, 20251 Hamburg, Germany

* Correspondence: tim.fieblinger@zmnh.uni-hamburg.de (T.F.); angela.cenci_nilsson@med.lu.se (M.A.C.)

Abstract: Non-apoptotic caspase-3 activation is critically involved in dendritic spine loss and synaptic dysfunction in Alzheimer's disease. It is, however, not known whether caspase-3 plays similar roles in other pathologies. Using a mouse model of clinically manifest Parkinson's disease, we provide the first evidence that caspase-3 is transiently activated in the striatum shortly after the degeneration of nigrostriatal dopaminergic projections. This caspase-3 activation concurs with a rapid loss of dendritic spines and deficits in synaptic long-term depression (LTD) in striatal projection neurons forming the indirect pathway. Interestingly, systemic treatment with a caspase inhibitor prevents both the spine pruning and the deficit of indirect pathway LTD without interfering with the ongoing dopaminergic degeneration. Taken together, our data identify transient and non-apoptotic caspase activation as a critical event in the early plastic changes of indirect pathway neurons following dopamine denervation.

Keywords: Parkinson's disease; striatum; caspase-3; spiny projection neurons; dendritic spines; long-term depression; Q-VD-OPh; mice

Citation: Fieblinger, T.; Li, C.; Espa, E.; Cenci, M.A. Non-Apoptotic Caspase-3 Activation Mediates Early Synaptic Dysfunction of Indirect Pathway Neurons in the Parkinsonian Striatum. *Int. J. Mol. Sci.* **2022**, *23*, 5470. https://doi.org/10.3390/ijms23105470

Academic Editors: Marcello Ciaccio and Alberto Pérez-Mediavilla

Received: 22 March 2022
Accepted: 12 May 2022
Published: 13 May 2022

1. Introduction

Caspases are a family of cysteine proteases classically associated with apoptosis (i.e., programmed cell death) [1–3]. However, accumulating evidence shows that caspases also fulfill non-apoptotic roles in both neurons and glial cells [4,5]. Studies in hippocampal neurons have shown that caspase-3 can mediate the pruning of axons, dendrites and spines in the absence of cell death [6–8]. Moreover, restricted activation of caspase-3 in dendrites has been linked to long-term depression (LTD) of Schaffer collateral-CA1 synapses [8]. In microglia and astrocytes, caspase activation regulates the transition to proinflammatory or reactive phenotypes [9–11], and cytosolic expression of active caspase-3 has been observed in various brain cells even under basal conditions [5,12]. In models of Alzheimer's disease, non-apoptotic caspase-3 activation has been shown to mediate hippocampal synaptic dysfunction, including altered AMPA receptor phosphorylation, loss of dendritic spines, and LTD deficits [6,13].

Loss of dendritic spines and deficits in LTD induction also occurs in striatal neurons in animal models of Parkinson's disease (PD), a neurodegenerative disorder with typical motor symptoms caused by the degeneration of nigrostriatal dopaminergic projections [14–21]. Using toxin-based models of PD, previous studies have reported high levels of caspase-3 activity in the striatum [22,23]. In contrast, a non-symptomatic genetic model of autosomal recessive PD (the *PINK1* knockout mouse) shows normal striatal levels of caspase-3 activity at baseline, but an activity-dependent reduction of caspase-3 activity appears to be causally linked to synaptic plasticity deficits in this model [24]. These studies warrant further investigations into the functional role of caspase-3 in the pathophysiology of PD.

In the present study, we set out to investigate whether caspase-3 plays a role in the loss of dendritic spines and the concomitant synaptic deficits that develop in striatal spiny projection neurons (SPNs) upon the loss of dopaminergic innervation. Loss of SPN spines and glutamatergic synapses have been well-documented in all animal models of PD exhibiting severe nigrostriatal dopamine (DA) depletion [15,16,19,25,26], and have moreover been detected post-mortem in PD patients (reviewed in [27]).

Striatal SPNs divide into two major groups, called direct and indirect pathway SPNs (dSPNs and iSPNs, respectively). These two populations differ in their respective projection targets, expression of DA receptors, physiological properties, and response to DA denervation [27–29]. In models of PD, spine pruning and a concomitant loss of LTD are more prominent in iSPNs than dSPNs [15–17,20]. Although marked pathophysiological adaptations affect dSPNs too [14,19,25,26], spine loss is delayed and more variable in these neurons [27,30]. Based on these considerations, we chose to focus our study on the spine loss and synaptic dysfunction affecting the iSPNs.

Our results reveal that caspase-3 is promptly and transiently activated in the striatum upon DA denervation, and that iSPNs show loss of dendritic spines and LTD disturbance at the same early time points. Treatment with Q-VD-OPh, a third-generation broad-spectrum caspase inhibitor with therapeutic potential [31], prevented the loss of dendritic spines and LTD in iSPNs without interfering with the ongoing degeneration of DA neurons. Our study, therefore, points to a connection between spine loss and early synaptic dysfunction of iSPNs and non-apoptotic activation of caspase-3 in the parkinsonian striatum.

2. Results

2.1. Transient and Early Non-Apoptotic Caspase-3 Activation in the DA-Denervated Striatum

To investigate the time course of caspase-3 activation, we performed 6-OHDA injections in the medial forebrain bundle (MFB) and euthanized the mice at 3 to 28 days post-lesion. Striatal sections were immunostained using an antibody recognizing both pro- and cleaved caspase-3, as in [23]. On the side contralateral to the lesion ("intact"), the levels of caspase-3 immunostaining were low, and comparable to those found in sham-operated mice (Figure 1A–F, left). In contrast, a large number of cell bodies and processes were immunoreactive for caspase-3 on the DA-denervated side ("lesion"), particularly at early time points (Figure 1A–F, right). The peak of caspase-3 expression in the DA-denervated striatum was detected at five days post-lesion (1.75 +/− 0.19-fold over intact side, Figure 1C,G), after which the signal steadily declined. At five days post-lesion, the caspase-3 immunostaining co-localized with the astrocyte markers GFAP and S100β (Figure 1H,I) and the microglial marker IBA1 (Figure 1J). To assess whether caspase-3 also was expressed in iSPNs, we labeled this neuron population using BAC-*adora2a*-Cre mice and a Cre-inducible AAV-eGFP vector. Upon examining these mice five days after a 6-OHDA lesion, we found labeling for caspase-3 in multiple puncta along iSPN dendrites (Figure 1K,L).

Why is caspase-3 elevated so shortly after the 6-OHDA lesion? Due to its known role in axon degeneration, a likely reason could be the involvement of caspase-3 in the breakdown of DAergic fibers that richly innervate the striatum. To test this hypothesis, we performed Western blot analysis of striatal tyrosine-hydroxylase (TH) levels, a widely used dopaminergic cell marker. Our analysis shows a rapid, exponential decay of striatal TH levels after the 6-OHDA injection, with the bulk already being lost three days post-lesion (0.29 +/− 0.03 and 0.20 +/− 0.04 relative to the intact side, at three and five days post-lesion, respectively; Figure 2A). This is in line with the report by Rentsch and colleagues, showing that striatal TH vanishes quickly even though the degeneration of dopaminergic neurons in the substantia nigra takes weeks to become established [32]. We then examined the same striatal samples for caspase-3 activation, measured as the ratio of cleaved caspase-3 fragments (17 and 19 kDA bands) over the pro-caspase-3 protein (35 kDa band). This ratio was significantly increased at five days (1.86 +/− 0.32 relative to intact side) but not three days post-lesion (1.21 +/− 0.12 relative to intact side; Figure 3B), indicating that the degeneration of TH axons precedes the peak of caspase-3 activation.

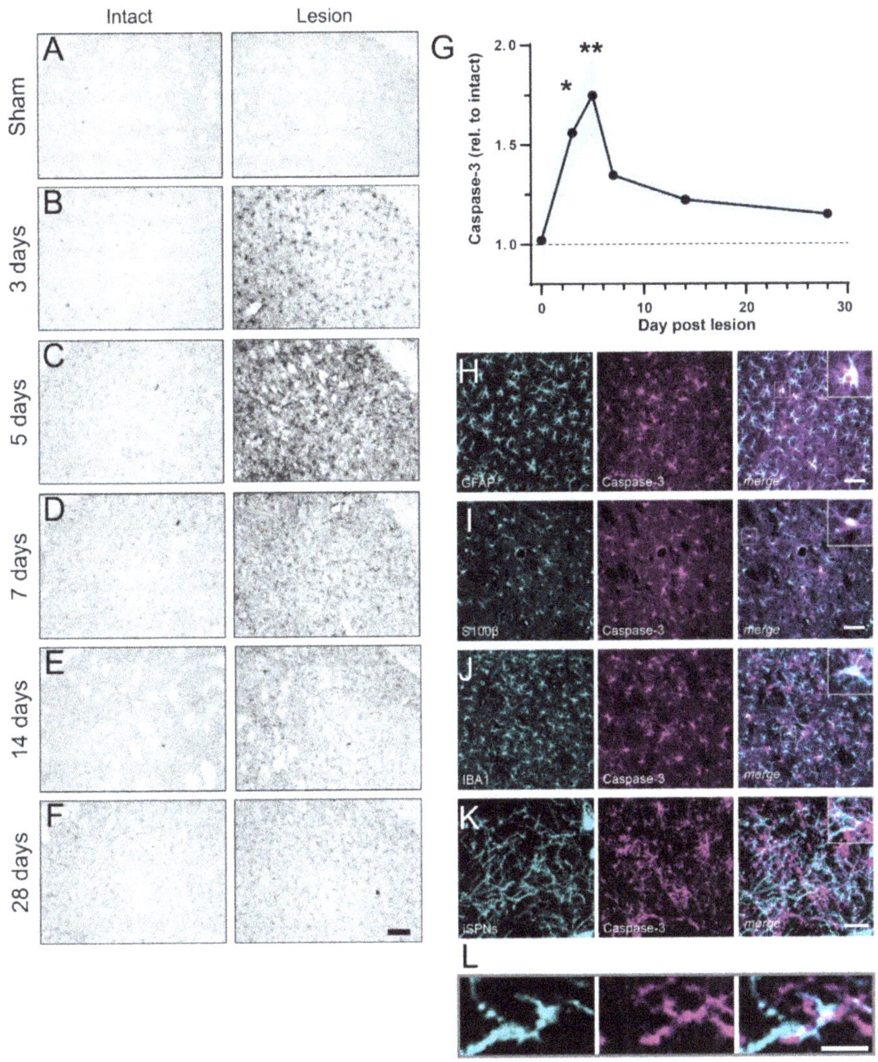

Figure 1. Caspase-3 is transiently upregulated in the DA-denervated striatum. (**A–F**) Striatal sections immunostained with a caspase-3 antibody depict the time course of caspase-3 activation in the DA-denervated striatum (right) relative to the control side (left) post 6-OHDA lesion. Scale bar: 100 μm. (**G**) Densitometric analysis reveals an increase of caspase-3 levels in the DA-denervated striatum at three- and five-days post-lesion as compared to sham-lesioned animals (represented by "Day 0" datapoint). * $p < 0.05$, ** $p < 0.01$ vs. Sham, ANOVA and post hoc Bonferroni test, N = 4–7. (**H–K**) Dual-antigen immunostaining demonstrates caspase-3 co-labeling with markers for astrocytes (GFAP and S100β; **H**,**I**), microglia (IBA1; **J**) and iSPN dendrites (eGFP; **K**). The latter were labeled using an AAV-based strategy (see methods). (**L**) Zoom-in of inset in (**K**) shows puncta of caspase-3 expression within iSPN dendrites. Scale bar: 20 μm (**H–J**), 10 μm (**K**), 3 μm (**L**).

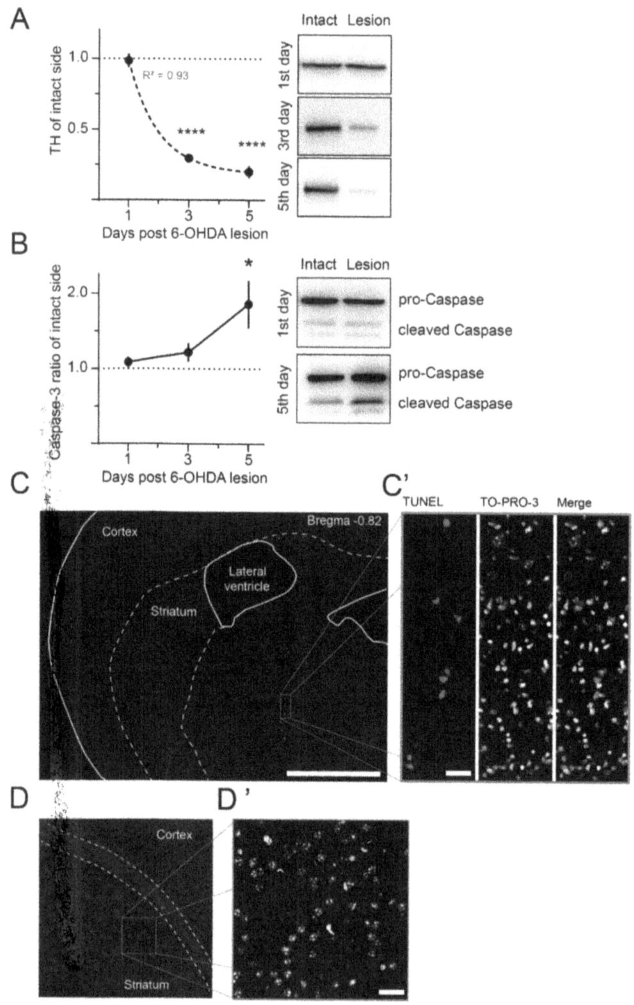

Figure 2. The peak of caspase-3 activation follows the loss of striatal DAergic innervation. (**A**) Western blot analysis of striatal TH levels shows an exponential decline of DAergic innervation between one and five days after the 6-OHDA lesion (dotted line depicts a one-phase decay fit with the given R^2 value). **** $p < 0.0001$ vs. one day post-lesion, ANOVA and post hoc Bonferroni test, N = 6–7. Right: Representative immunoblots with TH bands (60 kDa). (**B**) Caspase-3 activation was assessed as a ratio of the cleaved (19 and 17 kDa bands) over the uncleaved protein (35 kDa). The ratios are shown relative to the corresponding intact striatum and confirm a peak of caspase-3 activation at five days post-lesion. * $p < 0.05$ vs. one day post-lesion, ANOVA and post hoc Bonferroni test, N = 6–7. Right: Example of caspase-3-stained blots. (**C**,**D**) TUNEL staining was used to reveal apoptotic nuclei in sections through the caudal diencephalon (encompassing the MFB, **C**) and DA-denervated striatum (**D**). At the peak of caspase activation (five days post-6-OHDA lesion), no TUNEL positivity was found in the striatum (**D**,**D′**), whereas numerous cells displayed apoptotic features along the track of the 6-OHDA infusion (**C**,**C′**). (**C′**,**D′**) High magnification of (**C**,**D**), as indicated. TUNEL (magenta) is overlaid with TO-PRO-3 counterstain of neuronal nuclei (gray). Among the TUNEL-positive cells, some are colabeled with TO-PRO-3 and others are not, indicating different stages of apoptosis. Scale bars: 1 mm (**C**), 40 μm (**C′**) 10 μm (**D′**).

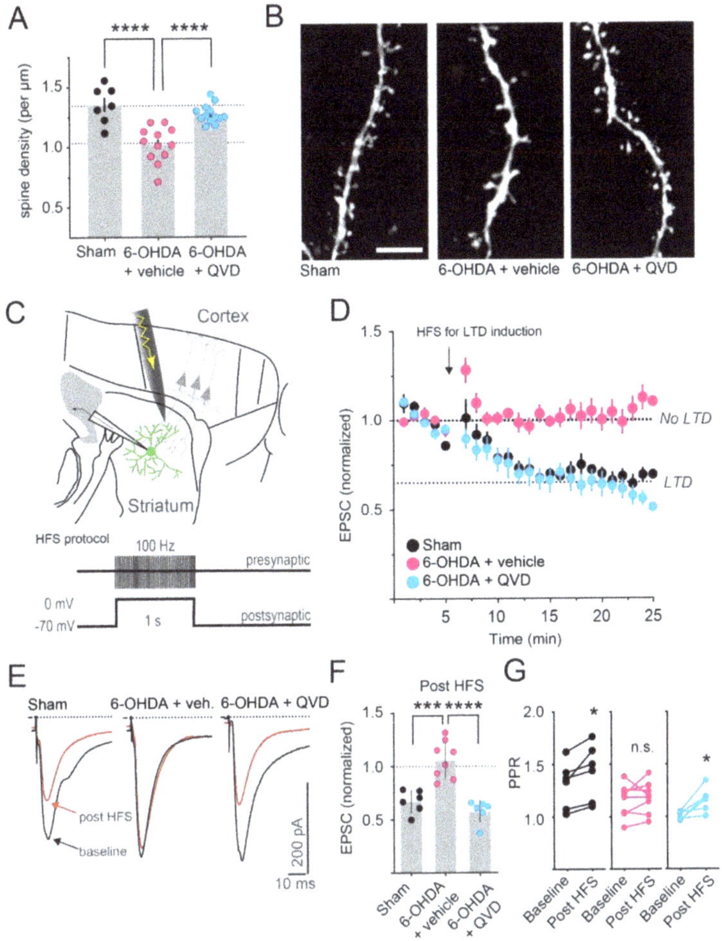

Figure 3. Early synaptic deficits after DA denervation are prevented by pharmacological caspase inhibition. (**A**) Dendritic spine density of iSPNs is reduced 5–6 days post-6-OHDA lesion. This loss is prevented by systemic treatment with the caspase inhibitor Q-VD-OPh ("QVD"). **** $p < 0.0001$ vs. 6-OHDA + vehicle, ANOVA and post hoc Bonferroni tests. N = 7–13 cells. (**B**) Two-photon images of iSPN dendrites, visualized after dye-filling the neurons through the patch pipette. Maximum-intensity projections of dendrites from controls and 6-OHDA-lesioned mice treated with vehicle or QVD are shown from left to right. Scale bar: 5 µm. (**C**) Sketch of the LTD recording paradigm. GFP-positive iSPNs were patched in the dorsolateral striatum and the stimulus electrode was placed near the border of the cortex. To induce LTD, high-frequent input stimulation (1 s at 100 Hz) was paired with postsynaptic depolarization. (**D**) Corticostriatal EPSCs are depressed using this protocol in control iSPNs (black). However, EPSC amplitudes are unchanged after HFS in 6-OHDA lesioned mice at 5–6 days after lesion (magenta). Co-treatment with QVD rescues this deficit, and LTD is readily induced (cyan). (**E**) Example traces from the recordings. Single EPSCs are shown before (black) and after (red) HFS. (**F**) Quantification shows loss and rescue of HFS-LTD. *** $p < 0.001$, **** $p < 0.0001$ vs. 6-OHDA + vehicle, ANOVA and post hoc Bonferroni test, N = 6-8 cells (**G**) After HFS-LTD induction, paired-pulse ratios (PPRs) are increased in control iSPNs (black) and unchanged in the DA-denervated striatum (magenta). Treatment with QVD rescues the LTD induction, with concomitant increase in PPRs (cyan). * $p < 0.05$, paired t-test, N = 6-8 cells.

When associated with apoptosis, caspase-3 activation shows a predominantly nuclear localization. However, in the DA-denervated striatum, caspase-3 immunostaining occurred mostly in filamentous structures. Accordingly, using terminal deoxynucleotidyl transferase-mediated biotinylated dUTP nick end labeling (TUNEL), we did not detect any apoptotic cells in DA-denervated striatal sections (Figure 2D,D′), whereas we found many intensely stained nuclei along the 6-OHDA injection track in the caudal diencephalon (Figure 2C,C′). The lack of striatal apoptosis is in keeping with previous studies finding a normal number of neurons in the MFB lesion model [22,33].

Taken together, these results indicate that the degeneration of dopaminergic fibers is promptly followed by prominent and transient non-apoptotic activation of caspase-3 in the striatum, with conspicuous upregulation of caspase-3 protein levels in both iSPNs and glial cells.

2.2. Caspase Activation Mediates Early Structural and Synaptic Changes in DA-Denervated iSPNs

Spine loss and synaptic deficits are prominent in hippocampal neurons in models of Alzheimer's disease, and caspase-3 is an important mediator of these changes [6,13]. Loss of dendritic spines and an inability to induce LTD are also observed in iSPNs after DA-denervation [14–20]. However, previous studies have not established when exactly these changes first appear. We therefore investigated if the loss of iSPN dendritic spines and LTD deficits already occur during the time window of marked caspase-3 activation.

At 5–6 days post-6-OHDA lesion, dendritic spine density was significantly reduced in iSPNs (1.36 +/− 0.06 and 1.03 +/− 0.04 spines per μm in sham and 6-OHDA + vehicle, respectively; Figure 3A,B). To evaluate if the activation of caspase-3 is causally linked to this spine loss, we co-treated 6-OHDA-lesioned mice with an effective dose of the systemically active pan-caspase-inhibitor Q-VD-OPh (10 mg/kg, s.c., twice daily on days 1-5 post-lesion; [34,35]). Interestingly, in the treated mice, iSPN spine density was not different from that in sham-lesioned controls (1.28 +/− 0.02 spines per μm; Figure 3A,B). Blocking caspase activation therefore prevents the loss of iSPN spines induced by DA denervation.

In PD models, the structural reorganization of iSPNs is accompanied by a deficit in LTD formation [14,17,18,20,21]. To induce LTD, we used an established protocol based on pairing high-frequency stimulation (HFS) of cortical afferents with postsynaptic depolarization (Figure 3C). In sham-lesioned mice, this protocol reduced evoked EPSCs in iSPNs by >30% (Figure 3D–F, black). This was accompanied by an increase in the paired-pulse ratio (PPR; 1.32 +/− 0.09 and 1.45 +/− 0.10 PPR at baseline and post HFS, respectively; Figure 3G). The PPR was measured by delivering two stimuli at a short interval (50 ms) and then dividing the amplitude of the second EPSC by the amplitude of the first. PPRs are a well-described indicator of synaptic release probabilities [36] and an increase in the PPR (indicating a reduced probability of presynaptic glutamate release) is expected for this form of LTD [37–39]. Interestingly, both LTD formation and the concomitant PPR increase were lost already 5–6 days after DA-denervation (1.06 +/− 0.06 EPSC amplitude relative to baseline; 1.16 +/− 0.05 and 1.18 +/− 0.05 PPR at baseline and post HFS, respectively; Figure 3D–G, magenta). However, upon cotreatment with the caspase-inhibitor Q-VD-OPh, HFS-induced LTD and concomitant PPR change did not differ from those in sham-lesioned controls (0.58 +/− 0.04 EPSC amplitude relative to baseline; 1.01 +/− 0.02 and 1.16 +/− 0.05 PPR at baseline and post HFS, respectively Figure 3D–G, cyan).

To determine whether Q-VD-OPh treatment had affected the extent of dopaminergic degeneration, we measured striatal TH levels by optical density analysis and the number of TH-positive neurons in the substantia nigra pars compacta (SNc) by unbiased stereological cell counting. In line with a previous study, we find that the loss of striatal TH-positive fibers precedes the loss of TH-positive cells in the SNc [32]. Additionally, striatal TH levels were reduced by over 90% regardless of the treatment (99 +/− 1.7 %, 4 +/− 0.6% and 3 +/− 0.5% of the intact side in Sham, 6-OHDA + vehicle and 6-OHDA + QVD, respectively; Figure 4A,C). Stereological counts of TH-positive cells in the SNc revealed a similar extent of DA neuron loss in 6-OHDA lesioned mice treated with Q-VD-OPh or vehicle (99 +/− 8%, 67 +/− 8% and 65 +/− 7% of the intact side in Sham, 6-OHDA + vehicle and 6-OHDA + QVD, respectively; Figure 4B,D). These data show that Q-VD-OPh treatment had not interfered with the process of dopaminergic degeneration induced by 6-OHDA.

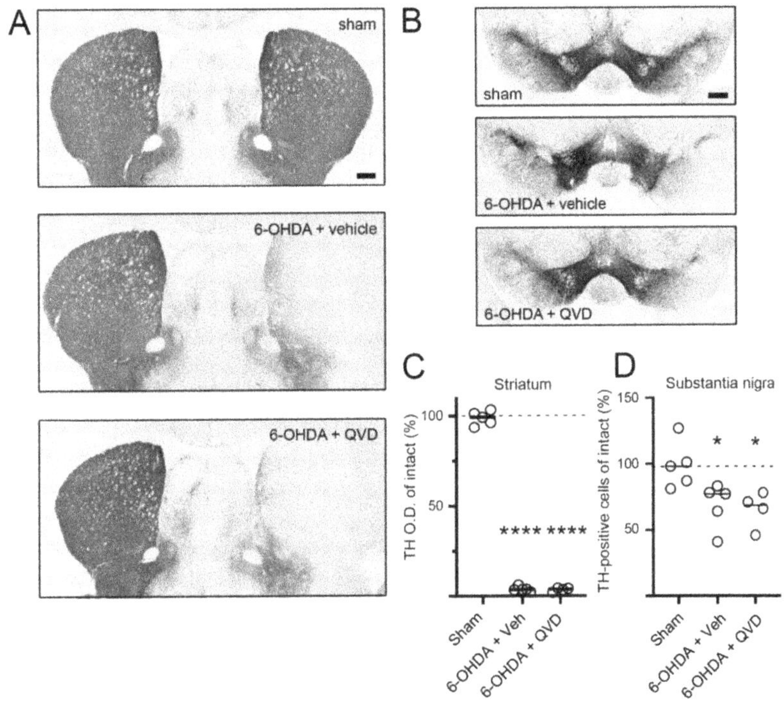

Figure 4. Q-VD-OPh treatment does not protect against 6-OHDA-induced dopaminergic degeneration. (**A**,**B**) Low-magnification images of striatal (**A**) and nigral (**B**) sections immunostained for TH, with the side ipsilateral to the 6-OHDA lesion on the right. Scale bar: 400 μm (**C**) Densitometric analysis shows that the loss of striatal DAergic innervation does not differ between vehicle- and QVD-treated animals. **** $p < 0.0001$ vs. sham, ANOVA and post hoc Bonferroni test, N = 4–5. (**D**) Stereological cell counts show that the 6-OHDA-induced loss of nigral DA neurons is not prevented by QVD treatment. * $p < 0.05$ vs. sham, ANOVA and post hoc Bonferroni test, N = 4–5.

Taken together, these results indicated that spine loss and the inability to form HFS-LTD affect iSPNs early after DA-denervation, coinciding with a period of high caspase-3 activity in the striatum. Interestingly, both alterations can be prevented by pharmacological treatment with a caspase inhibitor that does not modify the extent nor the pattern of nigrostriatal dopaminergic denervation.

3. Discussion

Our results show that transient, non-apoptotic activation of caspase-3 in the DA-denervated striatum temporally coincides with dendritic spine pruning and synaptic plasticity deficits in iSPNs. Treatment with a pan-caspase inhibitor prevented the loss of iSPN spines and preserved HFS-LTD, although it did not protect against the 6-OHDA-induced dopaminergic degeneration.

Previous studies addressing the role of caspase-3 in PD models have focused on its possible involvement in the apoptotic death of DA-producing neurons [23,40–47]. Thus, caspase-3 activation has been shown to mediate the apoptosis of nigral dopaminergic neurons in MPTP models [23,44,45,47], although results from 6-OHDA lesion models have varied between studies [40–43,46]. Few studies have investigated the expression and/or activity of caspase-3 in the striatum. One study reported that caspase-3 is apoptotically activated in the striatum of MPTP-treated mice seven days after the last MPTP dose [23], and a second study using a partial 6-OHDA lesion model observed expression of active caspase-3 in enkephalin-positive striatal neurons at four weeks post-lesion [22]. This upregulation of caspase-3 was reported to occur in the absence of cell loss, and its functional significance remained unclear [22]. We provide here the first evidence that non-apoptotic caspase-3 activation mediates early synaptic adaptations in iSPNs deprived of their dopaminergic afferents.

It is well established that, in both PD patients and DA-denervated animals, SPNs undergo a process of dendritic regression and loss of spines [16,48–52]. Previous studies in 6-OHDA-lesioned mice observed that the spine loss primarily occurs in iSPNs [15,16], or in the rare sub-population of SPNs expressing both the D1 and D2 receptors [53]. Although dSPNs undergo many modifications, spine pruning occurs only late in this cell population [30], presumably depending on a different mechanism. However, what does a loss of iSPN dendritic spines signify? An influential hypothesis states that the pruning of spines serves as a homeostatic response to rebalance the activity of iSPNs, which become hyperactive in the absence of D2 receptor stimulation [54]. Yet, maladaptive implications cannot be excluded since iSPN spine pruning concurs with loss of corticostriatal synapses and deficits in synaptic plasticity.

The present study provides valuable new insights into the temporal evolution and molecular mechanisms of iSPN spine loss. Our data show that spine pruning and LTD disruption are present in iSPNs already five days post lesion. This is a surprisingly early time point if one considers that the degeneration of nigral dopaminergic neurons and other SPN adaptations take weeks to manifest completely. The loss of iSPN dendritic spines is therefore very rapid, most likely directly triggered by the loss of DA afferents and the ensuing lack of D2 receptor stimulation. In line with this interpretation, iSPN dendritic spine density has been reported to decrease after pharmacological DA depletion with reserpine [15] or treatment with the D2 receptor-antagonist haloperidol, even when administered for only 5 days [55].

In a seminal study, Day et al. showed that selective elimination of iSPN spines in PD models is due to increased Ca^{2+} influx via CaV1.3 channels, which are positioned in spines. Accordingly, SPNs of CaV1.3-knockout mice have higher spine densities and are resistant to 6-OHDA-induced spine loss [15]. However, the mechanisms linking intraspine Ca^{2+} levels to spine removal have not been established. Based on the present results, we propose that caspase-3 activation provides a critical link, in a fashion similar to what has been described for hippocampal neurons. In these neurons, dendritic Ca^{2+} influx leads to sequential activation of caspase-9 and caspase-3, followed by spine pruning [7,8]. Caspase-3 could initiate spine removal by cleaving cytoskeletal components, such as actin [56] and actin-regulating proteins [57]. It also increases the activity of calcineurin [58], a Ca^{2+}-dependent phosphatase regulating cytoskeletal proteins [59]. Another potential role of caspase-3 may consist in promoting the release of chemoattractant factors necessary for phagocytic cells to engulf and remove damaged cellular components. This function of caspase-3 has been demonstrated in a model of dendritic pruning in *Drosophila* [60] and in various cellular models of apoptosis [61].

Besides the roles in neuronal plasticity, caspase-3 also activates microglia [11] and pushes astrocytes towards a reactive phenotype [9,10]. This glial activation can itself orchestrate the removal of dendritic spines [62]. We observed that many glia cells were labeled for caspase-3 in the DA-denervated striatum. Thus, a caspase-regulated glial contribution to the process of spine pruning seems likely. However, if spine loss was solely dependent on the lesion-induced glial activation, it would equally affect the two SPN populations, whereas it is well established that spine pruning selectively affects iSPNs for up to six weeks after 6-OHDA infusion, at least in the MFB lesion model [16,27]. Moreover, the observation that selective iSPN spine loss can be induced using pharmacological treatments that do not cause any significant glial activation [15,55] speaks in favor of an intraneuronal triggering mechanism.

The inability to induce HFS-LTD in iSPNs after DA denervation has been reported previously [14,17,18]. Here, we add that this deficit already occurs five days after the 6-OHDA lesion. In parallel with the preservation of dendritic spines, pharmacological caspase inhibition preserved the ability to induce LTD in iSPNs. HFS-LTD is mediated by postsynaptic endocannabinoid release and activation of presynaptic CB1 receptors (CB1Rs), which leads to an inhibition of transmitter release from the glutamatergic axon terminal. The CB1Rs are, however, only expressed on corticostriatal and not thalamostriatal terminals [39]. This is remarkable because we have previously shown that specifically corticostriatal synapses are pruned in iSPNs in this model of PD [16]. It is tempting to speculate that the selective loss of cortically innervated dendritic spines causes the disruption of this form of corticostriatal LTD by simply eliminating its structural substrate, and that Q-VD-OPh treatment preserves HFS-LTD in iSPNs by maintaining the integrity of their dendritic spines and the connected corticostriatal terminals. Interestingly, a previous ultrastructural study has revealed a marked loss of VGluT1-positive corticostriatal presynaptic terminals in MFB-lesioned rodents [63], and our previous study showed that denervation-induced iSPN spine pruning goes hand in hand with a loss of corticostriatal synapses [16]. Additionally, several studies show that input neurons from the motor cortex undergo various changes following DA-denervation [64–66]. The present set of data does not allow us to exclude that loss of iSPN spines and HFS-LTD are two independent consequences of DA denervation. The disadvantage of this viewpoint is, however, that it fails to readily explain why caspase-inhibition restores both phenomena, since our data show that DA innervation is not spared by such treatment.

We chose Q-VD-OPh because the more specific caspase-3 inhibitors like Z-DEVD-FMK do not cross the blood–brain barrier and would need to be injected directly into the brain. While being principally feasible, this approach is disadvantageous, as delivery of compounds directly into the striatum through cannulas is prone to induce tissue trauma and inflammatory responses that can confound the data. Q-VD-OPh on the other hand is a caspase inhibitor with therapeutic potential [67], and has been tested in several disease models. For example, it has been found to delay disease progression in AIDS models [68], reduce brain damage in models of stroke [69], and attenuate Alzheimer's pathology [35]. In the MPTP model of PD, systemic treatment with Q-VD-OPh reduces DA cell death and spares TH positive fibers in the striatum when low doses of MPTP are used [70], potentially by preventing caspase-3-mediated apoptosis of nigral dopaminergic neurons [71]. In our 6-OHDA model, however, Q-VD-OPh did not prevent the loss of dopaminergic neurons or TH fibers in the striatum. In agreement with a previous study [41], our results therefore indicate that the 6-OHDA-induced dopaminergic degeneration is largely independent of caspase-3-mediated apoptosis.

The functions of caspase-3 beyond apoptosis have started to be investigated relatively recently. After the pioneering results obtained in glutamatergic pyramidal neurons [6–8,13] we here provide the first evidence that also in the GABAergic, principal neurons of the striatum, caspase-3 serves non-apoptotic roles in mediating the pruning of dendritic spines and the associated synaptic alterations in a model of PD. Our results further show that loss of spines and dysfunctional LTD go hand in hand in iSPNs after DA denervation, which is reminiscent of the reports from other pathologies, such as hippocampal neurons in Alzheimer's disease models [6,13] and accumbal neurons in ethanol-dependent rats [72]. In future studies, it will be interesting to determine whether caspase-3 mediates the synaptic remodeling of SPNs also in non-pathological situations, such as during development or striatum-dependent learning tasks.

4. Materials and Methods

4.1. Animals

We used heterozygous BAC-*adora2a*-GFP and BAC-*adora2a*-Cre transgenic mice (GENSAT project) on a C57BL/6 background and their non-transgenic littermates (bred in the Animal Facility of Lund University, Biomedical Center). Mice were housed under a 12 h light/12 h dark cycle with free access to food and water. The animals were at least nine weeks old at the beginning of the experiments and a total of 80 mice of both sexes were used. Animal numbers and experimental layouts are provided in Figure S1. All experiments were approved by the Malmö-Lund Ethical Committee on Animal Research (ethical permit number M47-16, granted on 20 April 2016) and were conducted in adherence with the EU directive 2010/63/EU.

4.2. Stereotactic Surgeries

Unilateral 6-OHDA lesions and adeno-associated virus (AAV) injections were performed as described previously [73]. In brief, mice were anesthetized with a mixture of 4% isoflurane in air (Isoba vet, Apoteksbolaget) and placed in a stereotaxic frame (Kopf Instruments, Model 923-B mouse Gas Anethesia Head Holder, with 922 Ear Bars for mouse). 6-OHDA hydrochloride was dissolved in 0.02% ascorbic acid (3.2 μg free-base per μL) and 0.7 μL was injected at 0.2 μL/min in the median forebrain bundle at following coordinates: AP = −0.7, ML = −1.2, DV = −4.7, tooth bar: −4.0 [73]. For marking iSPNs, we injected AAV5-hSyn-DIO-EGFP-WPRE into the striatum at two injection sites (0.5 μL each): AP = +1, ML = −2.1, DV = −2.9; and AP = +0.3, ML = −2.3, DV = −3.0 [73]. After the surgery, the wound was closed with tissue glue and the animal received an s.c. injection of an analgesic (Marcaine, AstraZeneca; 2.5 mg/mL, 1 μL/g b.w.). To prevent dehydration, mice received s.c. injections of sterile glucose-ringer acetate (0.6 mL) immediately after the surgery. Further injections and dietary supplementations were given as necessary in the days post-lesion. Control mice received a sham-lesion surgery (capillary was inserted without delivering any injection). The success of the lesion was verified by TH-immunohistochemistry after the experiments.

4.3. Western Immunoblotting

Striatal samples were analyzed as described previously [74]. Animals were anesthetized with pentobarbital (500 mg/kg, i.p.) and briefly perfused with artificial cerebrospinal fluid (aCSF, see below). The brains were rapidly extracted and cut into thick sections on a Vibratome (Leica VT1200S) in ice-cold aCSF. The striatum was dissected out, transferred into tubes, rapidly frozen on crushed dry ice and stored at −80 °C until used. Samples were homogenized in RIPAlysis buffer, containing 65 mM Tris-base, 150 mM NaCl, 1% Triton-X, 0.25% sodium deoxycholate, 1 mM EDTA, and phosphatase and protease inhibitors ("phosSTOP" and "Complete, mini, EDTA-free," Roche Applied Science, Penzberg, Germany). Protein concentration was determined using the BCA Protein Assay Kit (Pierce #23225, Thermo Scientific, Uppsala, Sweden). Samples were run on a 7.5% SDS gel and transferred on PVD membranes. Membranes were blocked with 5% nonfat dry milk and incubated overnight with the following antibodies: anti-tyrosine hydroxylase (Pel-Freez, P40101), anti-caspase-3 (Cell Signaling, #9662, Lot 17, Danvers, MA, USA) and beta-actin (Sigma, A-3854). After washing, the membranes were incubated with HRP-linked secondary antibodies (anti-biotin HRP-linked antibody Cell Signaling, #7075; anti-rabbit-IgG HRP-linked antibody, Cell Signaling, #7074) and bands were visualized by chemiluminescence using an ECL kit (Pierce, #32106, Thermo Scientific). Images were acquired using a CCD camera (LAS1000 system, Fuji Films, Tokyo, Japan) and analyzed using ImageJ (National Institutes of Health, Bethesda, MD, USA). After image acquisition, membranes were stripped and re-blotted for loading controls.

4.4. Immunohistochemistry and Optical Density Analysis

Mice were anesthetized with sodium pentobarbital (240 mg/kg, i.p., Apoteksbolaget, Uppsala, Sweden) and transcardially perfused with 0.1 M phosphate-buffered saline (PBS, pH 7.4), followed by ice-cold 4% paraformaldehyde in PBS. Brains were post-fixed in the same solution overnight. Coronal sections of 30 μm thickness were cut while immersed in PBS at 4 °C using a vibratome (VT 1200S, Leica, Wetzlar, Germany). Sections were stored at −20 °C in a non-freezing solution (30% ethylene glycol and 30% glycerol in 0.1 M phosphate buffer).

Bright-field immunohistochemistry for TH and caspase-3 was performed according to previously described protocols [75], with minor modifications. Caspase-3 expression was studied using a primary antibody that recognizes both the pro- and cleaved caspase-3 (mouse anti-caspase-3, BD Transduction Laboratories, cat # BD 611049; 1:100). The immunostaining specificity was confirmed through omission of either primary or secondary antibodies, and through comparisons with the regional and cellular expression patterns produced by other caspase-3 antibodies (BD Transduction Laboratories #BD610322; Cell Signaling Technology, #9662; SCBT, #56053). Briefly, sections were rinsed in 0.02 M potassium phosphate-buffered saline (KPBS, pH 7.4) and transferred to a citrate buffer (pH 6.0) for antigen retrieval (carried out at 80 °C for 30 min). After three rinses with KPBS, sections were pretreated for 5 min in a solution of 3% hydrogen peroxide (H_2O_2) and 10% methanol to quench endogenous peroxidase activity. Sections were then pre-incubated for 1 h in a blocking buffer (5% normal horse serum in KPBS with 0.1% Triton-X, KPBS-T). This was followed by overnight incubation at 4 °C with the primary antibody. Next, sections were rinsed and incubated with biotinylated horse-anti-mouse secondary antibodies (Vector Laboratories, # BA2001, 1:200, Burlingame, CA, USA) for two hours, followed by incubation in an avidin-biotin-peroxidase solution (Vectastain Elite ABC, Vector Laboratories) for one hour at room temperature. The immunocomplexes were visualized using 3,3-diaminobenzidine (DAB) and H_2O_2 (0.05% and 0.04%, respectively). Finally, sections were rinsed in KPBS-T, mounted onto chromalum-coated slides, and coverslipped using DPX mounting medium. Images were taken on a Nikon Eclipse 80i microscope under a 20× objective (0.50 NA). Sections immunostained for TH (Pel-Freez, P40101-150; 1:1000) were digitized at low magnification (4× objective, NA 0.10) per our established methods [75,76]. Densitometric measurements of caspase-3 and TH immunostaining were carried out on predetermined regions of interest using ImageJ. Results from the side ipsilateral to the lesion were expressed as a percentage of the values measured on the contralateral side in each section.

Sections spanning the striatum were also processed for dual-antigen immunofluorescence to detect caspase-3 in iSPNs, microglia and astrocytes using the following primary antibodies: rabbit anti-caspase-3 (Cell Signaling Technology, #9662, 1:200) or mouse anti-caspase-3 (BD Transduction Laboratories, BD611049, 1:50), mouse anti-S100β (Sigma Aldrich, S2532, 1:100, St. Louis, MO, USA) and chicken anti-Iba1 (Synaptic Systems, #234009, 1:500, Göttingen, Germany). Sections were rinsed in Trizma buffered-saline (TBS) prior to a one hour pre-incubation in a blocking solution consisting of 5% normal serum (from goat or donkey) in TBS with 0.1% TritonX (TBS-T). Primary antibodies were diluted in the same blocking solution and sections were incubated overnight at 4 °C. On the second day, sections were incubated in secondary antibody solution for 2 h at room temperature. We used the following antibodies: Alexa Fluor 488-conjugated goat-anti-rabbit (Invitrogen, 1:200, Waltham, MA, USA); Alexa Fluor 647-conjugated donkey-anti-mouse (Invitrogen, 1:200) and Alexa Fluor 647-conjugated donkey anti-chicken (Jackson Immuno Research, 1:200, West Grove, PA, USA). Secondary antibodies were diluted in the same blocking solution as the primary antibody. Alexa FluorTM 488 Tyramide SuperBoostTM Kit (Invitrogen, B40922) was used to improve the signal-to-noise ratio for primary antibody rabbit anti-caspase-3. After secondary antibody incubation, sections were rinsed in TBS, mounted and coverslipped with polyvinyl alcohol mounting medium with DABCO (PVA-DABCO,

Sigma-Aldrich), and imaged using a confocal laser scanning microscope (LSM710 NLO, Zeiss, Jena, Germany).

4.5. Stereological Counts of Nigral DA Neurons

The number of TH positive cells was determined by unbiased stereology according to the optical fractionator method [77] using the protocol detailed previously [75,76]. In brief, four serial sections spanning the rostrocaudal extent of the substantia nigra pars compacta (from −2.9 to −3.6 mm posterior to bregma) were sampled in each animal. Analysis was performed using a Nikon 80i microscope with an x-y motorized stage controlled by NewCAST software (Visiopharm). The area of interest was outlined under 4× objective and counting of neurons was performed under 100× objective. The area sampling fraction and slice sampling fraction were set to 30% and 33.3%, respectively. The total number of TH-positive neurons in the substantia nigra pars compacta was estimated in each hemisphere using the optical fractionator formula, i.e., number of neurons = 1/ssf (slice sampling fraction) × 1/asf (area sampling fraction) × 1/tsf (thickness sampling fraction) × Σ(number of objects counted) [77].

4.6. Visualization of Apoptosis (TUNEL)

Terminal deoxynucleotidyl transferase-mediated biotinylated dUTP nick end labeling (TUNEL) was carried out in paraffin-embedded sections of 5 µm thickness, prepared from 3 mice that had been euthanized at 5 days post-lesion. The procedure was applied to both, striatal and midbrain sections encompassing the medial forebrain bundle (target of 6-OHDA infusion). We used a fluorescence-based TUNEL in situ cell death detection kit (FITC; Roche) according to the manufacturer's instructions and previous studies [24,78]. In brief, after preheating at 60 °C for 30 min, sections were rehydrated through a series of washing in xylene, absolute ethanol, graded ethanol and deionized distilled water. Following section rehydration, antigen retrieval in Proteinase K (20 µg/mL in 10 mM Tris/HCl, pH 7.4) was performed for 15 min at room temperature. Thereafter, the enzyme solution and label solution from the TUNEL kit were applied for 60 min at 37 °C. After washing three times with PBS the sections were counterstained with TO-PRO-3 Iodide 642/661 (Invitrogen, T3605) at a final concentration of 1 µM for 5 min at room temperature. Following a last washing step, sections were coverslipped with PVA-DABCO. Not all TUNEL-positive cells co-labeled with TO-PRO-3, which likely indicates different stages of apoptosis, as it has been shown that TO-PRO-3 labels early apoptotic and necrotic cells differentially [79].

4.7. In Vivo Caspase Inhibition with Q-VD-OPh

Q-VD-OPh (APExBio, Cat.No.: A1901) was dissolved first in DMSO and then diluted to the final concentration (2 mg/mL, 50% DMSO) with saline. Animals received injections of 10 mg/kg s.c., twice daily, with an injection volume of 5 mL/kg body weight.

4.8. Electrophysiology and Two-Photon Imaging

Acute brain slices were prepared as described previously [74,80]. In brief: mice were deeply anesthetized with Pentobarbital (65 mg/kg, i.p.) and shortly perfused with ice-cold aCSF containing (in mM): 124.0 NaCl, 3.0 KCl, 2.0 $CaCl_2$, 1.0 $MgCl_2$, 26.0 $NaCO_3$, 1.0 NaH_2PO_4 and 16.66 glucose. The osmolarity was 300–310 mOsm/L and the pH 7.4. Oxygenation and pH were maintained by gassing the aCSF with 5%/95% CO_2/O_2. Parasagittal slices (275 µm) were cut on a vibratome (VT1200s, Leica, Germany) and incubated at 34 °C for 30 min. Afterwards, the temperature was allowed to return to room temperature. LTD measurements were done at 30 °C.

The iSPNs in the dorsolateral striatum were identified by somatic eGFP expression. Patch pipettes (3–5 MOhm) were pulled from thick-walled borosilicate glass on a Sutter P-97 puller. Recordings were sampled at 10–20 kHz using a Multiclamp 700B amplifier (Molecular Devices, San Jose, CA, USA) and digitized (Digidata 1440, Molecular Devices,

USA). Data were analyzed in pClamp (v.10, Molecular Devices, USA). For voltage clamp recordings the internal solution contained (in mM): 120 $CsMeSO_3$, 5 NaCl, 10 TEA-Cl, 10 HEPES, 5 QX-314, 4 ATP-Mg and 0.3 GTP-Na. A stimulation electrode was placed in the dorsal striatum at the border with the corpus callosum. EPSCs were evoked at 0.05 Hz and cells were held at a potential of −70 mV in the presence of picrotoxin (50 μM). For LTD induction, synaptic inputs were stimulated with four trains (each 1 s at 100 Hz), spaced 10 s apart. During the stimulation, the cell was depolarized to 0 mV. All EPSCs were normalized to the average EPSC size of the respective baseline. Baseline EPSCS were not significantly different for the different groups: −241 +/− 15.6 pA, −269 +/− 25.4 pA and −386 +/− 77.9 pA, for Sham, 6-OHDA + vehicle and 6-OHDA + QVD, respectively (One-way ANOVA, $F(2, 17) = 2.728$, $p = 0.0938$; Bonferroni's multiple comparison test). For PPR measurements, two electric stimuli were given at an interstimulus interval (ISI) of 50 ms. To calculate the PPR, the amplitude of the second EPSC was divided by the amplitude of the first EPSC. Cortico-striatal synapses typically paired-pulse facilitate with short ISI, hence show a PPR > 1, and changes in PPRs are commonly interpreted as changes in presynaptic release probabilities [36–38]. Access resistance was continuously monitored and recordings were discarded if the access resistance at the end of the experiment was altered by >20%.

Two-photon imaging was performed on a Zeiss 710 NLO with a MaiThai laser (Spectra Physics, Sweden), as previously described [48,80]. To reveal SPN morphology, 50 μM of AlexaFluor-568 was added to the internal recording solution and excited at 780 nm through a 63× water-immersion objective (1.0 NA, Zeiss). For spine analysis, dendritic regions of interest were imaged at mid-distance from the soma (typically 50–100 μm), avoiding the aspiny proximal and the very distal regions (>100 μm). Dendritic stretches of 25–35 μm length were optically sectioned in a z-stack, with an x-y pixel size of $0.053 \times 0.053\ \mu m^2$ and z-sections were spaced 0.65 μm apart. Typically, 2–3 dendrites were analyzed per cell and dendritic spines were manually counted using ImageJ, in a blinded manner, as described previously [80].

4.9. Statistical Analysis

Statistical analysis was carried out using GraphPad Prism 8 and tests are specified in the respective figure legends and detailed results are provided in Table S1. A probability of $p < 0.05$ was considered statistically significant.

5. Concluding Remarks and Perspectives

This study presents three major findings: (1) the loss of dendritic spines and HFS-LTD in iSPNs occurs soon after striatal DA denervation, (2) caspase-3 is activated early and non-apoptotically in the same time frame, and (3) pharmacological caspase inhibition using a clinically relevant compound prevents the early synaptic alterations of iSPNs in a manner that does not require sparing or restoration of striatal DA fibers. These findings contribute new knowledge to the growing field of non-apoptotic caspase functions and an improved understanding of the striatal response to DA denervation. From a therapeutic perspective, our results point to a possible application of caspase inhibitors to prevent SPN alterations that are associated with the motor dysfunction of PD (reviewed in [27]). This suggestion, however, needs to be substantiated by new dedicated investigations.

Supplementary Materials: The following supporting information can be downloaded at: https://www.mdpi.com/article/10.3390/ijms23105470/s1.

Author Contributions: Conceptualization, T.F. and M.A.C.; formal analysis, T.F., C.L., E.E. and M.A.C.; investigation, T.F., C.L. and E.E.; data curation, T.F., C.L. and E.E.; writing—original draft preparation, T.F.; writing—review and editing, T.F., C.L. and M.A.C.; visualization, T.F. and C.L.; funding acquisition, T.F. and M.A.C. All authors have read and agreed to the published version of the manuscript.

Funding: This research was funded by project grants to M.A.C. from Swedish Government Funding for Clinical Research (Grant ALF project 43301), Swedish Research Council (Grant 2020-02696), Swedish Parkinson Foundation, and Swedish Brain Foundation. T.F. received grants for this project from The Craafoord Foundation and The Royal Physiographic Society of Lund, and held a Marie Skłodowska-Curie Individual Fellowship (MSCA-IF; SCAMPICITY, Grant agreement number 838736).

Institutional Review Board Statement: The animal study protocol was approved by the Malmö-Lund Ethical Committee on Animal Research (permit number M47-16, approved on 20 April 2016).

Informed Consent Statement: Not applicable.

Data Availability Statement: The presented data are available upon reasonable request from the corresponding authors.

Acknowledgments: The authors warmly thank Ann-Christin Lindh for excellent technical assistance and the Strategic Research Area MultiPark (Multidisciplinary Research on Parkinson's Disease), for infrastructure support).

Conflicts of Interest: The authors declare no conflict of interest.

Abbreviations

6-OHDA	6-hydroxy-dopamine
AAV	Adeno-associated virus
aCSF	Artificial cerebrospinal fluid
CCD	Charge-coupled device
DA	Dopamine
DAB	3,3′-diaminobenzidine
dSPN	Direct pathway spiny projection neuron
ECL	Electrochemiluminescence
EPSC	Excitatory postsynaptic potential
HFS	High-frequent stimulation
iSPN	Indirect pathway spiny projection neuron
KPBS	Potassium phosphate-buffered saline
LTD	Long-term depression
MFB	Median forebrain bundle
MPTP	1-methyl-4-phenyl-1,2,3,6-tetrahydropyridine
PBS	Phosphate-buffered saline
PD	Parkinson's disease
PPR	Paired pulse ratio
Q-VD-OPh	(3S)-5-(2,6-difluorophenoxy)-3-[[(2S)-3-methyl-2-(quinoline-2-carbonylamino) butanoyl]amino]-4-oxopentanoic acid
SPN	Spiny projection neuron
TBS	Trizma-buffered saline
TH	Tyrosine hydroxylase
TUNEL	Terminal deoxynucleotidyl transferase-mediated biotinylated dUTP nick end labeling

References

1. Li, J.; Yuan, J. Caspases in apoptosis and beyond. *Oncogene* **2008**, *27*, 6194–6206. [CrossRef] [PubMed]
2. Riedl, S.J.; Shi, Y. Molecular mechanisms of caspase regulation during apoptosis. *Nat. Rev. Mol. Cell Biol.* **2004**, *5*, 897–907. [CrossRef] [PubMed]
3. Shi, Y. Mechanisms of caspase activation and inhibition during apoptosis. *Mol. Cell* **2002**, *9*, 459–470. [CrossRef]
4. Hyman, B.T.; Yuan, J. Apoptotic and non-apoptotic roles of caspases in neuronal physiology and pathophysiology. *Nat. Rev. Neurosci.* **2012**, *13*, 395–406. [CrossRef] [PubMed]
5. Shen, X.; Venero, J.L.; Joseph, B.; Burguillos, M.A. Caspases orchestrate microglia instrumental functions. *Prog. Neurobiol.* **2018**, *171*, 50–71. [CrossRef]
6. D'Amelio, M.; Cavallucci, V.; Middei, S.; Marchetti, C.; Pacioni, S.; Ferri, A.; Diamantini, A.; De Zio, D.; Carrara, P.; Battistini, L.; et al. Caspase-3 triggers early synaptic dysfunction in a mouse model of Alzheimer's disease. *Nat. Neurosci.* **2011**, *14*, 69–76. [CrossRef]

7. Erturk, A.; Wang, Y.; Sheng, M. Local pruning of dendrites and spines by caspase-3-dependent and proteasome-limited mechanisms. *J. Neurosci.* **2014**, *34*, 1672–1688. [CrossRef]
8. Li, Z.; Jo, J.; Jia, J.M.; Lo, S.C.; Whitcomb, D.J.; Jiao, S.; Cho, K.; Sheng, M. Caspase-3 activation via mitochondria is required for long-term depression and AMPA receptor internalization. *Cell* **2010**, *141*, 859–871. [CrossRef]
9. Acarin, L.; Villapol, S.; Faiz, M.; Rohn, T.T.; Castellano, B.; Gonzalez, B. Caspase-3 activation in astrocytes following postnatal excitotoxic damage correlates with cytoskeletal remodeling but not with cell death or proliferation. *Glia* **2007**, *55*, 954–965. [CrossRef]
10. Aras, R.; Barron, A.M.; Pike, C.J. Caspase activation contributes to astrogliosis. *Brain Res.* **2012**, *1450*, 102–115. [CrossRef]
11. Burguillos, M.A.; Deierborg, T.; Kavanagh, E.; Persson, A.; Hajji, N.; Garcia-Quintanilla, A.; Cano, J.; Brundin, P.; Englund, E.; Venero, J.L.; et al. Caspase signalling controls microglia activation and neurotoxicity. *Nature* **2011**, *472*, 319–324. [CrossRef] [PubMed]
12. Hollville, E.; Deshmukh, M. Physiological functions of non-apoptotic caspase activity in the nervous system. *Semin. Cell Dev. Biol.* **2018**, *82*, 127–136. [CrossRef] [PubMed]
13. Park, G.; Nhan, H.S.; Tyan, S.H.; Kawakatsu, Y.; Zhang, C.; Navarro, M.; Koo, E.H. Caspase Activation and Caspase-Mediated Cleavage of APP Is Associated with Amyloid beta-Protein-Induced Synapse Loss in Alzheimer's Disease. *Cell Rep.* **2020**, *31*, 107839. [CrossRef] [PubMed]
14. Bagetta, V.; Picconi, B.; Marinucci, S.; Sgobio, C.; Pendolino, V.; Ghiglieri, V.; Fusco, F.R.; Giampa, C.; Calabresi, P. Dopamine-dependent long-term depression is expressed in striatal spiny neurons of both direct and indirect pathways: Implications for Parkinson's disease. *J. Neurosci.* **2011**, *31*, 12513–12522. [CrossRef]
15. Day, M.; Wang, Z.; Ding, J.; An, X.; Ingham, C.A.; Shering, A.F.; Wokosin, D.; Ilijic, E.; Sun, Z.; Sampson, A.R.; et al. Selective elimination of glutamatergic synapses on striatopallidal neurons in Parkinson disease models. *Nat. Neurosci.* **2006**, *9*, 251–259. [CrossRef] [PubMed]
16. Fieblinger, T.; Graves, S.M.; Sebel, L.E.; Alcacer, C.; Plotkin, J.L.; Gertler, T.S.; Chan, C.S.; Heiman, M.; Greengard, P.; Cenci, M.A.; et al. Cell type-specific plasticity of striatal projection neurons in parkinsonism and L-DOPA-induced dyskinesia. *Nat. Commun.* **2014**, *5*, 5316. [CrossRef]
17. Kreitzer, A.C.; Malenka, R.C. Endocannabinoid-mediated rescue of striatal LTD and motor deficits in Parkinson's disease models. *Nature* **2007**, *445*, 643–647. [CrossRef]
18. Lerner, T.N.; Kreitzer, A.C. RGS4 is required for dopaminergic control of striatal LTD and susceptibility to parkinsonian motor deficits. *Neuron* **2012**, *73*, 347–359. [CrossRef]
19. Suarez, L.M.; Solis, O.; Carames, J.M.; Taravini, I.R.; Solis, J.M.; Murer, M.G.; Moratalla, R. L-DOPA treatment selectively restores spine density in dopamine receptor D2-expressing projection neurons in dyskinetic mice. *Biol. Psychiatry* **2014**, *75*, 711–722. [CrossRef]
20. Thiele, S.L.; Chen, B.; Lo, C.; Gertler, T.S.; Warre, R.; Surmeier, J.D.; Brotchie, J.M.; Nash, J.E. Selective loss of bi-directional synaptic plasticity in the direct and indirect striatal output pathways accompanies generation of parkinsonism and l-DOPA induced dyskinesia in mouse models. *Neurobiol. Dis.* **2014**, *71*, 334–344. [CrossRef]
21. Shen, W.; Flajolet, M.; Greengard, P.; Surmeier, D.J. Dichotomous dopaminergic control of striatal synaptic plasticity. *Science* **2008**, *321*, 848–851. [CrossRef] [PubMed]
22. Ariano, M.A.; Grissell, A.E.; Littlejohn, F.C.; Buchanan, T.M.; Elsworth, J.D.; Collier, T.J.; Steece-Collier, K. Partial dopamine loss enhances activated caspase-3 activity: Differential outcomes in striatal projection systems. *J. Neurosci. Res.* **2005**, *82*, 387–396. [CrossRef] [PubMed]
23. Zhu, J.; Gao, W.; Shan, X.; Wang, C.; Wang, H.; Shao, Z.; Dou, S.; Jiang, Y.; Wang, C.; Cheng, B. Apelin-36 mediates neuroprotective effects by regulating oxidative stress, autophagy and apoptosis in MPTP-induced Parkinson's disease model mice. *Brain Res.* **2020**, *1726*, 146493. [CrossRef] [PubMed]
24. Imbriani, P.; Tassone, A.; Meringolo, M.; Ponterio, G.; Madeo, G.; Pisani, A.; Bonsi, P.; Martella, G. Loss of Non-Apoptotic Role of Caspase-3 in the PINK1 Mouse Model of Parkinson's Disease. *Int. J. Mol. Sci.* **2019**, *20*, 3407. [CrossRef]
25. Suarez, L.M.; Alberquilla, S.; Garcia-Montes, J.R.; Moratalla, R. Differential Synaptic Remodeling by Dopamine in Direct and Indirect Striatal Projection Neurons in Pitx3$^{-/-}$ Mice, a Genetic Model of Parkinson's Disease. *J. Neurosci.* **2018**, *38*, 3619–3630. [CrossRef]
26. Suarez, L.M.; Solis, O.; Aguado, C.; Lujan, R.; Moratalla, R. L-DOPA Oppositely Regulates Synaptic Strength and Spine Morphology in D1 and D2 Striatal Projection Neurons in Dyskinesia. *Cereb. Cortex* **2016**, *26*, 4253–4264. [CrossRef]
27. Fieblinger, T.; Cenci, M.A. Zooming in on the small: The plasticity of striatal dendritic spines in L-DOPA-induced dyskinesia. *Mov. Disord.* **2015**, *30*, 484–493. [CrossRef]
28. Gerfen, C.R.; Surmeier, D.J. Modulation of striatal projection systems by dopamine. *Annu. Rev. Neurosci.* **2011**, *34*, 441–466. [CrossRef]
29. Fieblinger, T. Striatal Control of Movement: A Role for New Neuronal (Sub-) Populations? *Front. Hum. Neurosci.* **2021**, *15*, 697284. [CrossRef]
30. Graves, S.M.; Surmeier, D.J. Delayed Spine Pruning of Direct Pathway Spiny Projection Neurons in a Mouse Model of Parkinson's Disease. *Front. Cell. Neurosci.* **2019**, *13*, 32. [CrossRef]

31. Caserta, T.M.; Smith, A.N.; Gultice, A.D.; Reedy, M.A.; Brown, T.L. Q-VD-OPh, a broad spectrum caspase inhibitor with potent antiapoptotic properties. *Apoptosis* **2003**, *8*, 345–352. [CrossRef] [PubMed]
32. Rentsch, P.; Stayte, S.; Morris, G.P.; Vissel, B. Time dependent degeneration of the nigrostriatal tract in mice with 6-OHDA lesioned medial forebrain bundle and the effect of activin A on L-Dopa induced dyskinesia. *BMC Neurosci.* **2019**, *20*, 5. [CrossRef] [PubMed]
33. Lindgren, H.S.; Rylander, D.; Ohlin, K.E.; Lundblad, M.; Cenci, M.A. The "motor complication syndrome" in rats with 6-OHDA lesions treated chronically with L-DOPA: Relation to dose and route of administration. *Behav. Brain Res.* **2007**, *177*, 150–159. [CrossRef] [PubMed]
34. Han, W.; Sun, Y.; Wang, X.; Zhu, C.; Blomgren, K. Delayed, long-term administration of the caspase inhibitor Q-VD-OPh reduced brain injury induced by neonatal hypoxia-ischemia. *Dev. Neurosci.* **2014**, *36*, 64–72. [CrossRef] [PubMed]
35. Rohn, T.T.; Kokoulina, P.; Eaton, C.R.; Poon, W.W. Caspase activation in transgenic mice with Alzheimer-like pathology: Results from a pilot study utilizing the caspase inhibitor, Q-VD-OPh. *Int. J. Clin. Exp. Med.* **2009**, *2*, 300–308.
36. Regehr, W.G. Short-term presynaptic plasticity. *Cold Spring Harb. Perspect. Biol.* **2012**, *4*, a005702. [CrossRef]
37. Choi, S.; Lovinger, D.M. Decreased probability of neurotransmitter release underlies striatal long-term depression and postnatal development of corticostriatal synapses. *Proc. Natl. Acad. Sci. USA* **1997**, *94*, 2665–2670. [CrossRef]
38. Gerdeman, G.L.; Ronesi, J.; Lovinger, D.M. Postsynaptic endocannabinoid release is critical to long-term depression in the striatum. *Nat. Neurosci.* **2002**, *5*, 446–451. [CrossRef]
39. Wu, Y.W.; Kim, J.I.; Tawfik, V.L.; Lalchandani, R.R.; Scherrer, G.; Ding, J.B. Input- and cell-type-specific endocannabinoid-dependent LTD in the striatum. *Cell Rep.* **2015**, *10*, 75–87. [CrossRef]
40. Dati, L.M.; Ulrich, H.; Real, C.C.; Feng, Z.P.; Sun, H.S.; Britto, L.R. Carvacrol promotes neuroprotection in the mouse hemiparkinsonian model. *Neuroscience* **2017**, *356*, 176–181. [CrossRef]
41. Ebert, A.D.; Hann, H.J.; Bohn, M.C. Progressive degeneration of dopamine neurons in 6-hydroxydopamine rat model of Parkinson's disease does not involve activation of caspase-9 and caspase-3. *J. Neurosci. Res.* **2008**, *86*, 317–325. [CrossRef] [PubMed]
42. Hanrott, K.; Murray, T.K.; Orfali, Z.; Ward, M.; Finlay, C.; O'Neill, M.J.; Wonnacott, S. Differential activation of PKC delta in the substantia nigra of rats following striatal or nigral 6-hydroxydopamine lesions. *Eur. J. Neurosci.* **2008**, *27*, 1086–1096. [CrossRef] [PubMed]
43. Hernandez-Baltazar, D.; Mendoza-Garrido, M.E.; Martinez-Fong, D. Activation of GSK-3beta and caspase-3 occurs in Nigral dopamine neurons during the development of apoptosis activated by a striatal injection of 6-hydroxydopamine. *PLoS ONE* **2013**, *8*, e70951. [CrossRef] [PubMed]
44. Koo, J.H.; Jang, Y.C.; Hwang, D.J.; Um, H.S.; Lee, N.H.; Jung, J.H.; Cho, J.Y. Treadmill exercise produces neuroprotective effects in a murine model of Parkinson's disease.e by regulating the TLR2/MyD88/NF-kappaB signaling pathway. *Neuroscience* **2017**, *356*, 102–113. [CrossRef] [PubMed]
45. Singh, A.; Verma, P.; Raju, A.; Mohanakumar, K.P. Nimodipine attenuates the parkinsonian neurotoxin, MPTP-induced changes in the calcium binding proteins, calpain and calbindin. *J. Chem. Neuroanat.* **2019**, *95*, 89–94. [CrossRef] [PubMed]
46. Stott, S.R.; Barker, R.A. Time course of dopamine neuron loss and glial response in the 6-OHDA striatal mouse model of Parkinson's disease. *Eur. J. Neurosci.* **2014**, *39*, 1042–1056. [CrossRef]
47. Yamada, M.; Kida, K.; Amutuhaire, W.; Ichinose, F.; Kaneki, M. Gene disruption of caspase-3 prevents MPTP-induced Parkinson's disease in mice. *Biochem. Biophys. Res. Commun.* **2010**, *402*, 312–318. [CrossRef]
48. Fieblinger, T.; Zanetti, L.; Sebastianutto, I.; Breger, L.S.; Quintino, L.; Lockowandt, M.; Lundberg, C.; Cenci, M.A. Striatonigral neurons divide into two distinct morphological-physiological phenotypes after chronic L-DOPA treatment in parkinsonian rats. *Sci. Rep.* **2018**, *8*, 10068. [CrossRef]
49. McNeill, T.H.; Brown, S.A.; Rafols, J.A.; Shoulson, I. Atrophy of medium spiny I striatal dendrites in advanced Parkinson's disease. *Brain Res.* **1988**, *455*, 148–152. [CrossRef]
50. Stephens, B.; Mueller, A.J.; Shering, A.F.; Hood, S.H.; Taggart, P.; Arbuthnott, G.W.; Bell, J.E.; Kilford, L.; Kingsbury, A.E.; Daniel, S.E.; et al. Evidence of a breakdown of corticostriatal connections in Parkinson's disease. *Neuroscience* **2005**, *132*, 741–754. [CrossRef]
51. Villalba, R.M.; Lee, H.; Smith, Y. Dopaminergic denervation and spine loss in the striatum of MPTP-treated monkeys. *Exp. Neurol.* **2009**, *215*, 220–227. [CrossRef] [PubMed]
52. Zaja-Milatovic, S.; Milatovic, D.; Schantz, A.M.; Zhang, J.; Montine, K.S.; Samii, A.; Deutch, A.Y.; Montine, T.J. Dendritic degeneration in neostriatal medium spiny neurons in Parkinson disease. *Neurology* **2005**, *64*, 545–547. [CrossRef] [PubMed]
53. Gagnon, D.; Petryszyn, S.; Sanchez, M.G.; Bories, C.; Beaulieu, J.M.; De Koninck, Y.; Parent, A.; Parent, M. Striatal Neurons Expressing D1 and D2 Receptors are Morphologically Distinct and Differently Affected by Dopamine Denervation in Mice. *Sci. Rep.* **2017**, *7*, 41432. [CrossRef] [PubMed]
54. Zhai, S.; Shen, W.; Graves, S.M.; Surmeier, D.J. Dopaminergic modulation of striatal function and Parkinson's disease. *J. Neural Transm.* **2019**, *126*, 411–422. [CrossRef]
55. Sebel, L.E.; Graves, S.M.; Chan, C.S.; Surmeier, D.J. Haloperidol Selectively Remodels Striatal Indirect Pathway Circuits. *Neuropsychopharmacology* **2017**, *42*, 963–973. [CrossRef]

56. Oo, T.F.; Siman, R.; Burke, R.E. Distinct nuclear and cytoplasmic localization of caspase cleavage products in two models of induced apoptotic death in dopamine neurons of the substantia nigra. *Exp. Neurol.* **2002**, *175*, 1–9. [CrossRef]
57. Sebbagh, M.; Renvoize, C.; Hamelin, J.; Riche, N.; Bertoglio, J.; Breard, J. Caspase-3-mediated cleavage of ROCK I induces MLC phosphorylation and apoptotic membrane blebbing. *Nat. Cell Biol.* **2001**, *3*, 346–352. [CrossRef]
58. Mukerjee, N.; McGinnis, K.M.; Park, Y.H.; Gnegy, M.E.; Wang, K.K. Caspase-mediated proteolytic activation of calcineurin in thapsigargin-mediated apoptosis in SH-SY5Y neuroblastoma cells. *Arch. Biochem. Biophys.* **2000**, *379*, 337–343. [CrossRef]
59. Spires-Jones, T.L.; Kay, K.; Matsouka, R.; Rozkalne, A.; Betensky, R.A.; Hyman, B.T. Calcineurin inhibition with systemic FK506 treatment increases dendritic branching and dendritic spine density in healthy adult mouse brain. *Neurosci. Lett.* **2011**, *487*, 260–263. [CrossRef]
60. Williams, D.W.; Kondo, S.; Krzyzanowska, A.; Hiromi, Y.; Truman, J.W. Local caspase activity directs engulfment of dendrites during pruning. *Nat. Neurosci.* **2006**, *9*, 1234–1236. [CrossRef]
61. Lauber, K.; Bohn, E.; Krober, S.M.; Xiao, Y.J.; Blumenthal, S.G.; Lindemann, R.K.; Marini, P.; Wiedig, C.; Zobywalski, A.; Baksh, S.; et al. Apoptotic cells induce migration of phagocytes via caspase-3-mediated release of a lipid attraction signal. *Cell* **2003**, *113*, 717–730. [CrossRef]
62. Siskova, Z.; Tremblay, M.E. Microglia and synapse: Interactions in health and neurodegeneration. *Neural Plast.* **2013**, *2013*, 425845. [CrossRef] [PubMed]
63. Zhang, Y.; Meredith, G.E.; Mendoza-Elias, N.; Rademacher, D.J.; Tseng, K.Y.; Steece-Collier, K. Aberrant restoration of spines and their synapses in L-DOPA-induced dyskinesia: Involvement of corticostriatal but not thalamostriatal synapses. *J. Neurosci.* **2013**, *33*, 11655–11667. [CrossRef] [PubMed]
64. Chen, L.; Daniels, S.; Kim, Y.; Chu, H.Y. Cell Type-Specific Decrease of the Intrinsic Excitability of Motor Cortical Pyramidal Neurons in Parkinsonism. *J. Neurosci.* **2021**, *41*, 5553–5565. [CrossRef]
65. Guo, L.; Xiong, H.; Kim, J.I.; Wu, Y.W.; Lalchandani, R.R.; Cui, Y.; Shu, Y.; Xu, T.; Ding, J.B. Dynamic rewiring of neural circuits in the motor cortex in mouse models of Parkinson's disease. *Nat. Neurosci.* **2015**, *18*, 1299–1309. [CrossRef] [PubMed]
66. Viaro, R.; Morari, M.; Franchi, G. Progressive motor cortex functional reorganization following 6-hydroxydopamine lesioning in rats. *J. Neurosci.* **2011**, *31*, 4544–4554. [CrossRef]
67. Brown, T.L. Q-VD-OPh, next generation caspase inhibitor. *Adv. Exp. Med. Biol.* **2004**, *559*, 293–300. [CrossRef]
68. Laforge, M.; Silvestre, R.; Rodrigues, V.; Garibal, J.; Campillo-Gimenez, L.; Mouhamad, S.; Monceaux, V.; Cumont, M.C.; Rabezanahary, H.; Pruvost, A.; et al. The anti-caspase inhibitor Q-VD-OPH prevents AIDS disease progression in SIV-infected rhesus macaques. *J. Clin. Investig.* **2018**, *128*, 1627–1640. [CrossRef]
69. Keoni, C.L.; Brown, T.L. Inhibition of Apoptosis and Efficacy of Pan Caspase Inhibitor, Q-VD-OPh, in Models of Human Disease. *J. Cell Death* **2015**, *8*, 1–7. [CrossRef]
70. Yang, L.; Sugama, S.; Mischak, R.P.; Kiaei, M.; Bizat, N.; Brouillet, E.; Joh, T.H.; Beal, M.F. A novel systemically active caspase inhibitor attenuates the toxicities of MPTP, malonate, and 3NP in vivo. *Neurobiol. Dis.* **2004**, *17*, 250–259. [CrossRef]
71. Turmel, H.; Hartmann, A.; Parain, K.; Douhou, A.; Srinivasan, A.; Agid, Y.; Hirsch, E.C. Caspase-3 activation in 1-methyl-4-phenyl-1,2,3,6-tetrahydropyridine (MPTP)-treated mice. *Mov. Disord.* **2001**, *16*, 185–189. [CrossRef]
72. Spiga, S.; Talani, G.; Mulas, G.; Licheri, V.; Fois, G.R.; Muggironi, G.; Masala, N.; Cannizzaro, C.; Biggio, G.; Sanna, E.; et al. Hampered long-term depression and thin spine loss in the nucleus accumbens of ethanol-dependent rats. *Proc. Natl. Acad. Sci. USA* **2014**, *111*, E3745–E3754. [CrossRef] [PubMed]
73. Alcacer, C.; Andreoli, L.; Sebastianutto, I.; Jakobsson, J.; Fieblinger, T.; Cenci, M.A. Chemogenetic stimulation of striatal projection neurons modulates responses to Parkinson's disease therapy. *J. Clin. Investig.* **2017**, *127*, 720–734. [CrossRef] [PubMed]
74. Fieblinger, T.; Sebastianutto, I.; Alcacer, C.; Bimpisidis, Z.; Maslava, N.; Sandberg, S.; Engblom, D.; Cenci, M.A. Mechanisms of dopamine D1 receptor-mediated ERK1/2 activation in the parkinsonian striatum and their modulation by metabotropic glutamate receptor type 5. *J. Neurosci.* **2014**, *34*, 4728–4740. [CrossRef] [PubMed]
75. Francardo, V.; Bez, F.; Wieloch, T.; Nissbrandt, H.; Ruscher, K.; Cenci, M.A. Pharmacological stimulation of sigma-1 receptors has neurorestorative effects in experimental parkinsonism. *Brain* **2014**, *137*, 1998–2014. [CrossRef] [PubMed]
76. Francardo, V.; Recchia, A.; Popovic, N.; Andersson, D.; Nissbrandt, H.; Cenci, M.A. Impact of the lesion procedure on the profiles of motor impairment and molecular responsiveness to L-DOPA in the 6-hydroxydopamine mouse model of Parkinson's disease. *Neurobiol. Dis.* **2011**, *42*, 327–340. [CrossRef] [PubMed]
77. West, M.J. Stereological methods for estimating the total number of neurons and synapses: Issues of precision and bias. *Trends Neurosci.* **1999**, *22*, 51–61. [CrossRef]
78. Ren, Q.; Hu, Z.; Jiang, Y.; Tan, X.; Botchway, B.O.A.; Amin, N.; Lin, G.; Geng, Y.; Fang, M. SIRT1 Protects Against Apoptosis by Promoting Autophagy in the Oxygen Glucose Deprivation/Reperfusion-Induced Injury. *Front. Neurol.* **2019**, *10*, 1289. [CrossRef]
79. Jiang, L.; Tixeira, R.; Caruso, S.; Atkin-Smith, G.K.; Baxter, A.A.; Paone, S.; Hulett, M.D.; Poon, I.K. Monitoring the progression of cell death and the disassembly of dying cells by flow cytometry. *Nat. Protoc.* **2016**, *11*, 655–663. [CrossRef]
80. Sebastianutto, I.; Cenci, M.A.; Fieblinger, T. Alterations of striatal indirect pathway neurons precede motor deficits in two mouse models of Huntington's disease. *Neurobiol. Dis.* **2017**, *105*, 117–131. [CrossRef]

International Journal of *Molecular Sciences*

MDPI

Article

Neuroprotective Effect of Luteolin-7-O-Glucoside against 6-OHDA-Induced Damage in Undifferentiated and RA-Differentiated SH-SY5Y Cells

Stephanie Cristine Hepp Rehfeldt [1,†], Joana Silva [2,†], Celso Alves [2], Susete Pinteus [2], Rui Pedrosa [3], Stefan Laufer [4,5] and Márcia Inês Goettert [1,4,*]

1 Cell Culture Laboratory, Graduate Program in Biotechnology, University of Vale do Taquari (Univates), Lajeado 95914-014, RS, Brazil; srehfeldt@universo.univates.br
2 Marine and Environmental Sciences Centre (MARE), Polytechnic of Leiria, 2520-630 Peniche, Portugal; joana.m.silva@ipleiria.pt (J.S.); celso.alves@ipleiria.pt (C.A.); susete.pinteus@ipleiria.pt (S.P.)
3 MARE—Marine and Environmental Sciences Centre, ESTM, Polytechnic of Leiria, 2520-614 Peniche, Portugal; rui.pedrosa@ipleiria.pt
4 Department of Pharmaceutical and Medicinal Chemistry, Institute of Pharmacy, Eberhard Karls Universität Tübingen, D 72076 Tubingen, Germany; stefan.laufer@uni-tuebingen.de
5 Tübingen Center for Academic Drug Discovery (TüCAD2), D 72076 Tubingen, Germany
* Correspondence: m.goettert@uni-tuebingen.de
† These authors contributed equally to this work.

Abstract: Luteolin is one of the most common flavonoids present in edible plants and its potential benefits to the central nervous system include decrease of microglia activation, neuronal damage and high antioxidant properties. The aim of this research was to evaluate the neuroprotective, antioxidant and anti-inflammatory activities of luteolin-7-O-glucoside (Lut7). Undifferentiated and retinoic acid (RA)-differentiated SH-SY5Y cells were pretreated with Lut7 and incubated with 6-hydroxydopamine (6-OHDA). Cytotoxic and neuroprotective effects were determined by MTT assay. Antioxidant capacity was determined by DPPH, FRAP, and ORAC assays. ROS production, mitochondrial membrane potential ($\Delta\Psi$m), Caspase–3 activity, acetylcholinesterase inhibition (AChEI) and nuclear damage were also determined in SH-SY5Y cells. TNF-α, IL-6 and IL-10 release were evaluated in LPS-induced RAW264.7 cells by ELISA. In undifferentiated SH-SY5Y cells, Lut7 increased cell viability after 24 h, while in RA-differentiated SH-SY5Y cells, Lut7 increased cell viability after 24 and 48 h. Lut7 showed a high antioxidant activity when compared with synthetic antioxidants. In undifferentiated cells, Lut7 prevented mitochondrial membrane depolarization induced by 6-OHDA treatment, decreased Caspase-3 and AChE activity, and inhibited nuclear condensation and fragmentation. In LPS-stimulated RAW264.7 cells, Lut7 treatment reduced TNF-α levels and increased IL-10 levels after 3 and 24 h, respectively. In summary, the results suggest that Lut7 has neuroprotective effects, thus, further studies should be considered to validate its pharmacological potential in more complex models, aiming the treatment of neurodegenerative diseases.

Keywords: 6-hydroxydopamine; apoptosis; mitochondrial membrane potential; neurodegenerative diseases; oxidative stress; cell culture techniques; neurodegenerative disorders; neuroprotective effect; biological products

Citation: Rehfeldt, S.C.H.; Silva, J.; Alves, C.; Pinteus, S.; Pedrosa, R.; Laufer, S.; Goettert, M.I. Neuroprotective Effect of Luteolin-7-O-Glucoside against 6-OHDA-Induced Damage in Undifferentiated and RA-Differentiated SH-SY5Y Cells. *Int. J. Mol. Sci.* **2022**, *23*, 2914. https://doi.org/10.3390/ijms23062914

Academic Editors: Marcello Ciaccio and Luisa Agnello

Received: 14 January 2022
Accepted: 3 March 2022
Published: 8 March 2022

1. Introduction

Neurodegenerative diseases (ND) are comprised of distinct and heterogeneous disorders characterized by the progressive and selective loss of neurons. Usually, the prevalence and symptoms worsening are intimately related with age, and as the global population gets older, the need for new ND therapeutic alternatives and deeper knowledge of their pathophysiology is urgent. The World Health Organization (WHO) estimates that by 2040, ND such as Alzheimer's disease (AD), and other types of dementia, or conditions that compromises motor function like Parkinson's disease (PD) or amyotrophic lateral sclerosis (ALS), will be the second most prevalent cause of death, after cardiovascular diseases [1]. However, due to the complexity and heterogeneity of NDs, most of the synthetic drugs evaluated in in vitro models and/or clinical trials end up failing [2], and thus further efforts must be conducted to develop an efficient disease-modifying treatment.

Luteolin, a phytochemical belonging to the flavone class of polyphenols, is one of the most common flavonoids present in edible plants. Its potential benefits to the CNS include the decrease of microglia activation and neuronal protection [3–5]. However, the glycosylated form of luteolin, known as cyranoside or luteolin-7-O-glucoside (Lut7) (PubChem ID 5280637), was reported as a selective JNK3 inhibitor, five times more selective than luteolin [6], which plays a key role in neurodegenerative diseases [7–11], suggesting that patients with neurodegenerative diseases might benefit from a natural or bio-inspired product-based therapy [12–16].

The effects of herbal extracts containing Lut7 as a majority compound are extensively explored in the literature. However, it is not possible to confirm that those reported effects are a result of the Lut7 activity or if other components are mediating the observed effects. Therefore, the aim of this study was to evaluate the antioxidant and anti-inflammatory activities of Lut7, as well as its neuroprotective effects in an in vitro human neurodegenerative model (SH-SY5Y cells induced with 6-OHDA) in both undifferentiated and differentiated cells.

2. Materials and Methods

2.1. Cell Lines and Reagents

Dulbecco's modified Eagle medium (DMEM) (D5523), F12 (N6760), heat-inactivated fetal bovine serum (FBS) (F4135), 6-hydroxydopamine hydrobromide (6-OHDA) (162957), 3-[4,5-dimethylthiazol-2]-2,5 diphenyltetrazolium bromide (MTT) (M5655), Penicillin (P3032), streptomycin (S9137), lipopolysaccharide (LPS) (from *Escherichia coli*, O111:B4, L2630), trypsin-EDTA (T4049), 2′,7′-dichlorofluorescin diacetate (DCFDA) (D6883), Caspase-3 Activity Fluorimetric kit (CASP3F), 2,4,6-Tris(2-pyridyl-s-triazine (TPTZ) (T125), FCCP, and oligomycin A were purchased from Sigma-Aldrich™ (St. Louis, MO, USA). DMEM (1200-058) used to culture RAW264.7 cell line and enzyme-linked immunosorbent assay (ELISA) kits for TNF-α, IL-6, and IL-10 were acquired from Gibco®, Invitrogen Life Science Technologies (Grand Island, NY, USA). All-trans-retinoic acid (ATRA) (SC200898) was purchased from Santa Cruz Biotechnology, (Dallas, TX, USA). Spectrophotometer analysis was performed using a SpectraMax® i3 microplate reader (Molecular Devices, San Jose, CA, USA). 5,5,6,6-tetrachloro-1,1,3,3-tetraethylbenzimidazolylcarbocyanine iodide (JC-1) staining was acquired from Molecular Probes (Eugene, OR, USA) and 4,6-diamidino-2-phenylindole (DAPI) staining was obtained from Applichem (Darmstadt, Germany). Photographs for the DAPI probe were taken with a fluorescence microscope Zeiss, model Axio Vert. A1, (Oberkochen, Germany). Lut7 (26-S) was obtained from Extrasynthese (Genay, Cedex, France). SH-SY5Y (ATCC® CRL-2266™) and RAW264.7 cell line (ATCC® TIB-71™) were acquired from American Type Culture Collection (ATCC). MAPK inhibitors (SP600125 and SB203580) were synthesized by Prof. Dr. Stefan Laufer research laboratory with a high purity grade ($\geq$95%).

2.2. Cell Culture Methods

2.2.1. SH-SY5Y Cell Line

Undifferentiated human SH-SY5Y neuroblastoma cells were cultured in DMEM mixed with F12 (1:1) and supplemented with 10% (*v/v*) FBS and 1% streptomycin/penicillin under controlled conditions in a 95% humidified atmosphere, at 37 °C and 5% CO_2. Culture medium was replaced every two days until the cells reached confluence 4–5 days after the initial seeding. For subculture, SH-SY5Y cells were dissociated with trypsin-EDTA, split into a 1:3 ratio. Cells were grown to 80% confluence before treatment. Culture conditions were performed according to ATCC recommendations.

2.2.2. SH-SY5Y Differentiation Protocol

The differentiation of SH-SY5Y cells was carried out in two steps. Firstly, SH-SY5Y cells were cultivated in media with 1% FBS supplemented with 10 μM of all-trans retinoic acid (ATRA) for 4 days. At the 5th day, the cell culture medium was replaced by fresh medium and cells were cultivated for 6 days.

2.2.3. RAW264.7 Cell Line

RAW264.7 cells were cultured in DMEM supplemented with 10% FBS (*v*/*v*) and 1% streptomycin/penicillin. The medium was replaced every 2 to 3 days. Sub-culturing was carried out with a cell scraper at a 1:4 split ratio. All procedures were performed according with ATCC recommendations.

2.3. Determination of Cell Viability and Neuroprotection Potential by MTT Assay

2.3.1. Evaluation of Lut7 Cytotoxicity

Cell viability was assessed using the colorimetric MTT assay [17] with slight modifications as described by Rehfeldt et al. [11]. SH-SY5Y (differentiated and undifferentiated) and RAW264.7 cells were seeded in 96-well dishes and left overnight for proper attachment. Cells were exposed to different concentrations (10–0.1 μM) of Lut7 over 24 or 48 h. MTT reagent was then added to each well at a final concentration of 5 mg/mL and the plate was placed in a humidified incubator at 37 °C with 5% CO_2 during 3 h. Formazan salts were dissolved in DMSO and the colorimetric determination of MTT reduction was estimated at 570 nm wavelength using the SpectraMax® i3 microplate reader. Untreated cells were used as control and considered as 100% viable.

2.3.2. Neuroprotection Potential

Neuroprotection effect was assessed using the colorimetric MTT assay with slight adaptations [11,17,18]. To investigate the neuroprotective potential of the compound, SH-SY5Y (differentiated and undifferentiated) cells were seeded at a density of 2×10^4 per well in a 96-well dish and cultivated overnight. Cells were then treated with different concentrations (1 or 0.1 μM) of Lut7 over 30 min before adding 6-OHDA (100 μM) stabilized with 0.02% of ascorbic acid to avoid its auto-oxidation. Cells' medium was then removed following 24 or 48 h of treatment and MTT (5 mg/mL) solution was added and the cells were incubated for 3 h. Following the MTT removal, DMSO was used to dissolve the formazan salts and the absorbance read at 570 nm.

2.4. Determination of Antioxidant Activity

The antioxidant activity of Lut7 was evaluated by the means of different methodologies, namely (a) 2,2-diphenyl-1-picrylhydrazyl (DPPH) radical scavenging ability [19]; (b) oxygen radical absorbance capacity (ORAC) [20]; and (c) ferric reducing antioxidant power (FRAP) assays [21] with slight adaptations [15]. Butylated hydroxytoluene (BHT) was used as a positive control for antioxidant activity.

2.5. Mechanisms of Cell Recovery after 6-OHDA-Induced Damage

2.5.1. Mitochondrial Membrane Potential (MMP) Assay

Cells were seeded in 96-well plates and left overnight in the incubator. SH-SY5Y cells were then treated with 6-OHDA (100 μM) for 6 h, in the absence or presence of Lut7 (0.1 or 1 μM). After exposure, cells were washed with PBS 1× and incubated with JC-1 at 37 °C for 30 min. Then, the reagent was gently removed, and cells were washed with PBS 1×. 100 μL/well PBS 1× was added and the fluorescence was measured at 490/595 nm (red fluorescence) and 490/530 nm (green fluorescence) of excitation and emission wavelengths, respectively [11,22]. FCCP (2.5 μM) plus oligomycin A (1 μg/mL) conjugate solution was used as positive control. Mitochondrial membrane potential was estimated by measuring the fluorescence of free JC-1 monomers (green) and JC-1 aggregates in mitochondria (red) and the results were expressed as the ratio of the monomers/aggregates of JC-1 in percentage of control.

2.5.2. ROS Production

The levels of reactive oxygen species (ROS) were determined using the 5-(and-6)-carboxy-2′, 7′-dichlorodihydrofluorescein diacetate (carboxy-H2DCFDA) probe as previously described [11,22] with slight modifications. Briefly, SH-SY5Y cells were treated with Lut7 at different concentrations (0.1 or 1 μM) and 6-OHDA (100 μM) over 6 h. Following treatment, cells were washed with PBS (1×) and20 μM carboxyH2DCFDA solution, previously dissolved in serum-free medium, and cells were incubated for 1 h at 37 °C. The fluorescence was read at 527 nm and 495 nm wavelength of emission and excitation, respectively.

2.5.3. Caspase 3 Activity

The enzyme was assessed using the Caspase-3 Activity Fluorometric kit, according to manufacturer's instructions. SH-SY5Y cells were cultured in 6-well plates and treated with 6-OHDA (100 μM) for 6 h in the presence or absence of Lut7 (0.1 or 1 μM). Caspase-3 activity was calculated from the slope of the linear phase of the fluorescence resulting from the rhodamine 110 accumulation and expressed in arbitrary fluorescence units per mg protein per minute (Δ fluorescence (a.u.)/mg protein/min) [22].

2.5.4. DAPI Staining

The nucleic condensation and/or fragmentation was determined by 4,6-diamidino-2-phenylindole (DAPI) staining. SH-SY5Y cells were cultured in 6-well plates and treated with 6-OHDA (100 μM) for 24 h in the presence or absence of Lut7 (0.1 or 1 μM). The cells were fixed in paraformaldehyde (4%) over 30 min. After this time, the solution was removed, and cells were incubated in Triton X-100 (0.1%) over 30 min. Then, Triton X-100 was removed, followed by the addition of DAPI (1 μg/mL) solution. After a 30 min reaction, DAPI was removed, and 1 mL PBS (pH 7.4) was added to each well. Then, the cells were observed using an Axio Vert. A1 fluorescence microscope (Zeiss, Oberkochen, Germany) [22].

2.5.5. AChE Activity

AChE activity was measured spectrophotometrically in a 96-well microplate according to a modified Ellman assay [23,24]. SH-SY5Y cells were cultured in 6-well plates overnight and treated with 6-OHDA (100 μM) for 6 h, in the absence or presence of Lut7 (0.1 or 1 μM). Then, cells were trypsinized with PBS (1×) + 0.1% Triton X-100. The supernatant was transferred to a 96-well microplate and incubated with DTNB (0.5 mM) and acetylcholine (Ach; 1 mM). The absorbance was measured at λ = 414 nm.

2.6. Cytokine Levels in RAW264.7 Cell Line

RAW264.7 cells were seeded at a density of 5×10^5 cells per well in a 24-dish plate. After adherence time, cells were pretreated for 1 h before LPS (1 μg/mL) treatment. The supernatant was collected at different times as described by Rehfeldt and co-workers [11]. To evaluate TNF-α release, samples of supernatant were collected after 3 and 24 h of treatment. For IL-6 analysis, samples were collected after 12 h of treatment, and for IL-10 after 48 h of treatment. All samples were frozen at −80 °C until further analysis. ELISA was performed according to the manufacturer's instructions. The absorbances were measured at 450 nm and 570 nm using a spectrophotometer (SpectraMax® i3). Values of 570 nm were subtracted from those of 450 nm to remove background interference. TNF-α, IL-6 and IL-10 standard curves were used to quantify the release from each cytokine by the cells.

2.7. Data and Statistical Analysis

The statistical analysis was performed on GraphPad Prism 6.0 software (GraphPad software, San Diego, CA, USA) using ANOVA. The results are expressed in mean ± standard error of the mean (SEM). Differences were considered significant at level of 0.001 (*** $p < 0.001$), 0.01 (** $p < 0.01$), and 0.05 (* $p < 0.05$). At least three independent experiments carried out in triplicate were performed.

3. Results

3.1. Cytotoxic and Neuroprotective Effect of Lut7 in Undifferentiated and RA-Differentiated SH-SY5Y Cells

The first set of experiments examined the impact of Lut7 on cell viability. The results shown in Figure 1A indicate that 10 μM Lut7 reduced SH-SY5Y cells viability by 33% ($p < 0.05$). At 1 μM and 0.1 μM, the compound did not significantly decrease the cells' viability, thus, these concentrations were selected for the neuroprotective assays.

The capacity of undifferentiated SH-SY5Y cells to recover from 6-OHDA stimuli was evaluated. Cells were pretreated with Lut7 at sub-toxic concentrations over 1 h before 6-OHDA treatment. The exposition of SH-SY5Y cells to 6-OHDA (100 μM) significantly reduced the cell viability after 24 h when compared to the untreated cells. However, when cells were treated with 0.1 μM Lut7, there was a 13% increase in cell viability when compared with 6-OHDA treatment (Figure 1B). However, this effect was not observed after 48 h (Figure 1C).

After differentiation, cells start to upregulate genes involved with antioxidant defense. This modified profile of gene expression reflects directly in the capacity of cells to recover from the oxidative stress caused by 6-OHDA. To confirm this higher resistance, one positive control with doxorubicin (10 μM) was included. The mechanism of action of doxorubicin is achieved by specifically blocking the activity of the enzyme topoisomerase II, which is involved in DNA replication during mitosis, and does not interfere in oxidative stress and, therefore, does not impact the damage caused by 6-OHDA. In this sense, SH-SY5Y cells were differentiated 10 days before neuroprotective assay. RA-differentiated SH-SY5Y cells pretreated with Lut7 at 1 and 0.1 μM increased cell viability in 27.4 and 27.1%, respectively ($p < 0.001$) (Figure 1D). On the other hand, after 48 h, RA-differentiated SH-SY5Y cells pretreated with 1 and 0.1 μM Lut7 increased cell viability in 112% ($p < 0.001$) and 67.5% ($p < 0.001$), respectively, when compared with 6-OHDA treatment (Figure 1E).

3.2. Lut7 Antioxidant Activity

The antioxidant activity of Lut7 was evaluated by three different chemical methods: (a) DPPH assay to determine the capacity of Lut7 scavenging potential; (b) FRAP to determine the Lut7 capacity to reduce ferric ions; (c) ORAC method to evaluate the presence of antioxidant molecules with the ability to neutralize the peroxyl radicals. The results are presented in Table 1 and Figure 2.

Table 1. Antioxidant activity of Lut7 and BHT.

	DPPH $^{(A)}$	FRAP $^{(B)}$	ORAC $^{(C)}$
Lut7	6.8 (0.76–0.9)	19,570.78 ± 291.48	8804.19 ± 409.99
BHT	>100	2821.50 ± 63.03	143.70 ± 23.36

BHT (butylated hydroxytoluene) was used as a standard. The values in the table represent the mean ± SEM from 3 independent experiments. $^{(A)}$ radical scavenging activity (EC_{50} μg/mL); $^{(B)}$ μM of $FeSO_4$ per gram of compound; $^{(C)}$ μM of Trolox equivalent (TE)/g of compound.

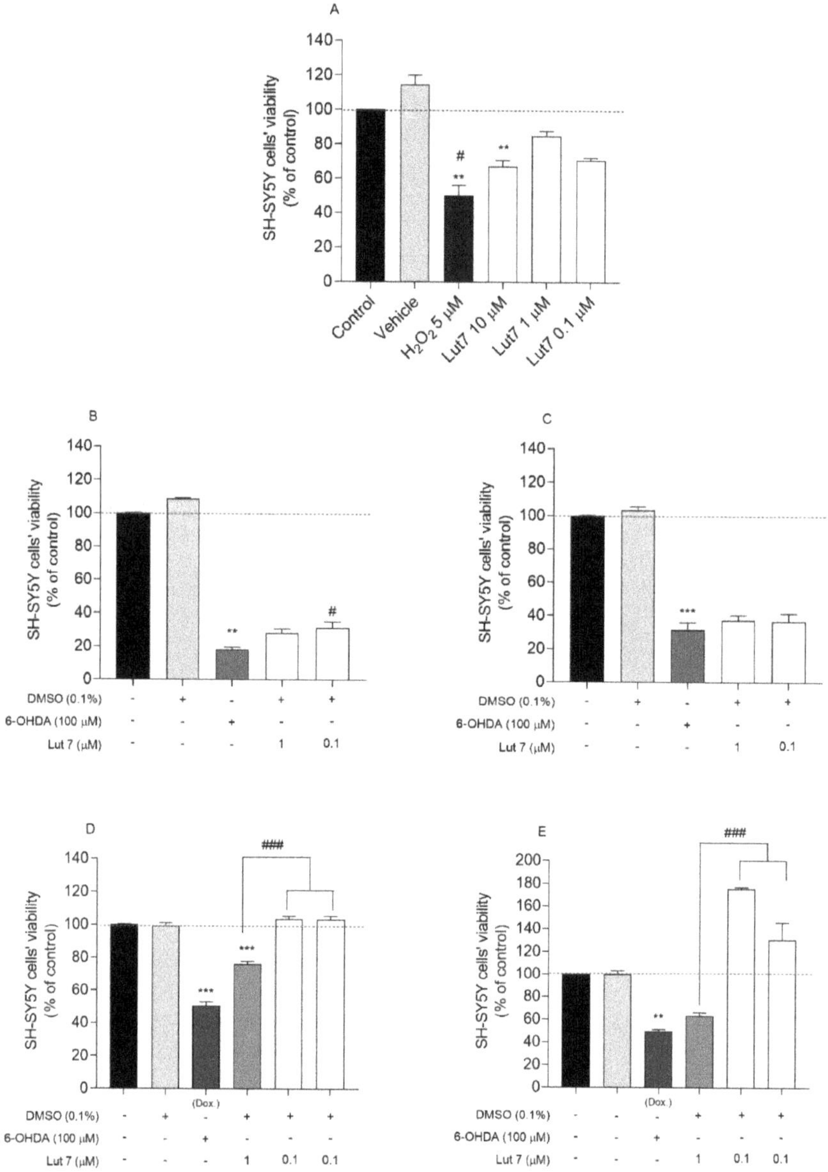

Figure 1. Cytotoxicity induced by Luteolin-7-O-glucoside (Lut7) in SH-SY5Y cells after 24 h of treatment (**A**); Neuroprotective effects of Lut7 against 6-OHDA induced neurotoxicity on undifferentiated SH-SY5Y cells: 24 h incubation (**B**); 48 h incubation (**C**). Neuroprotective effects of Lut7 against 6-OHDA induced neurotoxicity on RA-differentiated SH-SY5Y cells: 24 h incubation (**D**); 48 h incubation (**E**). The values in each column represent the mean ± SEM of at least three independent experiments carried out in triplicate. Statistical calculations were performed with ANOVA via the Tukey post hoc test and significant differences were considered for *** $p < 0.001$; ** $p < 0.01$ (vs. control); ### $p < 0.001$ and # $p < 0.05$ (vs. 6-OHDA). Doxorubicin was used as positive control; DMSO 0.1% was used as vehicle. Negative control (untreated cells) was considered to be 100% viable.

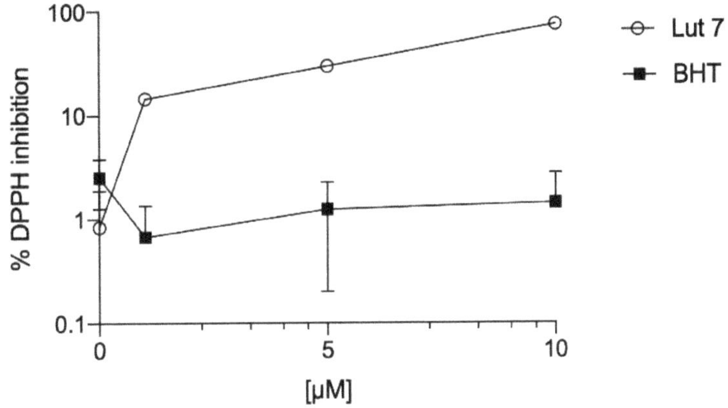

Figure 2. DPPH scavenging ability of Lut 7 and BHT—dose response analysis.

It was observed that Lut7 presented the highest potential of scavenging DPPH radical with an EC_{50} value of 6.80 μM when compared to the standard BHT (EC_{50} > 100 μM). In the ORAC method, Lut7 showed the highest antioxidant activity with 8804.19 ± 409.99 μmol of Trolox/g compound, when compared with BHT (143.70 ± 23.36 μmol of Trolox/g compound). Lut7 was also effective in reducing ferric ions (19,570.78 ± 291.48 μM $FeSO_4$/g of compound) when compared with the synthetic antioxidant.

3.3. *Lut7 Effects on Cellular Hallmarks Associated with ND*

Hallmarks of apoptotic cell death include the activation of caspases, the disruption of mitochondrial membrane potential and DNA fragmentation. These same events are also present during neurodegenerative diseases. To verify if the neuroprotective effect demonstrated by Lut7 on undifferentiated SH-SY5Y cells was associated with these hallmarks, different in vitro assays on cells treated with neurotoxin 6-OHDA in the absence or presence of Lut7 were carried out (Figure 3).

Firstly, the Caspase-3 activity was measured to understand if the Lut7 had capability to prevent the cell death mediated by apoptosis when exposed to the 6-OHDA neurotoxin. At 1 μM, Lut7 was able to decrease Caspase-3 activity by 57.6% (Figure 3A).

Secondly, the ability of Lut7 to prevent a condition of oxidative stress was evaluated through ROS quantification. After exposing SH-SY5Y cells to 6-OHDA (100 μM, 6 h) a two-fold increase in ROS levels was verified when compared to vehicle. However, Lut7 was not able to decrease ROS levels induced by 6-OHDA treatment (Figure 3B).

Thirdly, the MMP was determined to evaluate mitochondrial dysfunction and to understand if the neuroprotective effects of Lut7 were mediated by biological events that usually take place on mitochondria. Treatment with 6-OHDA at 100 μM for 6 h increased depolarization of SH-SY5Y cells mitochondrial membrane potential in 359.2% when compared to vehicle. On the other hand, the treatment with Lut7 (1 μM) exhibited a protective effect against cell depolarization induced by 6-OHDA treatment in 320% (139.2% vs. 459.2% of 6-OHDA). At 0.1 μM, Lut7 showed a similar effect decreasing the monomers/aggregates ratio of JC-1) in 192% (Figure 3C).

Considering the mechanism of action of many FDA-approved drugs to treat AD, the AChE activity was evaluated after Lut7 treatment. It is well documented that the AChE inhibition decreases the breakdown and promotes the accumulation of ACh, therefore, compensating the loss of functional cholinergic neurons and alleviating cognitive symptoms of AD. The results presented here showed that 6-OHDA treatment increased AChE activity in 43.47%. On the other hand, cells treated with 1 μM Lut7 reduced AChE activity in 77.49 ± 8.63% when compared with 6-OHDA treatment (Figure 3D). Cells pre-treated with

0.1 μM Lut7 followed by 6-OHDA exposure showed approximately 1-fold increase in AChE activity (242.67 ± 8.9%) when compared to the neurotoxin.

Finally, to understand if Lut7 had the ability to prevent the DNA fragmentation induced by 6-OHDA treatment, the integrity of SH-SY5Y DNA was observed following DAPI staining (Figure 4).

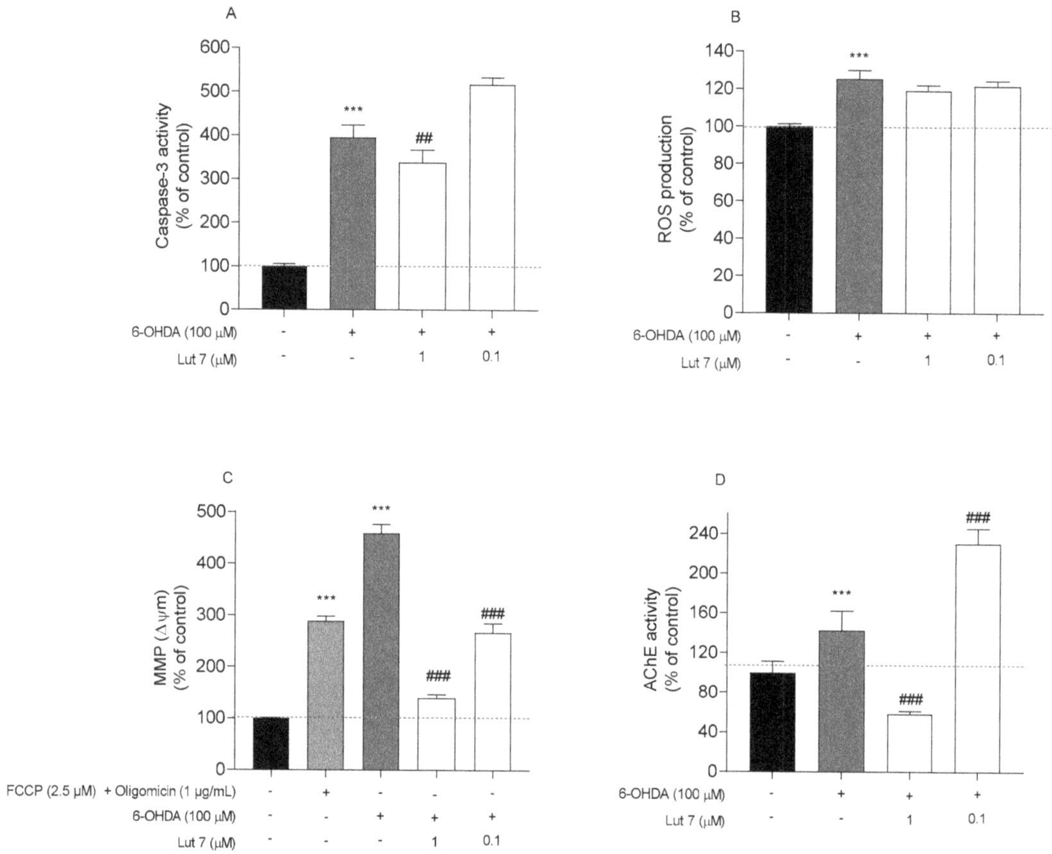

Figure 3. Effect of Luteolin-7-O-glucoside (Lut7) on SH-SY5Y cells after exposition to 6-OHDA (100 μM) over 6 h. (**A**) Effect on Caspase-3 activity (Δ Fluorescence (U.A)/mg of protein/min); (**B**) ROS production; (**C**) MMP (ratio of monomers/aggregates of JC-1); and (**D**) AChE activity (nmol/min/mg of protein). The values in each column represent the mean ± standard error of the mean (SEM) of at least 3 independent experiments carried out in triplicate. Statistical calculations were performed with ANOVA via the Tukey post hoc test and significant differences were considered for *** $p < 0.001$ (vs. control); ### $p < 0.001$ and ## $p < 0.01$ (vs. 6-OHDA). Negative control (untreated cells) was considered to be 100% viable and is represented by the red dashed line. DMSO 0.1% was used as a vehicle.

The exposition of SH-SY5Y cells to 100 μM 6-OHDA over 24 h led to nuclear condensation and fragmentation, which are characteristic features of apoptosis. However, it was possible to verify that Lut7 decrease the occurrence of those events induced by 6-OHDA.

3.4. Cytokines Levels on LPS-Induced RAW264.7 Cells Treated with Lut7

Cytokines play a crucial role in the inflammatory response. Before evaluating the effects of Lut7 treatment over cytokine release, its cytotoxic effects on the RAW267.4 cells viability were evaluated (Figure 5).

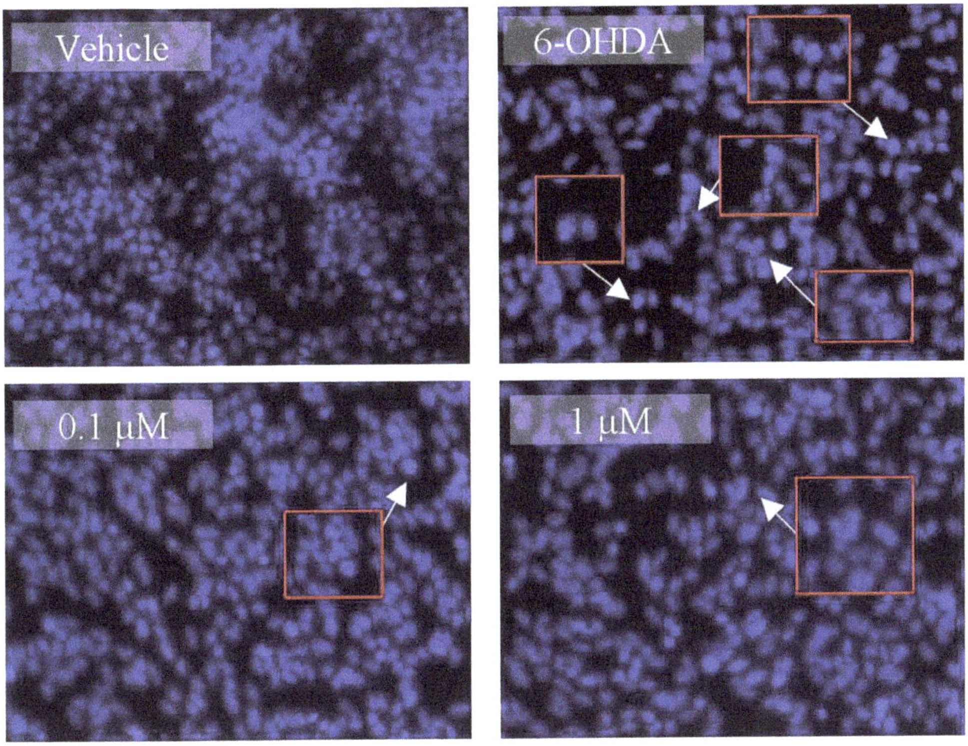

Figure 4. Nuclear morphology of SH-SY5Y cells stained with DAPI probe. SH-SY5Y cell stained with DAPI showing the anti-apoptotic effect of the Lut7 (0.1 or 1 μM) against neurotoxicity mediated by 6-OHDA (100 μM; 24 h). Arrows show nuclear abnormalities (fragmentation pattern), which is an indicator of apoptosis. Red Boxes represent the amplified zone where is visible nuclear changes.

Since at 10 μM, Lut7 reduced SH-SY5Y cells' viability, the cytotoxic effect of Lut7 in RAW264.7 cells was only tested at 1 and 0.1 μM. It was verified that at 0.1 and 1 μM, Lut7 did not induce cytotoxicity in RAW 264.7 cells (Figure 5A), and thus, these concentrations were used in the following assays. It was verified that, at both concentrations, Lut7 did not stimulate IL-6 release (Figure 5B). On the other hand, after 3 h, cells pretreated with 1 μM and 0.1 Lut7 decreased TNF-α levels in 30.97% and in 43.02%, respectively (Figure 5C). Lut7 treatment showed no effect over TNF-α levels after 24 h (Figure 5D). Concerning IL-10 release, which is delayed, the supernatant was collected after 24 and 48 h of treatment. Only after 24 h, Lut7 at 0.1 μM was able to stimulate the IL-10 release in 96.9% (Figure 5E). However, these effects were not maintained over 48 h (Figure 5F).

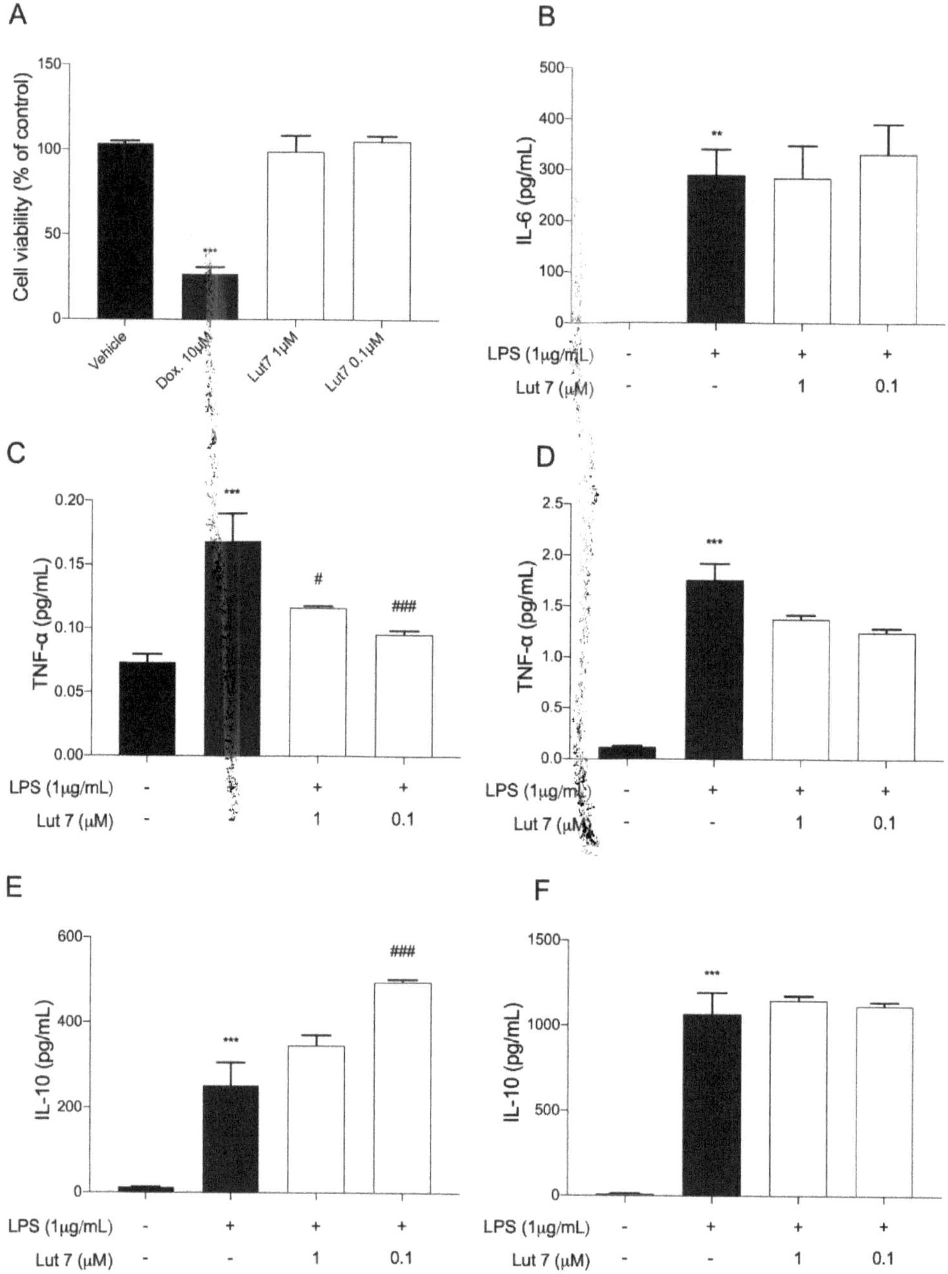

Figure 5. (**A**) Cytotoxicity of luteolin-7-O-glucoside (Lut7) on RAW264.7 cells viability after treatment for 24 h; (**B**) IL-6 levels were determined after 12 h of treatment; (**C**) TNF-α levels were determined after 3 h and (**D**) 24 h of treatment. IL-10 levels were estimated (**E**) after 24 h; and (**F**) 48 h of treatment. The values represent the mean ± SEM of at least three experiments carried out in triplicate. ** $p < 0.01$; *** $p < 0.001$; (vs. control); ### $p < 0.001$ and # $p < 0.05$ (vs. LPS). Doxorubicin was used as positive control; DMSO 0.1% was used as vehicle.

4. Discussion

Pre-clinical studies reported that luteolin, a flavonoid present in many fruits and vegetables, has anti-inflammatory and antioxidative properties [25–28]. However, the impact on neuroprotection using the RA-differentiated SH-SY5Y cell line has been poorly explored. Despite both differentiated and undifferentiated SH-SY5Y cells being used in experiments as suitable in vivo models of NDs, authors suggest that cells should be differentiated since undifferentiated cells are prevenient from a metastatic tumor and continuously undergo division, making it difficult to predict the effect of protective agents against neurotoxins. Indeed, differences in gene expression profiles, antioxidant capacity, synthesis of neurotransmitters and other phenotypic aspects have been observed between differentiated and undifferentiated cells [29–31]. Additionally, authors evaluated the enzymatic activity of AChE and choline acetyltransferase (ChAT) (cholinergic markers) in both differentiated and undifferentiated SH-SY5Y cells and demonstrated that cells show a different pattern of AChE activity when treated with RA [32]. Those findings suggest that the SH-SY5Y cell line may respond to the 6-OHDA stimuli differently, depending on whether they are differentiated or not.

In undifferentiated SH-SY5Y cells, our findings did not reveal a significant neuroprotection effect. These results are in line with a previous study performed with Lut7, in which the compound also did not show significant differences in SH-SY5Y cells' viability when treated with an amyloidogenic molecule [33]. On the other hand, on RA-differentiated cells, Lut7 showed a marked protective effect against 6-OHDA-induced damage, probably due to its increased ability to promote the expression of genes related to antioxidant defenses. This is consistent with previous studies where RA-induced differentiation of SH-SY5Y cells has been related with resistance to oxidants, possibly due to modulation of ROS production and oxidative stress responses [34–37]. This result ties well with a previous study wherein intensified oxidative phosphorylation in differentiated SH-SY5Y was observed [38].

The results herein presented show that Lut7 display high antioxidant capacity. This potential is especially relevant since neuronal cells present a high metabolic rate, continuously generating reactive oxygen species (ROS) during aerobic metabolism, as a result of electron transport chain (ETC) action during oxidative phosphorylation. As a consequence, brain tissue is particularly susceptible to oxidative stress [39,40]. Several events have been associated with neurodegeneration such as synaptic dysfunction, excitotoxicity, and oxidative stress. Indeed, because of its high metabolic rate combined with a limited capacity of cellular regeneration, the brain is particularly sensitive to oxidative damage.

The neurotoxin 6-OHDA is a potent inhibitor of complex I and causes direct oxidative damage through superoxide and hydrogen peroxide production and indirect damage after suffering auto-oxidation, generating even more ROS [41–44]. In order to evaluate ROS production and the capacity of Lut7 to promote cell recovery from 6-OHDA-induced damage, undifferentiated SH-SY5Y cells were used. Our findings revealed that Lut7 did not affect ROS production. Another study reported that Lut7 did not interfere on hepatitis B virus-induced intracellular ROS accumulation in HepG2 cells [45]. However, in contrast to our results, previous studies reported that Lut7 decreases ROS in many cell lines or in vivo models. Palombo et al. evaluated ROS production in IL-22 or IL-6-stimulated human keratinocytes (HEKn cells) and verified that at 20 μM, Lut7 treatment reduced ROS generation [46]. Similarly, in HUVEC cells (human umbilical vein endothelial cells) Lut7 at 20 μM reduced ROS generation and downregulated genes involved in inflammation [47]. The divergent results found in literature may derive from different cellular responses to different Lut7 concentrations, and/or different cell lines.

A correlation between mitochondrial membrane potential ($\Delta\Psi m$) and reactive oxygen species (ROS) production has been demonstrated [48–50]. Additionally, $\Delta\psi m$ depolarisation is usually correlated with neuronal death [51,52]. In this case, the present results indicate that Lut7 reverted the 6-OHDA-induced $\Delta\psi m$ depolarisation but it was not able to reduce the ROS production [45]. The MMP recovery mediated by Lut7 observed in SH-SY5Y cells after 6-OHDA-induced cell injury was similar to a previous study conducted with cisplatin in HK-2 cells (human proximal tubule cell line). According to Nho et al., (2018) Lut7 decreased cell death, promoted a recovery in MMP and abolished Caspase-3 activity [53]. In this study, Lut7 also decreased Caspase-3 activity in SH-SY5Y cells exposed to 6-OHDA. In H9c2 cells (rat cardiomyoblast cell line), Lut7 pretreatment reduced apoptosis, intracellular ROS, chromatin condensation and DNA damage, and reverted mitochondrial dysfunction induced by doxorubicin [54]. Another study reported similar results in H9c2 cells (reduction of apoptosis, ROS generation and mitochondrial dysfunction) but also showed downregulation of Caspase-3, p-ERK1/2, p-JNK and p-P38 inhibition, and p-ERK5 activation in angiotensin II-induced cells [55].

It was reported that mitochondrial dysfunction is mediated by JNK activation, while its inhibition prevent both the loss of $\Delta\psi m$ and apoptotic events [56–59]. One possible explanation for these findings is that JNK plays a significant role in apoptosis via the intrinsic pathway (also known as the 'mitochondrial pathway'), which is activated by extracellular or intracellular perturbations usually found in AD, such as oxidative stress. In response to a deleterious stimulus (such as ROS), JNK phosphorylates 14-3-3 protein, and induces the translocation of pro-apoptotic proteins (Bax and Bad) from the cytoplasm to the mitochondria, the major source of ROS in cells. The translocation of these pro-apoptotic proteins induces mitochondrial outer membrane permeabilization (MOMP), allowing the cytosolic release of pro-apoptogenic factors that normally reside in the mitochondrial intermembrane space, such as cytochrome c and Smac/DIABLO [60,61]. Cytochrome c then associates with Apaf-1, pro-Caspase 9 (CASP9), (and possibly other proteins) to form an apoptosome, which activates CASP9. When activated, CASP9 catalyzes the proteolytic activation of CASP3 and CASP7 (known as 'executioner caspases'), which handle cell demolition during intrinsic and extrinsic apoptosis pathways. However, DNA damage can also activate JNK. p53 is another JNK substrate that induces expression of pro-apoptotic genes (puma, fas and bax), leading to apoptosis in a mitochondrial-independent manner. On the other hand, p53 can trigger the MOMP as well in a transcription-independent manner by activating pro-apoptotic Bcl-2 proteins (Bax or Bak) or by inactivating anti-apoptotic Bcl-2 proteins (Bcl-2 and Bcl-X1) [62,63].

Recently, the role of JNK3 in Alzheimer's disease (AD) was reviewed [64]. It was demonstrated that synthetic JNK3 inhibitors have a promising future as therapeutic alternatives for AD treatment as they appear to attenuate many neurodegenerative-associated phenomena in different models. Interestingly, Lut7 also showed important JNK3 selectivity. The IC_{50} for JNK3 was reported to be as low as 2.45 ± 0.1 μM, while the IC_{50} for p38α was 87.1 ± 2.1 μM, indicating a 35-fold increase of selectivity to JNK3 over p38. Authors hypothesized that the selectivity for JNK3 is a result of the interaction of Lut7 and the residues Asn152, Gln155, Asn 194 and Ser 193 of JNK3 [6]. Therefore, according to our results, it is possible that the main mechanism of Lut7 in preventing mitochondrial-dependent apoptosis may be related with JNK3 inhibition.

Although many drugs have been evaluated as AD treatment in vivo, in vitro and in silico models, one of the best pharmacological alternatives by far, still consist in AChE inhibitors (AChEI), such as galantamine, rivastigmine or donepezil. It is believed that in early or mid-stages of AD, the increase of ACh induced by AChEI compensates the loss of cholinergic neurons and prolongs the cholinergic sinalization, improving cognitive symptoms, such as learning and memory impairment [65–67]. Indeed, AChEIs have been used for over 30 years to increase the levels of ACh in muscarinic and nicotinic receptors, and interestingly, galantamine, an AChEI which has been approved since 2001 by the FDA in the treatment of AD, was first isolated from botanical sources back in the 1950s [68], which reinforces the potential of natural products in providing efficient treatments to several diseases.

Here, we demonstrated that Lut7 is able to inhibit AChE activity in 6-OHDA-treated cells. Corroborating with our results, the role of flavonoids (including Lut7) as potential AChEI have been reviewed elsewhere [13,69,70]. The AChE inhibitory capacity of Lut7 has been also demonstrated by in vivo and in silico studies [71–73] despite other studies having reported a wide range of IC_{50} values, probably due to distinct methodologies. The in vitro study conducted by Liu et al. (2020) indicated an IC_{50} of 18.24 ± 2.33 μM [73], while the in vitro and in silico study of Sevindik et al. (2015) identified an IC_{50} of 1.65 μM to AChEI [72].

Both toxic protein accumulation and oxidative stress are main hallmarks of NDs and contribute to neuroinflammation, further worsening the disease. Previously, studies had already reported that luteolin suppressed the production of proinflammatory cytokines in macrophages by blocking kappa B (NF-κB), and activator protein 1 (AP1) nuclear signaling pathways, and inhibited the production of nitric oxide and proinflammatory eicosanoids. In addition, luteolin decreased the release of TNF-α and superoxide after LPS induced in microglial cell cultures, and reduced the production of LPS-induced IL-6 in cerebral microglia in vivo and in vitro [74,75]. In the CNS, it decreased inflammation and axonal damage by preventing monocyte migration through the blood–brain barrier (BBB) [3–5]. Since both microglia cells and RAW264.7 cells are capable of expressing major histocompatibility complex (MHC) antigens, as well as T and B cell markers and share other phenotypic traits and innate immunological functions with other mononuclear phagocytes, we also demonstrated that Lut7 can reduce TNF-α after 3 h, and increase IL-10 after 24 h. In the past two decades, neuroinflammation has been considered an important component of the NDs' pathogenesis. It is well established that the CNS is composed of distinct kinds of cells that perform specific roles in brain homeostasis and therefore, may contribute differently to the worsening of symptoms or progression of ND.

A previous study reported that pretreatment with Lut7 suppressed the induction of nitrite, ROS, PGE2, and TNF-α in a dose-dependent manner in IL-1β-stimulated rat primary chondrocytes [76], suggesting that Lut7 has a potent anti-inflammatory effect. Additionally, Lut7 inhibited the IL-1β-induced nuclear accumulation of NF-κB subunit p65 by suppressing phosphorylation and degradation of IκB-α and significantly inhibited the IL-1β-induced phosphorylation of ERK, JNK, and p38 MAPK in a dose-dependent manner [76].

Despite the promising potential of Lut7, there are some limitations in our study that should be addressed in future research. The primary focus of our study was to evaluate the neuroprotective, antioxidant and anti-inflammatory effect of Lut7 in in vitro models of neurodegenerative diseases, yet, further studies are necessary to attest the results here described. Therefore, evaluation of inflammatory mediators such as iNOS and COX-2, as well as the expression of transcription factors (e.g., Nrf2, AP1, NF-κB), antioxidant enzymes (e.g., SOD, GSH-Px), and other apoptosis-related proteins (e.g., Caspase-9 and -8, Bax, cytochrome C, JNK, p38) would be of extreme value to fully elucidate the anti-inflammatory, anti-apoptotic and antioxidant mechanisms of Lut7, especially based on co-culturing systems (e.g., neuron and microglia-derived cells co-culture, 3D) or more complex models (e.g., in vivo models) of neurodegenerative diseases.

5. Conclusions

In summary, our results suggested that Lut7 protected SH-SY5Y cell line against 6-OHDA-induced damage and protected differentiated SH-SY5Y cells against neurotoxicity induced by 6-OHDA over 48 h. Although we did not observe a reduction of ROS production, Lut7 protected SH-SY5Y cells against 6-OHDA-induced mitochondrial and nuclear damage and reduced Caspase-3 and AChE activity. In RAW264.7 cell line, Lut7 was able to decrease TNF-α and to increase IL-10 levels (Figure 6).

Several questions remain unanswered but we believe this is an interesting topic for future work and may be a good starting point for further investigation in in vivo models of neurodegenerative diseases to validate Lut7 pharmacological potential.

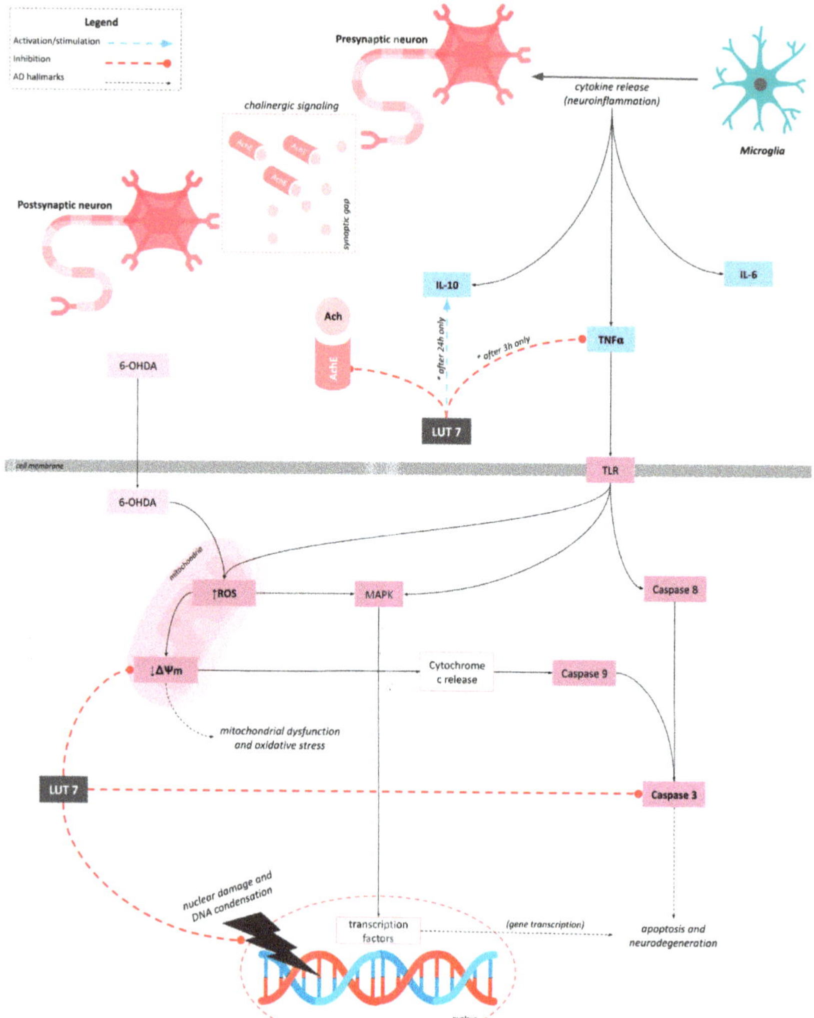

Figure 6. Proposed mechanism of action of Luteolin-7-O-glucoside (Lut7). Regarding the mitochondria, Lut7 prevented membrane depolarization induced by 6-OHDA and indirectly reduced mitochondrial dysfunction and oxidative stress. Lut7 also decreased Caspase-3 activity protecting cells against 6-OHDA-induced apoptosis. Additionally, at low concentrations, Lut7 was able to inhibit AChE activity, which may contribute to alleviating AD symptoms. In RAW264.7 cells, Lut7 was able to reduce TNF-α production (after 3 h), and induce IL-10 release (after 24 h), which may contribute to modulate the neuroinflammation.

Author Contributions: S.C.H.R., J.S., C.A. and S.P. did main experiments (antioxidant, cytotoxicity, neuroprotective, signaling pathways mechanisms and anti-inflammatory activity) and wrote the manuscript. M.I.G., S.L. and R.P. participated in designing and coordinating the study. All authors have read and agreed to the published version of the manuscript.

Funding: This study was funded in part by the Coordenação de Aperfeiçoamento de Pessoal de Nível Superior—Brasil (CAPES)—Finance Code 001, and Conselho Nacional de Pesquisa (CNPq) 420381/2018-0. This work was also supported by the Portuguese Foundation for Science and Technology(FCT) through the strategic project UID/04292/2020 granted to MARE—Marine and Environmental Sciences Centre, and UIDP/04046/2020 and UIDB/04046/2020 granted to BioISI—BioSystems and Integrative Sciences Institute, through POINT4PAC project (Oncologia de Precisão: Terapias e Tecnologias Inovadoras, SAICTPAC/0019/2015-LISBOA- 01-0145-FEDER-016405), through CROSS-ATLANTIC project (PTDC/BIA-OUT/29250/2017), co-financed by COMPETE (POCI-01-0145-FEDER029250) and through Molecules for Health project (PTDC/BIA-BQM/28355/2017). This work was also supported by FCT and CAPES cooperation agreement through project MArTics (FCT/DRI/CAPES 2019.00277.CBM). We acknowledge support by Open Access Publishing Fund of University of Tübingen.

Institutional Review Board Statement: Not applicable.

Informed Consent Statement: Not applicable.

Data Availability Statement: The data presented in this study are available on request from the corresponding author.

Acknowledgments: The authors are very grateful for the financial support of the Projects and Programs described in the funding section.

Conflicts of Interest: The authors declare no conflict of interest.

References

1. Gammon, K. Neurodegenerative disease: Brain windfall. *Nature* **2014**, *515*, 299–300. [CrossRef] [PubMed]
2. Rezak, M.; de Carvalho, M. Disease modification in neurodegenerative diseases: Not quite there yet. *Neurology* **2020**, *94*, 12–13. [CrossRef] [PubMed]
3. Bakhtiari, M.; Panahi, Y.; Ameli, J.; Darvishi, B. Protective effects of flavonoids against Alzheimer's disease-related neural dysfunctions. *Biomed. Pharmacother.* **2017**, *93*, 218–229. [CrossRef]
4. Heim, K.E.; Tagliaferro, A.R.; Bobilya, D.J. Flavonoid antioxidants: Chemistry, metabolism and structure-activity relationships. *J. Nutr. Biochem.* **2002**, *13*, 572–584. [CrossRef]
5. Rasool, M.; Malik, A.; Qureshi, M.S.; Manan, A.; Pushparaj, P.N.; Asif, M.; Qazi, M.H.; Qazi, A.M.; Kamal, M.A.; Gan, S.H.; et al. Recent updates in the treatment of neurodegenerative disorders using natural compounds. *Evid. Based Complement. Alternat. Med.* **2014**, *2014*, 979730. [CrossRef] [PubMed]
6. Goettert, M.; Schattel, V.; Koch, P.; Merfort, I.; Laufer, S. Biological evaluation and structural determinants of p38α mitogen-activated-protein kinase and c-Jun-N-terminal kinase 3 inhibition by flavonoids. *Chembiochem* **2010**, *11*, 2579–2588. [CrossRef] [PubMed]
7. Haeusgen, W.; Boehm, R.; Zhao, Y.; Herdegen, T.; Waetzig, V. Specific activities of individual c-Jun N-terminal kinases in the brain. *Neuroscience* **2009**, *161*, 951–959. [CrossRef]
8. Gourmaud, S.; Paquet, C.; Dumurgier, J.; Pace, C.; Bouras, C.; Gray, F.; Laplanche, J.-L.; Meurs, E.F.; Mouton-Liger, F.; Hugon, J. Increased levels of cerebrospinal fluid JNK3 associated with amyloid pathology: Links to cognitive decline. *J. Psychiatry Neurosci.* **2015**, *40*, 151–161. [CrossRef]
9. Takahashi, R.H.; Nagao, T.; Gouras, G.K. Plaque formation and the intraneuronal accumulation of β-amyloid in Alzheimer's disease. *Pathol. Int.* **2017**, *67*, 185–193. [CrossRef]
10. Gourmaud, S.; Thomas, P.; Thomasseau, S.; Tible, M.; Abadie, C.; Paquet, C.; Hugon, J. Brimapitide reduced neuronal stress markers and cognitive deficits in 5XFAD transgenic mice. *J. Alzheimers Dis.* **2018**, *63*, 665–674. [CrossRef]
11. Rehfeldt, S.C.H.; Laufer, S.; Goettert, M.I. A Highly Selective *In Vitro* JNK3 Inhibitor, FMU200, Restores Mitochondrial Membrane Potential and Reduces Oxidative Stress and Apoptosis in SH-SY5Y Cells. *Int. J. Mol. Sci.* **2021**, *22*, 3701. [CrossRef] [PubMed]
12. Dey, A.; Bhattacharya, R.; Mukherjee, A.; Pandey, D.K. Natural products against Alzheimer's disease: Pharmaco-therapeutics and biotechnological interventions. *Biotechnol. Adv.* **2017**, *35*, 178–216. [CrossRef] [PubMed]
13. Uddin, M.S.; Kabir, M.T.; Niaz, K.; Jeandet, P.; Clément, C.; Mathew, B.; Rauf, A.; Rengasamy, K.R.R.; Sobarzo-Sánchez, E.; Ashraf, G.M.; et al. Molecular Insight into the Therapeutic Promise of Flavonoids against Alzheimer's Disease. *Molecules* **2020**, *25*, 1267. [CrossRef]
14. Majolo, F.; Martins, A.; Rehfeldt, S.; Henriques, J.A.P.; Contini, V.; Goettert, M.I. Approaches for the treatment of neurodegenerative diseases related to natural products. In *Bioactive Natural Products*; Studies in Natural Products Chemistry; Elsevier: Amsterdam, The Netherlands, 2021; Volume 69, pp. 1–63; ISBN 9780128194874.
15. Silva, J.; Alves, C.; Martins, A.; Susano, P.; Simões, M.; Guedes, M.; Rehfeldt, S.; Pinteus, S.; Gaspar, H.; Rodrigues, A.; et al. Loliolide, a New Therapeutic Option for Neurological Diseases? In Vitro Neuroprotective and Anti-Inflammatory Activities of a Monoterpenoid Lactone Isolated from Codium tomentosum. *Int. J. Mol. Sci.* **2021**, *22*, 1888. [CrossRef]

16. Silva, J.; Alves, C.; Pinteus, S.; Susano, P.; Simões, M.; Guedes, M.; Martins, A.; Rehfeldt, S.; Gaspar, H.; Goettert, M.; et al. Disclosing the potential of eleganolone for Parkinson's disease therapeutics: Neuroprotective and anti-inflammatory activities. *Pharmacol. Res.* **2021**, *168*, 105589. [CrossRef] [PubMed]
17. Mosmann, T. Rapid colorimetric assay for cellular growth and survival: Application to proliferation and cytotoxicity assays. *J. Immunol. Methods* **1983**, *65*, 55–63. [CrossRef]
18. Silva, J.; Alves, C.; Pinteus, S.; Mendes, S.; Pedrosa, R. Neuroprotective effects of seaweeds against 6-hydroxidopamine-induced cell death on an in vitro human neuroblastoma model. *BMC Complement. Altern. Med.* **2018**, *18*, 58. [CrossRef] [PubMed]
19. Pinteus, S.; Silva, J.; Alves, C.; Horta, A.; Fino, N.; Rodrigues, A.I.; Mendes, S.; Pedrosa, R. Cytoprotective effect of seaweeds with high antioxidant activity from the Peniche coast (Portugal). *Food Chem.* **2017**, *218*, 591–599. [CrossRef]
20. Dávalos, A.; Gómez-Cordovés, C.; Bartolomé, B. Extending applicability of the oxygen radical absorbance capacity (ORAC-fluorescein) assay. *J. Agric. Food Chem.* **2004**, *52*, 48–54. [CrossRef]
21. Benzie, I.F.; Strain, J.J. The ferric reducing ability of plasma (FRAP) as a measure of "antioxidant power": The FRAP assay. *Anal. Biochem.* **1996**, *239*, 70–76. [CrossRef]
22. Ouazia, D.; Levros, L.C.; Rassart, É.; Desrosiers, R.R. The protein l-isoaspartyl (d-aspartyl) methyltransferase protects against dopamine-induced apoptosis in neuroblastoma SH-SY5Y cells. *Neuroscience* **2015**, *295*, 139–150. [CrossRef] [PubMed]
23. Ellman, G.L.; Courtney, K.D.; Andres, V.; Feather-Stone, R.M. A new and rapid colorimetric determination of acetylcholinesterase activity. *Biochem. Pharmacol.* **1961**, *7*, 88–95. [CrossRef]
24. Santillo, M.F.; Liu, Y. A fluorescence assay for measuring acetylcholinesterase activity in rat blood and a human neuroblastoma cell line (SH-SY5Y). *J. Pharmacol. Toxicol. Methods* **2015**, *76*, 15–22. [CrossRef] [PubMed]
25. Mansuri, M.L.; Parihar, P.; Solanki, I.; Parihar, M.S. Flavonoids in modulation of cell survival signalling pathways. *Genes Nutr.* **2014**, *9*, 400. [CrossRef] [PubMed]
26. Rahal, A.; Kumar, A.; Singh, V.; Yadav, B.; Tiwari, R.; Chakraborty, S.; Dhama, K. Oxidative stress, prooxidants, and antioxidants: The interplay. *Biomed Res. Int.* **2014**, *2014*, 761264. [CrossRef]
27. Nabavi, S.F.; Braidy, N.; Gortzi, O.; Sobarzo-Sanchez, E.; Daglia, M.; Skalicka-Woźniak, K.; Nabavi, S.M. Luteolin as an anti-inflammatory and neuroprotective agent: A brief review. *Brain Res. Bull.* **2015**, *119*, 1–11. [CrossRef]
28. Kim, S.; Chin, Y.; Cho, J. Protection of Cultured Cortical Neurons by Luteolin against Oxidative Damage through Inhibition of Apoptosis and Induction of Heme. *Biol. Pharm. Bull.* **2017**, *40*, 256–265. [CrossRef]
29. Luchtman, D.W.; Song, C. Why SH-SY5Y cells should be differentiated. *Neurotoxicology* **2010**, *31*, 164–165, author reply 165. [CrossRef]
30. Korecka, J.A.; van Kesteren, R.E.; Blaas, E.; Spitzer, S.O.; Kamstra, J.H.; Smit, A.B.; Swaab, D.F.; Verhaagen, J.; Bossers, K. Phenotypic characterization of retinoic acid differentiated SH-SY5Y cells by transcriptional profiling. *PLoS ONE* **2013**, *8*, e63862. [CrossRef]
31. Krishna, A.; Biryukov, M.; Trefois, C.; Antony, P.M.A.; Hussong, R.; Lin, J.; Heinäniemi, M.; Glusman, G.; Köglsberger, S.; Boyd, O.; et al. Systems genomics evaluation of the SH-SY5Y neuroblastoma cell line as a model for Parkinson's disease. *BMC Genom.* **2014**, *15*, 1154. [CrossRef]
32. de Medeiros, L.M.; De Bastiani, M.A.; Rico, E.P.; Schonhofen, P.; Pfaffenseller, B.; Wollenhaupt-Aguiar, B.; Grun, L.; Barbé-Tuana, F.; Zimmer, E.R.; Castro, M.A.A.; et al. Cholinergic Differentiation of Human Neuroblastoma SH-SY5Y Cell Line and Its Potential Use as an *In vitro* Model for Alzheimer's Disease Studies. *Mol. Neurobiol.* **2019**, *56*, 7355–7367. [CrossRef] [PubMed]
33. Iakovleva, I.; Begum, A.; Pokrzywa, M.; Walfridsson, M.; Sauer-Eriksson, A.E.; Olofsson, A. The flavonoid luteolin, but not luteolin-7-O-glucoside, prevents a transthyretin mediated toxic response. *PLoS ONE* **2015**, *10*, e0128222. [CrossRef] [PubMed]
34. Cheung, Y.-T.; Lau, W.K.-W.; Yu, M.-S.; Lai, C.S.-W.; Yeung, S.-C.; So, K.-F.; Chang, R.C.-C. Effects of all-trans-retinoic acid on human SH-SY5Y neuroblastoma as in vitro model in neurotoxicity research. *Neurotoxicology* **2009**, *30*, 127–135. [CrossRef] [PubMed]
35. Schneider, L.; Giordano, S.; Zelickson, B.R.; S Johnson, M.; A Benavides, G.; Ouyang, X.; Fineberg, N.; Darley-Usmar, V.M.; Zhang, J. Differentiation of SH-SY5Y cells to a neuronal phenotype changes cellular bioenergetics and the response to oxidative stress. *Free Radic. Biol. Med.* **2011**, *51*, 2007–2017. [CrossRef] [PubMed]
36. de Bittencourt Pasquali, M.A.; de Ramos, V.M.; Albanus, R.D.O.; Kunzler, A.; de Souza, L.H.T.; Dalmolin, R.J.S.; Gelain, D.P.; Ribeiro, L.; Carro, L.; Moreira, J.C.F. Gene Expression Profile of NF-κB, Nrf2, Glycolytic, and p53 Pathways During the SH-SY5Y Neuronal Differentiation Mediated by Retinoic Acid. *Mol. Neurobiol.* **2016**, *53*, 423–435. [CrossRef]
37. Kunzler, A.; Zeidán-Chuliá, F.; Gasparotto, J.; Girardi, C.S.; Klafke, K.; Petiz, L.L.; Bortolin, R.C.; Rostirolla, D.C.; Zanotto-Filho, A.; de Bittencourt Pasquali, M.A.; et al. Changes in Cell Cycle and Up-Regulation of Neuronal Markers During SH-SY5Y Neurodifferentiation by Retinoic Acid are Mediated by Reactive Species Production and Oxidative Stress. *Mol. Neurobiol.* **2017**, *54*, 6903–6916. [CrossRef]
38. Forster, J.I.; Köglsberger, S.; Trefois, C.; Boyd, O.; Baumuratov, A.S.; Buck, L.; Balling, R.; Antony, P.M.A. Characterization of Differentiated SH-SY5Y as Neuronal Screening Model Reveals Increased Oxidative Vulnerability. *J. Biomol. Screen.* **2016**, *21*, 496–509. [CrossRef]
39. Hamanaka, R.B.; Chandel, N.S. Mitochondrial reactive oxygen species regulate cellular signaling and dictate biological outcomes. *Trends Biochem. Sci.* **2010**, *35*, 505–513. [CrossRef]
40. Murphy, M.P. How mitochondria produce reactive oxygen species. *Biochem. J.* **2009**, *417*, 1–13. [CrossRef]

41. Glinka, Y.; Gassen, M.; Youdim, M.B. Mechanism of 6-hydroxydopamine neurotoxicity. *J. Neural. Transm. Suppl.* **1997**, *50*, 55–66. [CrossRef]
42. Glinka, Y.; Tipton, K.F.; Youdim, M.B. Nature of inhibition of mitochondrial respiratory complex I by 6-Hydroxydopamine. *J. Neurochem.* **1996**, *66*, 2004–2010. [CrossRef] [PubMed]
43. Glinka, Y.; Tipton, K.F.; Youdim, M.B. Mechanism of inhibition of mitochondrial respiratory complex I by 6-hydroxydopamine and its prevention by desferrioxamine. *Eur. J. Pharmacol.* **1998**, *351*, 121–129. [CrossRef]
44. Glinka, Y.Y.; Youdim, M.B. Inhibition of mitochondrial complexes I and IV by 6-hydroxydopamine. *Eur. J. Pharmacol.* **1995**, *292*, 329–332. [CrossRef]
45. Cui, X.-X.; Yang, X.; Wang, H.-J.; Rong, X.-Y.; Jing, S.; Xie, Y.-H.; Huang, D.-F.; Zhao, C. Luteolin-7-O-Glucoside Present in Lettuce Extracts Inhibits Hepatitis B Surface Antigen Production and Viral Replication by Human Hepatoma Cells in Vitro. *Front. Microbiol.* **2017**, *8*, 2425. [CrossRef]
46. Palombo, R.; Savini, I.; Avigliano, L.; Madonna, S.; Cavani, A.; Albanesi, C.; Mauriello, A.; Melino, G.; Terrinoni, A. Luteolin-7-glucoside inhibits IL-22/STAT3 pathway, reducing proliferation, acanthosis, and inflammation in keratinocytes and in mouse psoriatic model. *Cell Death Dis.* **2016**, *7*, e2344. [CrossRef]
47. De Stefano, A.; Caporali, S.; Di Daniele, N.; Rovella, V.; Cardillo, C.; Schinzari, F.; Minieri, M.; Pieri, M.; Candi, E.; Bernardini, S.; et al. Anti-Inflammatory and Proliferative Properties of Luteolin-7-O-Glucoside. *Int. J. Mol. Sci.* **2021**, *22*, 1321. [CrossRef]
48. Zorov, D.B.; Juhaszova, M.; Sollott, S.J. Mitochondrial reactive oxygen species (ROS) and ROS-induced ROS release. *Physiol. Rev.* **2014**, *94*, 909–950. [CrossRef]
49. Turrens, J.F. Mitochondrial formation of reactive oxygen species. *J Physiol.* **2003**, *552*, 335–344. [CrossRef]
50. Korshunov, S.S.; Skulachev, V.P.; Starkov, A.A. High protonic potential actuates a mechanism of production of reactive oxygen species in mitochondria. *FEBS Lett.* **1997**, *416*, 15–18. [CrossRef]
51. Connolly, N.M.C.; Theurey, P.; Adam-Vizi, V.; Bazan, N.G.; Bernardi, P.; Bolaños, J.P.; Culmsee, C.; Dawson, V.L.; Deshmukh, M.; Duchen, M.R.; et al. Guidelines on experimental methods to assess mitochondrial dysfunction in cellular models of neurodegenerative diseases. *Cell Death Differ.* **2018**, *25*, 542–572. [CrossRef]
52. Norat, P.; Soldozy, S.; Sokolowski, J.D.; Gorick, C.M.; Kumar, J.S.; Chae, Y.; Yağmurlu, K.; Prada, F.; Walker, M.; Levitt, M.R.; et al. Mitochondrial dysfunction in neurological disorders: Exploring mitochondrial transplantation. *NPJ Regen. Med.* **2020**, *5*, 22. [CrossRef] [PubMed]
53. Nho, J.-H.; Jung, H.-K.; Lee, M.-J.; Jang, J.-H.; Sim, M.-O.; Jeong, D.-E.; Cho, H.-W.; Kim, J.-C. Beneficial Effects of Cynaroside on Cisplatin-Induced Kidney Injury In Vitro and In Vivo. *Toxicol. Res.* **2018**, *34*, 133–141. [CrossRef] [PubMed]
54. Yao, H.; Shang, Z.; Wang, P.; Li, S.; Zhang, Q.; Tian, H.; Ren, D.; Han, X. Protection of Luteolin-7-O-Glucoside Against Doxorubicin-Induced Injury Through PTEN/Akt and ERK Pathway in H9c2 Cells. *Cardiovasc. Toxicol.* **2016**, *16*, 101–110. [CrossRef] [PubMed]
55. Chen, S.; Yang, B.; Xu, Y.; Rong, Y.; Qiu, Y. Protection of Luteolin-7-O-glucoside against apoptosis induced by hypoxia/reoxygenation through the MAPK pathways in H9c2 cells. *Mol. Med. Rep.* **2018**, *17*, 7156–7162. [CrossRef]
56. Wang, Y.; Guo, S.-H.; Shang, X.-J.; Yu, L.-S.; Zhu, J.-W.; Zhao, A.; Zhou, Y.-F.; An, G.-H.; Zhang, Q.; Ma, B. Triptolide induces Sertoli cell apoptosis in mice via ROS/JNK-dependent activation of the mitochondrial pathway and inhibition of Nrf2-mediated antioxidant response. *Acta Pharmacol. Sin.* **2018**, *39*, 311–327. [CrossRef]
57. Chauhan, D.; Li, G.; Hideshima, T.; Podar, K.; Mitsiades, C.; Mitsiades, N.; Munshi, N.; Kharbanda, S.; Anderson, K.C. JNK-dependent release of mitochondrial protein, Smac, during apoptosis in multiple myeloma (MM) cells. *J. Biol. Chem.* **2003**, *278*, 17593–17596. [CrossRef]
58. Che, X.-F.; Moriya, S.; Zheng, C.-L.; Abe, A.; Tomoda, A.; Miyazawa, K. 2-Aminophenoxazine-3-one-induced apoptosis via generation of reactive oxygen species followed by c-jun N-terminal kinase activation in the human glioblastoma cell line LN229. *Int. J. Oncol.* **2013**, *43*, 1456–1466. [CrossRef]
59. Fan, P.; Yu, X.-Y.; Xie, X.-H.; Chen, C.-H.; Zhang, P.; Yang, C.; Peng, X.; Wang, Y.-T. Mitophagy is a protective response against oxidative damage in bone marrow mesenchymal stem cells. *Life Sci.* **2019**, *229*, 36–45. [CrossRef]
60. Schroeter, H.; Boyd, C.S.; Ahmed, R.; Spencer, J.P.E.; Duncan, R.F.; Rice-Evans, C.; Cadenas, E. c-Jun N-terminal kinase (JNK)-mediated modulation of brain mitochondria function: New target proteins for JNK signalling in mitochondrion-dependent apoptosis. *Biochem. J.* **2003**, *372*, 359–369. [CrossRef]
61. Hanawa, N.; Shinohara, M.; Saberi, B.; Gaarde, W.A.; Han, D.; Kaplowitz, N. Role of JNK translocation to mitochondria leading to inhibition of mitochondria bioenergetics in acetaminophen-induced liver injury. *J. Biol. Chem.* **2008**, *283*, 13565–13577. [CrossRef]
62. Roos, W.P.; Kaina, B. DNA damage-induced cell death by apoptosis. *Trends Mol. Med.* **2006**, *12*, 440–450. [CrossRef] [PubMed]
63. Yue, J.; López, J.M. Understanding MAPK signaling pathways in apoptosis. *Int. J. Mol. Sci.* **2020**, *21*, 2346. [CrossRef] [PubMed]
64. Hepp Rehfeldt, S.C.; Majolo, F.; Goettert, M.I.; Laufer, S. c-Jun N-Terminal Kinase Inhibitors as Potential Leads for New Therapeutics for Alzheimer's Diseases. *Int. J. Mol. Sci.* **2020**, *21*, 9677. [CrossRef] [PubMed]
65. Kandimalla, R.; Reddy, P.H. Therapeutics of neurotransmitters in alzheimer's disease. *J. Alzheimers Dis.* **2017**, *57*, 1049–1069. [CrossRef] [PubMed]
66. Ali, T.B.; Schleret, T.R.; Reilly, B.M.; Chen, W.Y.; Abagyan, R. Adverse Effects of Cholinesterase Inhibitors in Dementia, According to the Pharmacovigilance Databases of the United-States and Canada. *PLoS ONE* **2015**, *10*, e0144337. [CrossRef] [PubMed]

67. Abdul Manap, A.S.; Wei Tan, A.C.; Leong, W.H.; Yin Chia, A.Y.; Vijayabalan, S.; Arya, A.; Wong, E.H.; Rizwan, F.; Bindal, U.; Koshy, S.; et al. Synergistic Effects of Curcumin and Piperine as Potent Acetylcholine and Amyloidogenic Inhibitors With Significant Neuroprotective Activity in SH-SY5Y Cells via Computational Molecular Modeling and in vitro Assay. *Front. Aging Neurosci.* **2019**, *11*, 206. [CrossRef]
68. Kalola, U.K.; Nguyen, H. Galantamine. In *StatPearls*; StatPearls Publishing: Treasure Island, FL, USA, 2021.
69. Khan, H.; Amin, S.; Kamal, M.A.; Patel, S. Flavonoids as acetylcholinesterase inhibitors: Current therapeutic standing and future prospects. *Biomed. Pharmacother.* **2018**, *101*, 860–870. [CrossRef]
70. Orhan, I.; Kartal, M.; Tosun, F.; Sener, B. Screening of various phenolic acids and flavonoid derivatives for their anticholinesterase potential. *Z. Naturforsch. C J. Biosci.* **2007**, *62*, 829–832. [CrossRef]
71. Istifli, E.S.; Sarıkürkcü, C. Assessment of apigenin-7-glucoside and luteolin-7-glucoside as multi-targeted agents against Alzheimer's disease: A molecular docking study. *Int. J. Plant Based Pharm.* **2021**, *1*, 56–64.
72. Sevindik, H.G.; Güvenalp, Z.; Yerdelen, K.Ö.; Yuca, H.; Demirezer, L.Ö. The discovery of potential anticholinesterase compounds from *Achillea millefolium* L. *Ind. Crops Prod.* **2015**, *76*, 873–879. [CrossRef]
73. Liu, M.-Y.; Zeng, F.; Shen, Y.; Wang, Y.-Y.; Zhang, N.; Geng, F. Bioguided Isolation and Structure Identification of Acetylcholinesterase Enzyme Inhibitors from *Drynariae rhizome*. *J. Anal. Methods Chem.* **2020**, *2020*, 2971841. [CrossRef]
74. Jang, S.; Kelley, K.W.; Johnson, R.W. Luteolin reduces IL-6 production in microglia by inhibiting JNK phosphorylation and activation of AP-1. *Proc. Natl. Acad. Sci. USA* **2008**, *105*, 7534–7539. [CrossRef] [PubMed]
75. Zhang, J.-X.; Xing, J.-G.; Wang, L.-L.; Jiang, H.-L.; Guo, S.-L.; Liu, R. Luteolin Inhibits Fibrillary β-Amyloid1-40-Induced Inflammation in a Human Blood-Brain Barrier Model by Suppressing the p38 MAPK-Mediated NF-κB Signaling Pathways. *Molecules* **2017**, *22*, 334. [CrossRef] [PubMed]
76. Lee, S.A.; Park, B.-R.; Moon, S.-M.; Hong, J.H.; Kim, D.K.; Kim, C.S. Chondroprotective Effect of Cynaroside in IL-1β-Induced Primary Rat Chondrocytes and Organ Explants via NF-κB and MAPK Signaling Inhibition. *Oxid. Med. Cell. Longev.* **2020**, *2020*, 9358080. [CrossRef] [PubMed]

International Journal of *Molecular Sciences*

MDPI

Article

Chicoric Acid Prevents Neuroinflammation and Neurodegeneration in a Mouse Parkinson's Disease Model: Immune Response and Transcriptome Profile of the Spleen and Colon

Ning Wang [1], Rui Li [2], Bainian Feng [3], Yuliang Cheng [1], Yahui Guo [1,*] and He Qian [1,*]

1 School of Food Science and Technology, Jiangnan University, Wuxi 214122, China; wangyuning31@163.com (N.W.); wxfoodcyl@126.com (Y.C.)
2 Cedars-Sinai Medical Center, Los Angeles, CA 90048, USA; lr2008525@163.com
3 School of Pharmaceutical Science, Jiangnan University, Wuxi 214122, China; fengbainian@jiangnan.edu.cn
* Correspondence: guoyahui@jiangnan.edu.cn (Y.G.); amtf168168@126.com (H.Q.)

Abstract: Chicoric acid (CA), a polyphenolic acid compound extracted from chicory and echinacea, possesses antiviral, antioxidative and anti-inflammatory activities. Growing evidence supports the pivotal roles of brain–spleen and brain–gut axes in neurodegenerative diseases, including Parkinson's disease (PD), and the immune response of the spleen and colon is always the active participant in the pathogenesis and development of PD. In this study, we observe that CA prevented dopaminergic neuronal lesions, motor deficits and glial activation in PD mice, along with the increment in striatal brain-derived neurotrophic factor (BDNF), dopamine (DA) and 5-hydroxyindoleacetic acid (5-HT). Furthermore, CA reversed the level of interleukin-17(IL-17), interferon-gamma (IFN-γ) and transforming growth factor-beta (TGF-β) of PD mice, implicating its regulatory effect on the immunological response of spleen and colon. Transcriptome analysis revealed that 22 genes in the spleen (21 upregulated and 1 downregulated) and 306 genes (190 upregulated and 116 downregulated) in the colon were significantly differentially expressed in CA-pretreated mice. These genes were functionally annotated with GSEA, GO and KEGG pathway enrichment, providing the potential target genes and molecular biological mechanisms for the modulation of CA on the spleen and gut in PD. Remarkably, CA restored some gene expressions to normal level. Our results highlighted that the neuroprotection of CA might be associated with the manipulation of CA on brain–spleen and brain–gut axes in PD.

Keywords: chicoric acid; Parkinson's disease; neuroinflammation; neurodegeneration; peripheral immune system; spleen; gut

Citation: Wang, N.; Li, R.; Feng, B.; Cheng, Y.; Guo, Y.; Qian, H. Chicoric Acid Prevents Neuroinflammation and Neurodegeneration in a Mouse Parkinson's Disease Model: Immune Response and Transcriptome Profile of the Spleen and Colon. *Int. J. Mol. Sci.* **2022**, *23*, 2031. https://doi.org/10.3390/ijms23042031

Academic Editors: Marcello Ciaccio, Luisa Agnello and Fabrizio Michetti

Received: 6 December 2021
Accepted: 21 January 2022
Published: 12 February 2022

1. Introduction

Parkinson's disease (PD) is the most prevalent neurodegenerative movement disorder caused by the progressive loss of dopaminergic neurons in the substantia nigra pars compacta region of the midbrain [1,2]. Neuroinflammation is well documented as the common pathological characteristic of PD. Notably, the neuroinflammatory phenotypes of PD can be modulated by peripheral immunoreactions through the molecular crosstalk between resident and blood-derived cellular components, suggesting that peripheral immunity is an active participant in the neuroinflammatory and neurodegenerative progression of PD [3–5].

The spleen and gut, two of the largest peripheral immunological organs, were proposed as the targets for anti-neurodegeneration owing to their modulatory roles in the immune system [4,6]. For example, increased splenic macrophages with activated M1 subtype could induce the systemic proinflammatory response, leading to the neurological damage and motor disorder in a mouse model of parkinsonism [7]. In addition, the

1-methyl-4-phenyl-1,2,3,6-tetrathydropyridine (MPTP)-induced mice treated with an immune modulator exhibited increased splenocytes and spleen size, which suppressed the neuroinflammatory response, motor dysfunctions and dopaminergic neuronal depletion, implicating a mediating role of the spleen in the immunological communication between central nervous system (CNS) and peripheral immune system of PD [8–10]. Increasing evidence has shown that gut microbial dysbiosis might be a major mediator for neuroinflammation in PD via gut microbial metabolites through the microbiota–gut–brain axis [11,12]. Microbial-dysbiosis-driven inflammation in the gut may lead to the hyperpermeability of the colon responsible for the leaky gut and the release of gut-derived toxins in PD [13]. Moreover, α-synuclein, a neuroinflammatory mediator in PD, might retrograde transport from the enteric nervous system to the CNS [12,14], suggesting that the gut plays a key role in maintaining the balance of the gut–brain axis in PD. Taken together, the spleen and gut are the vital peripheral organs that are essential for systemic immunity, are heavily involved in the pathomechanism and development of PD, and might be the potential targets aiming to combat PD neurodegeneration.

Chicoric acid (CA), a polyphenolic acid compound obtained from plants, such as chicory, purple coneflower, lettuce, dandelion, and other edible plants, exhibits antiviral, antioxidative and anti-inflammatory activities [15]. Previous studies reported that CA possessed anti-neuroinflammatory and immunoregulatory properties [16–18]. However, limited researches focus on the neuroprotective effects of CA on PD, and the evidence for the influence of CA on peripheral immune organs, such as the spleen and colon, in MPTP-intoxicated PD mice is lacking.

In this paper, we demonstrate that CA prevents the neurodegenerative progression in MPTP-induced PD mice. The mechanism exploration indicates the CA-regulated immune response as well as gene expression of the spleen and colon in PD. Moreover, target genes and molecular biological mechanisms for the modulation of CA on the spleen and gut in PD are obtained by transcriptome analysis. Our study provides the evidence that the neurorescue effects of CA are associated with the peripheral immune system in PD, especially the spleen and colon, which may present the potential target organs for PD therapy.

2. Results

2.1. CA Improves Motor Deficits in PD Mice

The pole test and traction test were conducted to assess the effects of CA on motor disorder caused by MPTP. Mice treated with MPTP displayed motor dysfunction including longer pole descent time in pole test (Figure 1c) and lower scores in traction test (Figure 1d). However, CA alleviated motor disorder significantly. Mice pretreated with CA shortened the time of downward climbing from top to the bottom of the pole in the pole test (Figure 1c) and exhibited a better performance with a higher score in the measurement of gripping the rope in the traction test (Figure 1d). The results revealed that CA possessed neuroprotective effects on the motor function of PD mice.

2.2. CA Promoted Dopaminergic Neuron Survival and Striatal TH Levels in PD Mice

The most prominent pathological feature of PD is the death of nigrostriatal dopaminergic (DA) neurons. Therefore, the survival of DA neurons in the substantia nigra (SN) was investigated by immunofluorescence (IF) staining, and the expression of striatal tyrosine hydroxylase (TH) was determined by Western blot. The results displayed that MPTP induced a decrease of TH-positive DA neurons in SN and CA pretreatment reversed the MPTP-mediated loss of DA neurons (Figure 2a,b). Western blotting analysis showed a lower expression of striatal TH in MPTP group, but CA exhibited an enhanced effect on striatal TH expression (Figure 2c,d). The above results indicated that MPTP destroyed DA neurons in SN and TH in striatum, while this lesion was attenuated by CA.

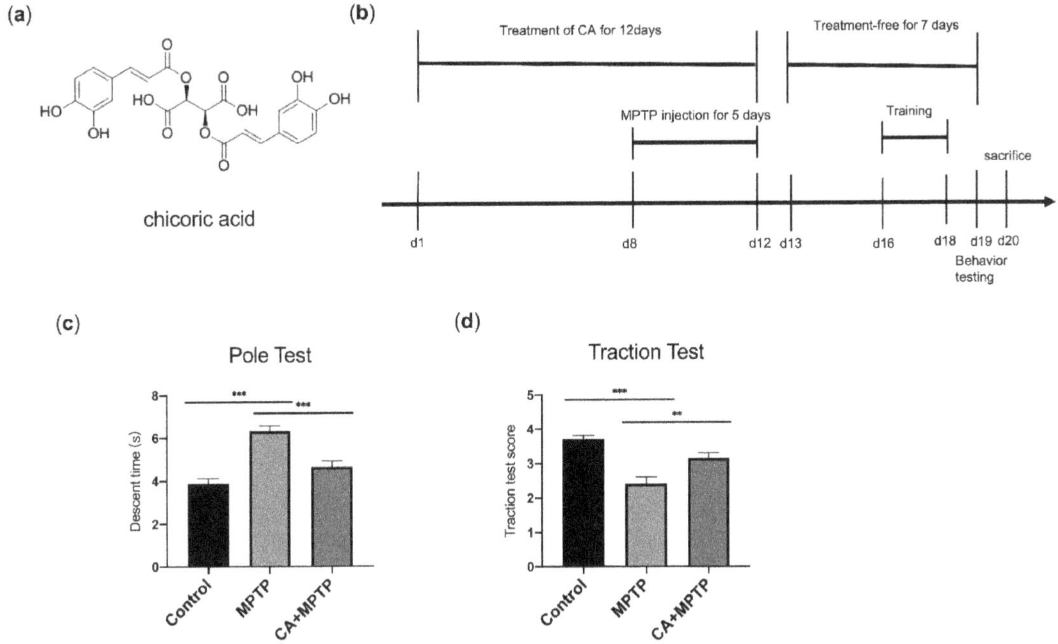

Figure 1. CA improved behavioral deficits in PD mice. (**a**) Chemical structure of CA. (**b**) Timeline for the experimental procedure. (**c**) Pole test: time to descend the pole. (**d**) Traction test: score of traction reflexes. Statistical comparison by one-way ANOVA with Tukey's post hoc test. Data represent means ± SEM. ** $p < 0.01$, *** $p < 0.001$. $n = 10$.

2.3. CA Improved the Reduction in Striatal Dopamine and Serotonin of PD Mice

Striatal neurotransmitters, dopamine (DA) and 5-hydroxytryptamine (5-HT), and their metabolites, including 3,4-dihydroxyphenylacetic (DOPAC), homovanillic acid (HVA) and 5-hydroxyindoleacetic acid (5-HIAA), were detected by high-performance liquid chromatography (HPLC) with fluorescence detection. The turnover of DA and 5-HT was used to evaluate their metabolism in PD represented as the ratio of (DOPAC+HVA)/DA and 5-HIAA/5-HT, respectively. As shown in Figure 2e–h, MPTP mice displayed a decrease of DA and 5-HT, as well as an increased ratio of (DOPAC+HVA)/DA and 5-HIAA/5-HT. CA increased the level of DA and 5-HT of PD mice, likewise the MPTP-mediated increment in the ratio of (DOPAC+HVA)/DA and 5-HIAA/5-HT were decreased by CA, indicating that CA alleviated the MPTP-induced reduction in striatal DA and 5-HT, and contributed to the restoration of their metabolite level.

2.4. CA Suppressed Glial-Mediated Neuroinflammation Accompanied by an Increment in Striatal Neurotrophic Factors

To investigate whether the neuroinflammation accompanied by glial activation occurs in PD mice, glial fibrillary acidic protein (GFAP) and ionized calcium binding adaptor molecule-1 (Iba-1), the markers of astrogliosis and microgliosis, were detected by double IF staining for TH and GFAP or Iba-1, respectively. Simultaneously, GFAP, Iba-1, cluster of differentiation molecule 11b (CD11b) (microglia marker) and BDNF in the striatum were examined by Western blotting to verify the changes in striatal protein and neurotrophics. The highest level of GFAP and Iba-1 in SN were observed in MPTP group along with the reduction in TH-positive cells (Figure 3a–d). CA-pretreated mice exhibited an inhibition of astrocyte and microglial activation shown as a decline in number of positive cells for GFAP and Iba-1, respectively (Figure 3a–d), as well as a higher level of TH-positive cells.

Western blotting analysis confirmed the neuroinflammatory response induced by MPTP, as evidenced by the enhanced expression of striatal GFAP, Iba-1, as well as CD11b, another microglia-activation marker (Figure 3e,g–i). Of note, these increased proteins were reduced by CA (Figure 3e,g–i). We further explored the effects of CA on striatal BDNF, an essential neurotrophic factor for DA neuronal survival. As displayed in Figure 3f,j, a severe depletion of BDNF appeared in MPTP group; CA partially but significantly restored the BDNF level.

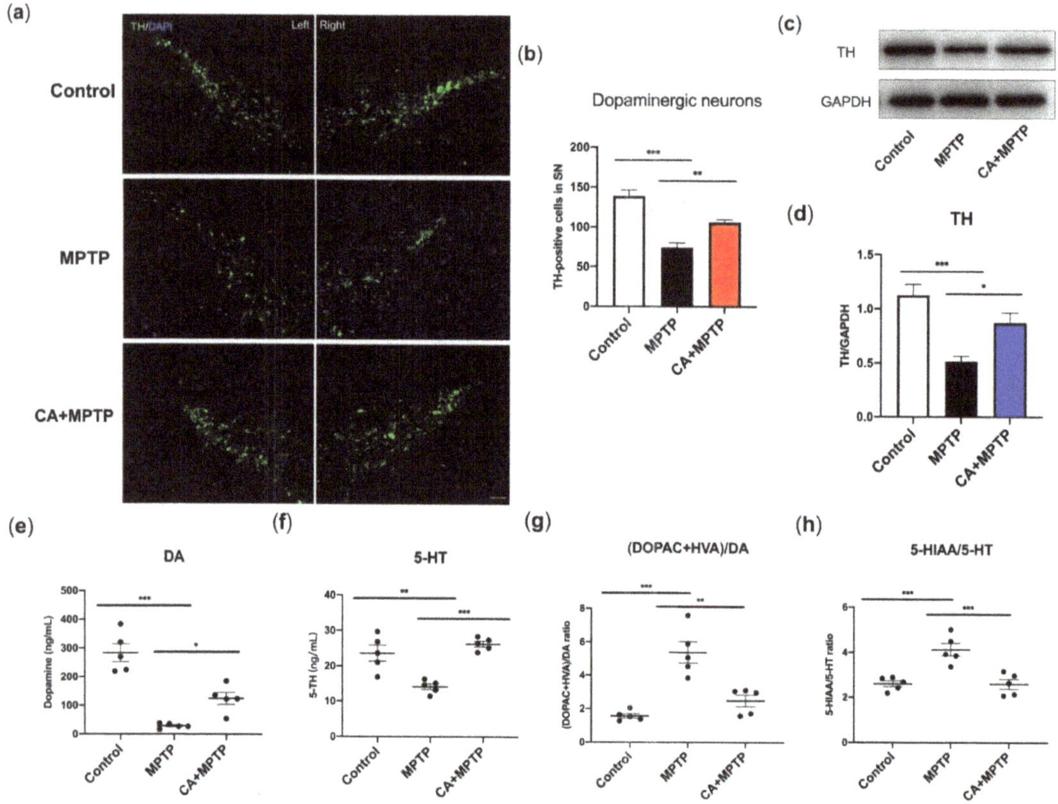

Figure 2. CA promoted the dopaminergic neuron survival, TH expression in striatum and the enhancement in neurotransmitters. (**a**) Representative IF staining for TH in the right and left SN, respectively. Scale bar: 100 μm. (**b**) Quantification for the number of TH-positive cells in left SN. (**c**) Representative bands of Western blotting for striatal TH expression. (**d**) Quantification for striatal TH expression, band intensity normalized to GAPDH. (**e**) The level of striatal DA. (**f**) The level of 5-TH. (**g**) The turnover rate of striatal DA ([DOPAC+HVA]/DA). (**h**) The turnover rate of striatal 5-HT (5-HIAA/5-HT). Statistical comparison by one-way ANOVA with Tukey's post hoc test. Data represent means ± SEM. * $p < 0.05$, ** $p < 0.01$, *** $p < 0.001$. $n = 4$ for IF, $n = 5$ for the measurement of neurotransmitter and Western blotting.

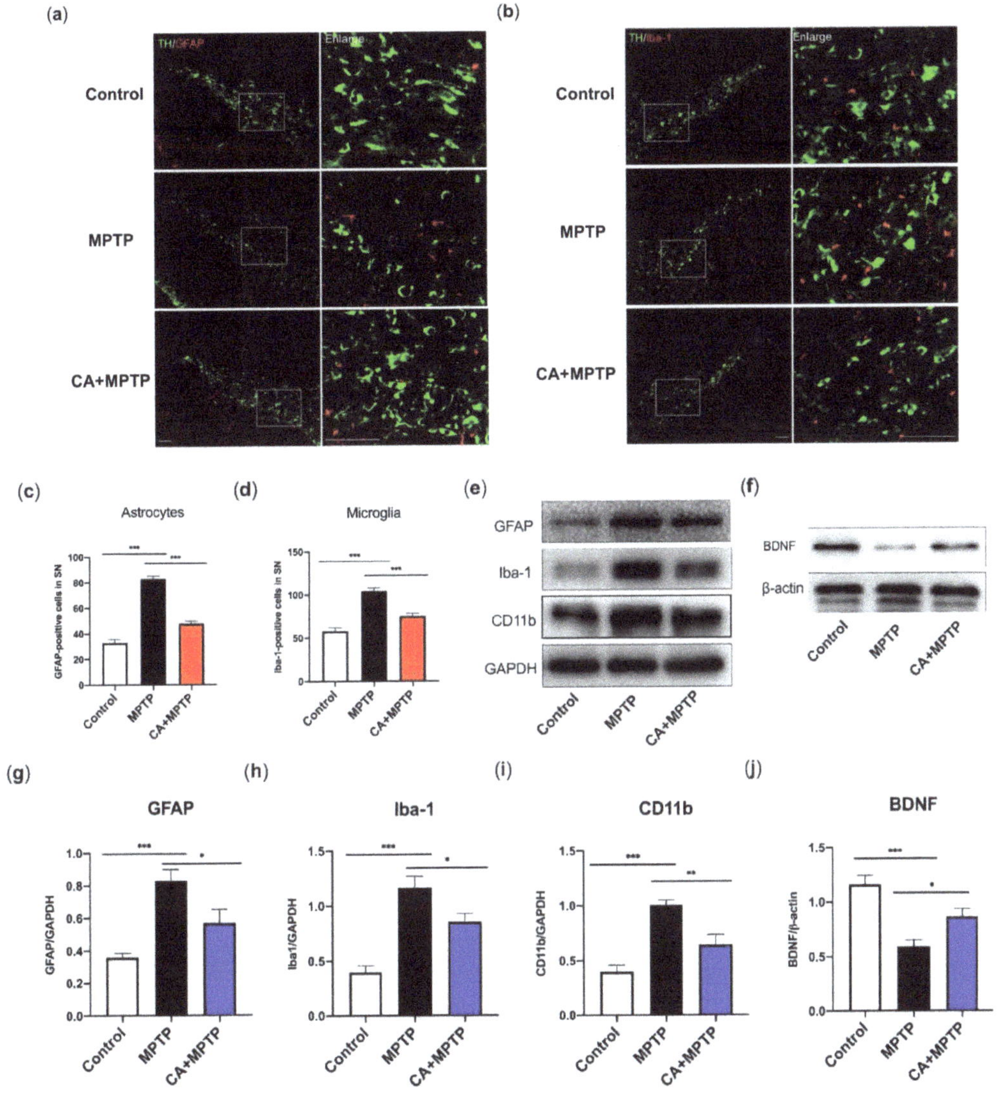

Figure 3. CA inhibited MPTP-induced microglial and astrocyte activation in SN and striatum, respectively, accompanied by the restoration of striatal BDNF. (**a**) Representative double-IF staining for TH (green) and GFAP (red) in SN. Scale bars: 100 μm. (**b**) Representative double-IF staining for TH (green) and Iba-1 (red) in SN. Scale bars: 100 μm. (**c**) Quantification for the number of GFAP-positive cells. (**d**) Quantification for the number of Iba-1-positive cells. (**e**,**f**) Representative bands of Western blotting for the expression of striatal GFAP, Iba-1, CD11b and BDNF. (**g**–**j**) Quantification for protein expression, band intensity normalized to GAPDH or β-actin. Statistical comparison by one-way ANOVA with Tukey's post hoc test. Data represent means ± SEM. * $p < 0.05$, ** $p < 0.01$, *** $p < 0.001$. $n = 4$ for IF, $n = 5$ for Western blotting.

2.5. CA Restored the Protein Levels of IL-17, IFN-γ and TGF-β in the Serum, Striatum, Spleen and Solon of PD Mice

In order to explore the influence of CA on the peripheral immune system, the protein expression of T helper 17 (Th17) cell-related cytokine (IL-17), T helper 1 (Th1)-related cytokine (IFN-γ) and regulatory T cells (Tregs)-related cytokine (TGF-β) in the serum, striatum, spleen and colon were determined by ELISA. As shown in Figure 4a–c, mice in MPTP group exhibited high levels of IL-17 and IFN-γ, and a low level of TGF-β in the serum, striatum, spleen and colon, respectively. Notably, CA pretreatment reduced the protein levels of IL-17 and IFN-γ, and increased TGF-β expression in the serum, striatum, spleen, and colon of PD mice.

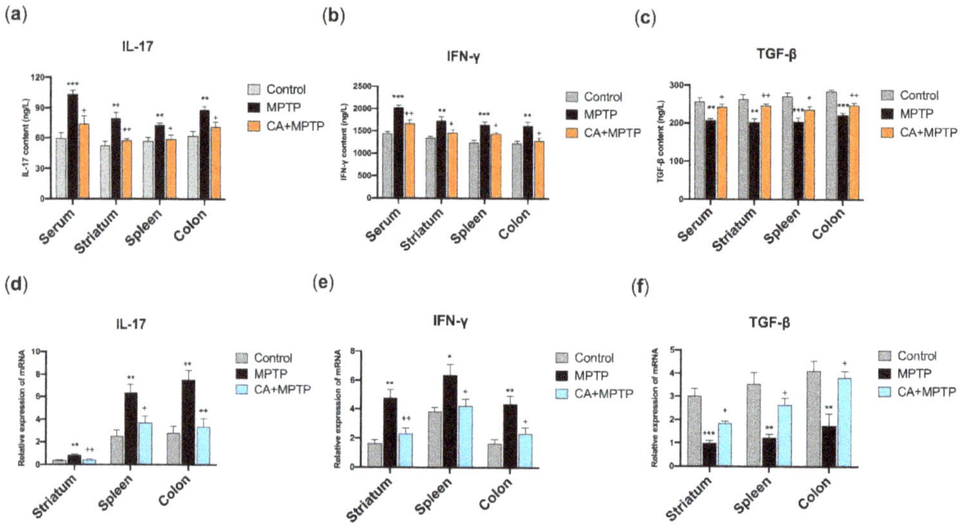

Figure 4. CA restored MPTP-mediated changes in IL-17, IFN-γ and TGF-β at the protein level and mRNA level in the serum, striatum, spleen, and colon. (**a**) The protein level of IL-17. (**b**) The protein level of IFN-γ. (**c**) The protein level of TGF-β. (**d**) The relative expression of IL-17 mRNA. (**e**) The relative expression of IFN-γ mRNA. (**f**) The relative expression of TGF-β mRNA. Statistical comparison by one-way ANOVA with Tukey's post hoc test. Data represent means ± SEM. * $p< 0.05$, ** $p< 0.01$, *** $p < 0.001$, vs. Control group; + $p < 0.05$, ++ $p < 0.01$, vs. MPTP group. $n = 4$ for ELISA, $n = 5$ for qPCR.

2.6. CA Restored the mRNA Levels of IL-17, IFN-γ and TGF-β in the Striatum, Spleen and Colon of PD Mice

To further verify the ELISA results, the mRNA expressions of IL-17, IFN-γ and TGF-β of the striatum, spleen and colon were examined by quantitative PCR (qPCR). The primers of the cytokines are presented in Table 1. As expected, MPTP mice showed an enhancement of mRNA expression of IL-17 and IFN-γ, and a decrease in TGF-β in the striatum, spleen and colon (Figure 4d–e), respectively. These changes in mRNA expression were attenuated by CA, as a downregulation of IL-17 and IFN-γ mRNA, and an upregulation of TGF-β mRNA were observed in the striatum, spleen and colon of mice in CA+MPTP group (Figure 4d–e).

Table 1. Sequences of PCR primers.

Gene	Forward Primers	Reverse Primers
IL-17	5′-TGGACTCTGAGCCGCAATG-3′	5′-GGCGGACAATAGAGGAAACG-3′
IFN-γ	5′-CATAGATGTGGAAGAAAAGAG-3′	5′-AGAGTCTGAGGTAGAAAGAGATA-3′
TGF-β	5′-CGAAGCGGACTACTATGCTA-3′	5′-GAATGTCTGACGTATTGAAGAA-3′
β-actin	5′-CCTCTATGCCAACACAGT-3′	5′-AGCCACCAATCCACACAG-3′

2.7. CA Modulated the Gene Expression of the Spleen in PD Mice Based on Transcriptome Analysis

The above results indicate that MPTP caused neuroinflammation and neurodegeneration, which was suppressed by CA markedly. In an attempt to explore the molecular mechanism role of the spleen and colon for the neuroprotection of CA in PD, the transcriptome profile of the spleen and colon was performed.

Differentially expressed genes (DEGs) were identified by the comparison between CA+MPTP and MPTP groups. The results show that a total of 22 DEGs in the spleen were expressed, of which 21 were upregulated and 1 was downregulated (Figure 5a). However, there were no significant differences of splenic gene expression between the CA+MPTP and control groups at the difference threshold of (|$\log_2$ Fold change (FC)|) > 1.2 and adjusted *p*-value (padj) < 0.05 (results not shown in figure). These results were also verified by the cluster analysis of splenic DGEs between the NC, MPTP and CA+MPTP groups (Figure 5c), as CA restored some splenic genes to normal levels in PD mice, suggesting that CA could improve neuronal survivals involving its regulation on splenic gene expression in MPTP mice.

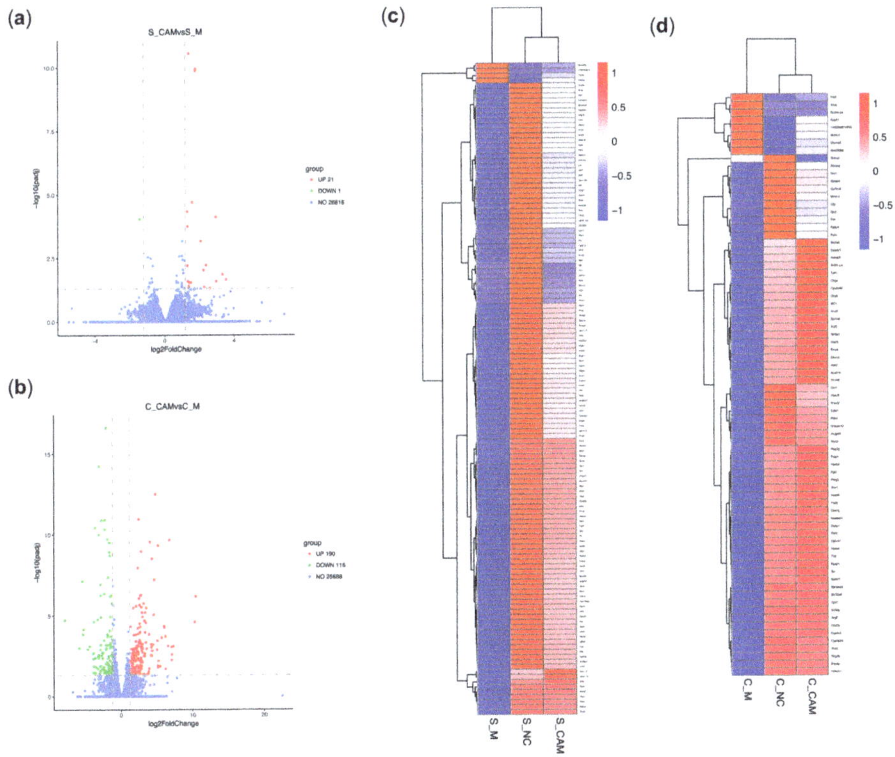

Figure 5. The influence of CA on splenic gene expression in PD mice based on RNA-Seq. (**a**) Volcano plot of splenic DEGs between the MPTP and CA+MPTP groups. (**b**) Volcano plot of colonic DEGs between the MPTP and CA+MPTP groups. The heatmaps showing the results of cluster analysis for DGEs between the control, MPTP and CA+MPTP groups in the spleen (**c**) and colon (**d**). CA pretreatment significantly restored the gene expression in the colon and spleen of MPTP mice; combining with the regulation of CA on IL-17, IFN-γ and TGF-β in PD mice, it was indicated that the neuroprotective effects of CA were closely related to the modulation of CA on peripheral immune system. Cut-off value of (|$\log_2$FC|) > 1.2 and padj < 0.05, $n = 3$ per group. S: spleen, C: colon. CAM: CA+MPTP group, M: MPTP group, NC: control group.

Next, the functional annotation of splenic DEGs between the CA+MPTP and MPTP groups was conducted using Gene Ontology (GO) and Kyoto Encyclopedia of Genes and Genomes (KEGG) enrichment with the significant cut-off value of padj < 0.05. The analysis of GO functional dissimilarity showed that the DEGs were mainly enriched in the cellular responses to dexamethasone and insulin stimulus, and the responses to dexamethasone, ketone, insulin, glucocorticoid and corticosteroid, as well as the immunoglobulin production and negative regulation of cytokine biosynthetic process (Table 2). The enrichKEGG analysis indicated that splenic DEGs were mainly involved in PPAR signaling pathway (Table 3). Gene Set Enrichment Analysis (GSEA) demonstrated that the expression of splenic genes induced by CA was positively enriched in epithelial mesenchymal transition, apoptosis, Kirsten rat sarcoma viral oncogene homolog (KRAS)-signaling, notch signaling, inflammatory response and TGF-β signaling (Figure 6a). The results reveal the molecular biological mechanisms for the regulation of CA on the spleen, which might be related to CA-mediated neuroprotection.

Table 2. List of GO terms enriched with splenic DEGs (padj < 0.05).

Category	GO ID	Description	Gene Name	Count	Up	Down	padj
BP	GO:0042036	negative regulation of cytokine biosynthetic process	Errfi1↑Muc16↑	2	2	0	0.037678798
BP	GO:0051384	response to glucocorticoid	Errfi1↑Dusp1↑Pck1↑	3	3	0	0.037678798
BP	GO:0031960	response to corticosteroid	Errfi1↑Dusp1↑Pck1↑	3	3	0	0.037678798
BP	GO:0032869	cellular response to insulin stimulus	Pdk4↑Errfi1↑Pck1↑	3	3	0	0.037678798
BP	GO:1901654	response to ketone	Errfi1↑Dusp1↑Pck1↑	3	3	0	0.037678798
BP	GO:0071549	cellular response to dexamethasone stimulus	Errfi1↑Pck1↑	2	2	0	0.037678798
BP	GO:0071548	response to dexamethasone	Errfi1↑Pck1↑	2	2	0	0.046519989
BP	GO:0002377	immunoglobulin production	Igkv9-124↑Igkv14-100↑Igkv4-91↑	3	3	0	0.046927892
BP	GO:0032868	response to insulin	Pdk4↑Errfi1↑Pck1↑	3	3	0	0.046927892

↑ upregulation.

Table 3. List of pathway enriched with splenic DEGs (padj < 0.05).

KEGG ID	Description	Gene Name	Count	Up	Down	padj
mmu03320	PPAR signaling pathway	Pck1↑Plin4↑	2	2	0	0.00949221

↑ upregulation.

2.8. CA Modulated the Gene Expression of the Colon in PD Mice Based on Transcriptome Analysis

Inflammatory processes participate in the initiation and development of PD due to their effects on the brain–gut axis, including the immune system [19,20]. We further investigated the influence of CA on the colon in PD at the transcriptome level. As displayed in Figure 5b, a total of 306 colonic DEGs in PD mice were expressed with CA pretreatment, of which 190 were upregulated and 116 were downregulated. Notably, compared to the control group, only one downregulated DEG in the CA+MPTP group was obtained under the screening threshold of ($|\log_2 FC|$) > 1.2 and padj < 0.05 (results not shown in the figures). Moreover, the results of cluster analysis for colonic DGEs between the NC, MPTP and CA+MPTP groups demonstrated that the MPTP-induced gene expression of the colon was significantly reversed by CA (Figure 5d). We thus inferred that CA pretreatment could modulate MPTP-induced gene expression in the colon, which was involved in the neuroprotection of CA.

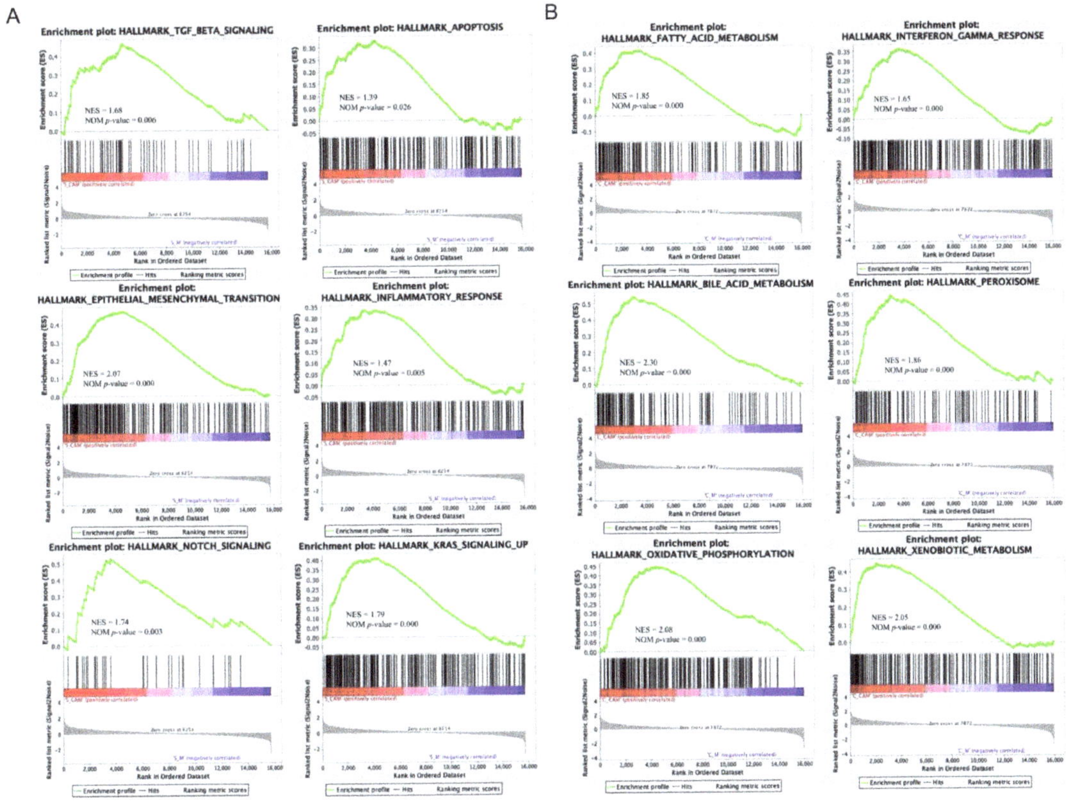

Figure 6. Enrichment plots from the gene set enrichment analysis (GSEA). (**A**) The significantly enriched hallmark terms associated with CA-mediated modulation on the spleen in PD mice. (**B**) The significantly enriched hallmark terms associated with CA-mediated modulation on the colon in PD mice. Cut-off value of | Normalized Enrichment Score (NES) | > 1.0 and nominal p-value < 0.05, n = 3 per group. S: spleen, C: colon. CAM: CA+MPTP group, M: MPTP group.

GO enrichment results show that DEGs between MPTP and CA+MPTP mice were largely involved in the processes of fatty acid metabolism, organic acid biosynthesis, lipid catabolism and drug catabolism, as well as in the cellular response to xenobiotic stimulus, and the activity of monooxygenase and oxidoreductase; the rest of the enrichGO terms are displayed in Table 4. Furthermore, KEGG pathway annotations showed that colonic DEGs were mainly enriched in retinol metabolism, steroid hormone biosynthesis, chemical carcinogenesis, bile secretion and serotonergic synapse; the rest of involved pathways are displayed in Table 5. Additionally, GSEA enrichment plots indicated that the expression of colonic genes induced by CA were mainly enriched in fatty acid metabolism, IFN-γ response, oxidative phosphorylation, bile acid metabolism, peroxisome and xenobiotic metabolism (Figure 6b). These results provided biological molecular mechanisms in the CA-mediated communication between the brain and gut.

Table 4. List of GO terms enriched with colonic DEGs (padj < 0.05).

Category	GO ID	Description	Gene Name	Count	Up	Down	padj
BP	GO:0006805	xenobiotic metabolic process	Acaa1b↑Cyp2c66↑Ugt1a1↑Lpo↑Cyp2c65↑Cyp2c55↑Nceh1↑ Ugt1a7c↑Cyp2d12↓Cyp2d9↓Srd5a2↓Cyp2c68↓Cyp2c69↓Cyp2f2↓	14	8	6	1.06×10^{-8}
BP	GO:0042737	drug catabolic process	Cyp2c66↑Cyp2c65↑Akr1c18↑Cyp4b1↑Adh5↑Cyp2c55↑Cubn↑ Aldh3b1↑Cyp2d12↓Cyp2d9↓Adh1↓Cyp2c68↓Nt5e↓Cyp2c69↓ Cyp2f2↓	15	8	7	1.79×10^{-7}
BP	GO:0042738	exogenous drug catabolic process	Cyp2c66↑Cyp2c65↑Cyp4b1↑Cyp2c55↑Cyp2d12↓Cyp2d9↓ Cyp2c68↓Cyp2c69↓Cyp2f2↓	9	4	5	1.79×10^{-7}
BP	GO:0006690	icosanoid metabolic process	Cyp2c66↑Cyp2c65↑Akr1c18↑Cyp2c55↑Tlr2↑Ggt1↑Cyp2d12↓ Cyp2d9↓Pla2g4f↓Cyp2c68↓Pla2g5↓Cyp2c69↓Cyp2f2↓	13	6	7	1.79×10^{-7}
BP	GO:0006631	fatty acid metabolic process	Acaa1b↑Ces1f↑Ppara↑Cyp2c66↑Lpin2↑Ces1d↑Pdk4↑Acsf2↑ Cyp2c65↑Akr1c18↑Lpl↑Cyp2c55↑Slc27a4↑Ggt1↑Ehhadh↑Abhd3↑ Adipoq↑Cyp2d12↓Cyp2d9↓Pla2g4f↓Cyp2c68↓Cyp2c69↓Cyp2f2↓	23	17	6	3.37×10^{-7}
BP	GO:0016042	lipid catabolic process	Acaa1b↑Ces1f↑Lpin2↑Ces1d↑Akr1c18↑Lpl↑Ces1g↑Aspg↑Nceh1↑ Slc27a4↑Ugt1a7c↑Ehhadh↑Abhd3↑Adipoq↑Pla2g4f↓Srd5a2↓ Pla2g2a↓Pla2g5↓Ces3a↓Hexb↓	20	14	6	7.85×10^{-7}
BP	GO:1901568	fatty acid derivative metabolic process	Cyp2c66↑Cyp2c65↑Akr1c18↑Cyp2c55↑Tlr2↑Ggt1↑Cyp2d12↓ Cyp2d9↓Pla2g4f↓Cyp2c68↓Pla2g5↓Cyp2c69↓Cyp2f2↓	13	6	7	2.80×10^{-6}
BP	GO:0046394	carboxylic acid biosynthetic process	Gapdh↑Hkdc1↑Ppara↑Pdk4↑Akr1c18↑Nags↑Lpl↑Malrd1↑ Aldh1a1↑Ggt1↑Abhd3↑Ugdh↑Tkfc↑Gpd1↑Rdh16↓Eno3↓Pla2g4f↓ Hif1a↓Hk2↓Rdh9↓Pla2g5↓	21	14	7	2.80×10^{-6}
BP	GO:0016053	organic acid biosynthetic process	Gapdh↑Hkdc1↑Ppara↑Pdk4↑Akr1c18↑Nags↑Lpl↑Malrd1↑ Aldh1a1↑Ggt1↑Abhd3↑Ugdh↑Tkfc↑Gpd1↑Rdh16↓Eno3↓Pla2g4f↓ Hif1a↓Hk2↓Rdh9↓Pla2g5↓	21	14	7	2.80×10^{-6}
BP	GO:0071466	cellular response to xenobiotic stimulus	Acaa1b↑Cyp2c66↑Ugt1a1↑Lpo↑Cyp2c65↑Cyp2c55↑Nceh1↑ Ugt1a7c↑Cyp2d12↓Cyp2d9↓Srd5a2↓Cyp2c68↓Cyp2c69↓Cyp2f2↓	14	8	6	2.80×10^{-6}
MF	GO:0016712	oxidoreductase activity, acting on paired donors, with incorporation or reduction of molecular oxygen, reduced flavin or flavoprotein as one donor, and incorporation of one atom of oxygen	Cyp2c66↑Cyp2c65↑Cyp2d26↑Cyp4b1↑Cyp3a44↑Cyp2c55↑ Cyp2d12↓Cyp2d9↓Cyp2c68↓Cyp2c69↓Cyp2f2↓	11	6	5	2.77×10^{-9}
MF	GO:0004497	monooxygenase activity	Tph1↑Fmo5↑Cyp2c66↑Cyp2c65↑Cyp2d26↑Akr1c18↑Cyp4b1↑ Cyp3a44↑Cyp2c55↑Akr1c19↑Cyp2d12↓Cyp2d9↓Cyp2c68↓ Cyp2c69↓Cyp2f2↓	15	10	5	2.02×10^{-8}

Table 4. *Cont.*

Category	GO ID	Description	Gene Name	Count	Up	Down	padj
MF	GO:0008395	steroid hydroxylase activity	Cyp2c66↑Cyp2c65↑Cyp3a44↑Cyp2c55↑Cyp2d12↓Cyp2d9↓Cyp2c68↓Cyp2c69↓Cyp2f2↓	9	4	5	8.37×10^{-7}
MF	GO:0052689	carboxylic ester hydrolase activity	Ces1f↑Ces1d↑Ces2b↑Car1↑Lpl↑Ces1g↑Aspg↑Nceh1↑Abhd3↑Bche↑Pla2g4f↓Pla2g2a↓Pla2g5↓Ces3a↓	14	10	4	8.37×10^{-7}
MF	GO:0016705	oxidoreductase activity, acting on paired donors, with incorporation or reduction of molecular oxygen	Tph1↑Fmo5↑Cyp2c66↑Cyp2c65↑Cyp2d26↑Akr1c18↑Cyp4b1↑Cyp3a44↑Cyp2c55↑Akr1c19↑Cyp2d12↓Cyp2d9↓Cyp2c68↓Cyp2c69↓Cyp2f2↓	15	10	5	2.51×10^{-6}
MF	GO:0008392	arachidonic acid epoxygenase activity	Cyp2c66↑Cyp2c65↑Cyp2c55↑Cyp2c68↓Cyp2c69↓Cyp2f2↓	6	3	3	5.79×10^{-6}
MF	GO:0008391	arachidonic acid monooxygenase activity	Cyp2c66↑Cyp2c65↑Cyp2c55↑Cyp2c68↓Cyp2c69↓Cyp2f2↓	6	3	3	1.12×10^{-5}
MF	GO:0046906	tetrapyrrole binding	Cyp2c66↑Lpo↑Cyp2c65↑Cyp2d26↑Cyp4b1↑Cyp3a44↑Cyp2c55↑Cubn↑Cyp2d12↓Cyp2d9↓Cyp2c68↓Cyp2c69↓Cyp2f2↓	13	8	5	1.56×10^{-5}
MF	GO:0005506	iron ion binding	Tph1↑Cyp2c66↑Cyp2c65↑Cyp2d26↑Cyp4b1↑Cyp3a44↑Cyp2c55↑Cyp2d12↓Cyp2d9↓Cyp2c68↓Nt5e↓Cyp2c69↓Cyp2f2↓	13	7	6	4.05×10^{-5}
MF	GO:0020037	heme binding	Cyp2c66↑Lpo↑Cyp2c65↑Cyp2d26↑Cyp4b1↑Cyp3a44↑Cyp2c55↑Cyp2d12↓Cyp2d9↓Cyp2c68↓Cyp2c69↓Cyp2f2↓	12	7	5	4.27×10^{-5}

↑upregulation
↓downregulation

Table 5. List of pathways enriched with colonic DEGs (padj < 0.05).

KEGG ID	Description	Gene Name	Count	Up	Down	padj
mmu00830	Retinol metabolism	Ugt2b5↑Cyp2c66↑Ugt1a1↑Cyp2c65↑Cyp3a44↑Aldh1a1↑Ugt2b36↑ Adh5↑Cyp2c55↑Gm15368↑Ugt1a7c↑Rdh16↓Adh1↓Cyp2c68↓Rdh9↓	15	11	4	1.76×10^{-10}
mmu00140	Steroid hormone biosynthesis	Ugt2b5↑Cyp2c66↑Ugt1a1↑Cyp2c65↑Cyp2d26↑Akr1c18↑Cyp3a44↑ Ugt2b36↑Cyp2c55↑Gm15368↑Ugt1a7c↑Cyp2d12↓Cyp2d9↓Srd5a2↓ Cyp2c68↓	15	11	4	1.99×10^{-10}
mmu05204	Chemical carcinogenesis	Ugt2b5↑Cyp2c66↑Ugt1a1↑Cyp2c65↑Cyp3a44↑Ugt2b36↑Adh5↑ Cyp2c55↑Gm15368↑Ugt1a7c↑Aldh3b1↑Adh1↓Cyp2c68↓	13	11	2	4.60×10^{-7}
mmu00591	Linoleic acid metabolism	Cyp2c66↑Cyp2c65↑Cyp3a44↑Cyp2c55↑Pla2g4f↓Cyp2c68↓Pla2g2a↓ Pla2g5↓	8	4	4	1.34×10^{-5}
mmu04976	Bile secretion	Ugt2b5↑Ugt1a1↑Slc51a↑Slc10a2↑Sct↑Ugt2b36↑Gm15368↑Nceh1↑ Ugt1a7c↑Adcy9↑	10	10	0	0.000109402
mmu00053	Ascorbate and aldarate metabolism	Ugt2b5↑Ugt1a1↑Ugt2b36↑Gm15368↑Ugt1a7c↑Ugdh↑	6	6	0	0.000126403
mmu00982	Drug metabolism - cytochrome P450	Ugt2b5↑Fmo5↑Ugt1a1↑Ugt2b36↑Adh5↑Gm15368↑Ugt1a7c↑ Aldh3b1↑Adh1↓	9	8	1	0.000135848
mmu00980	Metabolism of xenobiotics by cytochrome P450	Ugt2b5↑Ugt1a1↑Ugt2b36↑Adh5↑Gm15368↑Ugt1a7c↑Aldh3b1↑ Adh1↓Cyp2f2↓	9	7	2	0.000182884
mmu04726	Serotonergic synapse	Tph1↑Cyp2c66↑Cyp2c65↑Cyp2d26↑Slc18a1↑Cyp2c55↑Cyp2d12↓ Cyp2d9↓Pla2g4f↓Cyp2c68↓Htr4↓	11	6	5	0.000408708
mmu00590	Arachidonic acid metabolism	Cyp2c66↑Cyp2c65↑Cyp2c55↑Ggt1↑Pla2g4f↓Cyp2c68↓Pla2g2a↓Pla2g5↓	8	4	4	0.001096684
mmu00040	Pentose and glucuronate interconversions	Ugt2b5↑Ugt1a1↑Ugt2b36↑Gm15368↑Ugt1a7c↑Ugdh↑	6	6	0	0.001096684
mmu00983	Drug metabolism - other enzymes	Ces1f↑Ugt2b5↑Ugt1a1↑Ces1d↑Ces2b↑Ugt2b36↑Gm15368↑Ugt1a7c↑ Gm45727↑	9	9	0	0.0021122
mmu03320	PPAR signaling pathway	Acaa1b↑Ppara↑Lpl↑Angptl4↑Hmgcs2↑Slc27a4↑Ehhadh↑Adipoq↑	8	8	0	0.002234301
mmu00860	Porphyrin and chlorophyll metabolism	Ugt2b5↑Ugt1a1↑Ugt2b36↑Gm15368↑Ugt1a7c↑	5	5	0	0.008876674
mmu00010	Glycolysis / Gluconeogenesis	Hkdc1↑Adh5↑Pgm2↑Aldh3b1↑Eno3↓Adh1↓Hk2↓	7	4	3	0.008876674
mmu00592	alpha-Linolenic acid metabolism	Acaa1b↑Pla2g4f↓Pla2g2a↓Pla2g5↓	4	1	3	0.01005902
mmu00500	Starch and sucrose metabolism	Hkdc1↑Sis↑Pgm2↑Hk2↓	4	3	1	0.030913453
mmu00520	Amino sugar and nucleotide sugar metabolism	Hkdc1↑Pgm2↑Ugdh↑Hk2↓Hexb↓	5	3	2	0.036511025
mmu00052	Galactose metabolism	Hkdc1↑Sis↑Pgm2↑Hk2↓	4	3	1	0.041823524

↑upregulation
↓downregulation

3. Discussion

Previous studies reported that the CNS inflammation could be regulated by systemic immunity targeting the spleen and gut, implicating the immunological communication between the brain and periphery [5,6]. In the current study, a neuroinflammatory response was observed in PD mice, as the glia population increased after being MPTP challenged, which was characterized by higher levels of CD11b, Iba-1 and GFAP, the hallmarks of activated microglia and astrocytes. In addition, MPTP induced an increment in IL-17 and IFN-γ, and decrease in the TGF-β of PD mice. Since IL-17, IFN-γ and TGF-β are well known as the markers of Th17, Th1 and Treg [21], respectively, it was suggested that the immunoreactions of peripheral organs participated in the neuroinflammatory and neurodegenerative progress of PD. Of note, Th1 and Th17, the CD4 effector T cells, have been considered as the potential contributors to neuroinflammation in PD [10], and microgliosis can exacerbate the deterioration of DA neuron via a proinflammatory mediator, such as IFN-γ released from Th1 [22]. Consistent with this, our data reveal gliosis-mediated neuroinflammation and enhancement in IL-17 and IFN-γ, whereas CA reduced their production, and prevented MPTP-mediated microgliosis as well as astrogliosis along with increased BDNF, the improvement in DA neuronal survival and motor function, suggesting that CA acted as an immunological mediator of the peripheral immune system and inhibited neuroinflammation in PD. Compelling evidence showed that the immunologic modulation of Tregs had an inhibitory effect on neuroinflammation leading to the reduction in the motor disorder of PD [22], and TGF-β released from Tregs acted against the MPP+-induced DA neuronal death via TβR-I on microglia [23–25], which also supported our results that CA upregulated TGF-β expression accompanied by the decrease in gliosis. Furthermore, a previous demonstration that CA could promote neuronal survivals and prevent LPS-induced neuroinflammation in vitro and vivo [16] is also consistent with our results. Accordingly, it is reasonable to infer that the neuroprotection of CA might be associated with the immunological modulation of CA on the peripheral immune system by targeting the spleen and colon.

CA is a diester of caffeic acid with L-tartaric acid (2,3-dicaffeoyl-L-tartaric acid), and a previous study has reported an inhibitory activity of caffeic acid against intestinal inflammation [26]. Additionally, gut inflammation may initiate and aggravate neuroinflammatory reaction in PD [20]. Our results demonstrated CA significantly alleviated gut inflammation and neuroinflammation in PD mice. Thus, targeting gut inflammation might be a promising therapy to inhibit neuroinflammation in PD and phenolic acid compounds with the basic structure of caffeic acid could be the potential candidates.

To further explore the prominent role of the spleen and gut in PD, and the regulatory effects of CA on them, we investigated transcriptome-level changes in the spleen and colon between the MPTP and CA+MPTP groups to provide a holistic view of genetic networks and demonstrate the molecular mechanism of the spleen and gut for the neuroprotection of CA on PD.

The heatmaps showed a remarkable gene expression change in the spleen and colon of PD mice, but CA pretreatment blocked these changes and restored some gene expression to normal levels, suggesting a favorable modulation of CA on the spleen and colon in PD. Notably, these genes might have a direct correlation with the self-function and self-homeostasis of organs and might be the potential targets for the regulation of CA on the spleen and colon in PD.

GO terms enriched with splenic and colonic DEGs were mainly involved in biological processes and cellular components, such as cellular responses to dexamethasone and insulin stimulus, and the responses to dexamethasone, ketone, insulin, glucocorticoid and corticosteroid, and the immunoglobulin production and negative regulation of cytokine biosynthetic process for the spleen, as well as the processes of fatty acid metabolism, organic acid biosynthesis, lipid catabolism and drug catabolism, cellular response to xenobiotic stimulus, and the activity of monooxygenase and oxidoreductase for colon. The KEGG analysis indicated peroxisome proliferator-activated receptors (PPAR) signal pathway for the spleen, and retinol metabolism, steroid hormone biosynthesis, chemical carcinogenesis, bile secretion and serotonergic synapse pathway for the colon. Moreover, GSEA identified some biological functions, including epithelial mesenchymal transition, apoptosis, KRAS-signaling, notch signaling, inflammatory response and TGF-β signaling, for the spleen, and fatty acid metabolism, IFN-γ response, oxidative phosphorylation, bile acid metabolism, peroxisome and xenobiotic metabolism for the colon. Thus, it is suggested that CA might influence the self-function and self-homeostasis of the spleen and colon to impact the neurological function in PD via these molecular biological processes and signaling pathways.

In this work, we observed that ERBB receptor feedback inhibitor 1 (ERRfi1), dual specificity phosphatase 1(Dusp1) and Phosphoenolpyruvate Carboxykinase (Pck1) were up regulated in the spleen. ERRfi1, also known as MIG-6 and RALT, is an indirect regulator for immunoreactions, which could drive epidermal growth factor receptors (EGFR) internalized and degraded due to its selective inhibition on EGFR [27,28], and multiple immune-related signal pathways can be activated by EGFR, including PI3K/AKT, STAT-1/STAT-3 and Raf/Ras/MEK/ERK [29]. Moreover, Dusp1 is described as an element of innate immunity to suppress the MAPK/ERK signals, which is essential for restraining the activation of JNK and p38 pathways to trigger anti-inflammatory responses [30,31]. Consistently, the results in this study showed that CA inhibited the IFN-γ-mediated inflammation in the spleen possibly via the related immunological pathways by upregulating ERRfi1 and Dusp1. In addition, PCK1 is a metabolic enzyme that is indispensable for gluconeogenesis [32,33], which is consistent with our results that the splenic PCK1 was upregulated and gluconeogenesis-related pathways, including PPAR signaling, are activated by CA. We inferred that CA exerted a neuroprotective effect on PD involving the brain–spleen axis by regulating these genes.

In PD, the imbalance of the microbiome–gut–brain axis plays a pivotal role [11], and gut microbial dysbiosis could conduce to the changes of metabolites, such as SCFAs, which might be one of the major mediators of neuroinflammation and gut inflammation [12,20]. In our previous work, we observed that CA showed a modulatory effect on gut microbiota and reduced gut inflammation [34], indicating that the gut is a crucial participant in PD. In the current study, transcriptome analysis was performed to examine the impact of CA on the gut in PD mice. PI3k serves as a pivotal part of the central pathway bearing cell proliferation, metabolism and cell survival, whose results alternate in the autophagy disruption in neurodegenerative diseases, including Alzheimer's disease and PD [35]. In our study, gene PI3k of the colon was reversed to nearly to the normal level by CA, indicating that the regulation of CA on gene PI3k plays a key role in the gut function in PD. Furthermore, we found that the colonic dominant DEGs induced by CA in PD mice were mainly from cytochrome P450 (CYP) enzymes, and the inhibition and induction of CYP450 are the central mechanisms for pharmacokinetic interactions [36]. CYP450 pathway in vascular diseases provided anti-inflammatory oxylipins to prevent the inflammatory process [37]. We thus inferred that the anti-neuroinflammation of CA might be heavily involved in the inhibition of CA on gut inflammation via the regulation of CA on colonic CYPs. Notably, multiple pathways were activated by CA in the colon of PD mice, suggesting a modulatory role of CA on the gut–brain axis. For instance, retinol deficiency may lead to gut microbiota disorder that is proposed as a mediator of PD [38], and CA activated this pathway in PD via the regulation of related genes. In addition, steroid hormone is the initiator and maintainer for sexual differentiation and reproduction responsible for water and salt balance, metabolism and stress response [39]; in the current study, CA retained steroid hormone biosynthesis in the colon by regulating some gene expressions, including certain CYPs, which are the functional enzymes for biosynthesis of all steroid hormones. Moreover, the PPAR pathway, ligand-activated transcription factors, is capable of controlling energy homeostasis, its activation can boost fatty acid metabolism and enhance the glucose metabolism resulting from insulin sensitization [37,40]. Intriguingly, it was observed that the PPAR pathway was activated in both the spleen and colon in PD mice, suggesting that the CA-mediated PPAR pathway in the spleen and colon may play a key role in preventing the neurotoxicity of PD.

4. Materials and Methods

4.1. Materials

CA (≥98%) was purchased from Yiyan Bio-technology (Shanghai, China); its chemical structure is shown in Figure 1a. MPTP (M0896) was purchased from Sigma (Missouri, MO, USA). The primary antibodies used were mouse anti-TH (MAB318, Millipore, Burlington, MA, USA), rabbit anti- BDNF (ab108319, Abcam, UK), rabbit anti-GFAP (#80788, Cell Signaling, Beverly, MA, USA), rabbit anti-Iba-1 (#17198, Cell Signaling, Beverly, MA, USA), rabbit anti-CD11b (#17800, Cell Signaling, Beverly, MA, USA), mouse anti-β-actin (66009-1-Ig, Proteintech, Chicago, IL, USA) and rabbit anti-GAPDH (10494-1-AP, Proteintech, Chicago, IL, USA). The secondary antibodies used were FITC-conjugated goat anti-mouse IgG (A0568, Beyotime, Shanghai, China) and CY3-conjugated goat anti-rabbit IgG (A0516, Beyotime, Shanghai, China); goat anti-rabbit IgG (15015, Proteintech, Chicago, IL, USA) and goat anti-mouse IgG (15014, Proteintech, Chicago, IL, USA) conjugated to horseradish peroxidase (HRP). The materials for RNA-sequence are indicated in methods below.

4.2. Animals and Treatment

The male C57BL/6 mice used in this study were obtained from Beijing Weitonglihua Laboratory Animal Technology Institute (Shanghai, China). After acclimation under standard conditions at 20–22 °C on a 12–12 h light/dark cycle with food and water provided ad libitum, the mice (6–8 weeks old, 18 g ± 2 g) were randomly assigned to the control group, MPTP group and CA+MPTP group (n = 10 per group). For the CA treatment, mice were pretreated with CA (40 mg/kg) intragastrically daily for 12 days. For the MPTP injection, 2 h after CA administration, the mice received daily MPTP (30 mg/kg) injections intraperitoneally on the 8th day for 5 days. The mice in the control group were given the same volume of normal saline. After the last injection of MPTP, no further treatments were given to the animals for 7 days. The behavioral test was performed after 7 treatment-free days. Subsequently, the mice were sacrificed after being anesthetized with isoflurane (Yuyan, Shanghai, China); the serum, striatum, spleen and colon were harvested immediately, and the whole brain was obtained from the mice perfused transcardially with 0.01 M of phosphate-buffered saline (PBS) followed by 4% paraformaldehyde (PFA). This experimental timeline is illustrated in Figure 1b. All experimental procedures were approved by the Animal Ethics Committee of Jiangnan University (Wuxi, China).

4.3. Behavioral Tests

On the 5th day of the 7-day treatment-free period, the mice underwent behavioral training once per day for 3 days, and the formal behavioral experiments were conducted the next day. The behavioral test was carried out as previously described [41,42].

Pole test: a straight wooden pole with a diameter of 1 cm, a height of 50 and a spherical protuberance (diameter 2 cm) on the top was entangled with non-adhesive gauze. The mice were placed head-down on the top of the pole fixed in the home cage vertically, the total time of downward climbing from the top to the bottom of the pole was recorded for each animal. The test for each mouse was conducted three times at 10 min intervals, and the average time was obtained for the subsequent statistical analysis.

Traction test: the mice were placed on a straight and horizontal rope (diameter 5 mm) with the fore limbs gripping the rope; the placement of limbs was observed for 10 s and scored from 1 to 4. The scores were evaluated as: 1 for the mice gripping the rope with one front paw, 2 for the mice gripping the rope with both front paws, 3 for the mice gripping the rope with both front paws and 1 hind paw, and 4 for the mice gripping the rope with both front paws and both hind paws. The test for each mouse was repeated three times and the average score was obtained for statistical analyses.

4.4. Measurement of Neurotransmitters

Striatal neurotransmitters DA and 5-HT, and their metabolites including DOPAC, HVA and 5-HIAA were measured by HPLC (Waters 2475, Milford, MA, USA) as described previously (Ahmed et al., 2014; Rangasamy et al., 2019). The HPLC was equipped with a fluorescence detector and an Atlantis T3 column (150 mm × 4.6 mm, 5 µm, Waters); gradient elution was performed with the mobile phases composed of water, acetonitrile and 0.01 M PBS (pH 4.0). Striatum from half of the brain was homogenized in 0.1 M perchloric acid, and the resulting homogenates were centrifuged at 13,000× *g*, 4 °C for 15 min to collect the supernatant. A total of 25 µL supernatant of each sample were injected into the column. The standard solution was freshly prepared with DA, 5-HT and their metabolites DOPAC, HVA, 5-HIAA. Then, standard peaks were generated to calculate the concentrations of each neurotransmitter in each sample. Data were presented in ng/mL.

4.5. Immunofluorescence

The whole brain for frozen sectioning was post-fixed in 4% PFA at 4 °C for 24 h, then dehydrated in 20% and 30% sucrose successively at 4 °C based on the subsidence of the brain in the solution. Next, the dehydrated brain was embedded in an optimal cutting temperature compound (SAKURA, Torrance, CA, USA) for frozen sectioning. Each mouse brain was cryosectioned at 10µm by cryostat microtome (Leica CM1950, Wetzlar Germany) and the sections containing major part of substantia nigra pars compacta from bregma −2.92 mm to −3.52 mm were used for immunofluorescent staining.

Brain slices were subjected to antigen retrieval by being immersed in 0.01 M sodium citrate buffer (pH 6.0) (Solarbio, Beijing, China), and blocked with 5% goat serum (*v*/*v*) (Beyotime, Shanghai, China) diluted with PBS containing 0.2% Triton X-100 (*v*/*v*) (Beyotime, Shanghai, China) for 1 h at 37 °C. The slices were then incubated overnight at 4 °C with the primary antibodies against TH (1:1000), GFAP (1:2000) or Iba-1 (1:1000), and then incubated with the secondary antibodies conjugated to the FITC or Cy3 (1:1000) in the dark for 1 h at 37 °C. Finally, the brain slices were mounted with a mounting medium containing 4′,6-diamidino-2-phenylindole (DAPI) and imaged with a fluorescence microscope (ZEISS AXIO IMAGER, Oberkochen, Germany). The positive cells for TH, GFAP and Iba-1 were counted by a blind experimenter using Image J software (National Institutes of Health, Bethesda, MA, USA). A total of 6 brain sections containing major part of SN were quantified for each animal, and 5 animals in each group.

4.6. Western Blot Analysis

Striatum tissues were homogenized in ice-cold RIPA buffer (Beyotime, China) with an addition of a protease and phosphatase inhibitor cocktail (Solarbio, China), then centrifuged at 13,000 rpm, 4 °C for 10 min to collect supernatant. The concentration of proteins was determined using a BCA kit (BioSharp, Shanghai, China) following the manufacturer's instructions. The supernatant samples were boiled at 95 °C for 10 min with loading buffer. A total of 30µg of total proteins from each sample were separated by electrophoresis in 10% sodium dodecyl sulfate-polyacrylamide gel electrophoresis (SDS-PAGE), then transferred onto polyvinylidene fluoride (PVDF) membranes (Millipore, Burlington, MA, USA). After being blocked in 5% BSA at room temperature for 2 h, the membranes were probed overnight at 4 °C with primary antibodies against TH (1:1000), BDNF (1:1000), CD11b (1:1000), GFAP (1:1000), Iba-1 (1:500), GAPDH (1:4000) and β-actin (1:2000), then incubated at room temperature for 2 h with HRP-conjugated secondary antibodies: goat anti-rabbit IgG (1:8000) and goat anti-mouse IgG (1:4000). Finally, the protein bands were exposed to be visualized using the super enhanced chemiluminescent (ECL; Millipore, Burlington, MA, USA) and imaged by a Gel Image System (Bio-Rad, Hercules, CA, USA). The protein content was normalized to GAPDH or β-Actin, and the grey value was analyzed by Image J software.

4.7. Enzyme-Linked Immunosorbent Assay (ELISA)

The content of IL-17, IFN-γ and TGF-β in the serum, striatum, spleen and colon was detected by a commercial ELISA kit (Mouse IL-17 Kit, Mouse IFN-γ Kit and TGF-β) (Nanjing Senbeijia Bioengineering Institute, China). All experimental procedures were conducted according to the manufacturers' protocol. IL-17, IFN-γ and TGF-β concentration were expressed in ng/L protein.

4.8. qPCR for Cytokines IL-17, IFN-γ and TGF-β

Total RNA was extracted from the striatum, spleen and colon using the Trizol reagent (Invitrogen, Waltham, MA, USA) following the manufacturer's protocol. Then, RNA samples were reverse transcribed into complementary DNA (cDNA) using the PrimeScript RT Master Mix reverse transcription kit (TaKaRa, Shiga, Japan). The mRNA expression was quantified by qPCR with SYBR Green Master Mix (Roche, Mannheim, Germany) and CFX96TM Real-Time System (Bio-Rad, Hercules, CA, USA). Relative gene expression was calculated by the ΔΔCt method, and Ct value was normalized to β-actin.

4.9. The Isolation of Total RNA

RNAs samples were isolated from the spleen and colon tissue using TRIzol Reagent (Invitrogen, Waltham, MA, USA) following the manufacturer's protocol and treated with DNase using TURBO DNA-free kit (AM1907, Thermo Fisher Scientific (Waltham, MA, USA)). The quality and quantity of total RNAs were determined by the Agilent 2100 Bioanalyzer system. RNA samples whose RNA integrity numbers were greater than 8.0 were submitted to the following library preparation.

4.10. cDNA Library Construction and Sequencing

The transcriptome library was constructed using NEBNext® Ultra™ RNA Library Prep Kit (NEB, Ipswich, MA, USA) according to the manufacturer's recommendations. The oligo(dT)-enriched mRNA was fragmented, and the interrupted RNA fragments were converted into the first cDNA strand using random hexamer primers and reverse transcriptase, followed by the synthesis of the second cDNA strand using DNA Polymerase I and RNase H. Subsequently, the double-stranded cDNAs were end-repaired and ligated to adaptors. Next, the size-selected cDNA was enriched by polymerase chain reaction (PCR), and the PCR products were purified by AMPure XP system to establish the final transcriptome library. After being assessed for quality by the Agilent Bioanalyzer 2100 system, the library was sequenced on an Illumina Novaseq platform (Illumina, San Diego, CA, USA). Each group consists of three biological replicates each with three technical replicates. The gene expression data were submitted to the National Center of Biotechnology Information (NCBI) Sequence Read Archive (SRA) repository (https://submit.ncbi.nlm.nih.gov/about/sra/, accessed on 20 October 2021) under accession numbers SRX12648145, SRX12648146, SRX12648147, SRX12639649, SRX12639650, SRX12639651, SRX12648142, SRX12648143, SRX12648144, SRX12639641, SRX12639642, SRX12639645.

4.11. Sequencing Data Analysis

DEGs were measured by the DESeq2 R package (1.16.1) with $|\log_2\text{FoldChange}|$ ($|\log_2\text{FC}|$) > 1.2 and adjusted p-value (padj) < 0.05. The functional annotation of DEGs was performed with GO and pathway enrichment assessment; the functions enrichGO and enrichKEGG were analyzed using clusterProfiler package R under the R programing environment. GSEA was performed using the TCGA dataset with $|\text{NES}| > 1.0$ and nominal p-value < 0.05.

4.12. Statistical Analysis

Data are expressed as the mean ± SEM. Statistical analyses were conducted using GraphPad Prism Software 8.2.1. Differences were analyzed by one-way analysis of variance

(ANOVA) with Tukey's post hoc test for multiple groups comparisons. A *p*-value of <0.05 was set as the threshold for significance.

5. Conclusions

The present findings demonstrated that CA could prevent neuroinflammation and neurodegeneration via motor deficits, DA neuronal survivals in SN and glial reactions, along with the increased BDNF, DA and 5-HT of PD mice. The underlying mechanism might be associated with the modulatory role of CA on the immunological response and gene expression of the spleen and colon. This study provides the supportive evidence that the effects of CA on the spleen and colon may play a neuroprotective role in PD. However, further studies are needed to elucidate how a dysfunctional spleen and colon affect the brain–spleen and brain–gut axes in the pathology and development of PD.

Author Contributions: Conceptualization, N.W., B.F. and Y.C.; methodology, N.W. and R.L.; software, N.W. and R.L.; formal analysis, N.W.; investigation, N.W.; writing—original draft preparation, N.W.; writing—review and editing, Y.G.; supervision, H.Q. All authors have read and agreed to the published version of the manuscript.

Funding: This research received no external funding.

Institutional Review Board Statement: The animal study protocol was approved by the Animal Ethics Committee of Jiangnan University (Wuxi, China) (JN.No20200430c1081220[016], 30 April 2020).

Informed Consent Statement: Not applicable.

Data Availability Statement: All sequence data in this study are available in the National Center of Biotechnology Information (https://www.ncbi.nlm.nih.gov/bioproject/, accessed on 20 October 2021) under Accession No. PRJNA771869 and PRJNA771189.

Acknowledgments: This work was supported by the National Key Research and Development Program of China (No. 2017YFC1601704).

Conflicts of Interest: The authors declare no conflict of interest.

References

1. Olanow, C.W.; Tatton, W.G. Etiology and pathogenesis of Parkinson's disease. *Annu. Rev. Neurosci.* **1999**, *22*, 123–144. [CrossRef] [PubMed]
2. Kish, S.J.; Shannak, K.; Hornykiewicz, O. Uneven pattern of dopamine loss in the striatum of patients with idiopathic Parkinson's disease. Pathophysiologic and clinical implications. *N. Engl. J. Med.* **1988**, *318*, 876–880. [CrossRef] [PubMed]
3. Forloni, G.; La Vitola, P.; Cerovic, M.; Balducci, C. Inflammation and Parkinson's disease pathogenesis: Mechanisms and therapeutic insight. *Prog. Mol. Biol. Transl. Sci.* **2021**, *177*, 175–202. [CrossRef]
4. Mok, S.W.; Wong, V.K.; Lo, H.H.; de Seabra Rodrigues Dias, I.R.; Leung, E.L.; Law, B.Y.; Liu, L. Natural products-based polypharmacological modulation of the peripheral immune system for the treatment of neuropsychiatric disorders. *Pharmacol. Ther.* **2020**, *208*, 107480. [CrossRef] [PubMed]
5. Benkler, M.; Agmon-Levin, N.; Hassin-Baer, S.; Cohen, O.S.; Ortega-Hernandez, O.D.; Levy, A.; Moscavitch, S.D.; Szyper-Kravitz, M.; Damianovich, M.; Blank, M.; et al. Immunology, autoimmunity, and autoantibodies in Parkinson's disease. *Clin. Rev. Allergy Immunol.* **2012**, *42*, 164–171. [CrossRef] [PubMed]
6. Dantzer, R.; O'Connor, J.C.; Freund, G.G.; Johnson, R.W.; Kelley, K.W. From inflammation to sickness and depression: When the immune system subjugates the brain. *Nat. Rev. Neurosci.* **2008**, *9*, 46–57. [CrossRef]
7. Yan, A.; Zhang, Y.; Lin, J.; Song, L.; Wang, X.; Liu, Z. Partial depletion of peripheral M1 macrophages reverses motor deficits in MPTP-treated mouse by suppressing neuroinflammation and dopaminergic neurodegeneration. *Front. Aging Neurosci.* **2018**, *10*, 160. [CrossRef]
8. Olson, K.E.; Namminga, K.L.; Schwab, A.D.; Thurston, M.J.; Lu, Y.; Woods, A.; Lei, L.; Shen, W.; Wang, F.; Joseph, S.B.; et al. Neuroprotective activities of long-acting granulocyte-macrophage colony-stimulating factor (mPDM608) in 1-Methyl-4-Phenyl-1,2,3,6-Tetrahydropyridine-Intoxicated Mice. *Neurotherapeutics* **2020**, *17*, 1861–1877. [CrossRef]
9. Dutta, D.; Kundu, M.; Mondal, S.; Roy, A.; Ruehl, S.; Hall, D.A.; Pahan, K. RANTES-induced invasion of Th17 cells into substantia nigra potentiates dopaminergic cell loss in MPTP mouse model of Parkinson's disease. *Neurobiol. Dis.* **2019**, *132*, 104575. [CrossRef]
10. Chung, E.S.; Kim, H.; Lee, G.; Park, S.; Kim, H.; Bae, H. Neuro-protective effects of bee venom by suppression of neuroinflammatory responses in a mouse model of Parkinson's disease: Role of regulatory T cells. *Brain Behav. Immun.* **2012**, *26*, 1322–1330. [CrossRef]

11. Baizabal-Carvallo, J.F.; Alonso-Juarez, M. The link between gut dysbiosis and neuroinflammation in Parkinson's Disease. *Neuroscience* **2020**, *432*, 160–173. [CrossRef] [PubMed]
12. Sampson, T.R.; Debelius, J.W.; Thron, T.; Janssen, S.; Shastri, G.G.; Ilhan, Z.E.; Challis, C.; Schretter, C.E.; Rocha, S.; Gradinaru, V.; et al. Gut microbiota regulate motor deficits and neuroinflammation in a model of Parkinson's Disease. *Cell* **2016**, *167*, 1469–1480. [CrossRef] [PubMed]
13. Salat-Foix, D.; Tran, K.; Ranawaya, R.; Meddings, J.; Suchowersky, O. Increased intestinal permeability and Parkinson Disease patients: Chicken or Egg? *Can. J. Neurol. Sci.* **2012**, *39*, 185–188. [CrossRef] [PubMed]
14. Phillips, R.J.; Walter, G.C.; Wilder, S.L.; Baronowsky, E.A.; Powley, T.L. Alpha-synuclein-immunopositive myenteric neurons and vagal preganglionic terminals: Autonomic pathway implicated in Parkinson's disease? *Neuroscience* **2008**, *153*, 733–750. [CrossRef] [PubMed]
15. Peng, Y.; Sun, Q.C.; Park, Y. The bioactive effects of chicoric acid as a functional food ingredient. *J. Med. Food* **2019**, *22*, 645–652. [CrossRef] [PubMed]
16. Liu, Q.; Chen, Y.; Shen, C.; Xiao, Y.; Wang, Y.; Liu, Z.; Liu, X. Chicoric acid supplementation prevents systemic inflammation-induced memory impairment and amyloidogenesis via inhibition of NF-kappaB. *FASEB J.* **2017**, *31*, 1494–1507. [CrossRef]
17. Lee, N.Y.; Chung, K.S.; Jin, J.S.; Bang, K.S.; Eom, Y.J.; Hong, C.H.; Nugroho, A.; Park, H.J.; An, H.J. Effect of chicoric acid on mast cell-mediated allergic inflammation In Vitro and In Vivo. *J. Nat. Prod.* **2015**, *78*, 2956–2962. [CrossRef]
18. Kour, K.; Bani, S. Augmentation of immune response by chicoric acid through the modulation of CD28/CTLA-4 and Th1 pathway in chronically stressed mice. *Neuropharmacology* **2011**, *60*, 852–860. [CrossRef]
19. Dalile, B.; Van Oudenhove, L.; Vervliet, B.; Verbeke, K. The role of short-chain fatty acids in microbiota-gut-brain communication. *Nat. Rev. Gastroenterol. Hepatol.* **2019**, *16*, 461–478. [CrossRef]
20. Devos, D.; Lebouvier, T.; Lardeux, B.; Biraud, M.; Rouaud, T.; Pouclet, H.; Coron, E.; des Varannes, S.B.; Naveilhan, P.; Nguyen, J.M.; et al. Colonic inflammation in Parkinson's disease. *Neurobiol. Dis.* **2013**, *50*, 42–48. [CrossRef]
21. Yang, H.C.; Won, E.J.; Kim, M.J.; Sung, C.M.; Rhee, J.H.; Nam, K.I. Intralymphatic administration of metagonimus yokogawai-extracted protein attenuates experimental murine allergic rhinitis model. *Int. Arch. Allergy Immunol.* **2021**, *182*, 381–387. [CrossRef] [PubMed]
22. Huang, Y.; Liu, Z.; Cao, B.B.; Qiu, Y.H.; Peng, Y.P. Treg cells attenuate neuroinflammation and protect neurons in a mouse model of Parkinson's Disease. *J. Neuroimmune Pharm.* **2020**, *15*, 224–237. [CrossRef] [PubMed]
23. Liu, Z.; Chen, H.Q.; Huang, Y.; Qiu, Y.H.; Peng, Y.P. Transforming growth factor-beta 1 acts via T beta R-I on microglia to protect against MPP+-induced dopaminergic neuronal loss. *Brain Behav. Immun.* **2016**, *51*, 131–143. [CrossRef] [PubMed]
24. Liesz, A.; Suri-Payer, E.; Veltkamp, C.; Doerr, H.; Sommer, C.; Rivest, S.; Giese, T.; Veltkamp, R. Regulatory T cells are key cerebroprotective immunomodulators in acute experimental stroke. *Nat. Med.* **2009**, *15*, 192–199. [CrossRef] [PubMed]
25. Reynolds, A.D.; Banerjee, R.; Liu, J.N.; Gendelman, H.E.; Mosley, R.L. Neuroprotective activities of CD4+CD25+ regulatory T cells in an animal model of Parkinson's disease. *J. Leukocyte. Biol.* **2007**, *82*, 1083–1094. [CrossRef]
26. Zielinska, D.; Zielinski, H.; Laparra-Llopis, J.M.; Szawara-Nowak, D.; Honke, J.; Gimenez-Bastida, J.A. Caffeic acid modulates processes associated with intestinal inflammation. *Nutrients* **2021**, *13*, 554. [CrossRef]
27. Park, E.; Kim, N.; Ficarro, S.B.; Zhang, Y.; Lee, B.I.; Cho, A.; Kim, K.; Park, A.K.J.; Park, W.Y.; Murray, B.; et al. Structure and mechanism of activity-based inhibition of the EGF receptor by Mig6. *Nat. Struct. Mol. Biol.* **2015**, *22*, 703–711. [CrossRef]
28. Frosi, Y.; Anastasi, S.; Ballaro, C.; Varsano, G.; Castellani, L.; Maspero, E.; Polo, S.; Alema, S.; Segatto, O. A two-tiered mechanism of EGFR inhibition by RALT/MIG6 via kinase suppression and receptor degradation. *J. Cell Biol.* **2010**, *189*, 557–571. [CrossRef]
29. Chen, J.; Zeng, F.; Forrester, S.J.; Eguchi, S.; Zhang, M.Z.; Harris, R.C. Expression and function of the epidermal growth factor receptor in physiology and disease. *Physiol. Rev.* **2016**, *96*, 1025–1069. [CrossRef]
30. Wancket, L.M.; Frazier, W.J.; Liu, Y. Mitogen-activated protein kinase phosphatase (MKP)-1 in immunology, physiology, and disease. *Life Sci.* **2012**, *90*, 237–248. [CrossRef]
31. Abraham, S.M.; Clark, A.R. Dual-specificity phosphatase 1: A critical regulator of innate immune responses. *Biochem. Soc. Trans.* **2006**, *34*, 1018–1023. [CrossRef] [PubMed]
32. Geng, X.; Shen, J.; Li, F.; Yip, J.; Guan, L.; Rajah, G.; Peng, C.; DeGracia, D.; Ding, Y. Phosphoenolpyruvate carboxykinase (PCK) in the brain gluconeogenic pathway contributes to oxidative and lactic injury after stroke. *Mol. Neurobiol.* **2021**, *58*, 2309–2321. [CrossRef] [PubMed]
33. Ko, C.W.; Counihan, D.; Wu, J.; Hatzoglou, M.; Puchowicz, M.A.; Croniger, C.M. Macrophages with a deletion of the phosphoenolpyruvate carboxykinase 1 (Pck1) gene have a more proinflammatory phenotype. *J. Biol. Chem.* **2018**, *293*, 3399–3409. [CrossRef] [PubMed]
34. Wang, N.; Feng, B.; Hu, B.; Cheng, Y.; Guo, Y.; He, Q. Neuroprotection of chicoric acid in a mouse model of Parkinson's disease involves gut microbiota and TLR4 signaling pathway. *Food Funct.* 2022; *in press*. [CrossRef] [PubMed]
35. Brunet, A.; Datta, S.R.; Greenberg, M.E. Transcription-dependent and -independent control of neuronal survival by the PI3K-Akt signaling pathway. *Curr. Opin. Neurobiol.* **2001**, *11*, 297–305. [CrossRef]
36. Hakkola, J.; Hukkanen, J.; Turpeinen, M.; Pelkonen, O. Inhibition and induction of CYP enzymes in humans: An update. *Arch. Toxicol.* **2020**, *94*, 3671–3722. [CrossRef]
37. Wagner, N.; Wagner, K.D. The role of PPARs in Disease. *Cells* **2020**, *9*, 2367. [CrossRef]

38. Marie, A.; Darricau, M.; Touyarot, K.; Parr-Brownlie, L.C.; Bosch-Bouju, C. Role and mechanism of vitamin a metabolism in the pathophysiology of Parkinson's Disease. *J. Parkinsons Dis.* **2021**, *11*, 949–970. [CrossRef]
39. Schiffer, L.; Barnard, L.; Baranowski, E.S.; Gilligan, L.C.; Taylor, A.E.; Arlt, W.; Shackleton, C.H.L.; Storbeck, K.H. Human steroid biosynthesis, metabolism and excretion are differentially reflected by serum and urine steroid metabolomes: A comprehensive review. *J. Steroid. Biochem. Mol. Biol.* **2019**, *194*, 105439. [CrossRef]
40. Han, L.; Shen, W.J.; Bittner, S.; Kraemer, F.B.; Azhar, S. PPARs: Regulators of metabolism and as therapeutic targets in cardiovascular disease. Part I: PPAR-alpha. *Future Cardiol.* **2017**, *13*, 259–278. [CrossRef]
41. Cao, Q.; Qin, L.; Huang, F.; Wang, X.; Yang, L.; Shi, H.; Wu, H.; Zhang, B.; Chen, Z.; Wu, X. Amentoflavone protects dopaminergic neurons in MPTP-induced Parkinson's disease model mice through PI3K/Akt and ERK signaling pathways. *Toxicol Appl. Pharmacol.* **2017**, *319*, 80–90. [CrossRef] [PubMed]
42. Kuribara, H.; Higuchi, Y.; Tadokoro, S. Effects of central depressants on rota-rod and traction performances in mice. *Jpn. J. Pharmacol.* **1977**, *27*, 117–126. [CrossRef] [PubMed]

International Journal of
Molecular Sciences

MDPI

Article

The Anxiolytic Drug Buspirone Prevents Rotenone-Induced Toxicity in a Mouse Model of Parkinson's Disease

Sarah Thomas Broome and Alessandro Castorina *

Laboratory of Cellular and Molecular Neuroscience (LCMN), School of Life Sciences, Faculty of Science, University of Technology Sydney, Sydney, NSW 2007, Australia; sarah.j.thomasbroome@student.uts.edu.au
* Correspondence: Alessandro.Castorina@uts.edu.au

Abstract: A pharmacological and genetic blockade of the dopamine D3 receptor (D3R) has shown to be neuroprotective in models of Parkinson's disease (PD). The anxiolytic drug buspirone, a serotonin receptor 1A agonist, also functions as a potent D3R antagonist. To test if buspirone elicited neuroprotective activities, C57BL/6 mice were subjected to rotenone treatment (10mg/kg i.p for 21 days) to induce PD-like pathology and were co-treated with increasing dosages of buspirone (1, 3, or 10 mg/kg i.p.) to determine if the drug could prevent rotenone-induced damage to the central nervous system (CNS). We found that high dosages of buspirone prevented the behavioural deficits caused by rotenone in the open field test. Molecular and histological analyses confirmed that 10 mg/kg of buspirone prevented the degeneration of TH-positive neurons. Buspirone attenuated the induction of interleukin-1β and interleukin-6 expression by rotenone, and this was paralleled by the upregulation of arginase-1, brain-derived neurotrophic factor (BDNF), and activity-dependent neuroprotective protein (ADNP) in the midbrain, striatum, prefrontal cortex, amygdala, and hippocampus. Buspirone treatment also improved mitochondrial function and antioxidant activities. Lastly, the drug prevented the disruptions in the expression of two neuroprotective peptides, pituitary adenylate cyclase-activating polypeptide (PACAP) and vasoactive intestinal peptide (VIP). These results pinpoint the neuroprotective efficacy of buspirone against rotenone toxicity, suggesting its potential use as a therapeutic agent in neurodegenerative and neuroinflammatory diseases, such as PD.

Keywords: Parkinson's disease; neurodegeneration; rotenone; neuroinflammation; dopamine; dopamine-D3-receptor; buspirone

Citation: Thomas Broome, S.; Castorina, A. The Anxiolytic Drug Buspirone Prevents Rotenone-Induced Toxicity in a Mouse Model of Parkinson's Disease. *Int. J. Mol. Sci.* **2022**, *23*, 1845. https://doi.org/10.3390/ijms23031845

Academic Editors: Marcello Ciaccio and Luisa Agnello

Received: 11 January 2022
Accepted: 2 February 2022
Published: 6 February 2022

1. Introduction

Parkinson's disease (PD) is the most common neurodegenerative movement disorder, characterised by the progressive loss of dopaminergic neurons in the *substantia nigra pars compacta* (SNpc) and resulting in a deficit of dopamine (DA) in the *striatum* [1]. PD is a multi-factorial disease that arises from a complex interplay between several environmental factors, a genetic predisposition, and defective cellular processes [2,3]. Many attempts were made in rodents to reproduce some of the pathological domains of PD. For example, the environmental toxin, rotenone, is a popular PD-mimetic due to its ability to reproduce the major clinical and behavioural features of PD in rodents, including motor deficits, dopaminergic degeneration, mitochondrial impairment, and neuroinflammation [4,5].

Neuroinflammation describes the local immune response within the central nervous system (CNS), and is predominantly driven by resident microglia [6,7]. It is a major contributor to PD pathology and is consistently linked to disease progression and clinical severity [8]. Furthermore, it was shown that neuroinflammation alone is sufficient to promote the death of dopaminergic neurons, as shown in a study involving an injection of the inflammatory mimetic, LPS, into the rodent brain, which resulted in increased levels of inflammatory mediators prior to the loss of dopaminergic neurons [8]. Adding complexity to PD pathogenesis is the discovery that DA itself can control inflammation [9].

The discovery of functional dopamine receptors expressed on the surface of multiple immune cell subtypes, including microglia [10], suggests that DA itself can modulate at least certain immune responses [9]. This is clearly seen in the ability of the dopamine D3 receptor (D3R) to promote inflammation. It was shown that D3R-signalling promotes disease progression by favouring neuroinflammation and promoting the pathogenic CD4+ T cell response associated with PD [11,12]. Additionally, microglial activation is repressed in D3R-deficient mice [11]. Most importantly, compounds able to block D3R reduced CNS inflammation and, consequently, slowed the progression of PD [11,13].

Computational and neuroimaging studies have revealed that the anxiolytic drug buspirone, a partial 5-hydroxytryptamine receptor (5-HT1Ar) agonist, has a strong "off-target" function as a D3R antagonist [14–16]. In recent work, we demonstrated that either the genetic deletion of D3R or buspirone treatment reliably attenuated LPS-triggered inflammation in BV2 microglia [17]. It is well accepted that blocking microglial-induced inflammation reduces neurodegeneration and slows disease progression [18,19].

The emerging evidence extending the pharmacological properties of buspirone, together with the neuroprotective effects seen in response to D3R blockade, prompted us to investigate if buspirone can protect from dopaminergic degeneration by attenuating neuroinflammation in a rotenone mouse model of PD. We focused on the ability of buspirone to protect against rotenone-induced behavioural deficits, dopaminergic degeneration, mitochondrial dysfunction, and inflammation. To the best of our knowledge, this is the first study to characterise the neurochemical changes triggered by buspirone as an anti-inflammatory agent in several CNS structures. Our results indicate that, at high dosages, buspirone is able to prevent rotenone-induced deficits in locomotor and exploratory behaviour, protect dopaminergic neurons from degeneration, and reduce the expression of inflammatory mediators whilst heightening the expression of neurotrophic factors and protective neuropeptides in several CNS regions. Collectively, these findings support the idea that buspirone protects the CNS against rotenone intoxication in a mouse model of PD.

2. Results

2.1. Buspirone Prevents Rotenone-Induced Deficits in Locomotor and Exploratory Behaviour

Buspirone is clinically used as an anxiolytic and has been previously utilised as a positive control in the assessment of the exploratory behaviour of C57BL/6 mice in the open field (OF) [20] (Supplementary Figure S1). Rotenone (10 mg/kg) induced significant deficits in the exploratory behaviour of mice, as determined by a reduction in the number of entries and the time spent in the centre quadrant of the OF (inner white box on representative heat maps; Figure 1A). Notably, at day 21 there was a significant reduction in the number of entries in the centre of the OF (* $p < 0.05$; Figure 1B), which correlated with a reduction in the time spent in the centre at both day 14 (* $p < 0.05$) and day 21 (* $p < 0.05$) (Figure 1C). As expected, buspirone was able to prevent rotenone-induced deficits in exploratory behaviour (Figure 1B–C).

Drug-induced effects on locomotor behaviour were also appraised by comparing the total distance travelled (Figure 1D) and the average speed (Figure 1E). Rotenone significantly impaired locomotion after only 14 days of treatment. There was a significant reduction in the total distance travelled at day 14 (* $p < 0.05$ vs. baseline) and day 21 (* $p < 0.05$) (Figure 1D), concurrent with a reduction in the average speed at both day 14 (* $p < 0.05$) and 21(* $p < 0.05$) (Figure 1E). In rotenone-treated mice, 1 mg/kg of buspirone was unable to prevent deficits in locomotor behaviour. In fact, mice still displayed a significant reduction in the total distance travelled at both day 14 (* $p < 0.05$ vs. baseline) and day 21 (* $p < 0.05$ vs. baseline) (Figure 1D), which correlated with a slower average speed at day 21 (*** $p < 0.001$ vs. baseline; Figure 1E), compared to baseline measurements. However, higher concentrations of buspirone (3 and 10 mg/kg) prevented these locomotor deficits (Figure 1D,E).

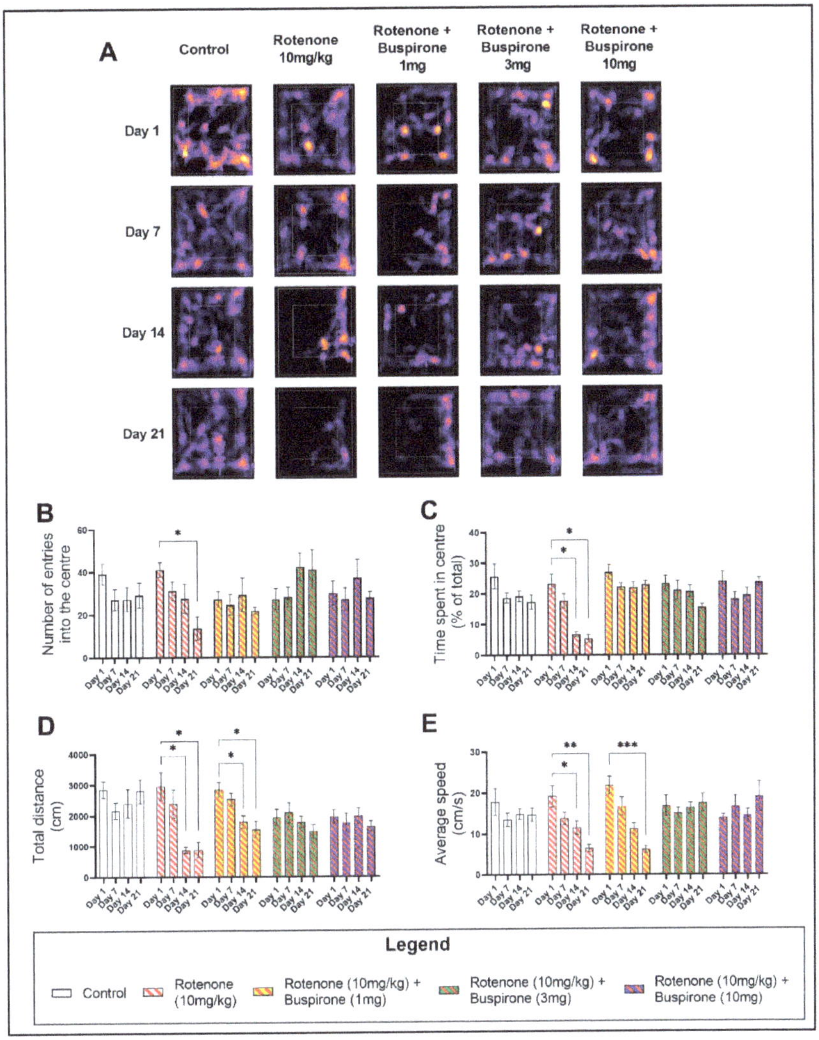

Figure 1. Buspirone prevents rotenone-induced locomotor and exploratory behavioural deficits. The open field (OF) test was used to assess the locomotor and exploratory behaviours of mice on a weekly basis for a total of 21 days. Representative heat maps generated from MouBeAt Software illustrate the locomotor pattern of mice during 5 min spent in the OF (**A**). Exploratory behaviour was determined by measuring the number of entries (**B**) and the total time (**C**) each mouse spent in the centre quadrant. Locomotor activity was assessed by comparing the total distance travelled (**D**) and average speed (**E**) of mice. Comparisons were made within the same treatment group compared to baseline measurements. Data shown represents 8–12 mice per group. * $p < 0.05$, ** $p < 0.01$, or *** $p < 0.001$ as determined by a one-way ANOVA followed by Dunnett's post-hoc test.

2.2. Buspirone Prevents the Loss of Dopaminergic Neurons in Rotenone-Treated Mice

To assess if buspirone protected dopaminergic neurons from the detrimental effects of rotenone, we analysed the expression two dopaminergic markers, tyrosine hydroxylase (TH) and dopamine transporter (DAT), in the midbrain and striatum, respectively (Figure 2). As expected, rotenone administration resulted in a significant reduction of TH mRNA (**** $p < 0.0001$ vs. vehicle; Figure 2A) and protein expression (**** $p < 0.0001$ vs. vehicle; Figure 2B,C) in the midbrain. Buspirone treatment upregulated the expression of TH transcripts in the midbrain, at all dosages, compared with rotenone-treated mice (**** $p < 0.0001$; Figure 2A). These results were associated with a dose-dependent increase in TH expression at the protein level, which was statistically significant at the highest dosage of buspirone (### $p < 0.001$ vs. rotenone-treated mice; Figure 2B). Immunohistochemical data corroborated these findings, as the significant reduction in the number of TH-positive cells in the SNpc (** $p < 0.01$ vs. vehicle) was prevented by buspirone in a dose-dependent manner, although results were significant only at the highest dosage tested (### $p < 0.001$, buspirone 10 mg/kg vs. rotenone-treated mice; Figure 2D,E).

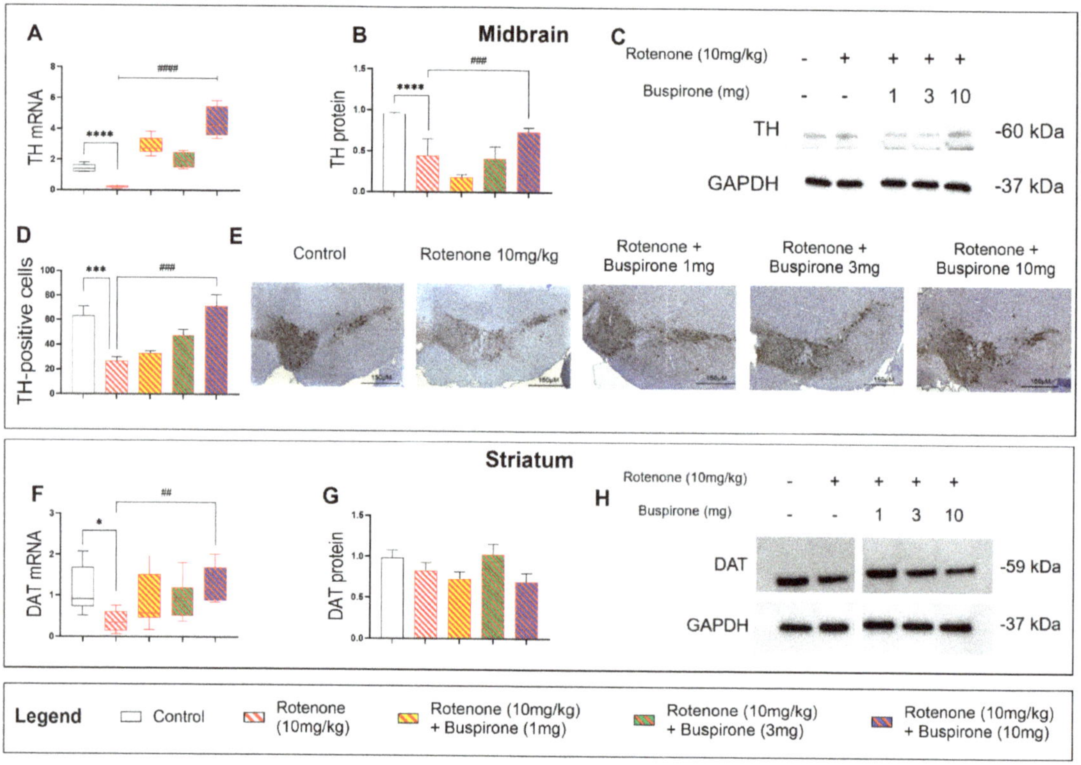

Figure 2. Buspirone at high dosages protects nigro-striatal dopaminergic neurons from degeneration. TH mRNA (**A**) and protein expression in the midbrain (**B**,**C**). Representative photomicrographs depicting TH immunoreactivity and semi-quantitative measurement of staining intensity in the *SNpc*, scale bar = 150 μM (**D**,**E**). Striatal DAT mRNA (**F**) and protein expression (**G**,**H**). Gene expression was measured by qRT-PCR and quantified using the ΔΔCt method after normalization to s18 (the housekeeping gene). qRT-PCR results are reported as mean fold changes with respect to vehicle-treated control mice. Protein expression was determined by Western blot and normalized to GAPDH (the loading control). Western blots are cropped and removed lanes depicting drug-treatment controls are included in Supplementary Figure S1. Densitometry results are presented as means ± S.E.M. TH-positive cells are reported as the percent of TH-positive cells normalized to the total number of nuclei ± S.E.M. Data represents 3–4 mice per group. * $p < 0.05$, *** $p < 0.001$, or **** $p < 0.0001$, ## $p < 0.01$, ### $p < 0.001$, or #### $p < 0.0001$, as determined by an ANOVA followed by Dunnett's post-hoc test. TH, tyrosine hydroxylase; DAT, dopamine transporter; GAPDH, glyceraldehyde 3-phosphate dehydrogenase; kDa, Kilodalton; s18, ribosomal protein s18 gene.

2.3. Buspirone Triggers Brain-Region-Specific Changes in the Expression of Glial Activation Markers

To determine if buspirone treatment prevented rotenone-induced activation of astrocytes and microglia, we measured the relative mRNA and protein expression of GFAP (astrocytic activation marker) and CD11b or Iba1 (microglial activation markers) in several brain regions (i.e., midbrain, striatum, prefrontal cortex, amygdala, and hippocampus). To our surprise, glial activation followed a region-specific pattern and was more pronounced in extra-nigral regions (Figure 3).

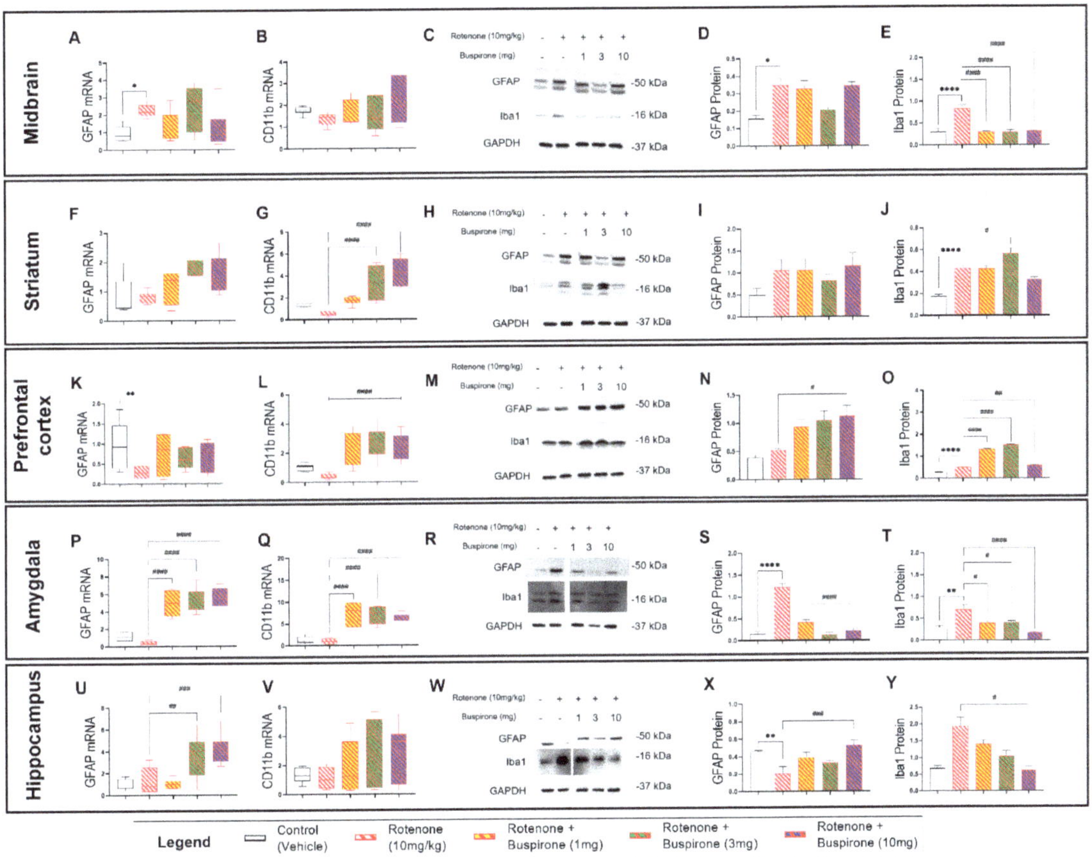

Figure 3. Buspirone triggers region-specific changes in the expression of glial activation markers. Real-time qPCR analyses of GFAP and CD11b in the midbrain (**A**,**B**), striatum (**F**,**G**), prefrontal cortex (**K**,**L**), amygdala (**P**,**Q**), and hippocampus (**U**,**V**) reported as mean fold changes calculated with the ΔΔCt method after normalization to s18 (the housekeeping gene). Western blot and densitometric analysis of GFAP and Iba1 in the midbrain (**C**–**E**), striatum (**H**–**J**), prefrontal cortex (**M**–**O**), amygdala (**R**–**T**), and hippocampus (**W**–**Y**). Western blots are cropped and removed lanes depicting drug-treatment controls are included in Supplementary Figure S3. Densitometric results are expressed as means ± S.E.M. All data represents 4–6 samples per treatment group. * $p < 0.05$, ** $p < 0.01$, or **** $p < 0.0001$, compared to vehicle-treated controls; or # $p < 0.05$, ## $p < 0.01$, ### $p < 0.001$, or #### $p < 0.0001$, compared to rotenone (10 mg/kg)-treated mice, as determined by an ANOVA followed by Sidak's post-hoc test. GFAP, Glial Fibrillary Acidic Protein; Iba1, ionized calcium binding adapter molecule 1; s18, ribosomal protein s18 gene; GAPDH, Glyceraldehyde 3-phosphate Dehydrogenase; kDa, Kilodalton.

In the midbrain, GFAP mRNA and protein expression were increased in rotenone-treated mice (* $p < 0.05$ vs. vehicle; Figure 3A,C,D); however, co-treatment with buspirone failed to prevent GFAP gene and protein induction at all the dosages tested ($p > 0.05$ vs. rotenone). Striatal GFAP transcripts and proteins were not significantly affected by rotenone exposure or buspirone co-treatment, although expression levels were slightly increased when compared with controls ($p > 0.05$ vs. vehicle; Figure 3F,H,I). In the prefrontal cortex, GFAP transcripts were significantly downregulated in response to rotenone administration (** $p < 0.01$ vs. vehicle; Figure 3K), but not in animals that were co-treated with buspirone, where levels were comparable to controls ($p > 0.05$ vs. vehicle). In contrast, GFAP protein levels were not diminished by rotenone and were significantly increased in animals that received 10 mg/kg of buspirone (# $p < 0.05$; Figure 3M,N). In the amygdala, GFAP mRNA expression was slightly reduced by rotenone exposure ($p > 0.05$ vs. vehicle). Surprisingly, buspirone co-treatment caused a strong increase in GFAP transcripts at all dosages (#### $p < 0.0001$ vs. rotenone; Figure 3P). Conversely, GFAP protein expression was remarkably increased in the amygdala or rotenone-intoxicated mice (**** $p < 0.0001$ vs. vehicle) and significantly reduced by buspirone co-administration (#### $p < 0.0001$ vs. rotenone; Figure 3R,S). Hippocampal GFAP transcripts were not affected by rotenone ($p > 0.05$ vs. vehicle); however, mRNA levels were significantly increased when animals were co-administered with the two highest dosages of buspirone (## $p < 0.01$ and ### $p < 0.0001$ vs. rotenone, respectively; Figure 3U). GFAP protein expression in the hippocampus did not match mRNA data. In fact, GFAP protein levels were significantly reduced upon rotenone treatment (** $p < 0.01$ vs. vehicle) and increased in response to the highest dosage of buspirone (### $p < 0.001$ vs. rotenone; Figure 3W,X).

Gene expression studies of the microglial marker CD11b across the different brain regions did not show any major effects of rotenone ($p > 0.05$ vs. vehicle; Figure 3B,G,L,Q,V), although some trends towards a reduction were seen in the striatum and prefrontal cortex. In contrast, buspirone co-treatment induced significant CD11b gene upregulation in the striatum (#### $p < 0.0001$ vs. rotenone at 3 and 10 mg/kg; Figure 3G), prefrontal cortex (#### $p < 0.0001$ vs. rotenone at 1, 3, and 10 mg/kg: Figure 3L), amygdala (#### $p < 0.0001$ vs. rotenone at 1, 3, and 10 mg/kg: Figure 3Q), and, to some extent, in the hippocampus, although not significantly ($p > 0.05$ vs. rotenone at 1, 3, and 10 mg/kg: Figure 3V).

Protein expression of the microglial marker Iba1 showed a robust induction in the CNS following rotenone exposure, and levels were in great part prevented upon buspirone co-administration (Figure 3E,J,O,T,Y). In more detail, rotenone significantly increased Iba1 protein expression in the midbrain (**** $p < 0.0001$ vs. vehicle; Figure 3C,E), striatum (**** $p < 0.0001$ vs. vehicle; Figure 3H,J), prefrontal cortex (**** $p < 0.0001$ vs. vehicle, Figure 3M,O), amygdala (** $p < 0.01$ vs. vehicle; Figure 3R,T), and, to a minor extent, in the hippocampus, although not significantly ($p > 0.05$ vs. vehicle; Figure 3W,Y).

In the midbrain, all dosages of buspirone prevented rotenone-induced Iba1 upregulations (#### $p < 0.0001$ vs. rotenone; Figure 3E). Within the striatum, Iba1 expression was not affected by either buspirone at 1 mg/kg or at 10 mg/kg; however, it was significantly increased at 3 mg/kg (# $p < 0.05$ vs. rotenone; Figure 3H,I). Buspirone also reduced Iba1 expression in the amygdala (# $p < 0.05$ at 1 and 3 mg/kg and #### $p < 0.0001$ at 10 mg/kg buspirone vs. rotenone; Figure 3R,T) and in the hippocampus, but only at the dosage of 10 mg/kg of buspirone (# $p < 0.05$ vs. rotenone; Figure 3Y).

2.4. Buspirone Reduces Neuroinflammation and Promotes the Expression of Neurotrophic Factors

In view of the effects of buspirone in the glial compartment, we sought to interrogate the mRNA expression of several pro-inflammatory cytokines (IL-1β and IL-6) and the anti-inflammatory marker (Arg1), as well as that of two neurotrophic factors (BDNF and ADNP) after 21-days of systemic rotenone administration with or without buspirone co-administration by qRT-PCR (Figure 4). The systemic administration of rotenone (10 mg/kg) significantly induced the expression of IL-1β in the midbrain (**** $p < 0.0001$; Figure 4A) and hippocampus (**** $p < 0.0001$; Figure 4E), but not in the striatum, prefrontal cortex, or amygdala ($p > 0.05$ vs. vehicle; Figure 4B–D). Similarly, the expression of IL-6 was significantly upregulated in the midbrain (**** $p < 0.0001$; Figure 4F), striatum (**** $p < 0.0001$; Figure 4G), amygdala (**** $p < 0.0001$; Figure 4I), and hippocampus (**** $p < 0.0001$; Figure 4J), but not in the prefrontal cortex ($p > 0.05$; Figure 4C).

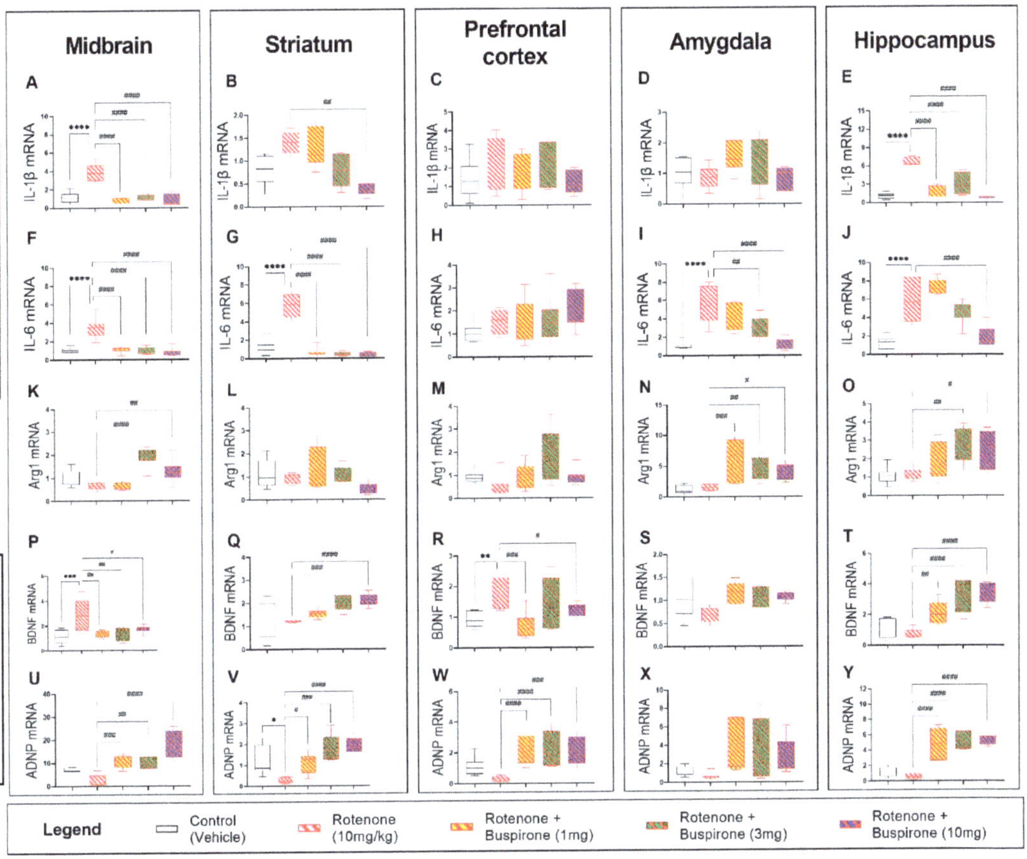

Figure 4. Buspirone reduces the expression of inflammatory cytokines and upregulates the expression of neurotrophic factors. Real-time qPCR analyses of the pro-inflammatory cytokines IL-1β (**A–E**) and IL-6 (**F–J**), the anti-inflammatory cytokine Arg1 (**K–O**), and the neurotrophic factors BDNF (**P–T**) and ADNP (**U–Y**) in the midbrain (box 1), striatum (box 2), prefrontal cortex (box 3), amygdala (box 4), and hippocampus (box 5) were reported as mean fold changes calculated with the ΔΔCt method after normalization to s18 (the housekeeping gene). All data represents 5–8 samples per treatment group. * $p < 0.05$, ** $p < 0.01$, *** $p < 0.001$, or **** $p < 0.0001$, compared to vehicle-treated controls; or # $p < 0.05$, ## $p < 0.01$, ### $p < 0.001$, or ####$p < 0.0001$, compared to rotenone (10 mg/kg)-treated mice, as determined by an ANOVA followed by Sidak's post-hoc test. IL-1β, interleukin-1beta; IL-6, interleukin-6; Arg1, arginase 1; BDNF, brain-derived neurotrophic factor; ADNP, activity-dependent neuroprotective protein; s18, ribosomal protein s18 gene; GAPDH, Glyceraldehyde 3-phosphate Dehydrogenase; kDa, Kilodalton.

Gene expression of the pro-inflammatory cytokines IL-1β and IL-6 were globally downregulated in response to buspirone co-treatment (Figure 4A–J). Both IL-1β (#### $p < 0.0001$; Figure 4A) and IL-6 (#### $p < 0.0001$; Figure 4F) were robustly downregulated in the midbrain compared with rotenone-treated mice. In the striatum, all buspirone dosages were able to reduce IL-6 expression (#### $p < 0.0001$; Figure 4G), but only the 10 mg/kg buspirone does was able to reduce IL-1β mRNA expression in the striatum (## $p < 0.01$ vs. rotenone-treated mice; Figure 4B). Similarly, in the hippocampus, 10 mg/kg of buspirone was required to fully prevent rotenone-driven upregulation of IL-6 expression (#### $p < 0.0001$; Figure 4J), whilst all buspirone dosages reliably attenuated IL-1β levels within the same brain region (#### $p < 0.0001$; Figure 4E). Both 3 and 10 mg/kg of buspirone significantly downregulated IL-6 expression in the amygdala (## $p < 0.01$ and #### $p < 0.0001$, respectively; Figure 4I). Buspirone treatment had no effects on IL-1β in the amygdala or on either pro-inflammatory cytokines in the prefrontal cortex ($p > 0.05$; Figure 4C,D,H).

Real-time qPCR analyses revealed that rotenone administration did not alter the expression of the anti-inflammatory marker Arg1 in the brain (Figure 4K–O). Conversely, both 3 and 10 mg/kg of buspirone significantly increased Arg1 mRNA in the midbrain (#### $p < 0.0001$ and ## $p < 0.01$ vs. rotenone, respectively; Figure 4K) and hippocampus (## $p < 0.01$ and # $p < 0.05$, respectively; Figure 4O). In the amygdala, all buspirone treatment groups demonstrated an increase in Arg1 mRNA expression in response to drug co-treatment (### $p < 0.001$, ## $p < 0.01$ and # $p < 0.05$ at 1, 3, and 10 mg/kg, respectively; Figure 4N), whereas it had no ameliorative effects in the striatum and prefrontal cortex ($p > 0.05$ vs. rotenone; Figure 4L,M).

The mRNA expression of the neurotrophic factors BDNF and ADNP was assessed in the CNS of mice exposed to rotenone and/or co-treated with buspirone. As shown, rotenone significantly increased BDNF transcripts in the midbrain and prefrontal cortex (*** $p < 0.001$ and ** $p < 0.01$ vs. vehicle, respectively; Figure 4P,R), but not in the striatum, amygdala, and hippocampus, where transcripts were marginally decreased ($p > 0.05$ vs. vehicle; Figure 4Q,S,T). The co-administration of buspirone to rotenone-treated mice partly reversed the effects of the toxicant on gene expression. In the midbrain, it significantly reversed BDNF gene induction at all the dosages tested (## $p < 0.01$ at 1 and 3 mg/kg and # $p < 0.05$ at 10 mg/kg; Figure 4P). Striatal BDNF mRNA were significantly increased when buspirone was used at 3 and 10 mg/kg (### $p < 0.001$ and #### $p < 0.0001$ vs. rotenone; Figure 4Q), an effect that was seen at all dosages in the hippocampus (## $p < 0.01$ at 1 mg/kg and #### $p < 0.0001$ vs. rotenone at 3 and 10 mg/kg; Figure 4T). In the prefrontal cortex, drug co-treatment reliably reduced BDNF gene expression (### $p < 0.001$ at 1 mg/kg and # $p < 0.05$ at 10 mg/kg, respectively; Figure 4R). In the amygdala, BDNF gene expression was not affected by rotenone or buspirone co-administration ($p > 0.05$; Figure 4S).

A global trend towards a reduction of ADNP transcripts was observed in the CNS of animals treated with rotenone, although this was significant only in the striatum (* $p < 0.05$ vs. vehicle; Figure 4U–Y). In contrast, buspirone robustly increased ADNP mRNA in the CNS regions tested. In the midbrain, buspirone increased ADNP (### $p < 0.001$, ## $p < 0.01$ and #### $p < 0.0001$ vs. rotenone at 1, 3 and 10 mg/kg; Figure 4U). Similarly, in the striatum, both 1, 3, and 10 mg/kg of buspirone significantly increased the expression of ADNP (# $p < 0.05$, ## $p < 0.01$, and #### $p < 0.0001$, respectively; Figure 4V). Additionally, all buspirone treatments promoted the upregulation of ADNP in the prefrontal cortex (#### $p < 0.0001$, #### $p < 0.0001$, and ### $p < 0.001$ at 1, 3, and 10 mg/kg buspirone, respectively; Figure 4W) and hippocampus (#### $p < 0.0001$; 1, 3, and 10 mg/kg buspirone, respectively; Figure 4Y), whereas only a trend was seen in the amygdala ($p > 0.05$ vs. rotenone; Figure 4X).

2.5. Buspirone Dampens Rotenone-Induced Oxidative Stress in Distinct Brain Regions

Rotenone is a known mitochondrial complex I inhibitor that is able to increase the levels of reactive oxygen species, promoting oxidative stress and mitochondrial damage. To determine if buspirone could prevent mitochondrial damage, we assessed mitochondrial function by analyzing OPA1 mRNA expression and the expression of the antioxidant enzyme SOD1, as an indirect measure of oxidative stress, throughout the brain (Figure 5).

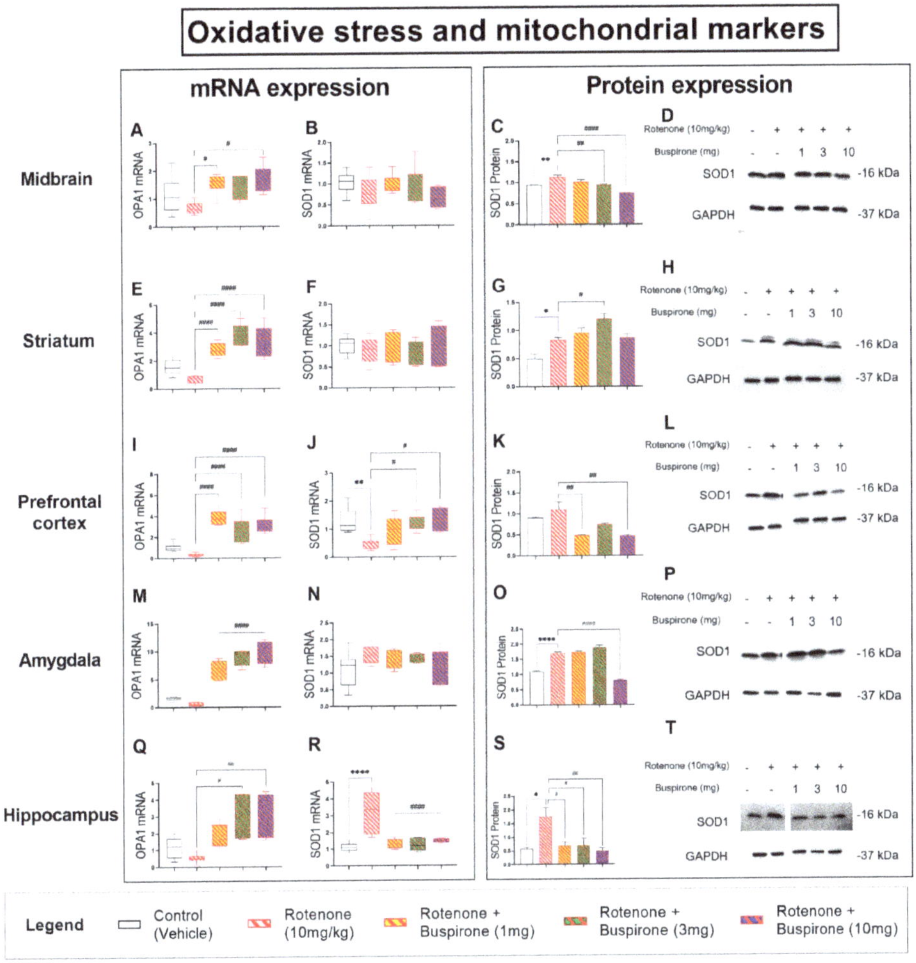

Figure 5. Buspirone dampens rotenone-induced oxidative stress in the brain. The mitochondrial marker, OPA1, was analysed to determine rotenone-induced mitochondrial damage and the antioxidant enzyme SOD1 was investigated as an indirect measure of oxidative stress. Real-time qPCR analyses of OPA1 and SOD1 mRNA expression in the midbrain (**A**,**B**), striatum (**E**,**F**), prefrontal cortex (**I**,**J**), amygdala (**M**,**N**), and hippocampus (**Q**,**R**) following the administration of rotenone (10 mg/kg) and/or buspirone (1, 3, or 10 mg) daily for 21 days. Fold-changes were calculated using the ΔΔCt method after normalization to s18 (ribosomal protein s18 gene), the housekeeping gene. Each data point represents the mean value from n = 5–8 mice per each group. Representative Western blots and densitometry of SOD1 protein expression in the midbrain (**C**,**D**), striatum (**G**,**H**), prefrontal cortex (**K**,**L**), amygdala (**O**,**P**), and hippocampus (**S**,**T**). Protein expression was normalized to GAPDH, the loading control. Western blots are cropped and removed lanes depicting drug-treatment controls are included in Supplementary Figure S5. Densitometric results are expressed as means ± S.E.M from n = 5–8 mice per each group. * $p < 0.05$, ** $p < 0.01$ or **** $p < 0.0001$, compared to vehicle-treated controls; or # $p < 0.05$, ## $p < 0.01$ or ####$p < 0.0001$, compared to rotenone (10 mg/kg)-treated mice, as determined by an ANOVA followed by Sidak's post-hoc test. OPA1, Mitochondrial dynamin like GTPase; SOD1, superoxide dismutase 1; s18, ribosomal protein s18 gene; GAPDH, glyceraldehyde 3-phosphate dehydrogenase; kDa, Kilodalton.

Analyses of OPA1 transcripts revealed that rotenone attenuated gene expression throughout the CNS, but not significantly ($p > 0.05$; Figure 5A,E,I,M,Q). Buspirone treatments promoted the expression of OPA1 mRNA, specifically in the midbrain (# $p < 0.05$ vs. rotenone at 1 and 10 mg/kg, Figure 5A), striatum (#### $p < 0.0001$; Figure 5E), prefrontal cortex (#### $p < 0.0001$; Figure 5I), amygdala (#### $p < 0.0001$; Figure 5M), and hippocampus (# $p < 0.05$ and ## $p < 0.01$ at 3 and 10 mg/kg; Figure 5Q).

Real-time qPCR analysis of SOD1 mRNA levels did not show any changes in the midbrain and striatum after treatment with rotenone and/or co-treatment with buspirone (Figure 5B,F). However, Western blots to measure SOD1 protein levels in these same brain regions showed significant increases following rotenone exposure (** $p < 0.01$ and * $p < 0.05$ vs. vehicle, respectively), which were prevented using buspirone (midbrain: ## $p < 0.01$ and #### $p < 0.0001$ vs. rotenone at 3 and 10 mg/kg; striatum: # $p < 0.05$ vs. rotenone; Figure 5C,D,G,H). In the prefrontal cortex, SOD1 mRNA expression was significantly reduced by rotenone (** $p < 0.01$ vs. vehicle) and rescued by 3 and 10 mg/kg buspirone (# $p < 0.05$ for both; Figure 5J). However, results in this brain region were not corroborated by protein expression analyses. Prefrontal SOD1 protein expression was slightly increased by rotenone ($p > 0.05$) and diminished following buspirone treatment (## $p < 0.01$ at both 1 and 10 mg/kg; Figure 5K,L). Similar to the midbrain and striatum, SOD1 mRNA were not affected by rotenone and/or buspirone treatments in the amygdala (Figure 5N). However, at the protein level, we observed a significant increase of SOD1 expression after rotenone exposure (**** $p < 0.0001$ vs. vehicle), which was prevented by co-treatment with the highest dosage of buspirone (#### $p < 0.0001$ vs. rotenone; Figure 5O,P). Lastly, hippocampal SOD1 mRNA levels were significantly increased by rotenone (**** $p < 0.0001$ vs. vehicle) and fully prevented by co-administration with buspirone at all dosages (#### $p < 0.0001$ vs. rotenone; Figure 5R). Transcript results were corroborated by protein findings, as SOD1 protein expression was induced by rotenone (* $p < 0.05$ vs. vehicle) and mitigated by buspirone (# $p < 0.05$ vs. rotenone at 1 and 3 mg/kg, ## $p < 0.01$ vs. rotenone at 10 mg/kg; Figure 5S,T).

2.6. Buspirone Prevents Dysregulations of the PACAP and VIP Neuropeptides in the Brains of Rotenone-Treated Mice

The neuropeptides PACAP and VIP are involved in a range of neuroprotective and immune modulatory functions in the CNS [21]. We have previously shown that the genetic blockade of the D3R promotes the expression of PACAP [22]. To determine if buspirone, a D3R antagonist, altered the expression of these peptides during rotenone intoxication, we analysed their gene and protein expression throughout the brain (Figure 6).

In the midbrain, rotenone exposure alone had no effects on PACAP mRNA ($p > 0.05$ vs. vehicle); however, the highest dosage of buspirone reduced gene expression (# $p < 0.05$ vs. rotenone; Figure 6A). Protein analyses did not show significant changes in PACAP protein expression in response to rotenone and/or buspirone co-treatment ($p > 0.05$; Figure 6C,D). In the striatum, PACAP mRNA were slightly reduced, and co-treatment with buspirone prevented these effects, although not significantly ($p > 0.05$ vs. rotenone; Figure 6F). In contrast, PACAP protein expression was robustly increased by rotenone in the striatum (**** $p < 0.0001$ vs. vehicle) and buspirone reduced the expression at 3 and 10 mg/kg (# $p < 0.05$ and #### $p < 0.0001$ vs. rotenone; Figure 6H,I). In the prefrontal cortex, PACAP transcripts were downregulated in response to rotenone (* $p < 0.05$ vs. vehicle). Co-treatment with buspirone significantly increased gene expression (# $p < 0.05$ vs. rotenone; Figure 6K). At the protein level, PACAP showed no changes after rotenone exposure; however, buspirone co-treatment significantly reduced PACAP protein expression (#### $p < 0.0001$ vs. rotenone; Figure 6M,N). In the amygdala, rotenone did not modify the mRNA expression of PACAP ($p > 0.05$ vs. vehicle); however, a significant induction was seen after 1 mg/kg buspirone co-administration (## $p < 0.01$ vs. rotenone; Figure 6P). In contrast, PACAP protein analyses showed that rotenone strongly increased protein expression (**** $p < 0.0001$ vs. vehicle) and levels were remarkably diminished in the amygdala of mice co-treated with both rotenone and buspirone (# $p < 0.05$ at 1 mg/kg and #### $p < 0.0001$ vs. rotenone at 3 and 10 mg/kg, respectively; Figure 6R,S). Hippocampal PACAP transcripts were only modestly reduced by rotenone ($p > 0.05$ vs. vehicle) and significantly increased in response to 3 mg/kg of buspirone (## $p < 0.01$ vs. rotenone; Figure 6U). At the protein level, PACAP was significantly reduced by rotenone (* $p < 0.05$ vs. vehicle) and reliably increased by buspirone (#### $p < 0.0001$ at 1 and 3 mg/kg, ## $p < 0.01$ vs. rotenone at 10 mg/kg; Figure 6W,X).

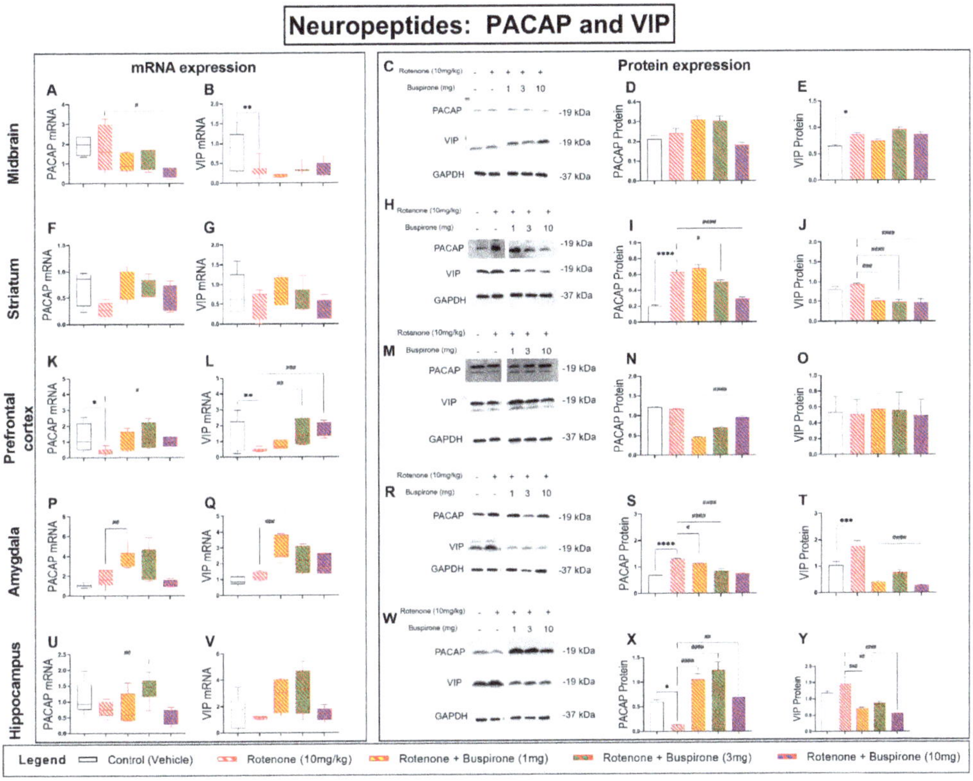

Figure 6. Buspirone prevents PACAP and VIP neuropeptide dysregulations in the brain of rotenone-treated mice. Real-time qPCR analyses of PACAP and VIP mRNA expression in the midbrain (**A**,**B**), striatum (**F**,**G**), prefrontal cortex (**K**,**L**), amygdala (**P**,**Q**), and hippocampus (**U**,**V**) following the administration of rotenone (10 mg/kg) and/or buspirone (1, 3, or 10 mg) daily for 21 days. Fold-changes were calculated using the ΔΔCt method after normalization to s18 (ribosomal protein s18 gene), the housekeeping gene. Each data point represents the mean value from $n = 5–8$ mice per each group. Representative Western blots and densitometry of PACAP and VIP protein expression in the midbrain (**C**–**E**), striatum (**H**–**J**), prefrontal cortex (**M**–**O**), amygdala (**R**–**T**), and hippocampus (**W**–**Y**). Protein expression was normalized to GAPDH, the loading control. Western blots are cropped and removed lanes depicting drug-treatment controls are included in Supplementary Figure S6. Densitometric results are expressed as means ± S.E.M from n = 5–8 mice per each group. * $p < 0.05$, ** $p < 0.01$, *** $p < 0.001$, or **** $p < 0.0001$, compared to vehicle-treated controls; or # $p < 0.05$, ## $p < 0.01$, ### $p < 0.001$, or #### $p < 0.0001$, compared to rotenone (10 mg/kg)-treated mice, as determined by an ANOVA followed by Sidak's post-hoc test. PACAP, pituitary adenylate cyclase-activating peptide; VIP, vasoactive intestinal peptide; s18, ribosomal protein s18 gene; GAPDH, glyceraldehyde 3-phosphate dehydrogenase; kDa, Kilodalton.

Analyses of VIP transcripts in the midbrain revealed a significant reduction in gene expression after rotenone exposure (** $p < 0.01$ vs. vehicle), which was not ameliorated by buspirone co-treatment ($p > 0.05$ vs. rotenone; Figure 6B). However, Western blots did not support these results, as VIP expression was upregulated by rotenone (* $p < 0.05$ vs. vehicle) and not further affected by buspirone ($p > 0.05$ vs. rotenone; Figure 6C,E). We did not observe any significant transcript changes in the striatum ($p > 0.05$; Figure 6G), whereas protein results showed that buspirone significantly reduced VIP expression (### $p < 0.001$ at 1 mg/kg, #### $p < 0.0001$ vs. rotenone at 3 and 10 mg/kg; Figure 6H,J). In the prefrontal cortex, VIP mRNA was reduced by rotenone (** $p < 0.01$ vs. vehicle) and returned to control levels upon buspirone co-treatment (## $p < 0.01$ and ### $p < 0.001$ at 3 and 10 mg/kg, respectively; Figure 6L). However, these results were not confirmed by Western blots, as VIP expression was unchanged in response to treatments ($p > 0.05$; Figure 6M,O). In the amygdala, VIP mRNA expression was not affected by rotenone treatment ($p > 0.05$ vs. vehicle); however, transcripts were upregulated by buspirone (### $p < 0.001$ vs. rotenone; Figure 6Q). Instead, VIP protein expression was upregulated by rotenone (*** $p < 0.001$ vs. vehicle) and robustly reduced upon buspirone co-administration (#### $p < 0.0001$ vs. rotenone at all dosages; Figure 6R,T). Finally, hippocampal VIP mRNA levels were unchanged after rotenone and/or buspirone co-treatment ($p > 0.05$; Figure 6V). However, VIP protein expression was reduced in the hippocampus of mice co-treated with rotenone and buspirone (### $p < 0.001$ at 1 mg/kg, ## $p < 0.01$ at 3 mg/kg and #### $p < 0.0001$ vs. rotenone at 10 mg/kg; Figure 6W,Y).

3. Discussion

Pharmacological blockade of the D3R has previously been shown to reduce inflammation and, consequently, slow disease progression in animal models of PD [11]. Interestingly, computational analyses have identified buspirone as a drug endowed with strong off-target function as D3R antagonist [16]. In addition, we previously demonstrated that buspirone treatment reliably attenuated LPS-triggered microglial polarization in BV2 cells, an effect that was not seen in D3R knockout (D3R$^{-/-}$) microglia [17]. This led us to test the efficacy of buspirone to reduce inflammation and promote neuroprotection in a rotenone mouse model of PD.

To our knowledge, this is the first study to investigate the ability of buspirone to prevent inflammation and promote neuroprotection in a model of PD. We report that buspirone prevented rotenone-induced deficits in locomotor and exploratory behaviour, and dose-dependently rescued dopaminergic degeneration in the *substantia nigra pars compacta*, midbrain, and striatum. Interestingly, buspirone also prevented glial activation in a brain-region-specific pattern. Concurrently, the drug robustly prevented the induction of pro-inflammatory cytokines by rotenone and increased the expression of neurotrophic and anti-inflammatory factors, suggesting that buspirone is able to afford some of its beneficial effects by modulating the immune response and by promoting the release of protective factors. Additionally, buspirone restored the levels of the mitochondrial marker OPA1 and the antioxidant enzyme SOD1, as well as prevented the disruptions to the two neuroprotective and immune modulatory peptides PACAP and VIP throughout the brain.

Several studies have described behavioural deficits in mice after systemic administration or rotenone [23–26]. These studies, consistent with current findings using 10 mg/kg of rotenone, indicate that a three-week induction regime using this dosage of pesticide is sufficient to attain locomotor impairments consistent with PD-like pathology. Previous studies on D3R$^{-/-}$ mice revealed that this dopamine receptor subtype mediates inhibitory effects on aversive learning [27] and both spontaneous and forced ethanol consumption in mice [16]. In this study, we did not observe any specific changes in general locomotor and exploratory behavior, nor in the expression of dopaminergic markers in mice treated with buspirone alone (Supplementary Figures S1 and S2), in contrast to previous reports from other laboratories using rats [28]. However, when the drug was co-administered with rotenone at the highest dosages (3 and 10 mg/kg), it successfully prevented the behavioural impairments caused by the insecticide. As such, it is conceivable that any preventative effect of buspirone on rotenone-induced behavioural impairments was likely to be attributed to its widespread restorative capacity in the CNS. Interestingly, similar anxiolytic and motor effects have been previously reported in mice treated with minocycline, a tetracycline antibiotic endowed with anti-inflammatory functions, further suggesting that agents able to attenuate inflammation may also be effective in preventing any associated behavioural deficits [20]. These effects are likely to be attributed to buspirone-mediated D3R antagonistic activity, as our results corroborate prior evidence of reduced anxiety-like behaviour in $D3R^{-/-}$ mice [29]. Unfortunately, buspirone's ameliorative effects on locomotor activity in rodents haven't been successfully translated into a clinical setting, as PD patients subjected to a buspirone drug treatment protocol showed poor outcomes in terms of anxiolytic response [30]. This species-specific diversity in drug response could be due to the different D3R vs. D2R distribution in the CNS between mice and humans. Additionally, given that buspirone also binds to D2R (with much lower affinity than D3R) to antagonize receptor activation [30], it is also possible that a drug-induced D2R blockade in humans (but not in rodents) may elicit some anti-psychotic effects that, however, are not accompanied by improvements on motor function, as shown in other studies [31]. It is the high homology between D2R and D3R that renders the development of specific D3R antagonists challenging and often results in mixed effects. Further refinement of the drug to enhance receptor selectivity may be a viable strategy to overcome this unwanted effect. However, it should be noted that the lack of motor improvements in PD patients is not necessarily paralleled by the lack of other critical disease-modifying effects of the drug, such as the reduced neuroinflammation and increased trophic support reported in this study. In addition, we cannot exclude that some of buspirone's inherent activities may also promote some degree of glial activation and stimulate the expression of trophic factors in certain CNS regions (Supplementary Figures S3–S6), whose significance warrants additional investigations. Furthermore, the neuroprotective effects can also be partly attributed to its canonical 5HT1a agonist activity, which may act synergistically to the D3R to improve the neurochemical disbalances seen in PD.

The landmark study by Elgueta and colleagues [11] demonstrated that pharmacological antagonism of D3R reduced the extent of dopaminergic degeneration in an MPTP model of PD [11]. Here, we report a similar dose-dependent protection of dopaminergic neurons in response to 10 mg/kg of buspirone. This finding corroborates the hypothesis that high dosages of buspirone are required to achieve enough D3R occupancy [32]. The protection of dopaminergic neurons is suggested to be secondary to buspirone's ability to reduce neuroinflammation. Accordingly, we analysed the expression of glial activation markers, inflammatory cytokines, and neurotrophic factors in distinct brain regions to determine how buspirone modulated the immune response in the brain.

Gao and colleagues [18] revealed that the presence of microglia greatly enhanced the neurodegenerative and neurotoxic effects of rotenone by increasing the susceptibility of neuron-enriched cultures to the toxin. In line with this study, we show that rotenone treatment induced the upregulation of microglial activation markers and pro-inflammatory cytokines. Previously, our laboratory has shown that D3R gene deletion and/or buspirone treatment prevented LPS-triggered microglial activation [17]. Here, we sought to determine if buspirone could exert similar effects in vivo. Therefore, we interrogated the expression of glial activation markers along with that of pro-inflammatory cytokines and neurotrophic factors. In contrast to our in vitro study, in vivo we observed a drug-induced reduction of Iba1 expression in the midbrain and amygdala, and an increase in the prefrontal cortex compared with mice receiving rotenone alone. Unexpectedly, these results were accompanied by a global increase in CD11b transcripts, which contrasts with the Iba1 response to the anxiolytic drug. These differences in the expression of glial activation markers could be due to the acquisition of differing microglial phenotypes during the course of PD. It is known that activated microglia exist in two main phenotypes, M1, considered to be pro-inflammatory, and M2, considered to be anti-inflammatory [33]. Depending on disease severity and the local CNS inflammatory state, microglia may present as heterogeneous subpopulations that may or may not benefit from the anti-inflammatory effects of the drug, although this remains to be ascertained. This theory would explain why we saw a region-specific variability in the expression of glial activation markers.

Buspirone also produced a robust downregulation of the pro-inflammatory cytokines IL-1β and IL-6. This was accompanied by a global upregulation of the anti-inflammatory marker Arg1, as well as a remarkable increase in the expression of neuroprotective and neurotrophic/growth factors, all known to exert a range of protective effects in the CNS [34]. Rotenone is known to reduce the expression of BDNF in vulnerable CNS regions [35]. This was partly seen in the hippocampus and amygdala, whereas levels were increased in the midbrain and prefrontal cortex. Irrespective of how BDNF transcripts were regulated by rotenone in the different CNS sites, buspirone prevented these changes. These results suggest that the drug is able to promote neuroplasticity, cell survival, and perhaps axonal growth during rotenone toxicity [36]. This is corroborated by the significant and robust induction of ADNP in all the brain regions we studied. ADNP is also essential in neuronal survival and brain formation, and is dysregulated in neurodegenerative and neuroinflammatory diseases [37]. The combined cytokine and neurotrophic profiles reported in this study indicate that buspirone favours the adoption of the M2 phenotype that promotes the dampening of neuroinflammation, CNS repair, and neuroprotection [38]. This is in agreement with our in vitro study that showed that $D3R^{-/-}$ and/or buspirone treatment inhibited M1 markers, including NO, NOS2, IL-1β, and TNF-α [17]. However, microglia are not the only glial cell type mediating the immune response. They work closely with astrocytes; however, more studies are needed to elucidate the role of buspirone on astrocytes. We report minimal changes in the expression of the astrocytic marker, GFAP. This correlates with analysis of GFAP-positive cells and morphology in $D3R^{-/-}$ mice that did not differ to wild-type animals [39], suggesting that the anti-inflammatory effect observed is primarily the result of buspirone's activity on microglia and, perhaps, on other immune cells. However, it was shown that the neuroprotective effects of 5HT1a agonism are mediated by astrocytes, which prevent the dopaminergic degeneration in parkinsonian mice [40]. This suggests that the 5HT1a agonist activity of buspirone may also contribute to the neuroprotective effects reported in this study. 5HT1a agonists, including buspirone, have been reported as effective neuroprotective agents in traumatic brain injury [41]. The role of the serotonin system in PD is being revealed, with some studies suggesting the 5HT1a receptor could be a target alone for improving both motor function [42] and microglial responses [43].

Rotenone toxicity is mediated by its inhibitory action on mitochondria, which ultimately results in oxidative stress [44]. To determine if the effects of buspirone were due to rotenone intoxication, we analysed the expression of OPA1, a mitochondrial marker, and SOD1, an antioxidant enzyme, as indirect evidence of oxidative stress. As expected, rotenone significantly reduced the expression of OPA1, indicating mitochondrial dysfunction. This was also verified by the upregulation of SOD1 throughout the various CNS regions. The widespread distribution of rotenone intoxication suggests this model may be capturing additional prodromal dysfunctions of the disease, as changes in extra-nigral regions correlated with changes in CNS regions that are associated with some of the non-motor symptoms that usually remain silent or do not clinically manifest until motor symptoms appear [45–47]. This could also explain the lack of glial activation, which is commonly observed in the late stages of the disease due to chronic neuroinflammation.

PACAP and VIP have both been shown to promote neuroprotection and reduce inflammation in several models of PD [48]. Notably, in primary rat mesencephalic neuron-glia cultures, PACAP prevented microglial polarization and protected against LPS-induced DAergic neurotoxicity [49]. This aligns with our recent study demonstrating the ability of these peptides to reduce LPS-induced inflammation and promote unique microglial phenotypes in vitro [50]. We have previously shown that D3R deletion enhances the expression of theses peptides [22]. The upregulation of neuroprotective growth factors and peptides by buspirone provides additional evidence of its neuroprotective activity and therapeutic potential in neurodegenerative diseases; however, further investigations on the exact role exerted by these protective agents with respect to the specific brain regions remains a topic of further investigation.

4. Materials and Methods

4.1. Animal Experiments

Seventy-two 7-week-old C57BL/6 male mice were purchased from ARC (Perth, WA, Australia) (Table 1). Mice were housed in individually ventilated cages (4 mice per cage), under normal 12:12 h light/dark cycle, with access to food and water ad libitum. All experiments were conducted in line with the Australian Code of Practice for the Care and Use of Animals for Scientific Purposes. All animal experiments were approved by the University of Technology Sydney Animal Care and Ethics Committee (ETH19-3322, approved on the 10 May 2019).

Table 1. Summary of treatment groups included in this study.

Group	Parkinson's Disease	Buspirone
1	Saline (vehicle)	Saline (vehicle)
2	Rotenone 10 mg/kg	Saline (vehicle)
3 [1]	Saline (vehicle)	Buspirone 1 mg/kg
4 [1]	Saline (vehicle)	Buspirone 3 mg/kg
5 [1]	Saline (vehicle)	Buspirone 10 mg/kg
6	Rotenone 10 mg/kg	Buspirone 1 mg/kg
7	Rotenone 10 mg/kg	Buspirone 3 mg/kg
8	Rotenone 10 mg/kg	Buspirone 10 mg/kg

[1] Buspirone control groups (groups 3–5) are included as Supplementary Figures S1–S6.

4.2. Experimental Protocol

Rotenone was prepared as a stock solution in 0.1% DMSO diluted in 0.1% saline. The 0.1% DMSO was added to Parkinson's disease saline treatment groups (Groups 1, 3–5) as a control. Buspirone was prepared as a stock solution in 0.1% saline. The experimental regime (Figure 7) lasted for 21 days, and involved daily intraperitoneal injections (i.p) of rotenone or saline, followed 2 h later by buspirone or saline, also administered by intraperitoneal injection. Mice were weighed daily and were subjected to the open field test every 7 days to assess locomotor and exploratory behavior. On day 22, mice were sacrificed and brains were collected. Brains were either snap frozen for molecular analysis or fixed in 4% paraformaldehyde for immunohistochemical analysis. Brains for molecular analysis were micro dissected into the following regions: prefrontal cortex, striatum, amygdala, hippocampus, and midbrain.

4.3. Open Field

The open field (OF) test was performed every 7 days. Baseline measurements were performed prior to any intervention. Animals were acclimated in the testing room for 30 min for habituation. The OF test was conducted in the dark using a square box made of grey Plexiglass plastic (30 × 30 × 30 cm). Individually, mice were placed in the centre of the OF and allowed to freely explore for 5 min while being recorded. The OF was cleaned thoroughly between each mouse to eliminate excretions and any odour cues using 70% ethanol and allowed to air dry. The FIJI/ImageJ Plugin Mouse Behavioural Analysis Toolbox (MouBeAT) software ver. 1.00 [51] was used to analyse videos. Locomotor activity was determined by the total distance travelled and average speed of the mice, as calculated by MouBeAt. Exploratory behavior was defined as the number of times a mouse entered the centre of the OF and the time spent in the centre.

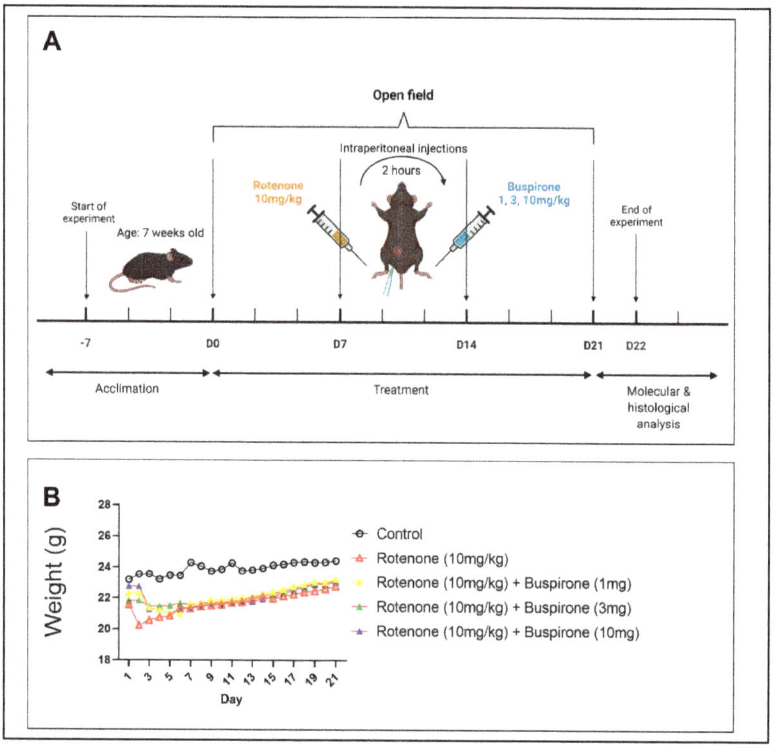

Figure 7. Experimental timeline. (**A**) Schematic of experimental timeline of injection regimen. Seven-week-old C57BL/6 male mice were allowed to acclimate for one week. Mice were exposed to daily intraperitoneal injections of rotenone (10 mg/kg), followed 2 h later by 1, 3, or 10 mg of buspirone for 21 days. At baseline, mice were subjected to the open field test for locomotor and exploratory behavioural assessment, and again on days 7, 14, and 21. On day 22, mice were euthanised and brains were collected for molecular and histological analysis. (**B**) Mice were weighed daily to monitor health and well-being. Data are represented as a daily average. N = 8–12 mice per group.

4.4. Real-Time Quantitative Polymerase Chain Reaction (RT-qPCR)

Total RNA was extracted using TRI-reagent and precipitated with 2-propanol following established protocols [52]. Single-stranded cDNA was synthesized using the Tetro cDNA synthesis kit (Bioline, Sydney, NSW, Australia) as per the manufacturer's instructions. Real-time qPCR was performed to analyse the mRNA levels of 12 genes (Table 2). The ribosomal protein 18S was used as the housekeeping gene. Each reaction consisted of 3 μL of cDNA (final concentration 100 ng), 5 μL iTaq Universal SYBR Green Master Mix (BioRad, VIC, Australia), and 0.8 μL of forward and reverse primers. To examine changes in expression, the mean fold changes of each sample was calculated using the ΔΔCt method as previously described [17]. PCR product specificity was evaluated by melting curve analysis, with each gene showing a single peak (data not shown).

Table 2. List of primer sets used in real-time qPCR analysis. Forward and reverse primers were selected from the 5′ and 3′ region of each gene. The expected length of each amplicon is indicated in the right column.

Accession #	Gene	Primer Sequence (5′-3′)	Length (bp)
NM_009377.3	Tyrosine hydroxylase (TH)	Fwd GCCCTACCAAGATCAAACCTAC Rev ATACGAGAGGCATAGTTCCTGA	93
NM_010020.3	Dopamine transporter (DAT)	Fwd ATGACATCAAGCAGATGACTGG Rev CACGACCACATACAGAAGGAAG	95
NM_001131020.1	Glial fibrillary acidic protein (GFAP)	Fwd GAGATTCGCACTCAATACGAGG Rev CTGCAAACTTAGACCGATACCA	79
NM_001082960.1	CD11b	Fwd GAGCAGGGGTCATTCGCTAC Rev GCTGGCTTAGATGCGATGGT	94
NM_008361.4	Interleukin-1β (IL-1β)	Fwd GCTACCTGTGTCTTTCCCGT Rev CATCTCGGAGCCTGTAGTGC	164
NM_007482.3	Arginase-1 (Arg1)	Fwd ACAAGACAGGGCTCCTTTCAG Rev TTAAAGCCACTGCCGTGTTC	105
NM_011434.2	Superoxidase dismutase (SOD1)	Fwd CAATGGTGGTCCATGAGAAACA Rev CCCAGCATTTCCAGTCTTTGTA	77
NM_001199177.1	Mitochondrial dynamin like GTPase (OPA1)	Fwd GCCCTTCTCTTGTTAGGTTCAC Rev ACACCTTCCTGTAATGCTTGTC	88
NM_007540.4	Brain derived neurotrophic factor (BDNF)	Fwd CGAGTGGGTCACAGCGGCAG Rev GCCCCTGCAGCCTTCCTTGG	160
NM_001310086.1	Activity-dependent neuroprotective protein (ADNP)	Fwd GTGACATTGGGTTGGAATACTGT Rev AGGTTTTGTCCGATAGTCCTGA	149
NM_016989.2	Pituitary adenylate cyclase-activating polypeptide (PACAP)	Fwd AGGCTTACGATCAGGACGGA Rev CTCCTGTCGGCTGGGTAGTA	121
NM_053991.1	Vasoactive intestinal peptide (VIP)	Fwd CCTGGCGATCCTGACACTCT Rev CTGCAGCCTGTCATCCAACC	100
NM_213557.1	18S ribosomal subunit (s18)	Fwd GGCGGAAAATAGCCTTCGCT Rev AGCCCTCTTGGTGAGGTCAA	101

4.5. Western Blot

Protein was extracted by homogenizing tissues in radioimmunoprecipitation assay (RIPA) buffer as previously described, and quantified using the bicinchoninic acid assay (Pierce BCA Protein Assay Kit, ThermoFisher Scientific, VIC, Australia). Next, 30 µg of protein was separated by SDS-polyacrylamide gel electrophoresis (SDS-PAGE) using 4–20% Mini-PROTEAN TGX Stain-Free Gels (15 well, BioRad, VIC, Australia). The Precision Plus Protein Prestained Standard in All Blue (BioRad, VIC, Australia) was included for comparison. Transfer to a PVDF membrane was performed using the semi-dry method (BioRad Trans-Blot Turbo Transfer System). Primary antibodies were incubated overnight in 5% skim milk in TBST blocking solution at 4 °C. Antibodies and dilutions are summarized in Table 3. Membranes were incubated in secondary antibody 1 hr at RT. Western blots were visualized using chemiluminescence BioRad Clarity Western ECL Blotting Substrate Solution. Images were acquired using the BioRad ChemiDoc MP System. Images were analysed using Fiji ImageJ and ratios normalized to GAPDH which was used as a loading control.

4.6. Immunohistochemistry

Brains were fixed in 4% paraformaldehyde at 4 °C for 48 h before dehydration and embedding in paraffin. Next, 5 µM thick coronal sections were cut with a microtome and mounted on glass slides. The sections were deparaffinised in xylene and rehydrated through decreasing concentrations of ethanol. Dopaminergic neurons were labeled with TH. The immunoreactivity of TH was visualized using the Rabbit-specific HRP/DAB (ABC) Detection IHC Kit (ab64261, Abcam, VIC, Australia) according to the manufacturer's protocol. Negative controls were created for all regions analysed by not incubating with the primary antibody. Hematoxylin was used as a counterstain to visualize nuclei. Sections were dehydrated in increasing concentrations of ethanol and xylene before being mounted.

Images were taken on a ZEISS AxioScan.Z1 (Carl Zeiss Australasia, NSW, Australia) at ×20 magnification. Fiji ImageJ was used for the semi-quantification of DAB-positive cells normalized to nuclei (hematoxylin stain) [53].

Table 3. Antibodies used in Western blots and immunohistochemistry.

Antibody	Dilution	Source (Cat. #)
Tyrosine hydroxylase (TH)	1:200 (WB) 1:500 (IHC-P)	Abcam (ab112)
Dopamine transporter (DAT)	1:1000	Abcam (ab128848)
Glial fibrillary acidic protein (GFAP)	1:1000	Abcam (ab68428)
Ionized calcium binding protein (Iba1)	1:1000	Sigma Aldrich (SAB2702364)
Mitochondrial dynamin like GTPase (OPA1)	1:1000	GeneTex (GTX129917)
Superoxidase dismutase (SOD1)	1:1000	GeneTex (GTX100554)
Pituitary adenylate cyclase-activating polypeptide (PACAP)	1:1000	GeneTex (GTX37576)
Vasoactive intestinal peptide (VIP)	1:1000	GeneTex (GTX129461)
Glyceraldehyde-3-phosphate dehydrogenase (GAPDH)	1:1000	BioRad (VPA00187)
Goat anti-Rabbit IgG HRP	1:10,000	BioRad (STAR208P)
Goat anti-Mouse IgG (H + L)-HRP	1:10,000	BioRad (1706516)

WB, Western blot; IHC-P, immunohistochemistry on paraffin embedded tissue.

4.7. Statistical Analysis

All data are reported as the mean ± SEM. Statistical analyses were calculated using GraphPad Prism software v. 9.0.2. (GraphPad Software, La Jolla, CA, USA). Comparisons between two or more groups were analysed by ANOVA followed by Dunnett's or Sidak's post-hoc tests, as appropriate. p-values ≤ 0.05 were considered statistically significant.

5. Conclusions

In conclusion, our results corroborate the idea that D3R receptor blockage may serve as a neuroprotective and anti-inflammatory mechanism to attenuate PD burden. Buspirone effectively prevented rotenone-induced behavioural deficits and dose-dependently protected dopaminergic neurons from the detrimental effects of rotenone intoxication. Buspirone mitigated rotenone toxicity throughout the brain, mainly by alleviating mitochondrial dysfunction, reducing inflammation, and promoting the expression of neuroprotective factors, including anti-inflammatory cytokines, neurotrophic and growth factors, and neuropeptides. Overall, this evidence suggests that buspirone may be a viable candidate for drug repurposing, as its anti-inflammatory and neuroprotective effects may prevent disease severity and progression in patients with neurodegenerative/neuroinflammatory conditions such as PD.

Supplementary Materials: The following supporting information can be downloaded at: https://www.mdpi.com/article/10.3390/ijms23031845/s1.

Author Contributions: Conceptualization, A.C.; methodology, S.T.B. and A.C.; validation, S.T.B. and A.C.; formal analysis, S.T.B. and A.C.; investigation, S.T.B.; data curation, S.T.B. and A.C.; writing—original draft preparation, S.T.B.; writing—review and editing, A.C.; supervision, A.C. All authors have read and agreed to the published version of the manuscript.

Funding: This research received no external funding.

Institutional Review Board Statement: The study was conducted according to the guidelines of the Declaration of Helsinki, and approved by the Institutional Review Board (or Ethics Committee) of University of Technology Sydney (protocol code: ETH19-3322, date of approval: 10/05/2019).

Informed Consent Statement: Not applicable.

Acknowledgments: The authors would like to acknowledge Fiona Ryan for her support and guidance during animal experiments. The authors acknowledge the use of the equipment AxioSlide Scanner in the Microbial Imaging Facility (MIF) at the iThree institute in the Faculty of Science, the University of Technology Sydney.

Conflicts of Interest: The authors declare no conflict of interest.

References

1. Braak, H.; Ghebremedhin, E.; Rub, U.; Bratzke, H.; Del Tredici, K. Stages in the development of Parkinson's disease-related pathology. *Cell Tissue Res.* **2004**, *318*, 121–134. [CrossRef] [PubMed]
2. Hunter, R.L.; Dragicevic, N.; Seifert, K.; Choi, D.Y.; Liu, M.; Kim, H.C.; Cass, W.A.; Sullivan, P.G.; Bing, G. Inflammation induces mitochondrial dysfunction and dopaminergic neurodegeneration in the nigrostriatal system. *J. Neurochem.* **2007**, *100*, 1375–1386. [CrossRef] [PubMed]
3. Gamber, K.M. Animal Models of Parkinson's Disease: New models provide greater translational and predictive value. *BioTechniques* **2016**, *61*, 210–211. [CrossRef]
4. Radad, K.; Al-Shraim, M.; Al-Emam, A.; Wang, F.; Kranner, B.; Rausch, W.D.; Moldzio, R. Rotenone: From modelling to implication in Parkinson's disease. *Folia Neuropathol.* **2019**, *57*, 317–326. [CrossRef] [PubMed]
5. Betarbet, R.; Sherer, T.B.; MacKenzie, G.; Garcia-Osuna, M.; Panov, A.V.; Greenamyre, J.T. Chronic systemic pesticide exposure reproduces features of Parkinson's disease. *Nat. Neurosci.* **2000**, *3*, 1301–1306. [CrossRef]
6. Gelders, G.; Baekelandt, V.; Van der Perren, A. Linking Neuroinflammation and Neurodegeneration in Parkinson's Disease. *J. Immunol. Res.* **2018**, *2018*, 4784268. [CrossRef]
7. DiSabato, D.J.; Quan, N.; Godbout, J.P. Neuroinflammation: The devil is in the details. *J. Neurochem.* **2016**, *139* (Suppl. S2), 136–153. [CrossRef]
8. Castano, A.; Herrera, A.J.; Cano, J.; Machado, A. Lipopolysaccharide intranigral injection induces inflammatory reaction and damage in nigrostriatal dopaminergic system. *J. Neurochem.* **1998**, *70*, 1584–1592. [CrossRef]
9. Thomas Broome, S.; Louangaphay, K.; Keay, K.A.; Leggio, G.M.; Musumeci, G.; Castorina, A. Dopamine: An immune transmitter. *Neural Regen. Res.* **2020**, *15*, 2173–2185. [CrossRef]
10. McKenna, F.; McLaughlin, P.J.; Lewis, B.J.; Sibbring, G.C.; Cummerson, J.A.; Bowen-Jones, D.; Moots, R.J. Dopamine receptor expression on human T- and B-lymphocytes, monocytes, neutrophils, eosinophils and NK cells: A flow cytometric study. *J. Neuroimmunol.* **2002**, *132*, 34–40. [CrossRef]
11. Elgueta, D.; Aymerich, M.S.; Contreras, F.; Montoya, A.; Celorrio, M.; Rojo-Bustamante, E.; Riquelme, E.; Gonzalez, H.; Vasquez, M.; Franco, R.; et al. Pharmacologic antagonism of dopamine receptor D3 attenuates neurodegeneration and motor impairment in a mouse model of Parkinson's disease. *Neuropharmacology* **2017**, *113*, 110–123. [CrossRef] [PubMed]
12. Elgueta, D.; Contreras, F.; Prado, C.; Montoya, A.; Ugalde, V.; Chovar, O.; Villagra, R.; Henriquez, C.; Abellanas, M.A.; Aymerich, M.S.; et al. Dopamine Receptor D3 Expression Is Altered in $CD4^+$ T-Cells From Parkinson's Disease Patients and Its Pharmacologic Inhibition Attenuates the Motor Impairment in a Mouse Model. *Front. Immunol.* **2019**, *10*, 981. [CrossRef] [PubMed]
13. Yang, P.; Perlmutter, J.S.; Benzinger, T.L.S.; Morris, J.C.; Xu, J. Dopamine D3 receptor: A neglected participant in Parkinson Disease pathogenesis and treatment? *Ageing Res. Rev.* **2020**, *57*, 100994. [CrossRef] [PubMed]
14. Raber, J.; Wienclaw, R.A.; Cataldo, L.J. Buspirone. In *The Gale Encyclopedia of Mental Health*; Gale: Detroit, MI, USA, 2012; pp. 264–267.
15. Kim, S.W.; Fowler, J.S.; Skolnick, P.; Muench, L.; Kang, Y.; Shea, C.; Logan, J.; Kim, D.; Carter, P.; King, P.; et al. Therapeutic doses of buspirone block D3 receptors in the living primate brain–CORRIGENDUM. *Int. J. Neuropsychopharmacol.* **2014**, *17*, 1354. [CrossRef]
16. Leggio, G.M.; Camillieri, G.; Platania, C.B.; Castorina, A.; Marrazzo, G.; Torrisi, S.A.; Nona, C.N.; D'Agata, V.; Nobrega, J.; Stark, H.; et al. Dopamine D3 receptor is necessary for ethanol consumption: An approach with buspirone. *Neuropsychopharmacology* **2014**, *39*, 2017–2028. [CrossRef] [PubMed]
17. Thomas Broome, S.; Fisher, T.; Faiz, A.; Keay, K.A.; Musumeci, G.; Al-Badri, G.; Castorina, A. Assessing the Anti-Inflammatory Activity of the Anxiolytic Drug Buspirone Using CRISPR-Cas9 Gene Editing in LPS-Stimulated BV-2 Microglial Cells. *Cells* **2021**, *10*, 1312. [CrossRef]
18. Gao, H.-M.; Hong, J.-S.; Zhang, W.; Liu, B. Distinct Role for Microglia in Rotenone-Induced Degeneration of Dopaminergic Neurons. *J. Neurosci.* **2002**, *22*, 782–790. [CrossRef]
19. Subhramanyam, C.S.; Wang, C.; Hu, Q.; Dheen, S.T. Microglia-mediated neuroinflammation in neurodegenerative diseases. *Semin. Cell Dev. Biol.* **2019**, *94*, 112–120. [CrossRef]

20. Liu, H.Y.; Yue, J.; Hu, L.N.; Cheng, L.F.; Wang, X.S.; Wang, X.J.; Feng, B. Chronic minocycline treatment reduces the anxiety-like behaviors induced by repeated restraint stress through modulating neuroinflammation. *Brain Res. Bull.* **2018**, *143*, 19–26. [CrossRef]
21. Waschek, J.A. VIP and PACAP: Neuropeptide modulators of CNS inflammation, injury, and repair. *Br. J. Pharmacol.* **2013**, *169*, 512–523. [CrossRef]
22. Marzagalli, R.; Leggio, G.M.; Bucolo, C.; Pricoco, E.; Keay, K.A.; Cardile, V.; Castorina, S.; Salomone, S.; Drago, F.; Castorina, A. Genetic blockade of the dopamine D3 receptor enhances hippocampal expression of PACAP and receptors and alters their cortical distribution. *Neuroscience* **2016**, *316*, 279–295. [CrossRef] [PubMed]
23. Dhanalakshmi, C.; Janakiraman, U.; Manivasagam, T.; Justin Thenmozhi, A.; Essa, M.M.; Kalandar, A.; Khan, M.A.; Guillemin, G.J. Vanillin Attenuated Behavioural Impairments, Neurochemical Deficts, Oxidative Stress and Apoptosis Against Rotenone Induced Rat Model of Parkinson's Disease. *Neurochem. Res.* **2016**, *41*, 1899–1910. [CrossRef] [PubMed]
24. Zhang, Y.; Guo, H.; Guo, X.; Ge, D.; Shi, Y.; Lu, X.; Lu, J.; Chen, J.; Ding, F.; Zhang, Q. Involvement of Akt/mTOR in the Neurotoxicity of Rotenone-Induced Parkinson's Disease Models. *Int. J. Environ. Res. Public Health* **2019**, *16*, 3811. [CrossRef] [PubMed]
25. Martin, L.J.; Pan, Y.; Price, A.C.; Sterling, W.; Copeland, N.G.; Jenkins, N.A.; Price, D.L.; Lee, M.K. Parkinson's disease alpha-synuclein transgenic mice develop neuronal mitochondrial degeneration and cell death. *J. Neurosci.* **2006**, *26*, 41–50. [CrossRef] [PubMed]
26. Inden, M.; Kitamura, Y.; Takeuchi, H.; Yanagida, T.; Takata, K.; Kobayashi, Y.; Taniguchi, T.; Yoshimoto, K.; Kaneko, M.; Okuma, Y.; et al. Neurodegeneration of mouse nigrostriatal dopaminergic system induced by repeated oral administration of rotenone is prevented by 4-phenylbutyrate, a chemical chaperone. *J. Neurochem.* **2007**, *101*, 1491–1504. [CrossRef]
27. D'Amico, A.G.; Scuderi, S.; Leggio, G.M.; Castorina, A.; Drago, F.; D'Agata, V. Increased hippocampal CREB phosphorylation in dopamine D3 receptor knockout mice following passive avoidance conditioning. *Neurochem. Res.* **2013**, *38*, 2516–2523. [CrossRef]
28. Prut, L.; Belzung, C. The open field as a paradigm to measure the effects of drugs on anxiety-like behaviors: A review. *Eur. J. Pharmacol.* **2003**, *463*, 3–33. [CrossRef]
29. Leggio, G.M.; Torrisi, S.A.; Castorina, A.; Platania, C.B.; Impellizzeri, A.A.; Fidilio, A.; Caraci, F.; Bucolo, C.; Drago, F.; Salomone, S. Dopamine D3 receptor-dependent changes in alpha6 GABAA subunit expression in striatum modulate anxiety-like behaviour: Responsiveness and tolerance to diazepam. *Eur. Neuropsychopharmacol.* **2015**, *25*, 1427–1436. [CrossRef]
30. Schneider, R.B.; Auinger, P.; Tarolli, C.G.; Iourinets, J.; Gil-Diaz, M.C.; Richard, I.H. A trial of buspirone for anxiety in Parkinson's disease: Safety and tolerability. *Park. Relat. Disord.* **2020**, *81*, 69–74. [CrossRef]
31. Maramai, S.; Gemma, S.; Brogi, S.; Campiani, G.; Butini, S.; Stark, H.; Brindisi, M. Dopamine D3 Receptor Antagonists as Potential Therapeutics for the Treatment of Neurological Diseases. *Front. Neurosci.* **2016**, *10*, 451. [CrossRef]
32. Le Foll, B.; Payer, D.; Di Ciano, P.; Guranda, M.; Nakajima, S.; Tong, J.; Mansouri, E.; Wilson, A.A.; Houle, S.; Meyer, J.H.; et al. Occupancy of Dopamine D_3 and D_2 Receptors by Buspirone: A [^{11}C]-(+)-PHNO PET Study in Humans. *Neuropsychopharmacology* **2016**, *41*, 529–537. [CrossRef] [PubMed]
33. Chauhan, P.; Sheng, W.S.; Hu, S.; Prasad, S.; Lokensgard, J.R. Differential Cytokine-Induced Responses of Polarized Microglia. *Brain Sci.* **2021**, *11*, 1482. [CrossRef] [PubMed]
34. Castorina, A.; Vogiatzis, M.; Kang, J.W.M.; Keay, K.A. PACAP and VIP expression in the periaqueductal grey of the rat following sciatic nerve constriction injury. *Neuropeptides* **2019**, *74*, 60–69. [CrossRef] [PubMed]
35. Mendonca, I.P.; de Paiva, I.H.R.; Duarte-Silva, E.P.; de Melo, M.G.; da Silva, R.S.; Oliveira, W.H.; da Silveira Andrade da Costa, B.L.; Peixoto, C.A. Metformin and fluoxetine improve depressive-like behavior in a murine model of Parkinsons disease through the modulation of neuroinflammation, neurogenesis and neuroplasticity. *Int. Immunopharmacol.* **2021**, *102*, 108415. [CrossRef] [PubMed]
36. Zhao, X.; Kong, D.; Zhou, Q.; Wei, G.; Song, J.; Liang, Y.; Du, G. Baicalein alleviates depression-like behavior in rotenone- induced Parkinson's disease model in mice through activating the BDNF/TrkB/CREB pathway. *Biomed. Pharmacother.* **2021**, *140*, 111556. [CrossRef]
37. Ivashko-Pachima, Y.; Gozes, I. Activity-dependent neuroprotective protein (ADNP)-end-binding protein (EB) interactions regulate microtubule dynamics toward protection against tauopathy. *Prog. Mol. Biol. Transl. Sci.* **2021**, *177*, 65–90. [CrossRef]
38. Burke, N.N.; Kerr, D.M.; Moriarty, O.; Finn, D.P.; Roche, M. Minocycline modulates neuropathic pain behaviour and cortical M1-M2 microglial gene expression in a rat model of depression. *Brain Behav. Immun.* **2014**, *42*, 147–156. [CrossRef]
39. Wang, J.; Lai, S.; Li, G.; Zhou, T.; Wang, B.; Cao, F.; Chen, T.; Zhang, X.; Chen, Y. Microglial activation contributes to depressive-like behavior in dopamine D3 receptor knockout mice. *Brain Behav. Immun.* **2020**, *83*, 226–238. [CrossRef]
40. Miyazaki, I.; Asanuma, M.; Murakami, S.; Takeshima, M.; Torigoe, N.; Kitamura, Y.; Miyoshi, K. Targeting 5-HT_{1A} receptors in astrocytes to protect dopaminergic neurons in Parkinsonian models. *Neurobiol. Dis.* **2013**, *59*, 244–256. [CrossRef]
41. Cheng, J.P.; Leary, J.B.; Sembhi, A.; Edwards, C.M.; Bondi, C.O.; Kline, A.E. 5-hydroxytryptamine$_{1A}$ (5-HT_{1A}) receptor agonists: A decade of empirical evidence supports their use as an efficacious therapeutic strategy for brain trauma. *Brain Res.* **2016**, *1640*, 5–14. [CrossRef]
42. Jiang, X.; Liang, P.; Wang, K.; Jia, J.; Wang, X. Serotonin 1A receptor agonist modulation of motor deficits and cortical oscillations by NMDA receptor interaction in parkinsonian rats. *Neuropharmacology* **2022**, *203*, 108881. [CrossRef]

43. Krabbe, G.; Matyash, V.; Pannasch, U.; Mamer, L.; Boddeke, H.W.; Kettenmann, H. Activation of serotonin receptors promotes microglial injury-induced motility but attenuates phagocytic activity. *Brain Behav. Immun.* **2012**, *26*, 419–428. [CrossRef] [PubMed]
44. Miyazaki, I.; Isooka, N.; Imafuku, F.; Sun, J.; Kikuoka, R.; Furukawa, C.; Asanuma, M. Chronic Systemic Exposure to Low-Dose Rotenone Induced Central and Peripheral Neuropathology and Motor Deficits in Mice: Reproducible Animal Model of Parkinson's Disease. *Int. J. Mol. Sci.* **2020**, *21*, 3254. [CrossRef] [PubMed]
45. Das, T.; Hwang, J.J.; Poston, K.L. Episodic recognition memory and the hippocampus in Parkinson's disease: A review. *Cortex* **2019**, *113*, 191–209. [CrossRef] [PubMed]
46. Carey, G.; Gormezoglu, M.; de Jong, J.J.A.; Hofman, P.A.M.; Backes, W.H.; Dujardin, K.; Leentjens, A.F.G. Neuroimaging of Anxiety in Parkinson's Disease: A Systematic Review. *Mov. Disord.* **2021**, *36*, 327–339. [CrossRef] [PubMed]
47. Schapira, A.H.V.; Chaudhuri, K.R.; Jenner, P. Non-motor features of Parkinson disease. *Nat. Rev. Neurosci.* **2017**, *18*, 435–450. [CrossRef]
48. Yu, R.; Li, J.; Lin, Z.; Ouyang, Z.; Huang, X.; Reglodi, D.; Vaudry, D. TAT-tagging of VIP exerts positive allosteric modulation of the PAC1 receptor and enhances VIP neuroprotective effect in the MPTP mouse model of Parkinson's disease. *Biochim. Biophys. Acta Gen. Subj.* **2020**, *1864*, 129626. [CrossRef]
49. Yang, S.; Yang, J.; Yang, Z.; Chen, P.; Fraser, A.; Zhang, W.; Pang, H.; Gao, X.; Wilson, B.; Hong, J.S.; et al. Pituitary adenylate cyclase-activating polypeptide (PACAP) 38 and $PACAP_4$–6 are neuroprotective through inhibition of NADPH oxidase: Potent regulators of microglia-mediated oxidative stress. *J. Pharmacol Exp. Ther.* **2006**, *319*, 595–603. [CrossRef]
50. Karunia, J.; Niaz, A.; Mandwie, M.; Thomas Broome, S.; Keay, K.A.; Waschek, J.A.; Al-Badri, G.; Castorina, A. PACAP and VIP Modulate LPS-Induced Microglial Activation and Trigger Distinct Phenotypic Changes in Murine BV2 Microglial Cells. *Int. J. Mol. Sci.* **2021**, *22*, 10947. [CrossRef]
51. Bello-Arroyo, E.; Roque, H.; Marcos, A.; Orihuel, J.; Higuera-Matas, A.; Desco, M.; Caiolfa, V.R.; Ambrosio, E.; Lara-Pezzi, E.; Gomez-Gaviro, M.V. MouBeAT: A New and Open Toolbox for Guided Analysis of Behavioral Tests in Mice. *Front. Behav. Neurosci.* **2018**, *12*, 201. [CrossRef]
52. Mandwie, M.; Karunia, J.; Niaz, A.; Keay, K.A.; Musumeci, G.; Rennie, C.; McGrath, K.; Al-Badri, G.; Castorina, A. Metformin Treatment Attenuates Brain Inflammation and Rescues PACAP/VIP Neuropeptide Alterations in Mice Fed a High-Fat Diet. *Int. J. Mol. Sci.* **2021**, *22*, 13660. [CrossRef] [PubMed]
53. Crowe, A.R.; Yue, W. Semi-quantitative Determination of Protein Expression using Immunohistochemistry Staining and Analysis: An Integrated Protocol. *Bio. Protoc.* **2019**, *9*, e3465. [CrossRef] [PubMed]

International Journal of *Molecular Sciences*

MDPI

Article

From Molecules to Behavior in Long-Term Inorganic Mercury Intoxication: Unraveling Proteomic Features in Cerebellar Neurodegeneration of Rats

Leonardo Oliveira Bittencourt [1,†], Victória Santos Chemelo [1,†], Walessa Alana Bragança Aragão [1], Bruna Puty [1], Aline Dionizio [2], Francisco Bruno Teixeira [1], Mileni Silva Fernandes [2], Márcia Cristina Freitas Silva [1], Luanna Melo Pereira Fernandes [3], Edivaldo Herculano Corrêa de Oliveira [4], Marilia Afonso Rabelo Buzalaf [2], Maria Elena Crespo-Lopez [5], Cristiane do Socorro Ferraz Maia [3] and Rafael Rodrigues Lima [1,*]

1 Laboratory of Functional and Structural Biology, Institute of Biological Sciences, Federal University of Pará, Belém 66075-110, PA, Brazil; leo.bittencourt25@gmail.com (L.O.B.); vicchemelo@gmail.com (V.S.C.); walessa.aragao@gmail.com (W.A.B.A.); brunaputy@gmail.com (B.P.); teixeira.f.bruno@gmail.com (F.B.T.); marciaf@ufpa.br (M.C.F.S.)
2 Department of Biological Sciences, Bauru Dental School, University of São Paulo, Bauru 17012-901, SP, Brazil; alinesdionizio@usp.br (A.D.); mi_biol@yahoo.com.br (M.S.F.); mbuzalaf@fob.usp.br (M.A.R.B.)
3 Laboratory Pharmacology of Inflammation and Behavior, Institute of Health Sciences, Federal University of Pará, Belém 66075-110, PA, Brazil; luannafe@hotmail.com (L.M.P.F.); crismaia@ufpa.br (C.d.S.F.M.)
4 Laboratory of Cytogenetics, Environmental Session, Evandro Chagas Institute, Ananindeua 66093-020, PA, Brazil; ehco@ufpa.br
5 Laboratory of Molecular Pharmacology, Institute of Biological Sciences, Federal University of Pará, Belém 66075-110, PA, Brazil; maria.elena.crespo.lopez@gmail.com
* Correspondence: rafalima@ufpa.br
† These authors contributed equally to this work.

Citation: Bittencourt, L.O.; Chemelo, V.S.; Aragão, W.A.B.; Puty, B.; Dionizio, A.; Teixeira, F.B.; Fernandes, M.S.; Silva, M.C.F.; Fernandes, L.M.P.; de Oliveira, E.H.C.; et al. From Molecules to Behavior in Long-Term Inorganic Mercury Intoxication: Unraveling Proteomic Features in Cerebellar Neurodegeneration of Rats. *Int. J. Mol. Sci.* **2022**, *23*, 111. https://doi.org/10.3390/ijms23010111

Academic Editors: Marcello Ciaccio and Luisa Agnello

Received: 24 October 2021
Accepted: 22 November 2021
Published: 22 December 2021

Abstract: Mercury is a severe environmental pollutant with neurotoxic effects, especially when exposed for long periods. Although there are several evidences regarding mercury toxicity, little is known about inorganic mercury (IHg) species and cerebellum, one of the main targets of mercury associated with the neurological symptomatology of mercurial poisoning. Besides that, the global proteomic profile assessment is a valuable tool to screen possible biomarkers and elucidate molecular targets of mercury neurotoxicity; however, the literature is still scarce. Thus, this study aimed to investigate the effects of long-term exposure to IHg in adult rats' cerebellum and explore the modulation of the cerebellar proteome associated with biochemical and functional outcomes, providing evidence, in a translational perspective, of new mercury toxicity targets and possible biomarkers. Fifty-four adult rats were exposed to 0.375 mg/kg of $HgCl_2$ or distilled water for 45 days using intragastric gavage. Then, the motor functions were evaluated by rotarod and inclined plane. The cerebellum was collected to quantify mercury levels, to assess the antioxidant activity against peroxyl radicals (ACAPs), the lipid peroxidation (LPO), the proteomic profile, the cell death nature by cytotoxicity and apoptosis, and the Purkinje cells density. The IHg exposure increased mercury levels in the cerebellum, reducing ACAP and increasing LPO. The proteomic approach revealed a total 419 proteins with different statuses of regulation, associated with different biological processes, such as synaptic signaling, energy metabolism and nervous system development, e.g., all these molecular changes are associated with increased cytotoxicity and apoptosis, with a neurodegenerative pattern on Purkinje cells layer and poor motor coordination and balance. In conclusion, all these findings feature a neurodegenerative process triggered by IHg in the cerebellum that culminated into motor functions deficits, which are associated with several molecular features and may be related to the clinical outcomes of people exposed to the toxicant.

Keywords: mercury; central nervous system; proteomic; motor functions

1. Introduction

Mercury is a hazardous toxic pollutant distributed in the environment and is considered a severe public health concern [1,2]. Humans are subjected to the mercurial compound by different sources because of anthropogenic activities, such as occupational exposure and environmental contamination by illegal gold mining that contaminates fish and seafood [3,4]. Elemental mercury (Hg^0) and methylmercury (MeHg) are considered the main species of exposure in occupational and environmental (via seafood) outbreaks, respectively [3,4]. However, even in the case of MeHg exposure, consequences would additionally be due to inorganic mercury (IHg), since the latter specie was detected in both contaminated food and cells of central nervous system (CNS) origin [5,6].

Considering the presence of IHg in fish, this becomes a serious public health problem since those who consume them, especially populations whose food bases are these items [7], are subject to prolonged exposure. Some evidence points out that commercially available predatory species of fish have significant levels of IHg, reaching mean values of 0.15 μg/g of IHg [6]. Moreover, although IHg species present lower liposolubility than organic ones, they also cause damage. Due to its toxicokinetic characteristics, after absorption, kidneys [8] and the blood and cells from the human central nervous system (CNS) are also affected by IHg poisoning [9–11].

Neurological manifestations are the primary symptomatology of chronic mercury poisoning, including cognitive impairments, tremors, and ataxia [4]. Tremors and ataxia are symptoms associated with loss of motor control and higher brain function [12], and when it comes to motor functions, the cerebellum is a pivotal structure of the CNS that plays an essential role in cognitive and motor skills [12,13]. The organ is a dense network of reciprocal crossed cerebellum-cerebral connections, interconnected with supratentorial motor areas, paralimbic, and associated with cortices pathways [12,13]. Moreover, damages in cerebellum interrupt the complex cerebellar circuitry between the neurons of the cerebellar cortex which brings clinical implications, including multiple system atrophy, spinocerebellar ataxia, hypoxia in Purkinje cells, associated with neuropathologies, such as autism spectrum disorder, attention deficit-hyperactivity disorder, and developmental dyslexia [14,15].

In this way, seeking to investigate even more the molecular features associated with IHg-induced neurotoxicity, our group has been researching the effects of IHg exposure in rats under a model that has shown to be capable of generating outcomes similar to those observed in humans, from systemic to regional analyses [16–21]. Moreover, some of these studies showed damages to essential areas to motor function, such as the motor cortex [19–21] and spinal cord [18]. However, the relationship between long-term exposure to IHg and its influence on the biochemistry and functionality of the cerebellum still lacks evidence. Thus, this study aimed to investigate the global proteomic profile underlying IHg-induced neurotoxicity associated with the motor dysfunction outcome, including molecular, biochemical, and morphological approaches.

2. Results

2.1. Long-Term Exposure to IHg Did Not Impair Mass Body Gain of Adult Rats

The exposure to IHg did not cause any body weight change over time ($p > 0.05$), and no loss of animals was observed (Figure 1).

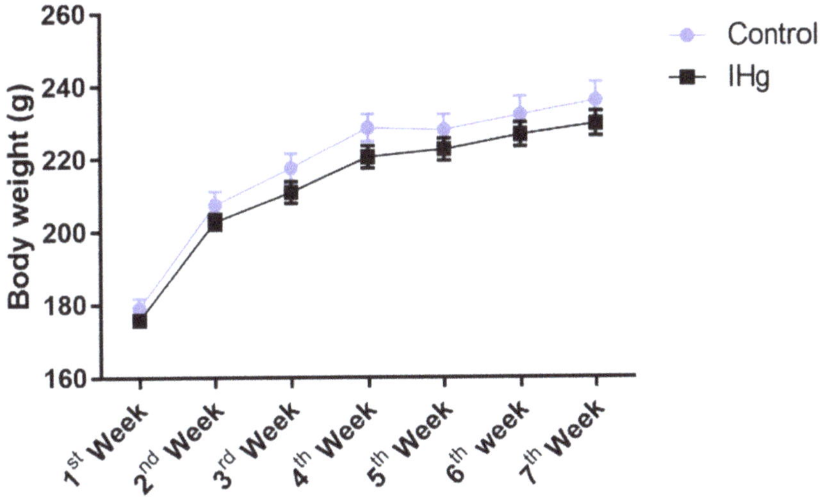

Figure 1. Body mass (g) register of adult rats exposed to 0.375 mg/kg of $HgCl_2$ for 45 days. Results are expressed as mean ± S.E.M. Two-way ANOVA with Sidak's post-test, $p > 0.05$.

2.2. Prolonged IHg Exposure Increased the Total Mercury Content in Cerebellar Parenchyma of Adult Rats

The total mercury levels were significantly higher (p = 0.0003) in the IHg group (0.0394 ± 0.004 mg/kg) in comparison with the control group (0.0120 ± 0.0001703 mg/kg), as observed in Figure 2.

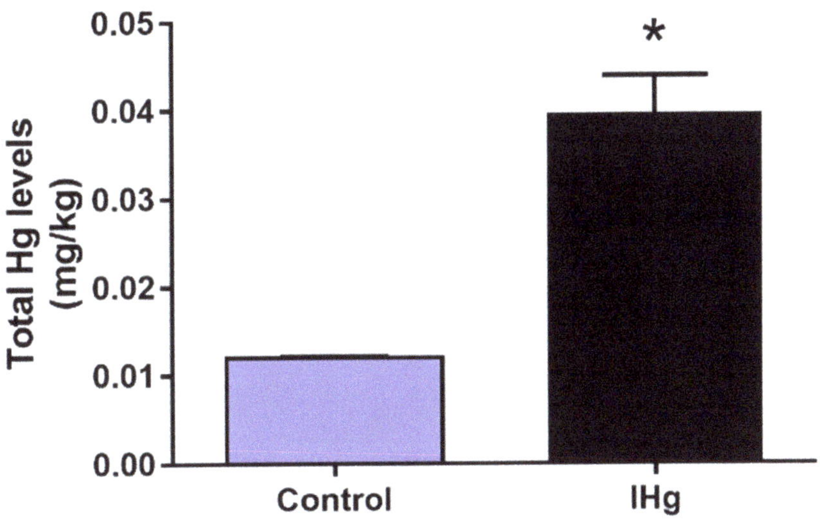

Figure 2. Total mercury (Hg) content in the cerebellum of adult rats exposed to 0.375 mg/kg of $HgCl_2$ for 45 days. Results are expressed as mg/kg (mean ± S.E.M.), * $p < 0.05$, t-Student test (n = 10).

2.3. IHg Long-Term Exposure Triggered Oxidative Stress in the Cerebellum of Adult Rats

The IHg reduced the cerebellar ACAP in 66.15% (±7%) in comparison with the control group (100 ± 6.67%; p = 0.0002) and in parallel, the LPO increased 41.70% (±5.75) when compared with the control (100 ± 3.63%) (p = 0.023) (Figure 3).

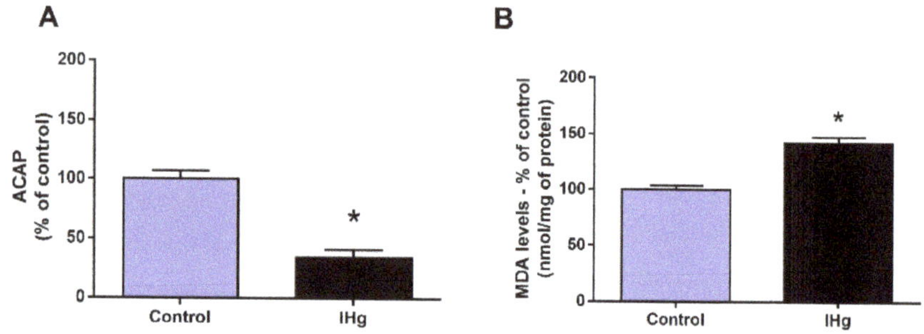

Figure 3. Oxidative biochemistry analyses of adult rats' cerebellum exposed to 0.375 mg/kg of $HgCl_2$ for 45 days. In (**A**) analyses of Antioxidant Capacity Against Peroxyl Radicals (ACAPs) and (**B**) lipid peroxidation (LPO). Data presented as percentage (%) of control (mean ± S.E.M.). * $p < 0.05$, *t*-Student test (n = 10).

2.4. The Cerebellar Proteomic Profile Was Significantly Modulated after Exposure to IHg Rats

The proteomic approach revealed the modulation of 419 proteins in the cerebellum of rats exposed to IHg. Among them, 24 proteins with exclusive expression in the IHg-exposed group and 166 with unique expression in the control group. Moreover, 196 proteins were found down-regulated and 33 up-regulated. Table 1 shows some proteins highlighted in the discussion, and the complete data are available in Supplementary Table S1 (See Supplementary Materials).

The bioinformatic analysis showed 31 biological processes associated with the proteins modulated in the cerebellum of exposed rats (Figure 4). Among them, the five most impaired processes were neuron projection development (19.2%), purine ribonucleotide biosynthetic process (6.7%), ADP metabolic process (6.2%), cerebellum development (3.9%), and regulation of synaptic vesicle exocytosis (3.9%).

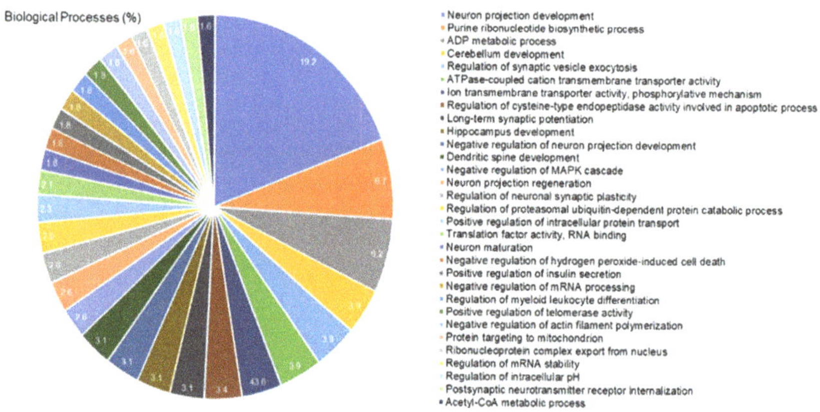

Figure 4. Functional distribution of proteins identified with differential expression in the cerebellum of rats exposed to IHg vs. control group. Categories of proteins based on gene ontology annotation of biological process. Terms significant (Kappa Score = 0.4) and distribution according to the percentage of the number of genes. UNIPROT provided proteins access number. The gene ontology was evaluated according to the ClueGo® plugin of Cytoscape® software 3.7.

Table 1. Identified proteins with expression significantly altered in the cerebellum of rats of the IHg group versus control group.

[a] Accession ID	Protein Name Description	PLGS Score	Fold Change
P34926	Microtubule-associated protein 1A	71.4	4.35
P10888	Cytochrome c oxidase subunit 4 isoform 1, mitochondrial	146.14	1.46
P0DP29	Calmodulin-1	1825.33	1.39
P0DP30	Calmodulin-2	1825.33	1.39
P0DP31	Calmodulin-3	1867.79	1.38
Q63345	Myelin-oligodendrocyte glycoprotein	290.08	1.30
P21707	Synaptotagmin-1	113.9	1.20
P07825	Synaptophysin	286.54	1.16
P68035	Actin, alpha cardiac muscle 1	10,468.68	−0.94
P39052	Dynamin-2	127.72	−0.92
P63039	60 kDa heat shock protein, mitochondrial	870.92	−0.91
P08413	Calcium/calmodulin-dependent protein kinase type II subunit beta	450.44	−0.90
P11730	Calcium/calmodulin-dependent protein kinase type II subunit gamma	323.39	−0.90
P00406	Cytochrome c oxidase subunit 2	786.54	−0.89
P0DMW0	Heat shock 70 kDa protein 1A	485.48	−0.87
Q06647	ATP synthase subunit O, mitochondrial	905.81	−0.86
P11275	Calcium/calmodulin-dependent protein kinase type II subunit alpha	220.44	−0.85
P12075	Cytochrome c oxidase subunit 5B, mitochondrial	942.79	−0.84
P15791	Calcium/calmodulin-dependent protein kinase type II subunit delta	215.3	−0.84
P60711	Actin, cytoplasmic 1	18,625.63	−0.83
P35704	Peroxiredoxin-2	2627.63	−0.82
P09951	Synapsin-1	1081.01	−0.82
P02688	Myelin basic protein	13,414.88	−0.76
P82995	Heat shock protein HSP 90-alpha	1727.77	−0.73
P07895	Superoxide dismutase [Mn], mitochondrial	514.65	−0.70
Q5U300	Ubiquitin-like modifier-activating enzyme 1	207.98	−0.69
P21575	Dynamin-1	876.32	−0.69
Q08877	Dynamin-3	202.61	−0.69
P31399	ATP synthase subunit d, mitochondrial	589.98	−0.66
P10719	ATP synthase subunit beta, mitochondrial	11,814.81	−0.65
O35244	Peroxiredoxin-6	2447.06	−0.64
P31016	Disks large homolog 4	86.82	-
Q62671	E3 ubiquitin-protein ligase UBR5	48.7	-
O88600	Heat shock 70 kDa protein 4	62.06	-
Q66HA8	Heat shock protein 105 kDa	45.92	-
Q63560	Microtubule-associated protein 6	72.62	-
P07722	Myelin-associated glycoprotein	55.31	-
P34064	Proteasome subunit alpha type-5	83.39	-
P61959	Small ubiquitin-related modifier 2	152.37	-
Q5XIF4	Small ubiquitin-related modifier 3	152.37	-
B2RYG6	Ubiquitin thioesterase OTUB1	121.2	-
Q7M767	Ubiquitin-conjugating enzyme E2 variant 2	735.2	-

+372 proteins with different status of regulation. [a] Accession ID according to Uniport.org database. Positive and negative values of fold change indicate up- and down-regulated proteins, respectively. Sign of—indicates exclusive expression in the control group, i.e., absent in the IHg group. Results of the comparison between the IHg group *versus* the control group.

2.5. *The IHg Exposure Increased Cell Death by Cytotoxicity and Apoptosis in the Cerebellum of Adult Rats*

After 45 days of exposure, the cell death markers of apoptosis increased around 41.70% (±7.35%) in comparison with the control (100 ± 1.81%; $p = 0.0002$), and in parallel, the marker of cytotoxicity increased around 63.10% (±4.19%) in comparison with the control (100 ± 1.66%; $p < 0.05$), as observed in Figure 5.

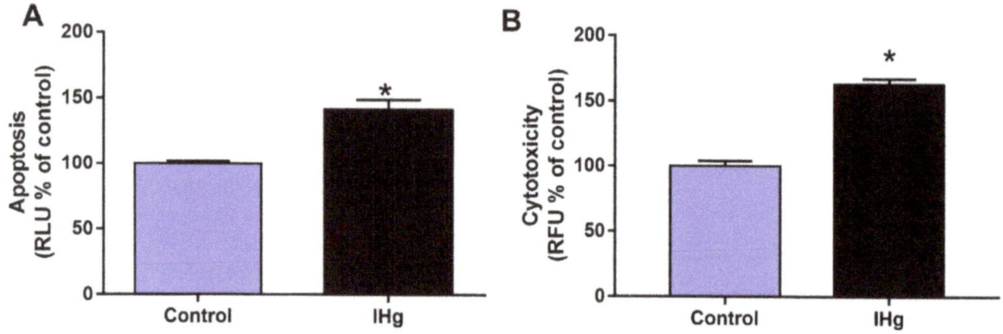

Figure 5. Nature of cell death analyses in the cerebellum of adult rats exposed to 0.375 mg/kg of $HgCl_2$ for 45 days. The apoptosis analysis (**A**) was initially represented as relative luminescence unit (RLU), and cytotoxicity analysis (**B**) was expressed initially as relative fluorescence unit (RFU). Data are presented as percentage (%) of control (mean ± S.E.M.). * $p < 0.05$, *t*-Student test (n = 10).

2.6. The IHg Intoxication Reduces Purkinje Cells Density in Adult Rats

The histological analysis showed a significant reduction in Purkinje cells number in rats (20.22 ± 1.593 cells/mm) in comparison with the control (28.67 ± 2.198 cells/mm; $p = 0.011$) (Figure 6).

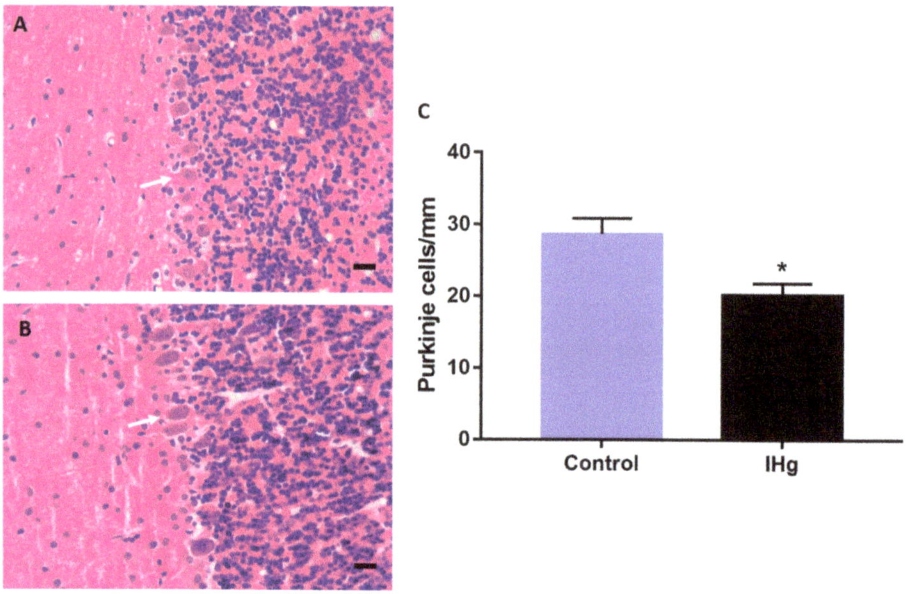

Figure 6. Morphological analysis of the cerebellum of rats exposed to 0.375 mg/kg of $HgCl_2$ for 45 days. In (**A**,**B**), representative photomicrographs of control and exposed groups, respectively. In (**C**), results are expressed as mean ± S.E.M and * $p < 0.05$, *t*-Student test (n = 12). Scale bar: 20 μm.

2.7. The Successive IHg-Induced Damages on the Cerebellum of Rats Led to a Poor Motor Function

The exposed animals showed a worse motor performance on the inclined plane than the control group ($p = 0.0129$). The control group showed a score of 76.2° (± 0.51°), while the exposed group 73.5° (±0.83°) (Figure 7A). Moreover, the cerebellar functions associated with the rotarod performance were affected by IHg long-term exposure. The latency to the first fall in the exposed group was shorter than the control group in all test sessions ($p < 0.0001$, Figure 7B).

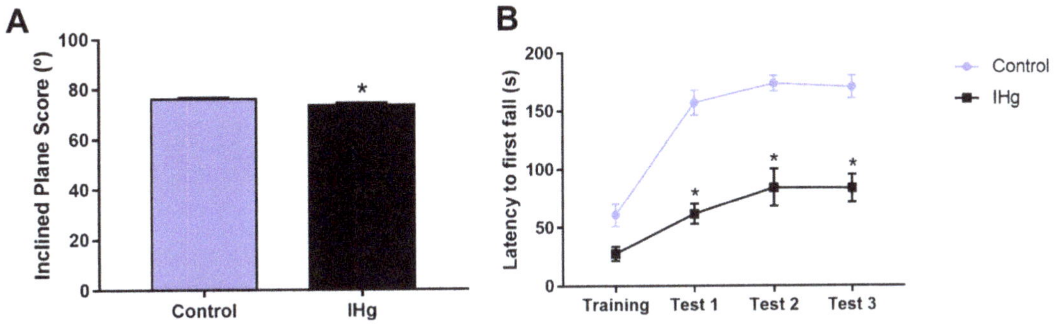

Figure 7. Inclined plane and rotarod performances of rats exposed to 0.375 mg/kg of $HgCl_2$ for 45 days. In (**A**), an average of the falling degree of five test sessions in inclined plane. In (**B**), latency (seconds) to the first fall in rotarod. Results are expressed as mean ± S.E.M. * $p < 0.05$, *t*-Student test (**A**) or one-way ANOVA with repeated measures (**B**) (n = 20).

3. Discussion

This study is the first investigation that brings to the literature evidence regarding the cerebellar global proteomic profile of rats exposed to IHg that underlies the IHg-induced neurotoxicity. Our results demonstrate the IHg exposure triggers oxidative stress and cell death by cytotoxicity and apoptosis, besides the reduction in Purkinje cells density in the cerebellum of rats. Those changes drove to impairments on cerebellar-related motor function as an inclined-plan and rotarod performance. In addition, the cerebellar proteome of IHg exposed rats revealed the modulation of proteins associated with cerebellar development, proteostasis, synaptic activity, and metabolic functions, which may be underlying the impairments on cerebellar-related motor function of adult rats.

IHg is known for its lower liposolubility, hence the lower capacity of crossing biological barriers as cell membranes, blood-brain-barrier, and placental barrier [22]. The detrimental effects IHg are directly associated with the oxidation state of the ion, i.e., mercurous (Hg^+) ion causes minor damage in comparison with mercuric ions (Hg^{2+}) [22], in which the latter was the chemical form used in this study. Although inorganic species have unfavorable physical-chemical characteristics to biomagnify and bioaccumulate, the presence of IHg in contaminated fish commercially available and consumed by humans was already evidenced [6]. In the past few years, our group has shown that several other organs are susceptible to IHg toxicity, such as blood [16], salivary glands [23], and the central nervous structures such as the hippocampus, motor cortex, and spinal cord [17–21] after the exposure to 0.375 mg/kg per day of $HgCl_2$. Then, despite all the toxicokinetic features, the long-term exposure to IHg increased the levels of total mercury in cerebellar parenchyma of adult rats about 4-fold higher than non-exposed animals significantly.

The literature shows that after systemic distribution, IHg can trigger cellular damage such as inhibition of microtubule assembly [24], DNA damage [25,26], mitochondrial impairments [27,28], and oxidative stress [17,29], which in turn, can trigger or aggravate those previously mentioned targets of mercury-induced toxicity. Oxidative stress is caused after an imbalance of cellular redox status, when there is an overproduction of reactive oxygen species (ROS) in detriment to the lower antioxidant capacity [29], being an essential mechanism of mercury damage as previously reviewed [30]. In addition, these biochemical impairments are also visualized in a systemic perspective once we observed oxidative stress triggering in peripheral blood of rats exposed to IHg under the same experimental protocol [16].

To investigate whether this dose of IHg would be capable of reproducing the classic features of mercury mechanisms of damage, we assessed the redox status by the cerebellar capacity of neutralizing peroxyl radicals and the lipid peroxidation. The peroxyl radical affects cells by oxidizing the DNA [31] and initializing the lipid peroxidation of cell and organelle membranes [32]. There is an essential link between peroxyl radicals and lipid peroxidation, which is why we also evaluated the malondialdehyde levels, that indicate the oxidation of polyunsaturated fatty acids present in cell and organelle membranes [33]. Our results showed a decrease in ACAP and increased levels of malondialdehyde, indicating a process of oxidative stress in the cerebellum of rats exposed to IHg, which is consistent with the literature on mercury toxicity, and possibly associated with the down-regulation of Peroxiredoxin 2 (P35704) and 6 (O35244), and Superoxide dismutase (P07895) as shown in the proteomics, suggesting that the decrease on enzymatic antioxidant competence against ROS, such as superoxide anions, peroxides and peroxide nitrites derived from IHg-induced toxicity increased lipid peroxidation [34,35].

Oxidative stress impairs several biological processes, such as mitochondrial function, energy metabolism, cell death, synaptic plasticity, and protein homeostasis (proteostasis) [36–39]. As part of the energy imbalance, the reduction in ATP levels may lead to a transient downregulation of global translation processes to generate new proteins, and, indeed, in time, an oxidative stress state; this is a natural strategy to prevent protein unfolding and aggregation [40,41]. Corroborating this fact, the proteomic approach showed a significant reduction in translation processes in the cerebellum of rats exposed to IHg, as shown by the 362 proteins down-regulated or exclusively found in the control group.

In addition, the adequate regulation of the ubiquitin-proteasome and molecular chaperones, such as heat shock proteins (HSP), is involved in proteome protection [40,42]. The former is a proteolytic system that degrades oxidized proteins, working as quality control, and the latter, mainly the HSP, is related to protein maturation, re-folding, and degradation [41,43]. The proteomic approach showed the down-regulation of Ubiquitin-like modifier-activating enzyme 1 (Q5U300) and exclusive regulation of E3 ubiquitin-protein ligase UBR5 (Q62671), proteasome subunit alpha type-5 (P34064), small ubiquitin-related modifier 2 and 3 (P61959 and Q5XIF4, respectively), ubiquitin thioesterase OTUB1 (B2RYG6) and ubiquitin-conjugating enzyme E2 variant 2 (Q7M767) in the control group (i.e., absent in exposed group), revealing an evident impairment on the ubiquitin-proteasome system. Furthermore, we also observed the exclusive regulation of HSP 70 kDa protein 4 and 105 kDa (O88600 and Q66HA8, respectively) in the control group and down-regulation of several HSP with different molecular weights as 60 kDa (P63039), 70 kDa (P0DMW0), 90 (P82995). It is worth mentioning that HSP with molecular weights between 40 and 105 kDa are ATP-dependent [44], and interestingly, our proteomic approach also demonstrated the down-regulation of several ATP-synthase subunits (e.g., Q06647, P31399, P10719), suggesting not only a cellular energy metabolism failure but also a compromise on HSP functioning. In fact, HSPs have been previously proposed as important therapeutic tools for many diseases [45], and here we provide evidence on their possible role as early biomarkers of neurological damage.

Besides the oxidation of essential macromolecules as proteins and lipids, we also had the firing of cell death pathways after mitochondrial impairment [46]. In this way, cytochrome c plays a crucial role in apoptosis once its release from mitochondria to cytoplasm promotes the generation of the apoptosome and further activation of pro-caspases 3 and 7 [47]. The proteomic analysis showed the modulation of some cytochrome c oxidase subunits, such as subunits 4 (P10888, up-regulated), 2 (P00406, down-regulated), and 5B (P12075, down-regulated). Corroborating these findings, we also observed increased apoptosis markers (caspase 3/7) in the IHg-exposed group. Regarding the result of cytotoxicity, it ratifies the outcome triggered by apoptotic mechanisms once the method used assesses the activity of a protease released from cells when there is a compromised membrane.

In addition to that, a reduction in Purkinje cell density was observed. These cells play a pivotal role in motor functions, so we elected them to perform the histological analysis. Indeed, the cerebellum presents several structures and different cells; however, the Purkinje cells are integrators and effectors in cerebellar neuronal communication and are important to sensorimotor calibration and motor learning. Each cell receives inputs from several parallel fibers, besides inputs and synaptic contacts from climbing fibers to regulate the movements [48–50]. In this way, our data indicate that the increase in apoptosis and cytotoxicity markers, besides the oxidative stress, may reduce Purkinje cell numbers.

Besides the structural component, it is important to highlight the role of proteins related to synaptic communication in neural functions. Firstly, we must clarify that synaptic transmission has a complex dynamic involving the formation, transport, docking of vesicles, the neurotransmitters, the rearrangement of the pre- and post-synaptic neuronal cytoskeleton and the calcium homeostasis [51,52]. Then, the first set of proteins that we observed significantly changed was calmodulin unities 1 (P0DP29), 2 (P0DP30), and 3 (P0DP31) found to be up-regulated, while calcium-calmodulin-dependent protein kinase type II subunit(s) alpha (P11275), beta (P08413), gamma (P11730), and delta (P15791) were down-regulated. The second set of proteins are constituents of the cytoskeleton, such as actin subunits (e.g., P68035 and P60711) down-regulation; microtubule-related proteins, as expected due to the mercury toxicity mechanisms, such as Dynamin 1, 2, and 3 (P21575, P39052, and Q08877, respectively, down-regulated); and several tubulin subunits. Calcium-calmodulin kinase II proteins play several biological roles, such as neurotransmitter synthesis and release, in cognitive and motor functions [53,54], while actin and microtubule-related proteins are important for cytosolic transport vesicles, exocytosis, and endocytosis of vesicles [55].

Comprising the third set of proteins involved in synaptic transmission, we highlight the vesicle constituents, such as synaptotagmin-1 (P21707, up-regulated) and synapsin-1 (P09951, down-regulated). However, four proteins must be emphasized: synaptophysin (P07825; up-regulated); microtubule-associated proteins 1A (P34926; up-regulated) and 6 (Q63560; absent in exposed group) and Disks large homolog 4 (P31016, absent in exposed group). These proteins represent one of the most studied synaptic complexes because they are components of synaptic vesicles [56], besides the pre-and post-synaptic platforms [44,57]. This modulation on the synaptosome suggests an impact on cerebellar functions caused by IHg long-term exposure, which may have led to the loss of balance and motor control demonstrated by the behavioral assessment.

For cerebellar functions and development, structural proteins play a crucial role in the action potential and synaptic communication. Considering that, we observed the down-regulation of myelin basic protein (P02688), the up-regulation of myelin-oligodendrocyte glycoprotein (Q63345), and exclusive regulation of myelin-associated glycoprotein (P07722) in the control group, indicating damages to the myelin-sheath structure. P02688 is an abundant protein in CNS, and it acts to mediate the cytosolic adhesion of multilayered compact myelin [58]. Besides that, P07722 and Q63345 are proteins associated with the myelin maintenance and axon-glial interplay [59,60]. Thus, modifications on the regulation status of those proteins may compromise the cerebellar functioning by myelin decompaction and axonal regeneration, which impair the action potentials of neurons, and hence, the cerebellum-related motor functions.

In this way, it is evident that IHg triggers considerable damages to the cerebellum, since in the proposed model, the increase in total mercury levels in cerebellar parenchyma was associated with oxidative stress triggering and a neurodegenerative pattern, possibly associated with the cytotoxic and apoptotic mechanism. Besides that, the global proteomic profile revealed several proteins that underlie these successive damages, culminating in the motor dysfunction of rats, as evidenced by rotarod and inclined plane assessment. The motor function depends on several anatomical regions in CNS, and the cerebellum plays a crucial role in balance, motor refinement, and even cognitive aspects [61]. For that reason, we used rotarod and inclined plane tests to assess the rats' abilities and validate the functional damages triggered by IHg long-term exposure. The results showed an evident impairment in balance and motor refinement as an outcome of IHg exposure and these data reinforce the understanding that IHg affects multiple motor-related regions from CNS.

4. Materials and Methods

4.1. Ethical Statement and Experimental Protocol

All experiments were performed following bioethical instructions after authorization from the Ethics Committee on Animal Experimentation from the Federal University of Pará under protocol NO 9228050418 and followed the ARRIVE 2.0 guideline (see Supplementary Table S2).

Fifty-four male Wistar rats (*Rattus novergicus*), weighing about 200 g and 90 days old, were used in this study. The animals were divided into two groups (control and exposed) by simple randomization and kept in plastic cages (4 animals per cage). The animals were maintained at a temperature of 25 °C with a 12 h dark/light cycle (lights on at 7 a.m.). The exposed group received a single concentration of 0.375 mg/kg/day for 45 days as $HgCl_2$ (Sigma-Aldrich, San Louis, MO, USA) solubilized in distilled water and administered by oral gavage. This dosage was previously established and seen as capable of triggering damage in CNS structures [17–21]; besides, the dose was adjusted weekly according to the body weight. The control group received only distilled water by intragastric gavage for the same period and proportional volume. The animals would be excluded in case of malnutrition and the presence of bruises on the body. During the experimental period there were no deaths and exclusions.

4.2. Behavioral Assessment

After 24 h from the last exposure to IHg, behavioral tests were performed. Afterward, ten random animals were conducted to the assay room with attenuation of noise levels and low illumination (12 lux) and acclimated for at least one hour before the behavioral tests to assess the motor functions associated with the cerebellum as primary outcomes. The animals were identified as sequential numbers by L.O.B and V.S.C., and the researchers who performed the behavioral assessment were blinded regarding this information.

4.2.1. Inclined Plane Test

The animals' ability to maintain its postural stability, a crucial cerebellar skill, was assessed with the inclined plane test. The rats were positioned individually in an apparatus consisting of an inclined plane with adjustable angulation. The angle is raised every 5 degrees (from 0° to 90°) every 5 s, which can be used as an index of climbing strength. The animals were subjected to five test sessions in which the maximum inclination that the animal was able to maintain itself for 5 s was recorded as the final degree. The average angle of the five sessions was analyzed [62,63].

4.2.2. Rotarod Test

To evaluate the motor coordination and balance through a forced task, we performed the rotarod test, consisting of an apparatus (Insight Scientific Equipment, Ribeirão Preto, Brazil) of a grooved metal roll (8 cm in diameter) with separated compartments for each animal. The animals were initially trained to stay on the rotating rod at 15 revolutions per minute (rpm) for 3 min (training phase). Then, we assessed the animal's ability to remain on the roll for three successive sessions of 3 min each at 15 rpm with an interval of 60 s [64] to evaluate the latency to first fall.

4.3. Sample Collection and Preparation

After behavioral assays, the animals were deeply anesthetized with a solution of xylazine hydrochloride (30 mg/kg) and ketamine chloride (180 mg/kg) (i.p.). Then, after total loss of paw and corneal reflexes, the animals were euthanized. The cerebellums were collected after brain removal, and some specimens were gently washed in cold saline solution. After, they were stored in microtubes, frozen in liquid nitrogen, and kept at

−80 °C until further procedures, while others were dissociated by collagenase as detailed below, both to assess secondary outcomes.

Before the oxidative biochemistry analyses, the samples were thawed and resuspended (1:1, *w*/*v*) in Tris-HCl buffer (20 mM, pH 7.4, at 4 °C) and sonically disaggregated. Then, each sample was divided into two aliquots and frozen. For mercury measurements and proteomic analysis, the entire samples were frozen until further analyses. The evaluation of apoptosis and cytotoxicity required tissue dissociation right after the sample collection, executed with collagenase in concentrations of 2 mg/mL (20 min) and 4 mg/mL (40 min) at 37 °C. Details are described in the respective topics below.

4.4. Mercury Measurements

The samples were digested in a nitric acid, perchloric acid, and sulfuric acid solution (1:1:5). The samples were heated, and the Hg^{2+} present in the samples was converted into Hg^0 after stannous chloride addition. The equipment used was the Semi-Automatic Mercury Analyzer-Hg 201 (Sanso Seisakusho Co. Ltd., Tokyo, Japan) as previously reproduced by our group [16,65–67], following the protocol of Suzuki et al. [68] and. The total mercury content was expressed as mg/kg.

4.5. Oxidative Biochemistry Analyses

4.5.1. Antioxidant Capacity against Peroxyl Radicals (ACAPs)

This analysis was proceeded based on Amado et al. [68]. One aliquot of the homogenate was centrifuged at 14,000 rpm at 4 °C for 10 min to collect the supernatant. From the supernatant, 200 µL was incubated with a peroxyl radical generator (2,2′-azobis-2-methylpropionamidine dihydrochloride (ABAP; 4 mM; Sigma-Aldrich, San Louis, MO, USA) in triplicate to verify the oxidation of ABAP and generate fluorescence. After the reaction with ABAP, the generated fluorescence was measured on a fluorimeter (Victor X3, Perkin Elmer, Waltham, MA, USA) at a controlled temperature of 35 °C. After the first reading to determine the fluorescence value, a total volume of 10 µL of 2′,7′ H2DCF-DA (40 nM) was added to the microplate, with cyclic readings every 5 min for 60 min. High relative areas under the curve established by a second-degree polynomial curve indicate a low antioxidant capacity to neutralize peroxyl radicals. The inverse relative difference between the area with and without ABAP was considered a measure of antioxidant capacity. The results were later expressed as the percentage of control.

4.5.2. Lipid Peroxidation (LPO)

For LPO, 5 µL from total homogenate was diluted with Tris-HCl buffer (1:30, *v*:*v*) followed by quantifying total protein content by Bradford's method [69] for data normalization. The remaining lysate was centrifuged at 5600 rpm for 10 min at 4 °C, and the supernatant was collected to determine LPO [70]. A volume of 325 µL of 10.3 mM NMFI diluted in methanol (1:3, *v*:*v*) and 75 µL of methanesulfonic acid were added to 100 µL of the standard malondialdehyde (MDA) solutions or samples and heated at 45 °C for 40 min. Then, absorbances were measured at 570 nm, and the results were expressed as nmol of MDA/mg of protein and then converted to the percentage of control.

4.6. Proteomic Approach and Bioinformatic Analyses

A total of six samples from each group were used. A pool of two samples was performed, and the analysis was carried out in biological triplicate. The detailed protocol is available elsewhere [18,71,72]. The samples were cryofractured using a cryogenic mill and liquid nitrogen followed by the extraction of proteins with lysis buffer (urea 7 M, thiourea 2 M, diluted in ammonium bicarbonate) in constant stirring at 4 °C; then the samples were centrifuged for 30 min at 14,000 rpm at 4 °C. Subsequently, 50 µg of protein (determined by Bradford's method) were diluted in 50 µL of ammonium bicarbonate (50 mM), and each sample 10 µL ammonium bicarbonate (50 mM) and 25 µL Rapigest (0.2%) (Waters Co., Manchester, UK) were added and incubated at 37 °C for 30 min. Subsequently, 2.5 µL of dithiothreitol (100 mM) was added and incubated at 37 °C for 60 min, followed by 2.5 μL of 300 mM iodoacetamide (BioRad, Hercules, CA, USA) addition and incubation for 30 min at room temperature and in the dark. Then, the protein digestion was preceded by adding 10 μL of trypsin (Thermo Fisher, Waltham, MA, USA) for 14 h at 37 °C. On the following day, 10 μL of 5% trifluoroacetic acid (Sigma-Aldrich, St. Louis, MO, USA) was added for 90 min at 37 °C and centrifuged at 14,000 rpm at 6 °C for 30 min. Afterwards, the supernatants were collected and purified using C18 spin columns (Thermo Fisher, Waltham, MA, USA). The samples were resuspended in 12 μL of ADH (1 pmol·μL^{-1}) + 108 μL of 3% acetonitrile (Sigma-Aldrich, St. Louis, MO, USA) and 0.1% formic acid (Thermo Fisher, Waltham, MA, USA).

The reading and identification of the peptides were performed using a nanoAcquity UPLCXevo QTof MS system (Waters, Manchester, UK) and the Protein Lynx Global Server (PLGS) software, applying the Monte-Carlo algorithm and identifying the peptides by Uniprot database for *Rattus norvegicus*. Our analyses considered $p < 0.05$ for down-regulated proteins and $1 - p > 0.95$ for up-regulated proteins. After compiling the results, we performed the bioinformatic analysis using the Cytoscape v. 3.8 software with the ClueGO plugin to determine the functional categories of proteins based on gene ontology annotations of biological processes.

4.7. Apoptosis and Cytotoxicity Assays

After dissociation, 100 μL of the homogenate containing the dissociated cells were added to a 96-well microplate with 100 μL of the reagent from the Caspase-Glo 3/7 Assay System, a luminescent assay to measure caspase-3/7 activities and CytoTox-Glo (Promega System, Delft, The Netherlands), which uses a luminogenic peptide substrate to measure dead cell protease activity. Readings were performed on GloMax equipment (Promega System, Delft, The Netherlands) according to the manufacturer's recommendations. The results were expressed in the relative luminescence unit (RLU) for cytotoxicity and relative fluorescence unit (RFU) for apoptosis and then converted into a percentage of control.

4.8. Morphological Analysis

A set of animals was deeply anesthetized with a solution of xylazine hydrochloride (30 mg/kg) and ketamine chloride (180 mg/kg) (i.p.). Then, after total loss of paw and corneal reflexes, the animals were perfused with heparinized saline solution (0.9%) and fixed with 4% paraformaldehyde. The samples were collected and post-fixed in Bouin's solution for six hours and then embedded in paraplast (Sigma-Aldrich, St. Louis, MO, USA). Coronal sections (7 μm) were obtained with a manual microtome and stained by hematoxylin and eosin routine staining.

The Purkinje cell density was evaluated using a grid attached to the eyepiece of the microscope, with dimensions corresponding to an area of 0.0625 mm^2 to count the cells in three different fields from each section. The cells are displayed in a linear form, and then, after counting the cells, a proportional area was considered of grids filled with the cells. The representative photomicrographs were taken by a DS-Fi3 microscope camera attached to the Nikon Eclipse Ci H550s bright field microscope.

4.9. Statistical Analyses

After tabulation of the results obtained, the values were analyzed through the GraphPad Prism 7.0 software. The normality of data distribution was tested by the Shapiro–Wilk method, and to analyze the difference between groups, we used Student's t-test (inclined plane score, total mercury determination, oxidative biochemistry assays, cell death nature evaluation, and morphological analysis). We also used a two-way ANOVA 2 way with Sidak's post-test to analyze the significant difference between groups in body weight register during the experimental period and rotarod test. It was adopted $p < 0.05$ as significance in all statistical methods. For proteomic data, the statistical analysis was performed using the equipment's software as described above. All methodological steps are summarized in Figure 8.

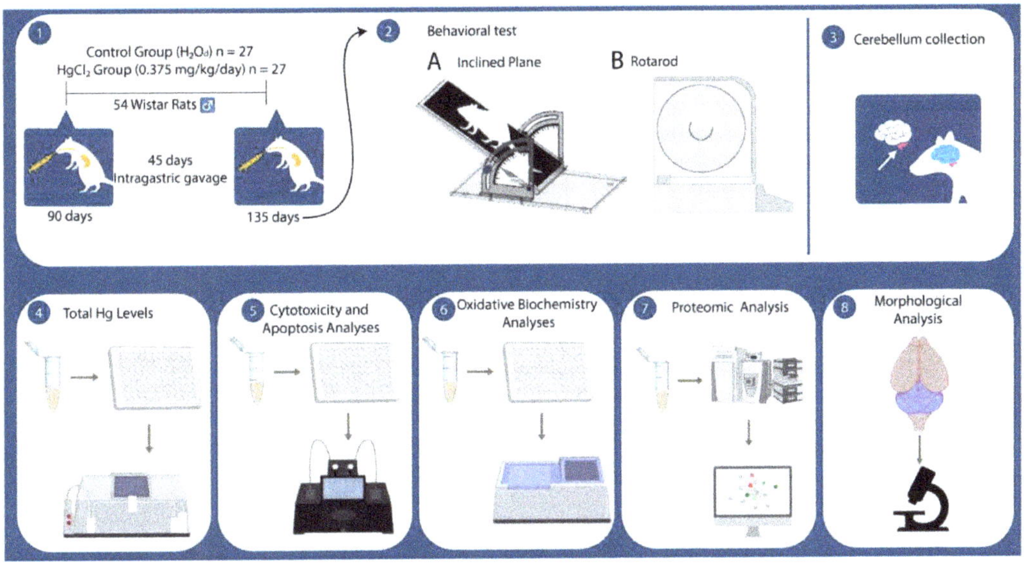

Figure 8. Schematic figure of all methodological steps. In (**1**), the experimental protocol of $HgCl_2$ long-term exposure in adult Wistar rats, followed by motor functions assessment (**2**) by inclined plane (**A**) and rotarod (**B**) tests and then, cerebellum collection (**3**); in (**4**), quantification of total mercury (Hg) levels by atomic absorption spectrometry; in (**5**), oxidative biochemistry analyses by the determination of lipid peroxidation (LPO) levels and antioxidant activity against peroxyl radicals (ACAPs); in (**6**), proteomic analysis by mass spectrometry with bioinformatics analysis though Cytoscape software. In (**7**), evaluation of cell death nature by cytotoxicity and apoptosis assays and in (**8**), Purkinje cell counting by bright field microscopy.

5. Conclusions

In conclusion, the results pointed out that IHg long-term exposure increases the mercury bioavailability in the cerebellum parenchyma and triggers several molecular, biochemical, and morphological impairments associated with the poor motor functions related to the cerebellum. It triggers oxidative stress and cell death by cytotoxicity and apoptosis, modulates several proteins from biological processes that may be potential new targets of researches designed to elucidate the mechanisms of damage.

Supplementary Materials: The following are available online at https://www.mdpi.com/article/10.3390/ijms23010111/s1.

Author Contributions: Conceptualization: L.O.B., F.B.T., R.R.L.; Methodology: F.B.T., L.O.B., V.S.C., A.D., M.S.F., M.C.F.S., E.H.C.d.O., W.A.B.A.; Formal analysis: L.O.B., R.R.L., L.M.P.F.; Investigation: L.O.B., V.S.C., E.H.C.d.O., C.d.S.F.M., R.R.L.; Resources: E.H.C.d.O., M.E.C.-L., R.R.L., M.A.R.B.; Data curation: L.O.B., V.S.C., A.D., M.C.F.S., L.M.P.F.; Writing—Original draft preparation: L.O.B., V.S.C., B.P.; Writing, review and editing: L.O.B., B.P., M.E.C.-L., R.R.L.; Visualization: R.R.L., B.P.; Supervision: R.R.L.; Project administration: R.R.L.; Funding acquisition: R.R.L. All authors have read and agreed to the published version of the manuscript.

Funding: L.O.B. and V.S.C. received CNPq—Brazilian Ministry of Science, Technology and Innovation scholarships. This paper had partial financial support from Fundação de Amparo à Pesquisa do Estado de São Paulo (FAPESP) and Coordenação de Aperfeiçoamento de Pessoal de Nível Superior, from Brazilian Ministry of Education (finance code 001). The APC payment was supported by Pró-reitoria de Pesquisa e Pós-graduação (PROPESP) from Federal University of Pará (UFPA).

Institutional Review Board Statement: All experiments were performed following bioethical instructions after authorization from the Ethics Committee on Animal Experimentation from the Federal University of Pará under protocol NO 9228050418 and followed the ARRIVE 2.0 guideline.

Data Availability Statement: All data are available within the article and in Supplementary Materials.

Acknowledgments: We are grateful to CNPq, CAPES and PROPESP for all the fellowship in the development of this research.

Conflicts of Interest: The authors declare no conflict of interest. The funders had no role in the design of the study; in the collection, analyses, or interpretation of data; in the writing of the manuscript, or in the decision to publish the results.

Abbreviations

IHg	Inorganic Mercury
CNS	Central Nervous System
i.p.	Intraperitoneal
w/v	weight/volume
ABAP	2,2′-azobis-2-methylpropionamidine dihydrochloride
LPO	Lipid Peroxidation
MDA	Malondialdehyde
UPLC	Ultra performance liquid chromatography
PLGS	Protein Lynx Global SERVER
RLU	Relative Luminescence Unit
RFU	Relative Fluorescence Unit
ROS	Reactive Oxygen Species
HSP	Heat Shock Proteins

References

1. Ha, E.; Basu, N.; Bose-O'Reilly, S.; Dórea, J.G.; McSorley, E.; Sakamoto, M.; Chan, H.M. Current progress on understanding the impact of mercury on human health. *Environ. Res.* **2017**, *152*, 419–433. [CrossRef]
2. Orr, S.E.; Barnes, M.C.; Joshee, L.; Uchakina, O.; McKallip, R.J.; Bridges, C.C. Potential mechanisms of cellular injury following exposure to a physiologically relevant species of inorganic mercury. *Toxicol. Lett.* **2019**, *304*, 13–20. [CrossRef]
3. Crespo-Lopez, M.E.; Augusto-Oliveira, M.; Lopes-Araújo, A.; Santos-Sacramento, L.; Yuki Takeda, P.; Macchi, B.M.; do Nascimento, J.L.M.; Maia, C.S.F.; Lima, R.R.; Arrifano, G.P. Mercury: What can we learn from the Amazon? *Environ. Int.* **2021**, *146*, 106223. [CrossRef] [PubMed]
4. Santos-Sacramento, L.; Arrifano, G.P.; Lopes-Araújo, A.; Augusto-Oliveira, M.; Albuquerque-Santos, R.; Takeda, P.Y.; Souza-Monteiro, J.R.; Macchi, B.M.; do Nascimento, J.L.M.; Lima, R.R.; et al. Human neurotoxicity of mercury in the Amazon: A scoping review with insights and critical considerations. *Ecotoxicol. Environ. Saf.* **2021**, *208*, 111686. [CrossRef]
5. Nevado, J.J.B.; Martín-Doimeadios, R.C.R.; Moreno, M.J.; do Nascimento, J.L.M.; Herculano, A.M.; Crespo-López, M.E. Mercury speciation analysis on cell lines of the human central nervous system to explain genotoxic effects. *Microchem. J.* **2009**, *93*, 12–16. [CrossRef]
6. Rodríguez Martín-Doimeadios, R.C.; Berzas Nevado, J.J.; Guzmán Bernardo, F.J.; Jiménez Moreno, M.; Arrifano, G.P.; Herculano, A.M.; do Nascimento, J.L.; Crespo-López, M.E. Comparative study of mercury speciation in commercial fishes of the Brazilian Amazon. *Environ. Sci. Pollut. Res. Int.* **2014**, *21*, 7466–7479. [CrossRef]
7. Machado, C.L.R.; Crespo-Lopez, M.E.; Augusto-Oliveira, M.; Arrifano, G.P.; Macchi, B.M.; Lopes-Araújo, A.; Santos-Sacramento, L.; Souza-Monteiro, J.R.; Alvarez-Leite, J.I.; Souza, C.B.A. Eating in the Amazon: Nutritional Status of the Riverine Populations and Possible Nudge Interventions. *Foods* **2021**, *10*, 1015. [CrossRef]
8. Park, J.D.; Zheng, W. Human exposure and health effects of inorganic and elemental mercury. *J. Prev. Med. Public Health Yebang Uihakhoe Chi* **2012**, *45*, 344–352. [CrossRef]
9. Ahmad, S.; Mahmood, R. Mercury chloride toxicity in human erythrocytes: Enhanced generation of ROS and RNS, hemoglobin oxidation, impaired antioxidant power, and inhibition of plasma membrane redox system. *Environ. Sci. Pollut. Res. Int.* **2019**, *26*, 5645–5657. [CrossRef]
10. Pamphlett, R.; Kum Jew, S. Inorganic mercury in human astrocytes, oligodendrocytes, corticomotoneurons and the locus ceruleus: Implications for multiple sclerosis, neurodegenerative disorders and gliomas. *Biometals Int. J. Role Met. Ions Biol. Biochem. Med.* **2018**, *31*, 807–819. [CrossRef] [PubMed]
11. Risher, J.F.; De Rosa, C.T. Inorganic: The other mercury. *J. Environ. Health* **2007**, *70*, 9–16; discussion 40.
12. Mariën, P.; Borgatti, R. Language and the cerebellum. *Handb. Clin. Neurol.* **2018**, *154*, 181–202. [CrossRef]
13. Schmahmann, J.D.; Guell, X.; Stoodley, C.J.; Halko, M.A. The Theory and Neuroscience of Cerebellar Cognition. *Annu. Rev. Neurosci.* **2019**, *42*, 337–364. [CrossRef]

14. Stoodley, C.J. The Cerebellum and Neurodevelopmental Disorders. *Cerebellum* **2016**, *15*, 34–37. [CrossRef]
15. Koeppen, A.H. The neuropathology of the adult cerebellum. *Handb. Clin. Neurol.* **2018**, *154*, 129–149. [CrossRef] [PubMed]
16. Dos Santos Chemelo, V.; Bittencourt, L.O.; Aragão, W.A.B.; Dos Santos, S.M.; Souza-Rodrigues, R.D.; Ribeiro, C.; Monteiro, M.C.; Lima, R.R. Long-Term Exposure to Inorganic Mercury Leads to Oxidative Stress in Peripheral Blood of Adult Rats. *Biol. Trace Elem. Res.* **2021**, *199*, 2992–3000. [CrossRef]
17. Aragão, W.A.B.; Teixeira, F.B.; Fagundes, N.C.F.; Fernandes, R.M.; Fernandes, L.M.P.; da Silva, M.C.F.; Amado, L.L.; Sagica, F.E.S.; Oliveira, E.H.C.; Crespo-Lopez, M.E.; et al. Hippocampal Dysfunction Provoked by Mercury Chloride Exposure: Evaluation of Cognitive Impairment, Oxidative Stress, Tissue Injury and Nature of Cell Death. *Oxidative Med. Cell. Longev.* **2018**, *2018*, 7878050. [CrossRef] [PubMed]
18. Corrêa, M.G.; Bittencourt, L.O.; Nascimento, P.C.; Ferreira, R.O.; Aragão, W.A.B.; Silva, M.C.F.; Gomes-Leal, W.; Fernandes, M.S.; Dionizio, A.; Buzalaf, M.R.; et al. Spinal cord neurodegeneration after inorganic mercury long-term exposure in adult rats: Ultrastructural, proteomic and biochemical damages associated with reduced neuronal density. *Ecotoxicol. Environ. Saf.* **2020**, *191*, 110159. [CrossRef] [PubMed]
19. Teixeira, F.B.; Fernandes, R.M.; Farias-Junior, P.M.; Costa, N.M.; Fernandes, L.M.; Santana, L.N.; Silva-Junior, A.F.; Silva, M.C.; Maia, C.S.; Lima, R.R. Evaluation of the effects of chronic intoxication with inorganic mercury on memory and motor control in rats. *Int. J. Environ. Res. Public Health* **2014**, *11*, 9171–9185. [CrossRef]
20. Teixeira, F.B.; de Oliveira, A.C.A.; Leão, L.K.R.; Fagundes, N.C.F.; Fernandes, R.M.; Fernandes, L.M.P.; da Silva, M.C.F.; Amado, L.L.; Sagica, F.E.S.; de Oliveira, E.H.C.; et al. Exposure to Inorganic Mercury Causes Oxidative Stress, Cell Death, and Functional Deficits in the Motor Cortex. *Front. Mol. Neurosci.* **2018**, *11*, 125. [CrossRef]
21. Teixeira, F.B.; Leão, L.K.R.; Bittencourt, L.O.; Aragão, W.A.B.; Nascimento, P.C.; Luz, D.A.; Braga, D.V.; Silva, M.; Oliveira, K.R.M.; Herculano, A.M.; et al. Neurochemical dysfunction in motor cortex and hippocampus impairs the behavioral performance of rats chronically exposed to inorganic mercury. *J. Trace Elem. Med. Biol. Organ Soc. Miner. Trace Elem.* **2019**, *52*, 143–150. [CrossRef]
22. Bridges, C.C.; Zalups, R.K. Transport of inorganic mercury and methylmercury in target tissues and organs. *J. Toxicol. Environ. Health. Part B Crit. Rev.* **2010**, *13*, 385–410. [CrossRef]
23. Aragão, W.A.B.; da Costa, N.M.M.; Fagundes, N.C.F.; Silva, M.C.F.; Alves-Junior, S.M.; Pinheiro, J.J.V.; Amado, L.L.; Crespo-López, M.E.; Maia, C.S.F.; Lima, R.R. Chronic exposure to inorganic mercury induces biochemical and morphological changes in the salivary glands of rats. *Met. Integr. Biometal Sci.* **2017**, *9*, 1271–1278. [CrossRef]
24. Bonacker, D.; Stoiber, T.; Wang, M.; Böhm, K.J.; Prots, I.; Unger, E.; Thier, R.; Bolt, H.M.; Degen, G.H. Genotoxicity of inorganic mercury salts based on disturbed microtubule function. *Arch. Toxicol.* **2004**, *78*, 575–583. [CrossRef] [PubMed]
25. Ben-Ozer, E.Y.; Rosenspire, A.J.; McCabe, M.J., Jr.; Worth, R.G.; Kindzelskii, A.L.; Warra, N.S.; Petty, H.R. Mercuric chloride damages cellular DNA by a non-apoptotic mechanism. *Mutat. Res.* **2000**, *470*, 19–27. [CrossRef]
26. Silva-Pereira, L.C.; Cardoso, P.C.; Leite, D.S.; Bahia, M.O.; Bastos, W.R.; Smith, M.A.; Burbano, R.R. Cytotoxicity and genotoxicity of low doses of mercury chloride and methylmercury chloride on human lymphocytes in vitro. *Braz. J. Med. Biol. Res. Rev. Bras. De Pesqui. Med. E Biol.* **2005**, *38*, 901–907. [CrossRef]
27. Weinberg, J.M.; Harding, P.G.; Humes, H.D. Mitochondrial bioenergetics during the initiation of mercuric chloride-induced renal injury. I. Direct effects of in vitro mercuric chloride on renal mitochondrial function. *J. Biol. Chem.* **1982**, *257*, 60–67. [CrossRef]
28. Araragi, S.; Kondoh, M.; Kawase, M.; Saito, S.; Higashimoto, M.; Sato, M. Mercuric chloride induces apoptosis via a mitochondrial-dependent pathway in human leukemia cells. *Toxicology* **2003**, *184*, 1–9. [CrossRef]
29. Pizzino, G.; Irrera, N.; Cucinotta, M.; Pallio, G.; Mannino, F.; Arcoraci, V.; Squadrito, F.; Altavilla, D.; Bitto, A. Oxidative Stress: Harms and Benefits for Human Health. *Oxidative Med. Cell. Longev.* **2017**, *2017*, 8416763. [CrossRef] [PubMed]
30. Crespo-López, M.E.; Macêdo, G.L.; Pereira, S.I.; Arrifano, G.P.; Picanço-Diniz, D.L.; do Nascimento, J.L.; Herculano, A.M. Mercury and human genotoxicity: Critical considerations and possible molecular mechanisms. *Pharmacol. Res.* **2009**, *60*, 212–220. [CrossRef]
31. Lim, P.; Wuenschell, G.E.; Holland, V.; Lee, D.H.; Pfeifer, G.P.; Rodriguez, H.; Termini, J. Peroxyl radical mediated oxidative DNA base damage: Implications for lipid peroxidation induced mutagenesis. *Biochemistry* **2004**, *43*, 15339–15348. [CrossRef] [PubMed]
32. Phaniendra, A.; Jestadi, D.B.; Periyasamy, L. Free radicals: Properties, sources, targets, and their implication in various diseases. *Indian J. Clin. Biochem. IJCB* **2015**, *30*, 11–26. [CrossRef] [PubMed]
33. Moldogazieva, N.T.; Mokhosoev, I.M.; Mel'nikova, T.I.; Porozov, Y.B.; Terentiev, A.A. Oxidative Stress and Advanced Lipoxidation and Glycation End Products (ALEs and AGEs) in Aging and Age-Related Diseases. *Oxidative Med. Cell. Longev.* **2019**, *2019*, 3085756. [CrossRef]
34. Fukai, T.; Ushio-Fukai, M. Superoxide dismutases: Role in redox signaling, vascular function, and diseases. *Antioxid. Redox Signal.* **2011**, *15*, 1583–1606. [CrossRef]
35. Perkins, A.; Nelson, K.J.; Parsonage, D.; Poole, L.B.; Karplus, P.A. Peroxiredoxins: Guardians against oxidative stress and modulators of peroxide signaling. *Trends Biochem. Sci.* **2015**, *40*, 435–445. [CrossRef]
36. Guo, C.; Sun, L.; Chen, X.; Zhang, D. Oxidative stress, mitochondrial damage and neurodegenerative diseases. *Neural Regen. Res.* **2013**, *8*, 2003–2014. [CrossRef]
37. Trostchansky, A.; Quijano, C.; Yadav, H.; Kelley, E.E.; Cassina, A.M. Interplay between Oxidative Stress and Metabolism in Signalling and Disease. *Oxidative Med. Cell. Longev.* **2016**, *2016*, 3274296. [CrossRef] [PubMed]
38. Mattson, M.P.; Liu, D. Energetics and oxidative stress in synaptic plasticity and neurodegenerative disorders. *Neuromolecular Med.* **2002**, *2*, 215–231. [CrossRef] [PubMed]

39. Korovila, I.; Hugo, M.; Castro, J.P.; Weber, D.; Höhn, A.; Grune, T.; Jung, T. Proteostasis, oxidative stress and aging. *Redox Biol.* **2017**, *13*, 550–567. [CrossRef]
40. Dahl, J.U.; Gray, M.J.; Jakob, U. Protein quality control under oxidative stress conditions. *J. Mol. Biol.* **2015**, *427*, 1549–1563. [CrossRef]
41. Reichmann, D.; Voth, W.; Jakob, U. Maintaining a Healthy Proteome during Oxidative Stress. *Mol. Cell* **2018**, *69*, 203–213. [CrossRef]
42. Shang, F.; Taylor, A. Ubiquitin-proteasome pathway and cellular responses to oxidative stress. *Free Radic. Biol. Med.* **2011**, *51*, 5–16. [CrossRef]
43. Miller, D.J.; Fort, P.E. Heat Shock Proteins Regulatory Role in Neurodevelopment. *Front. Neurosci.* **2018**, *12*, 821. [CrossRef]
44. Takei, Y.; Kikkawa, Y.S.; Atapour, N.; Hensch, T.K.; Hirokawa, N. Defects in Synaptic Plasticity, Reduced NMDA-Receptor Transport, and Instability of Postsynaptic Density Proteins in Mice Lacking Microtubule-Associated Protein 1A. *J. Neurosci. Off. J. Soc. Neurosci.* **2015**, *35*, 15539–15554. [CrossRef] [PubMed]
45. Almeida, M.B.; do Nascimento, J.L.; Herculano, A.M.; Crespo-López, M.E. Molecular chaperones: Toward new therapeutic tools. *Biomed. Pharmacother. Biomed. Pharmacother.* **2011**, *65*, 239–243. [CrossRef] [PubMed]
46. Kannan, K.; Jain, S.K. Oxidative stress and apoptosis. *Pathophysiology* **2000**, *7*, 153–163. [CrossRef]
47. Garrido, C.; Galluzzi, L.; Brunet, M.; Puig, P.E.; Didelot, C.; Kroemer, G. Mechanisms of cytochrome c release from mitochondria. *Cell Death Differ.* **2006**, *13*, 1423–1433. [CrossRef]
48. Smeyne, R.J.; Chu, T.; Lewin, A.; Bian, F.; Sanlioglu, S.; Kunsch, C.; Lira, S.A.; Oberdick, J. Local control of granule cell generation by cerebellar Purkinje cells. *Mol. Cell. Neurosci.* **1995**, *6*, 230–251. [CrossRef]
49. Medina, J.F. The multiple roles of Purkinje cells in sensori-motor calibration: To predict, teach and command. *Curr. Opin. Neurobiol.* **2011**, *21*, 616–622. [CrossRef]
50. Lamarão-Vieira, K.; Pamplona-Santos, D.; Nascimento, P.C.; Corrêa, M.G.; Bittencourt, L.O.; Dos Santos, S.M.; Cartágenes, S.C.; Fernandes, L.M.P.; Monteiro, M.C.; Maia, C.S.F.; et al. Physical Exercise Attenuates Oxidative Stress and Morphofunctional Cerebellar Damages Induced by the Ethanol Binge Drinking Paradigm from Adolescence to Adulthood in Rats. *Oxidative Med. Cell. Longev.* **2019**, *2019*, 6802424. [CrossRef]
51. Jockusch, B.M.; Rothkegel, M.; Schwarz, G. Linking the synapse to the cytoskeleton: A breath-taking role for microfilaments. *Neuroreport* **2004**, *15*, 1535–1538. [CrossRef] [PubMed]
52. Südhof, T.C. Calcium control of neurotransmitter release. *Cold Spring Harb. Perspect. Biol.* **2012**, *4*, a011353. [CrossRef]
53. Yamauchi, T. Neuronal Ca^{2+}/calmodulin-dependent protein kinase II–discovery, progress in a quarter of a century, and perspective: Implication for learning and memory. *Biol. Pharm. Bull.* **2005**, *28*, 1342–1354. [CrossRef]
54. Picconi, B.; Gardoni, F.; Centonze, D.; Mauceri, D.; Cenci, M.A.; Bernardi, G.; Calabresi, P.; Di Luca, M. Abnormal Ca^{2+}-calmodulin-dependent protein kinase II function mediates synaptic and motor deficits in experimental parkinsonism. *J. Neurosci. Off. J. Soc. Neurosci.* **2004**, *24*, 5283–5291. [CrossRef]
55. Chung, C.; Barylko, B.; Leitz, J.; Liu, X.; Kavalali, E.T. Acute dynamin inhibition dissects synaptic vesicle recycling pathways that drive spontaneous and evoked neurotransmission. *J. Neurosci. Off. J. Soc. Neurosci.* **2010**, *30*, 1363–1376. [CrossRef] [PubMed]
56. Kwon, S.E.; Chapman, E.R. Synaptophysin regulates the kinetics of synaptic vesicle endocytosis in central neurons. *Neuron* **2011**, *70*, 847–854. [CrossRef]
57. Béïque, J.C.; Andrade, R. PSD-95 regulates synaptic transmission and plasticity in rat cerebral cortex. *J. Physiol.* **2003**, *546*, 859–867. [CrossRef]
58. Boggs, J.M. Myelin basic protein: A multifunctional protein. *Cell. Mol. Life Sci. CMLS* **2006**, *63*, 1945–1961. [CrossRef]
59. Pronker, M.F.; Lemstra, S.; Snijder, J.; Heck, A.J.; Thies-Weesie, D.M.; Pasterkamp, R.J.; Janssen, B.J. Structural basis of myelin-associated glycoprotein adhesion and signalling. *Nat. Commun.* **2016**, *7*, 13584. [CrossRef]
60. Quarles, R.H. Myelin-associated glycoprotein (MAG): Past, present and beyond. *J. Neurochem.* **2007**, *100*, 1431–1448. [CrossRef] [PubMed]
61. Koziol, L.F.; Budding, D.; Andreasen, N.; D'Arrigo, S.; Bulgheroni, S.; Imamizu, H.; Ito, M.; Manto, M.; Marvel, C.; Parker, K.; et al. Consensus paper: The cerebellum's role in movement and cognition. *Cerebellum* **2014**, *13*, 151–177. [CrossRef] [PubMed]
62. Oliveira, G.B.; Fontes Ede, A., Jr.; de Carvalho, S.; da Silva, J.B.; Fernandes, L.M.; Oliveira, M.C.; Prediger, R.D.; Gomes-Leal, W.; Lima, R.R.; Maia, C.S. Minocycline mitigates motor impairments and cortical neuronal loss induced by focal ischemia in rats chronically exposed to ethanol during adolescence. *Brain Res.* **2014**, *1561*, 23–34. [CrossRef]
63. Lopes, G.O.; Martins Ferreira, M.K.; Davis, L.; Bittencourt, L.O.; Bragança Aragão, W.A.; Dionizio, A.; Rabelo Buzalaf, M.A.; Crespo-Lopez, M.E.; Maia, C.S.F.; Lima, R.R. Effects of Fluoride Long-Term Exposure over the Cerebellum: Global Proteomic Profile, Oxidative Biochemistry, Cell Density, and Motor Behavior Evaluation. *Int. J. Mol. Sci.* **2020**, *21*, 7297. [CrossRef] [PubMed]
64. Leão, L.K.R.; Bittencourt, L.O.; Oliveira, A.C.; Nascimento, P.C.; Miranda, G.H.N.; Ferreira, R.O.; Nabiça, M.; Dantas, K.; Dionizio, A.; Cartágenes, S.; et al. Long-Term Lead Exposure Since Adolescence Causes Proteomic and Morphological Alterations in the Cerebellum Associated with Motor Deficits in Adult Rats. *Int. J. Mol. Sci.* **2020**, *21*, 3571. [CrossRef] [PubMed]
65. Santana, L.; Bittencourt, L.O.; Nascimento, P.C.; Fernandes, R.M.; Teixeira, F.B.; Fernandes, L.M.P.; Freitas Silva, M.C.; Nogueira, L.S.; Amado, L.L.; Crespo-Lopez, M.E.; et al. Low doses of methylmercury exposure during adulthood in rats display oxidative stress, neurodegeneration in the motor cortex and lead to impairment of motor skills. *J. Trace Elem. Med. Biol. Organ Soc. Miner. Trace Elem.* **2019**, *51*, 19–27. [CrossRef]
66. Freire, M.A.M.; Santana, L.N.S.; Bittencourt, L.O.; Nascimento, P.C.; Fernandes, R.M.; Leão, L.K.R.; Fernandes, L.M.P.; Silva, M.C.F.; Amado, L.L.; Gomes-Leal, W.; et al. Methylmercury intoxication and cortical ischemia: Pre-clinical study of their comorbidity. *Ecotoxicol. Environ. Saf.* **2019**, *174*, 557–565. [CrossRef]

67. Nascimento, P.C.; Ferreira, M.K.M.; Balbinot, K.M.; Alves-Júnior, S.M.; Viana Pinheiro, J.J.; Silveira, F.M.; Martins, M.D.; Crespo-Lopez, M.E.; Lima, R.R. Methylmercury-Induced Toxicopathologic Findings in Salivary Glands of Offspring Rats After Gestational and Lactational Exposure. *Biol. Trace Elem. Res.* **2021**, *199*, 2983–2991. [CrossRef]
68. Amado, L.L.; Garcia, M.L.; Ramos, P.B.; Freitas, R.F.; Zafalon, B.; Ferreira, J.L.R.; Yunes, J.S.; Monserrat, J.M. A method to measure total antioxidant capacity against peroxyl radicals in aquatic organisms: Application to evaluate microcystins toxicity. *Sci. Total Environ.* **2009**, *407*, 2115–2123. [CrossRef]
69. Bradford, M.M. A rapid and sensitive method for the quantitation of microgram quantities of protein utilizing the principle of protein-dye binding. *Anal. Biochem.* **1976**, *72*, 248–254. [CrossRef]
70. Esterbauer, H.; Cheeseman, K.H. Determination of aldehydic lipid peroxidation products: Malonaldehyde and 4-hydroxynonenal. *Methods Enzymol.* **1990**, *186*, 407–421. [CrossRef]
71. Bittencourt, L.O.; Dionizio, A.; Nascimento, P.C.; Puty, B.; Leão, L.K.R.; Luz, D.A.; Silva, M.C.F.; Amado, L.L.; Leite, A.; Buzalaf, M.R.; et al. Proteomic approach underlying the hippocampal neurodegeneration caused by low doses of methylmercury after long-term exposure in adult rats. *Met. Integr. Biometal Sci.* **2019**, *11*, 390–403. [CrossRef] [PubMed]
72. Eiró, L.G.; Ferreira, M.K.M.; Bittencourt, L.O.; Aragão, W.A.B.; Souza, M.P.C.; Silva, M.C.F.; Dionizio, A.; Buzalaf, M.A.R.; Crespo-López, M.E.; Lima, R.R. Chronic methylmercury exposure causes spinal cord impairment: Proteomic modulation and oxidative stress. *Food Chem. Toxicol. Int. J. Publ. Br. Ind. Biol. Res. Assoc.* **2020**, *146*, 111772. [CrossRef] [PubMed]

International Journal of *Molecular Sciences*

MDPI

Review

Focus on the Small GTPase Rab1: A Key Player in the Pathogenesis of Parkinson's Disease

José Ángel Martínez-Menárguez [1,*], Emma Martínez-Alonso [1], Mireia Cara-Esteban [2] and Mónica Tomás [2]

1 Department of Cell Biology and Histology, Biomedical Research Institute of Murcia (IMIB-Arrixaca-UMU), Medical School, University of Murcia, Campus Mare Nostrum (CMN), 30100 Murcia, Spain; emma@um.es

2 Department of Human Anatomy and Embryology, Medical School, Universitat de Valencia, 46010 Valencia, Spain; micaes2@alumni.uv.es (M.C.-E.); monica.tomas@uv.es (M.T.)

* Correspondence: jamartin@um.es

Abstract: Parkinson's disease (PD) is the second most frequent neurodegenerative disease. It is characterized by the loss of dopaminergic neurons in the substantia nigra and the formation of large aggregates in the survival neurons called Lewy bodies, which mainly contain α-synuclein (α-syn). The cause of cell death is not known but could be due to mitochondrial dysfunction, protein homeostasis failure, and alterations in the secretory/endolysosomal/autophagic pathways. Survival nigral neurons overexpress the small GTPase Rab1. This protein is considered a housekeeping Rab that is necessary to support the secretory pathway, the maintenance of the Golgi complex structure, and the regulation of macroautophagy from yeast to humans. It is also involved in signaling, carcinogenesis, and infection for some pathogens. It has been shown that it is directly linked to the pathogenesis of PD and other neurodegenerative diseases. It has a protective effect against α–σψν toxicity and has recently been shown to be a substrate of LRRK2, which is the most common cause of familial PD and the risk of sporadic disease. In this review, we analyze the key aspects of Rab1 function in dopamine neurons and its implications in PD neurodegeneration/restauration. The results of the current and former research support the notion that this GTPase is a good candidate for therapeutic strategies.

Keywords: Rab1; GTPases; Parkinson's disease; secretory pathway; Golgi fragmentation; autophagy; α-synuclein; LRRK2

Citation: Martínez-Menárguez, J.Á.; Martínez-Alonso, E.; Cara-Esteban, M.; Tomás, M. Focus on the Small GTPase Rab1: A Key Player in the Pathogenesis of Parkinson's Disease. *Int. J. Mol. Sci.* **2021**, 22, 12087. https://doi.org/10.3390/ijms222112087

Academic Editors: Luisa Agnello and Marcello Ciaccio

Received: 30 September 2021
Accepted: 6 November 2021
Published: 8 November 2021

1. Introduction to PD Pathogenesis

PD is the second most frequent neurodegenerative disease, affecting around 4 million people around the world, with a prevalence and incidence rate of 108–257/ and 11–19/100,000 per year, respectively [1]. Even though a detailed analysis of the disease and almost all of its variations has been provided before, only palliative treatments for it exist. Unfortunately, diagnosis is almost exclusively based on symptomatology, while the only accurate diagnosis can be made by the post-mortem analysis of the brain [2,3]. Apart from classic motor symptoms such as rigidity or akinesia, non-motor symptoms have also been described, including dementia and other specific cognitive disorders, as well as gastrointestinal problems [4–6].

This disease is characterized by the loss of dopaminergic neurons in the substantia nigra pars compacta (SNpc), which affects many cerebral circuits controlling motor movements and other functions [7]. The cytopathological hallmark of the survival nigral neurons is the formation of insoluble deposits, called Lewy bodies (LBs), as well as the less studied Lewy neurites and pale bodies [8]. A recent electron microscopic analysis showed that these bodies contain filamentous materials, lipids, and altered organelles, such as mitochondria, lysosomes, and autophagosomes [9]. Although more than one hundred proteins have been identified, they are primarily composed of insoluble forms of α-synuclein (α-syn) [10]. This small protein (14 kDa) is detected in many compartments, such as cytosol, the nucleus, and mitochondria, although it is functionally associated with synaptic vesicles in presynaptic terminals. It is encoded by the *SNCA* gene and belongs to a well-conserved family of synaptic proteins formed by β-, γ-, and α-syn [11]; however, only the α isoform is able to form toxic fibrillary structures. Its exact function is unclear, although it has been suggested that it regulates synaptic vesicles' exocytosis and recycling and the formation of fusion pores [12]. Cytosolic α-syn is able to sense membrane curvature and bind negatively charged phospholipids present in the membranes of synaptic vesicles [13]. It has been proposed that it acts as a chaperone for the assembly of SNARE complexes (complexes mediating the specific fusion of compartments) at the terminal end, regulating the release of dopamine and avoiding non-specific interactions [14]. This protein is a major player in PD pathogenesis [10]. Monomers of α-syn can assemble to form beta-sheet-rich oligomers, protofibrils, and amyloid fibrils, which can aggregate to form LBs. These can alter many cellular processes, such as secretory and endocytic trafficking, autophagy, mitochondrial and nuclear functions, and proteasome degradation, which may induce cell stress and death [15,16]. The internalization of aggregates of misfolded α-syn might contribute to the spread of the pathology between neurons in a mechanism resembling prions [17]. Age is a key factor in α-syn aggregation and propagation [18].

The reason that dopaminergic neurons are particularly affected remains unclear; multiple alterations related to mutations in key proteins involved in protein homeostasis, mitochondria function, lysosome/autophagy pathways, dopamine synthesis, metabolism, the immune system, and inflammation have been proposed [19,20]. Many experimental data strongly support the notion that defective intracellular transport could be a major cause of nigral neuron death [21–23]. The axon of each nigral dopamine neuron travels a very long way to the striatum (the mesostriatal pathway); there, it branches extensively and establishes thousands or even millions of synapses with medium spiny neurons [24]. Thus, these neurons need to maintain "well-oiled" transport machinery to maintain a high flux of secretory proteins (and concomitantly a recycling apparatus), and its failure may trigger cell death.

The cause of sporadic PD is unknown, but 5–15% of cases are familial. Multiple mutations present in family members with PD are known, including autosomal dominant (*SNCA*, *LRRK2*, *VPS35*) and recessive (*PARK2*, *PINK1*, *DJ-1*) forms [25], although many additional risky loci have been identified [26,27]. The overexpression (duplication or triplication) or mutation of the *SNCA* gene is known to be a cause of familial PD [25,28,29]. Most *SNCA* point mutations are associated with the early onset of disease. These mutations and the greater production of α-syn result in aggregation, leading to the formation of LBs. The *SNCA* locus is also a risk for sporadic PD.

However, the most common causes of familial PD are mutations in leucine-rich repeat kinase 2 (*LRRK2*), and this is a strong risk factor for sporadic disease [30]. This is a large multi-domain protein (280 kDa) with a GTPase domain, serine/threonine kinase domains, and other domains involved in protein–protein interactions [31]. This kinase interacts with many molecules, including Rab GTPases. Mutations in these domains may lead to a hyperactive protein with toxic effects. LRRK2 plays multiple roles in cells, including in vesicular trafficking, endosomal transport, autophagy, cytoskeleton dynamics, ciliogenesis, neurite growth, mitochondrial morphology, and mitochondrial calcium regulation [31–35]. In nigral neurons, it is specifically involved in vesicle formation and docking, as well as ER-to-Golgi and Golgi-to-membrane transport and the internalization of dopamine receptors at the synapse [36,37].

VPS35 is also involved in transport, specifically in endosome–TGN retrograde transport mediated by the retromer complex. A PD-related mutation of this retromer subunit enhances the LRRK2-dependent phosphorylation of Rabs [38]. Mutations in the *PARK2* (parkin), *PINK1*, and *DJ-1* genes affect mitochondrial functions/homeostasis [39].

Many intracellular processes are altered in dopaminergic neurons, which can lead to their death [40]. All/most of these processes are connected and several molecules can be key regulators and monitors all of these routes. Thus, knowledge of these molecules is very important in order to understand neurodegeneration and could be the target of future therapeutic strategies. Rab GTPases are important regulators of membrane traffic. In the present review, we analyzed one of these small GTPases, Rab1. This protein has long been known as a regulator of the early steps of secretory traffic. It was a surprise when it was found to be related to α-syn toxicity. Furthermore, it was demonstrated to play a key role in autophagy and many other processes, such as unfolding protein response (UPR), signaling and carcinogenesis [41], glucose homeostasis [42], and infection with some parasites (such as Legionella pneumophila) [43,44].

Surviving nigral neurons have been found to overexpress this GTPase in a cellular model of PD and human samples [45,46]. Now, it is evident that it is a key molecule in PD and other neurodegenerative diseases. Thus, the alteration of Rab1-dependent ER–Golgi transport is a pathogenic mechanism involved in common types of familiar amyotrophic lateral sclerosis [47]. One of these disease-related mutants (FUS) was found to impair autophagy and, importantly, Rab1 overexpression plays a protective role [48]. Rab1 is also involved in the transport and processing of amyloid precursor protein (APP) in Alzheimer's disease [49]. Here, we analyze the molecular/functional aspects of the molecular biology of Rab1 directly related to PD.

2. Rab GTPases

Rab proteins are the largest subfamily of GTPases, with more than 60 members of approximately 25 kDa, sharing around 30% sequence identity. Their functions include the maintenance of the shape and dynamics of intracellular membranes and the establishment of membrane identity [50]. Rabs are involved in the formation, cytoskeleton-dependent transport, docking, and fusion of transport vesicles. To fulfill their functions, Rabs interact with many molecules, called effectors, which include coat proteins, tethering factors, cytoskeleton elements, etc. They were first discovered in nervous tissue, where vesicular transport is essential for dendrite and axon organization as well as synaptic transmission [51]. Each Rab or group of Rabs localizes in a particular membrane and regulates a specific transport step, including bidirectional ER–Golgi trafficking, intra-Golgi transport, endolysosomal routes, and autophagy [52,53]. In addition, they also regulate signal transduction, cell survival, and development [54].

Rabs are considered molecular switches and cycle between an active GTP-bound form and an inactive GDP-bound form [55] (Figure 1). In their active form, they are able to interact with many effector proteins, regulating multiple cellular processes. Their activity is regulated by guanine nucleotide exchange factors (GEFs), which promote the exchange of GDP to GTP and GTPase-activating proteins (GAPs); these activate the intrinsic low GTPase capability of Rabs. Inactive GDP-bound Rabs are found soluble in the cytosol, whereas, in their GTP-bound active forms, they are located in cellular membrane joints by geranylgeranyl groups. This binding is regulated by GDP dissociation inhibitor (GDI), which removes inactive GDP-bound Rabs from membranes. It has been proposed that GDI displacement factors (GDF) at the membrane inhibit this complex and maintain the association of Rabs with membranes. The association with membranes is also sensible to lipid composition, as it senses the curvature elastic energy stored in the membrane [56]. Rab activity is also finetuned by phosphorylation. This post-translational modification can affect the mode of interaction between Rabs and their partners [57]. Interestingly, a serine/threonine kinase that is specific for some Rabs is the above-mentioned LRRK2.

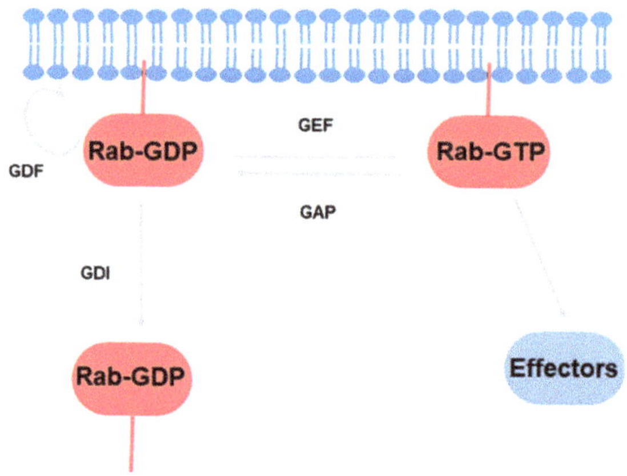

Figure 1. Regulation of GTP–GTP cycle of Rab GTPases by guanine nucleotide exchange factors (GEFs), GTPase-activating proteins (GAPs), GDP dissociation inhibitor (GDI), and GDI displacement factors (GDF). Membrane-associated GTP-bound Rabs are able to interact with effectors.

Rabs participate in many aspects of neuron life, including development, the polarized growth of neurites, endocytosis at the neuron end, axonal transport, the exocytosis of synaptic vesicles, and the dynamics of neurotransmitter receptors [58,59]. A few Rabs are selectively enriched in neurons, i.e., Rab3A (synaptic vesicle exocytosis), Rab8 (dendrite-specific transport), and Rab23 (neuron development and endosomal transport), all of which are associated with the post-Golgi transport step [60]. Most neurodegenerative diseases, including PD, are related to vesicular transport alterations mediated by Rabs [53,61,62]. Several Rabs (Rab8B, Rab11A, Rab13, and Rab39B) regulate the toxic aggregation of α-syn [63]. The LRRK2-dependent phosphorylation of Rab35 has been found to be involved in α-syn propagation between neurons [64]. The overexpression of Rab7 reduces α-syn aggregates by enhancing autophagy [65]. Wild-type α-syn physically interacts with Rab3A/B/C, Rab4B, Rab6A, Rab8A, Rab15, and Rab35 [66]. The PD-related mutant A30P α-syn interacts with Rab3A, Rab5, and Rab8, therefore affecting synaptic vesicle trafficking, endocytosis, and α-syn transport [67]. In a few cases, Rabs are the direct cause of the disease [51]. This is the case for Rab39B, which is involved in α-syn homeostasis, endosomal traffic, neurite outgrowth, synaptic maturation, and autophagy [50]. Rab39B mutation causes a rare form of early-onset PD with an α-syn pathology [68].

3. Rab1

Rab1 is one of the five Rabs present in all eukaryotes and is considered a housekeeping Rab. In yeast, Rab1 is encoded by the YTP1 gene. Two isoforms have been described, Rab1A and 1B, which share 75% and 66% amino acid identity with Ytp1, respectively, with 92% identity between them [69,70]. Both isoforms are expressed in all tissues, but there are important differences in their levels of expression [71]. The functional dissimilarities between these isoforms have not been exhaustively analyzed but there are differences in the interactions with coated vesicles and subcellular distribution (see below) and the role in the secretory flux of some proteins [72].

Rab1 interacts with many effectors related to membrane traffic, including p115 [73], GM130 [74,75], golgin 84 [76], and giantin [77]. Another effector of unknown function is Iporin, a ubiquitous protein that is highly expressed in the brain and that also interacts with GM130 [78]. Additionally, it interacts with members of the MICAL family and could form a link between membrane traffic and the cytoskeleton [79]. Rab1A and B interact with Golgi phosphoprotein 3 (GOLPH3), a protein involved in post-Golgi transport that is considered a Golgi oncoprotein and associated with several types of cancer [80].

As with all Rabs, GEFs, GAPs, GDI, and GDF regulate the association of Rab1 with membranes [55]. Two well-conserved GEFs for Rab proteins have been identified; these are named Transport Protein Particle (TRAPP) III and II (TRAPPI is now considered to be the core subcomplex of the other two). These regulate the early and post-Golgi transport steps, respectively. The multicomplex TRAPIII specifically activates Rab1, although TRAPPII has some activity on this Rab of unknown physiological significance [81]. In yeast, the membrane anchoring of TRAPIII is mediated by the Trs85 subunit, which is also necessary for Ypt1 activation [82]. The Bet5, Trs31, and Bet3 subunits are also important for this activity [83]. In mammals, TRAPPC4 (the ortholog of yeast Trs23) is a core component of the TRAPP complexes and is also essential for the GEF activity of Rab1 [55]. TRAPPC8 (which shares some characteristics with yeast Trs85) is also important for Rab1 activation [81]. TRAPPIII is located in COPII vesicles and Golgi. Its location in COPII vesicles depends in the interaction of the TRAPP core and the COPII component Sec23 [84,85]. TRAPPIII is also located in Atg9 vesicles and pre-autophagosomal structures in a process involving the Trs85 subunit [86] (see below for its role in autophagy). Interestingly, variants of TRAPPC4 are associated with human neurological disorders [87].

GAPs for Ypts are named GYPs in yeast and TBC1 domain proteins in other species due to all of them sharing this domain [88]. The yeast Gyp8 is a GAP for Ypt1, which is present in the ER. Gyp1, another GAP for Ypt1, is located in the early Golgi [89]. Their location in membranes depends on Arf1 GTPase, which recruits Gyp1 inactivating Ypt1 [90]. This sequential gradient of activation/inactivation of GTPases is important for maintaining compartment identity, ensuring cisternae maturation, and guiding the secretory traffic [91,92]. Gyp1 also plays a role in autophagy. It regulates the disassembly of Ytp1–Atg1 complexes, as well as the interaction of cargo receptor and Atg8 proteins in selective autophagy [93]. In humans, 38 Rab GAPs have been identified, with TBC1D20 (Gyp8 in yeast) being the specific (but not exclusive) GAP for Rab1. It regulates the exit from the ER of the secretory cargo, as well as the Golgi architecture [94]. It is also involved in the maturation of autophagosomes [95]. Interestingly, this study showed that TBC1D20-deficient mice display an altered autophagy flux in neurons, resulting in adult-onset motor dysfunction.

Less is known about the role of GDF and GDI in the location of Rab1. Prenylated Rab acceptor 1 (PRA1) (the mammalian homolog of yeast Yip3) has GDF activity for Rab1 (and other Rabs) [96] and is involved in its association with membranes. Interestingly, PRA1 interacts directly with α-syn [97]. Several well-conserved isoforms of GDI have been described that bind a wide spectrum of Rab proteins, with α-GDI being restricted to the brain [98]. GDI is involved in the retrieval of Rab1 from Golgi membranes [55].

4. The Role of Rab1 in the Secretory Pathway

It is well known that newly synthetized proteins and lipids move from the ER to the Golgi in COPII-coated vesicles, while retrograde transport is mediated by COPI-coated vesicles. The role of COPI-coated vesicles in the anterograde (i.e., ERGIC to Golgi and intra-Golgi transport) is under discussion. In contrast with yeast, in mammals, there is an intermediate compartment between the two organelles called ERGIC. The origin and role of these membranes are not completely understood. It is likely that it is formed by the fusion of ER-derived vesicles, which then move in block to the Golgi. Rab1 locates to the early membranes of the secretory pathway—more specifically, ERGIC and the cis-Golgi cisternae—but not in the ER [99]. In vivo experiments have shown that Rab1A is located in dynamic tubules with bidirectional movements, whereas Rab1B is concentrated in vesicular stationary elements [100], although these differences might be cell-type-specific. Rab1A-positive tubules exclude COPI and p58 (an ERGIC marker) [101], whereas Rab1B elements allow COPI budding [102].

Early studies support the notion that Rab1/Ypt1 is involved in the bidirectional traffic between ER and Golgi traffic (Figures 2A and 3), and its location is highly dynamic in living cells [55]. It has been implicated in regulating COPII vesicle formation and subsequent tethering and fusion with Golgi/ERGIC membranes [83,103–105]. Thus, GDP-bound

Rab1A and Rab1B (inactive forms) inhibit ER to Golgi transport [106]. Rab1B interacts with the COPII components Sec23, Sec24, and Sec31 [107]. Recent studies support the notion that cargo leaves the ER in Rab1-dependent carriers, whereas COPII components remain in stable ERES even when the cargo is released [108]. The Rab1 present in COPII vesicles may recruit p115, tethering these vesicles to the ERGIC membranes; similarly, Ypt1p interacts with Uso1, the yeast homolog of p115 [103]. The Rab1–p115 complex may anchor these elements to the cis-Golgi by interacting with GM130-GRASP65 [74]. The complex Rab1–p115 might be also involved in the tethering of COPI vesicles via giantin [77]. GM130, giantin, and p115 belong to the golgin family, which are Golgi matrix proteins characterized by extensive coiled-coil domains that, together with Rabs, organize Golgi membranes and transport [109]. Rab1 also interacts with another golgin, golgin-84, which is involved in the lateral linking of Golgi stacks, forming a continuous ribbon [110]. This interaction is necessary to maintain the Golgi structure and protein transport [111].

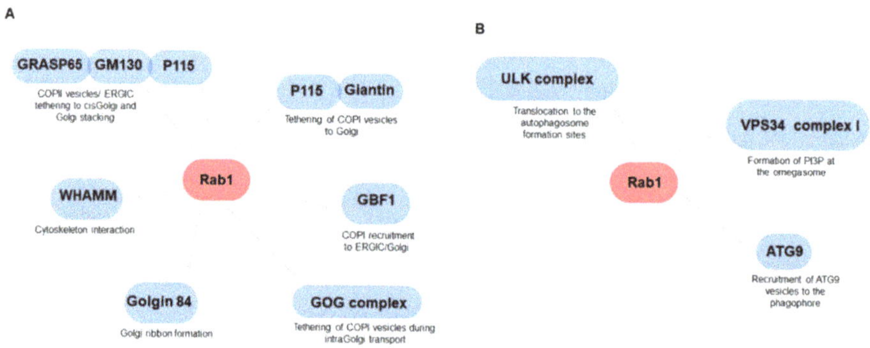

Figure 2. Main effectors of Rab1 involved in secretory traffic (**A**) and autophagy (**B**).

Ypt1 is also involved in COPI-dependent retrograde transport from the Golgi [112]. Ypt1 interacts with COPI [113]. In contrast with Rab1A, Rab1B regulates COPI coat recruitment to ERGIC and Golgi membranes because it is necessary for the recruitment of Golgi-specific brefeldin A-resistance guanine nucleotide exchange factor 1 (GBF1), which is the GEF that is specific for Arf1 [102], a GTPase involved in the recruitment of COPI components to membranes. In addition, Ypt1/Rab1 interacts with different subunits of the conserved oligomeric Golgi (COG) complex [114], which is involved in the tethering of vesicles during intra-Golgi retrograde transport.

Rab1 has recently been detected in the membrane of immature secretory granules, where it is recruited by Rab11 and might be involved in their maturation [115]. Whether this a general mechanism or restricted to the salivary glands of Drosophila remains to be clarified.

Rab1 also regulates the association of transport carriers and the cytoskeleton. Rab1 at the ERGIC can recruit WASP Homolog Associated with Actin, Golgi Membranes and Microtubules (WHAMM), which stimulates Arp2/3-mediated actin polymerization and interacts with microtubules, inducing membrane deformation [116]. The interaction of WHAMM with these cytoskeleton elements mediates membrane tubule elongation and formation, respectively [117]. Rab1 is also involved in the recruitment of the motor protein kinesin KifC1 to endocytic vesicles [118]. In addition, an effector of Rab1, GM130, is involved in microtubule polarization at the cis-Golgi [119]. Thus, Rab1 is a regulator of Golgi morphology and positioning.

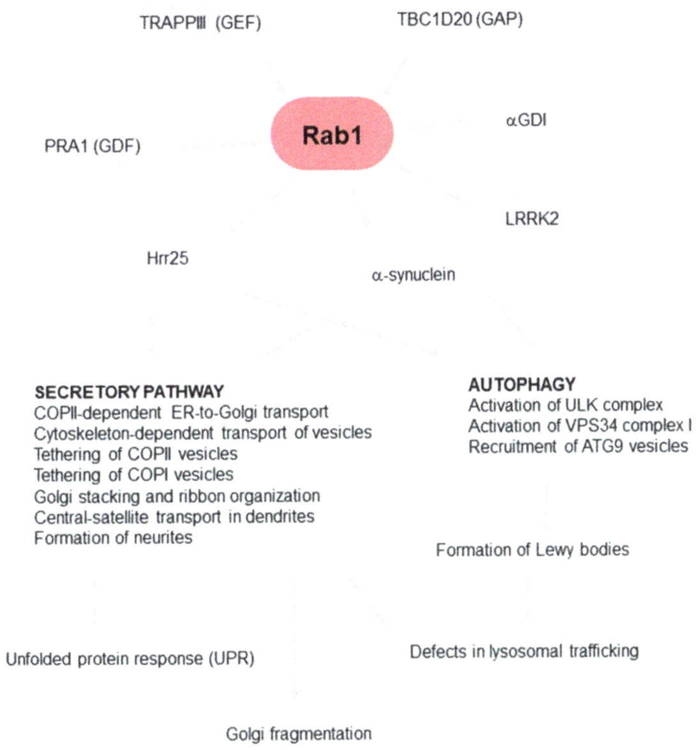

Figure 3. Overview of the roles and regulation of Rab1 and how it may be affected in PD. Activity and association with membranes of Rab1 are regulated by TRAPPIII (a guanine nucleotide exchange factor, GEF), TBC1D120 (a GTPase-activating protein, GAP), PRA1 (a GDI displacement factor, GDF), and αGDI (a GDP dissociation inhibitor, GDI). This activity is also regulated by the PD-related protein LRRK2. Rab1 regulates the early steps of the secretory pathway and autophagy through many effectors (see the full text for details). In addition, it regulates Hrr25, a member of the casein kinase 1 family, which also regulates these pathways. The expression of mutant forms of α-synuclein or overexpression of the wild type impair the secretory pathway, resulting in the activation of the UPR, the fragmentation of the Golgi ribbon, and the failure of the lysosomal route. These PD-associated alterations also affect autophagy. PD-related mutants of LRRK2 may enhance the activity of this GTPase, impairing secretory and autophagy pathways. Unbalanced secretory/autophagic/endolysosomal routes may lead to the formation of Lewy bodies.

Rab1 may also have specific roles in neuron trafficking. Most dendrites (80%) lack a Golgi outpost but contain ERGIC elements in close contact with early endosomes, forming a so-called satellite [120]. It has been proposed that, at this level, proteins and lipids are delivered from the ER to the synaptic spine membrane, using these two compartments by bypassing the Golgi [121]. Rab1-positive ERGIC tubules move to the growth cones of neurite-like processes in differentiated PC12 cells, a common model used for studying neuronal differentiation [101]. Thus, Rab1 may have a specific role in neuron development. In vivo experiments have shown that Rab1 tubules move bidirectionally between the central body and neurites connecting central Golgi and peripheric ERES. Rab1 is also involved in the formation of cell protrusions (i.e., neurites). It forms part of a conserved signaling pathway where the KDEL receptor stimulates the Golgi-resident monomeric Gαo, which uses Rab1 (and 3) for the promotion of secretion to form protrusions [122].

There is a link between Rab1-dependent traffic and ER homeostasis. Protein aggregation induced by the pesticide rotenone decreases the levels of Rab1 in hippocampal neurons, which may impair secretory traffic and, as a consequence, trigger an unfolded protein response (UPR) [123]. In yeast, Ypt1 plays an important role in the regulation of UPR by stabilizing unprocessed HAC1 RNA under normal growth conditions, avoiding UPR activation when there are no stress conditions [124]. In fact, Ypt1 can switch molecularly and functionally from GTPase to a molecular chaperone under heat stress conditions, enhancing the resistance of the cell under these conditions [125].

5. Rab1 as a Regulator of Autophagy

Macro-autophagy is an important mechanism for the maintenance of cellular homeostasis through the degradation of damaged organelles or protein aggregates, the recycling of materials in the absence of nutrients or stress conditions, and even the elimination of pathogens. Apart from macroautophagy, the lysosome is also involved in two other mechanisms of autophagy: chaperone-mediated autophagy and microautophagy. In the first type, proteins bearing the KFERQ sequence motif are recognized by heat shock cognate 71 kDa protein (HSC70) and target lysosomal-associated membrane protein 2 (LAMP2) [126,127]. In the second type, a small portion of the cytoplasm is engulfed by invaginations of the lysosomal membranes, leading to the formation of small internal vesicles [128].

Macro-autophagy (hereafter autophagy when not specified) involves the formation of double-membrane autophagosomes sequestering cytoplasmic portions that will fuse with lysosomes. Although autophagy protects cells against stress, an excess may induce cell death. It has been considered a bulk mechanism, but it can be also a highly selective process where specific receptors recognize specific cargo [129]. This is the case for the degradation of fibrillar aggregates (aggrephagy) or mitochondria (mitophagy), among others. In non-dividing cells such as neurons, it is an important mechanism for maintaining cellular homeostasis. However, few autophagosomes are found in normal neurons, and special mechanisms, such as engulfing Golgi cisternae, may cooperate in degrading materials [130]. The consequence of the failure of autophagy is the accumulation of damaged organelles and proteins, which may lead to neurodegeneration [131].

The initial elements of autophagy are small structures called phagophores or isolation membranes. In yeast, autophagosome formation occurs in a specific place in close proximity to the ER and the vacuole named phagophore assembly sites (PAS) or pre-autophagosomal structures, where specific molecules are selectively recruited when autophagy is activated [132]. In mammals, phagophores grow in multiples places in close proximity to the ER, especially in the ring-like subdomains of this organelle, called omegasomes, which are enriched in phosphatidylinositol-3-phosphate (PI3P) and specific PI3P-binding proteins [133–135]. Direct continuities between the phagophore (and autophagosome) and the ER have been described [136], as well as tubular connections between the omegasome and the phagophore [137]. ER exit sites, ER–mitochondria contact sites, and ERGIC can also be the places of initiation. Different membranes have been proposed as sources for the nascent autophagosome, including the ER, mitochondria, Golgi complex, endocytic elements, and plasma membrane. The phagophore grows, curves around a portion of the cytoplasm, and finally closes, forming a double-membrane organelle called the autophagosome. Finally, it fuses with the lysosome and the inner membrane and the content is degraded, releasing simple molecules such as amino acids, lipids, and sugars [138]. Autophagosomes can also fuse with late endosomes, forming amphisomes, which also fuse with lysosomes [139].

Many aspects of the complex molecular machinery involved in this process are well known (for a detailed explanation, see [140,141]). This process is regulated by evolutionarily conserved autophagy-related (ATG) proteins. Twenty of these ATG proteins constitute the core machinery and can be grouped into different functional unities [142]. Initiation requires two kinase complexes: the unc-51 like kinase (ULK) complex (Atg1 complex in yeast) and the phosphatidylinositol 3-kinase class 3 (PI3KC3)/vacuolar protein sorting 34 (VPS34) complex. The ULK complex is the first inductor of the autophagic process. Upstream nutrient/energy-sensing kinases (e.g., the mammalian target of rapamycin complex 1, mTORC1) activate this complex, which is mostly cytosolic, and recruit to phosphatidylinositol synthase-enriched subdomains of the ER [143]. The activation of ULK phosphorylates downstream effectors, including ATG9, in vesicles [144] to induce autophagosome formation. ATG9 is the only membrane protein and, in mammals, is under steady-state conditions in the TGN/late endosomes; however, after autophagic activation, ATG9 vesicles move close to the autophagosome in their formation (the Atg9 compartment in yeast) [145]. ATG9 vesicles seem to provide membranes for the formation of the autophagosome, but also supply important regulators such as lipid-metabolizing enzymes. In fact, its role as a lipid scramblase has recently been demonstrated [146]. Soon after ULK, the PI3KC3/VPS34 lipid kinase complex is recruited, inducing the production of PI3P (at low levels in the ER in basal conditions), which recruits autophagy effectors that drive the formation of the omegasome: DFCP1 and WIPI proteins [133,147]. WIPI proteins recruit two ubiquitin-like conjugation systems that work in cascade: ATG7-ATG3 (acting as E1- and E2-like enzymes) and ATG12- 5/ATG16L1 (acting as E3-like enzyme) [148]. These systems enable the lipidation of microtubule-associated protein light chain 3 (LC3) isoforms, members of the Atg8 family of proteins. These ubiquitin-like proteins are usually unlipidated and cytosolic (LC3-I) but become covalently bound to the amino acid of phosphatidylethanolamine in the membrane of the phagophore (LC3-II). The last one is usually used as a marker of autophagic flux. The insertion of lipidated Atg8 proteins and ATG9-containing vesicles (which putatively supply lipids from different membranes) is responsible for the growth of the phagophore and its closure to form the autophagosome [149,150]. Atg8 proteins work as adaptors to recruit other factors that contribute to the membrane expansion of the autophagosomal membrane and the recruitment of selective autophagy receptors.

Autophagy operates as an intracellular membrane pathway. Thus, it is evident that it shares some regulatory mechanisms with other transport routes, including Rabs. Rab1/Ypt1 is an important regulator of autophagy, operating at early steps [151–156] (Figures 2B and 3). In addition to Rab1, other Rabs are also known to be regulators of autophagy, including Rab33B (autophagosome elongation), Rab7 (fusion autophagosome-lysosome), Rab32 (autophagosome formation), Rab5 (autophagosome closure), Rab24 (co-localized with LC3), and Rab2 (autophagosome and autolysosome formation) [157,158].

In yeast, Ypt1 is activated by the Trs85 subunit of TRAPIII and regulates autophagy by recruiting the Atg1 complex to the PAS [159]. In mammals, Rab1 regulates the translocation of the ULK complex to the autophagosome formation sites [160]. In addition, it also regulates PI3P production at the omegasome. Thus, Rab1 is a specific activator of the VPS34 complex I (involved in autophagic sorting) but not II (involved in endocytic sorting) [161].

Ypt1 is activated by the TRAPPIII subunit Trs85 and also interacts with Atg11, which may be involved in the recruitment of Atg9 vesicles to PAS [154,162]. In mammals, Rab1 has been found in ATG9 vesicles and LC3-positive autophagosomes [152,163]. During starvation, ATG9 moves from Rab11-positive TGN/recycling endosomes to Rab1-positive ERGIC/Golgi membranes. This process is controlled by the interaction of TBC1D14 with the TRAPPC8 subunit (the mammalian ortholog of Trs85) of TRAPPIII [164]. There are several differences in the roles of Rab1 isoforms [165]. The depletion of Rab1A induces the mis-location of ATG9 and decreases omegasome formation. This route is independent of ER–Golgi transport. Thus, Rab1A specifically inhibits the very early steps of autophagy. In contrast, the depletion of Rab1B and Rab2 (another Rab involved in ER–Golgi transport) has the opposite effect of increasing autophagy. The significance of these differences remains to be established.

It has been proposed that there is a functional and spatial link between the secretory pathway and autophagy at the early stages, and the key component of this cross route may be Rab1. Under starvation conditions, there is a reduction in protein secretion and secretory compartments are redirected to the autophagy pathway. Autophagosomes are spatially and functionally linked to ERES [166]. Moreover, COPII vesicles and ERGIC can be sources of membranes for autophagy. Under stress conditions, ER-derived COPII vesicles, instead of being directed to the Golgi, can be a source of membranes for autophagosomes [85,167,168]. However, ERGIC can be the source of COPII vesicles after the activation of autophagy. ERGIC is a source of the small vesicles involved in LC3 lipidation, a process that requires the activation of phosphotidylinositol-3 kinase (PI3K) and the recruitment of COPII proteins in this compartment [169,170]. This process requires the remodeling of ERES, which is specifically mediated by some specific components of the UKL complex [171]. A critical step of this process is the phosphorylation of the COPII component Sec23B by ULK [172]. The ULK-dependent phosphorylation of COPII subunits also regulates ERES morphology and causes a reduction in secretion during active autophagy [173]. As an activator of ULK and an abundant component of the ERGIC, Rab1 may regulate this mechanism under stress conditions. TRAPPIII, the GEF for Rab1, binds the COP II coat subunit Sec23 at the PAS, which is necessary for autophagy [85]. In addition, Ypt1/Rab1 directly recruits the casein kinase 1 (CK1δ) family member Hrr25, a Golgi-localized serine–threonine kinase that regulates ER–Golgi traffic and autophagy, to COPII vesicles, activating its kinase activity and regulating the traffic of these vesicles in these two pathways [174,175].

6. Rab1, Secretory Pathway, and PD

Rab1A and Rab1B are expressed in SN, although at a lower level than other Rabs, such as Rab3A and Rab3C [71]. Interestingly, the levels of this Rab rose by 50% in human PD samples [46], as well as in a cellular model of PD [45]. Whether this overexpression is an attempt by the cell to rescue normality or is the result of cellular damage is not known.

The first connection between Rab1 and PD was observed in a yeast model [176]. Researchers have found that the main toxic effect of wild-type and PD-related mutant α-syn is due to the alteration of ER–Golgi transport and the best suppressor of this toxicity was found to be Ypt1. A few other proteins inhibit this toxicity, and, interestingly, all of them are involved in the same transport step, including Gyp8, which is the GAP for Ypt1. This effect is not restricted to this simple cellular model; is has also been shown that Rab1 protects against the mammalian dopaminergic neurons' death induced by α-syn. For the first time, intracellular trafficking, and, more specifically, ER–Golgi trafficking, has been related to the pathogenesis of PD. In yeast, only this Rab rescues α-syn. However, Rabs associated with later transport steps in neuronal models of PD (i.e., Rab3A and Rab8) have been shown to protect against α-syn toxicity [177]. In mammalian cells, the overexpression of wild-type or A53T mutant α-syn inhibits COPII vesicles' docking and fusion with early Golgi by the inhibition of the SNARE complex assembly [178]. The pathological accumulation of α-syn affects hydrolase trafficking and therefore lysosomal activity, and this dysfunction can be improved by Rab1 overexpression [179]. This alteration is due to the aberrant interaction of α-syn with GM130, which disrupts Rab1-dependent ER–Golgi traffic and the Golgi structure.

7. Relationship between PD-Associated Golgi Fragmentation and Rab1

GC is a cellular organelle that acts as the central station of intracellular transport. It is involved in the post-translational modification (mainly adding sugar moieties) of proteins and lipids synthesized in the ER. A network of tubule–vesicular elements associated with the trans side, the TGN, is involved in the classification, packaging, and delivery of these products to the final destination in or out of the cell. It is especially developed in cells with synthetic activity, as is the case of cells of neurons. It is commonly found in a perinuclear position, showing a half-moon disposal. Under the electron microscope, it appears as stacked cisternae, classified according to their position in the organelle as cis, medial, and trans cisternae. In mammalian cells, Golgi stacks are laterally linked, forming a so-called Golgi ribbon [180–182].

Interestingly, the Golgi complex of neurons appears to be fragmented in most neurodegenerative diseases, including AD and PD [45,46,183]. The Golgi ribbon is broken and stacks appear dispersed throughout the cytoplasm. Even though it has been demonstrated that the GC remains functional after fragmentation, the ribbon structure is implicated in high-level functions, such as mitosis entry, apoptosis regulation, stress response, axodendritic polarity, and cell migration [184,185]. Therefore, Golgi fragmentation significantly alters neural physiology, inducing traffic failure and thereby causing a deficiency in the dopaminergic transport in PD, resulting in the limited capture of this neurotransmitter in vesicles and its limited accumulation in the cytosol. Toxic dopamine could aggregate and modify α-syn and increase its toxicity, which could lead to the formation of LBs and, finally, cell death [186].

The maintenance of the GA cytoarchitecture depends on many structural (including golgins and Golgi reassembling stacking proteins, GRASPs) and regulating proteins (Rabs, SNARE, etc.) and the cytoskeleton. The alteration of these elements may be the cause of this fragmentation. In addition, the alteration in intracellular transport caused by AG fragmentation has been shown to cause deficits in cell degradation activity due to the deficiency of lysosomes with proteases such as cathepsin D, which facilitates the degradation of α-syn [187]. Previous studies support the notion that Golgi fragmentation occurs before the formation of α-syn aggregates [45]. In fact, it is the alteration of Rab1-dependent ER-to-Golgi transport that causes imbalances in the traffic into and out of the GC. Rab1 is also involved in intra-Golgi transport, although it has not been detected in medial/trans cisternae. It is important for the Golgi structure, because the downregulation or microinjection of mutants induces Golgi alterations [188–190]. The increased amount of Rab1a observed in survival nigral cells is theorized to be an attempt to restore the Golgi's structure during fragmentation. Thus, Rab1 rescues Golgi fragmentation in dopamine cells overexpressing α-syn and, interestingly, also improves the control of motor function in hemiparkinsonian rats [183]. The alteration of Rab1-dependent ER–Golgi transport seems to be a common cause of Golgi fragmentation in several neurodegenerative diseases [185].

8. Rab1, Autophagy, and PD

Altered autophagy seems to be a common feature of neurological diseases. The deletion of genes coding for essential proteins in autophagosome formation, such as Atg5 [191] or Atg7 [192], causes neurodegeneration. In AD brains, autophagosomes and cathepsin-containing autophagolysosomes are common in neurite processes, including synaptic ends [193]. Autophagic vacuoles containing cytoplasmic materials are also common in nigral neurons affected by PD [194]. Interestingly, Lewy bodies are not observed in cells with high amounts of these vacuoles. Lewy bodies and neurites contain autophagic markers, and the lipidated form of LC3 (LC3-II) is highly expressed in PD nigral samples [195,196]. Impaired autophagy may contribute to the development of sporadic PD. Thus, a deficit of ATG7 in dopamine neurons induces high alterations in these cells but delays neurodegeneration and the appearance of late-onset locomotor deficits in mice models [197]. The whole-brain loss of this ATG protein induces the presynaptic accumulation of α-syn and LRRK2.

LRRK2 and α-syn have been extensively studied in relation to their involvement with aberrant autophagy. Mutations in these PD-related genes affect autophagic function [198]. Wild-type α-syn contains a KFERQ sequence motif, so it is degraded by chaperone-mediated autophagy. Conversely, autophagy is the main mechanism for the degradation of fibrillary α-syn, while the proteasome system is preferred for phosphorylated α-syn oligomers [199]. Some PD-related mutations affect this pathway. Thus, A53T and A30P mutations block chaperone-mediated autophagy but enhance autophagy, which may be a compensatory response [200]. Moreover, autophagy is increased in samples of PD patients (with or without A53T mutations) [201]. In contrast, another study supports the notion that the overexpression of the A53T mutant inhibits autophagy, resulting in the accumulation of this defective protein [202]. In addition, α-syn can affect autophagosome formation by altering the actin cytoskeleton. It induces an excess of stabilization, which is dependent on Arp2/3, which was found to impair autophagosome maturation and mitophagy in a Drosophila model of PD [203].

The wild-type form of LRRK2 is degraded by chaperone-mediated autophagy or the ubiquitin–proteasome system but not autophagy [204]. LRRK2 is a negative regulator of autophagic activity and the expression of PD-related mutations and induces abnormal autophagy [205]. The G2019S mutation, the most common PD-related mutation that increases kinase activity, induces autophagy abnormalities and affects neurite development [206,207]. This mutation induces defective autophagy by affecting autophagosome transport and maturation [208]. Many Rabs are substrates of LRRK2 (Rab3A/B/C, Rab3D, Rab8A, Rab8B, Rab10, Rab12, Rab29/RAB7L1, Rab35, and Rab43) [209], and all PD-related mutations enhance Rab phosphorylation [210]. Interestingly, it was recently demonstrated that Rab1 is also an endogenous substrate of this kinase [211] and might be the mechanism used for autophagy regulation.

A relationship between Rab1, autophagy, and neurodegenerative diseases has been described [160]. A GGGGCC hexanucleotide repeat expansion in the C9orf72 gene is the most common genetic cause of amyotrophic lateral sclerosis and frontotemporal dementia (C9ALS/FTD) and is also associated with sporadic forms. C9orf72 is a Rab1 effector that regulates the translocation of the ULK complex in the initial steps of autophagy. This link is also evident in PD. The overexpression of α-syn inhibits autophagy by inhibiting Rab1A, which results in the mislocalization of Atg9 and a reduced number of omegasomes [165]. Moreover, the overexpression of Rab1A rescues the effects of the overexpression of α-syn.

9. Concluding Remarks

Rab1 controls the early steps of the secretory pathway and autophagy, as summarized in Figure 1. The overexpression of this GTPase by survival nigral neurons can have the effect of minimizing PD abnormalities. An elevation of the secretory flux may help to restore the reduction in dopamine traffic to the terminal end and the lysosomal defects caused by the alteration of α-syn-dependent ER to Golgi trafficking [176]. In addition, it also may restore the Golgi ribbon architecture, which affects polarized transport in neurons [185]. In fact, Rab1 overexpression was found to ameliorate motor symptoms in a PD model [183]. The activation of autophagy has been proposed as a treatment for PD. However, the results found are contradictory. The autophagy-stimulating antibiotic rapamycin prevents the loss of dopamine neurons [212]. Impaired autophagy induces progressive dopamine cell loss but, conversely, enhances dopamine neurotransmission and ameliorates motor symptoms [21]. It is possible that unbalanced autophagy could result in cell stress and, finally, death [213]. The overexpression of Rab1 (at least Rab1A) enhances autophagy and restores the defects produced by α-syn [165]. In fact, our group found that it reduced the number of α-syn aggregates in a cellular model of PD (Figure 4). Recent research has focused on α-syn as a therapeutic target for PD in an attempt to reduce the levels of production and aggregation or increase the degradation [214]. Rab1, as a regulator of α-syn toxicity and substrate of LRRK2, may also be a good candidate. Rab1 expression/function (and their binding partners) may be altered using small molecules that interact with selective regions of this protein that physically interact with regulators/effectors [215]. The search for therapeutic agents modulating Rab1-dependent α-syn toxicity is currently underway [216]. Alternatively, the function of Rab1 may be modified by targeting the regulatory proteins, including GEF, GAPs, and GDIs. Several compounds have been tested in attempts to inhibit the disease-dependent activation of members of the Ras superfamily of small GTPases (including Rabs), although these investigations are only at the initial laboratory/preclinical stage [217]. This inhibition can be achieved by the inhibition of GEF activity, GEF–GTPase interaction, or GAP activity or through the stabilization of the GTPase–GDI complex, among other strategies. However, it should be taken into consideration that most of these regulatory proteins are not specific for a single GTPase. Given that most data support the notion that Rab1 activation has a positive effect on dopamine neurons, it is necessary to look for specific Rab-activating compounds. As indicated above, the prenylation of Rab by isoprenoid groups is necessary for membrane location and interaction with effectors, meaning that this molecular mechanism can be also used as a regulator of Rab function. Isoprenoids are produced by the mevalonate pathway, which is also involved in cholesterol biosynthesis. Some inhibitors of this route have been successfully used as therapeutic approaches for type 2 diabetes, given that Rab GTPases (including Rab1) are necessary for glucose homeostasis [218]. Statins, which are well-known blockers of cholesterol generation, are inhibitors of hydroxymethyl-3-methylglutaryl coenzyme A, a critical enzyme of this pathway. The effects of statins in PD patients have been tested, although the results gained are conflicting [219]. Again, this approach does not seem to be as specific as needed. New experiments in cellular and animal models will be necessary to clarify the role of Rab1 in the development of PD and elucidate how specific compounds affect α-syn aggregation, Golgi structure, secretory traffic, and autophagy in dopamine neurons. This small GTP is a good example of how basic research may help us to understand complex pathologies and search for new therapeutic approaches.

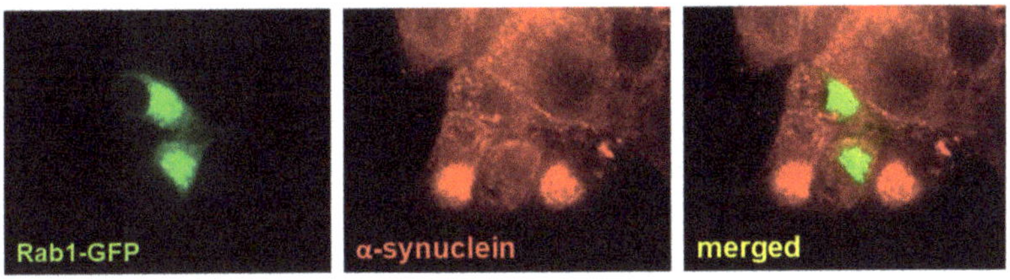

Figure 4. Rab1 overexpression reduces α-synuclein aggregates. These aggregates were induced in neuron-like differentiated PC12 cells by treatment with METH as a model of PD. Cells overexpressing Rab1-GFP have a reduced number of aggregates.

Author Contributions: J.Á.M.-M., E.M.-A., M.C.-E., and M.T. wrote and revised the manuscript. All authors have read and agreed to the published version of the manuscript.

Funding: This work was supported by grants from the Universidad de Valencia (UV-INV-AE11-41831) and Consellería de Cultura, Educación y Ciencia, Spain (GV/2013/093).

Institutional Review Board Statement: Not applicable.

Informed Consent Statement: Not applicable.

Data Availability Statement: Not applicable.

Acknowledgments: We apologize to authors whose primary research has not been cited in this review. The experiment shown in Figure 4 was performed by Wilson O. Rendón as part of his Ph.D Thesis for the University of Murcia under the supervision of José Ángel Martínez-Menárguez, Emma Martínez-Alonso and Mónica Tomás.

Conflicts of Interest: The authors declare no conflict of interest.

References

1. Balestrino, R.; Schapira, A. Parkinson disease. *Eur. J. Neurol.* **2020**, *27*, 27–42. [CrossRef] [PubMed]
2. Kalia, L.V.; Lang, A. Evolving basic, pathological and clinical concepts in PD. *Nat. Rev. Neurol.* **2016**, *12*, 65–66. [CrossRef] [PubMed]
3. Visanji, N.P.; Mollenhauer, B.; Beach, T.G.; Adler, C.H.; Coffey, C.S.; Kopil, C.M.; Dave, K.D.; Foroud, T.; Chahine, L.; Jennings, D. The Systemic Synuclein Sampling Study: Toward a biomarker for Parkinson's disease. *Biomark. Med.* **2017**, *11*, 359–368. [CrossRef] [PubMed]
4. Bordelon, Y.; Keener, A. Parkinsonism. *Semin. Neurol.* **2016**, *36*, 330–334. [CrossRef]
5. Sveinbjornsdottir, S. The clinical symptoms of Parkinson's disease. *J. Neurochem.* **2016**, *139*, 318–324. [CrossRef]
6. Hayes, M.T. Parkinson's Disease and Parkinsonism. *Am. J. Med.* **2019**, *132*, 802–807. [CrossRef]
7. Galvan, A.; Wichmann, T. Pathophysiology of parkinsonism. *Clin. Neurophysiol.* **2008**, *119*, 1459–1474. [CrossRef]
8. Kanazawa, T.; Adachi, E.; Orimo, S.; Nakamura, A.; Mizusawa, H.; Uchihara, T. Pale neurites, premature α-synuclein aggregates with centripetal extension from axon collaterals. *Brain Pathol.* **2012**, *22*, 67–78. [CrossRef]

9. Shahmoradian, S.H.; Lewis, A.J.; Genoud, C.; Hench, J.; Moors, T.E.; Navarro, P.P.; Castaño-Díez, D.; Schweighauser, G.; Graff-Meyer, A.; Goldie, K.N.; et al. Lewy pathology in Parkinson's disease consists of crowded organelles and lipid membranes. *Nat. Neurosci.* **2019**, *22*, 1099–1109. [CrossRef]
10. Mehra, S.; Sahay, S.; Maji, S.K. Alpha-Synuclein misfolding and aggregation: Implications in Parkinson's disease pathogenesis. *Biochim. Biophys Acta Proteins Proteom* **2019**, *1867*, 890–908. [CrossRef]
11. Stefanis, L. Alpha-Synuclein in Parkinson's disease. *Cold Spring Harb. Perspect. Med.* **2012**, *4*, a009399.
12. Oliveira, L.M.A.; Gasser, T.; Edwards, R.; Zweckstetter, M.; Melki, R.; Stefanis, L.; Lashuel, H.A.; Sulzer, D.; Vekrellis, K.; Halliday, G.M.; et al. Alpha-synuclein research: Defining strategic moves in the battle against Parkinson's disease. *NPJ Parkison's Dis.* **2021**, *7*, 1–23. [CrossRef]
13. Liu, C.; Zhao, Y.; Xi, H.; Jiang, J.; Yu, Y.; Dong, W. The Membrane Interaction of Alpha-Synuclein. *Front. Cell. Neurosci.* **2021**, *15*. [CrossRef]
14. Burré, J.; Sharma, M.; Tsetsenis, T.; Buchman, V.; Etherton, M.R.; Südhof, T.C. α-Synuclein Promotes SNARE-Complex Assembly in Vivo and in Vitro. *Science* **2010**, *329*, 1663–1667. [CrossRef]
15. Wang, T.; Hay, J.C. Alpha-synuclein Toxicity in the Early Secretory Pathway: How It Drives Neurodegeneration in Parkinsons Disease. *Front. Neurosci.* **2015**, *12*, 9–433. [CrossRef]
16. Wong, Y.C.; Krainc, D. α-synuclein toxicity in neurodegeneration: Mechanism and therapeutic strategies. *Nat. Med.* **2017**, *23*, 1–13. [CrossRef]
17. Jan, A.; Gonçalves, N.P.; Vaegter, C.B.; Jensen, P.H.; Ferreira, N. The Prion-Like Spreading of Alpha-Synuclein in Parkinson's Disease: Update on Models and Hypotheses. *Int. J. Mol. Sci.* **2021**, *22*, 8338. [CrossRef]
18. Berge, N.V.D.; Ferreira, N.; Mikkelsen, T.W.; Alstrup, A.K.O.; Tamgüney, G.; Karlsson, P.; Terkelsen, A.J.; Nyengaard, J.R.; Jensen, P.H.; Borghammer, P. Ageing promotes pathological alpha-synuclein propagation and autonomic dysfunction in wild-type rats. *Brain* **2021**, *144*, 1853–1868. [CrossRef]
19. Agarwal, D.; Sandor, C.; Volpato, V.; Caffrey, T.M.; Monzón-Sandoval, J.; Bowden, R.; Alegre-Abarrategui, J.; Wade-Martins, R.; Webber, C. A single-cell atlas of the human substantia nigra reveals cell-specific pathways associated with neurological disorders. *Nat. Commun.* **2020**, *11*, 1–11. [CrossRef]
20. Bu, M.; Farrer, M.J.; Khoshbouei, H. Parkinsons Dynamic control of the dopamine transporter in neurotransmission and homeostasis. *NPJ Parkinson's Dis.* **2021**, *7*, 22.
21. Hunn, B.H.M.; Vingill, S.; Threlfell, S.; Alegre-Abarrategui, J.; Magdelyns, M.; Deltheil, T.; Bengoa-Vergniory, N.; Oliver, P.L.; Cioroch, M.; Doig, N.M.; et al. Impairment of Macroautophagy in Dopamine Neurons Has Opposing Effects on Parkinsonian Pathology and Behavior. *Cell Rep.* **2019**, *29*, 920–931.e7. [CrossRef]
22. Abeliovich, A.; Gitler, A.D. Defects in trafficking bridge Parkinson's disease pathology and genetics. *Nature* **2016**, *539*, 207–216. [CrossRef]
23. Singh, P.K.; Muqit, M.M.K. Parkinson's: A Disease of Aberrant Vesicle Trafficking. *Annu. Rev. Cell Dev. Biol.* **2020**, *36*, 237–264. [CrossRef]
24. Matsuda, W.; Furuta, T.; Nakamura, K.; Hioki, H.; Fujiyama, F.; Arai, R.; Kaneko, T. Single Nigrostriatal Dopaminergic Neurons Form Widely Spread and Highly Dense Axonal Arborizations in the Neostriatum. *J. Neurosci.* **2009**, *29*, 444–453. [CrossRef]
25. Hardy, J.; Lewis, P.; Revesz, T.; Lees, A.; Paisan-Ruiz, C. The genetics of Parkinson's syndromes: A critical review. *Curr. Opin. Genet. Dev.* **2009**, *19*, 254–265. [CrossRef]
26. Chang, D.; Nalls, M.A.; Hallgrímsdóttir, I.B.; Hunkapiller, J.; Van Der Brug, M.; Cai, F.; Kerchner, G.A.; Ayalon, G.; Bingol, B.; Sheng, M.; et al. A meta-analysis of genome-wide association studies identifies 17 new Parkinson's disease risk loci. *Nat. Genet.* **2017**, *49*, 1511–1516. [CrossRef]
27. Gialluisi, A.; Reccia, M.G.; Modugno, N.; Nutile, T.; Lombardi, A.; Di Giovannantonio, L.G.; Pietracupa, S.; Ruggiero, D.; Scala, S.; Gambardella, S.; et al. Identification of sixteen novel candidate genes for late onset Parkinson's disease. *Mol. Neurodegener.* **2021**, *16*, 1–18. [CrossRef]
28. Lee, V.M.Y.; Trojanowski, J.Q. Mechanisms of Parkinson's disease linked to pathological α-synuclein: New targets for drug discovery. *Neuron* **2006**, *51*, 33–38. [CrossRef]
29. Ray, B.; Mahalakshmi, A.M.; Tuladhar, S.; Bhat, A.; Srinivasan, A.; Pellegrino, C.; Kannan, A.; Bolla, S.R.; Chidambaram, S.B.; Sakharkar, M.K. "Janus-Faced" α-Synuclein: Role in Parkinson's Disease. *Front. Cell Dev. Biol.* **2021**, *9*, 673395. [CrossRef]
30. Alessi, D.R.; Sammler, E. LRRK2 kinase in Parkinson's disease. *Science* **2018**, *360*, 36–37. [CrossRef]
31. Bonet-Ponce, L.; Cookson, M.R. LRRK2 recruitment, activity, and function in organelles. *FEBS J.* **2021**. [CrossRef] [PubMed]
32. Berwick, D.C.; Heaton, G.R.; Azeggagh, S.; Harvey, K. LRRK2 Biology from structure to dysfunction: Research progresses, but the themes remain the same. *Mol. Neurodegener.* **2019**, *14*, 1–22. [CrossRef] [PubMed]
33. Madureira, M.; Connor-Robson, N.; Wade-Martins, R. "LRRK2: Autophagy and Lysosomal Activity". *Front. Neurosci.* **2020**, *14*, 498. [CrossRef] [PubMed]
34. Taylor, M.; Alessi, D.R. Advances in elucidating the function of leucine-rich repeat protein kinase-2 in normal cells and Parkinson's disease. *Curr. Opin. Cell Biol.* **2020**, *63*, 102–113. [CrossRef] [PubMed]
35. Follett, J.; Farrer, M.J. LRRK2; a dynamic regulator of cellular trafficking. *Brain Res.* **2021**, *1761*, 147394. [CrossRef]
36. Rassu, M.; Del Giudice, M.G.; Sanna, S.; Taymans, J.M.; Morari, M.; Brugnoli, A.; Frassineti, M.; Masala, A.; Esposito, S.; Galioto, M.; et al. Role of LRRK2 in the regulation of dopamine receptor trafficking. *PLoS ONE* **2017**, *12*, e0179082. [CrossRef]

37. Cho, H.J.; Yu, J.; Xie, C.; Rudrabhatla, P.; Chen, X.; Wu, J.; Parisiadou, L.; Liu, G.; Sun, L.; Ma, B.; et al. Leucine-rich repeat kinase 2 regulates Sec16A at ER exit sites to allow ER-Golgi export. *EMBO J.* **2014**, *33*, 2314–2331. [CrossRef]
38. Mir, R.; Tonelli, F.; Lis, P.; Macartney, T.; Polinski, N.K.; Martinez, T.N.; Chou, M.Y.; Howden, A.J.M.; König, T.; Hotzy, C.; et al. The Parkinson's disease VPS35[D620N] mutation enhances LRRK2-mediated Rab protein phosphorylation in mouse and human. *Biochem. J.* **2018**, *475*, 1861–1883. [CrossRef]
39. Li, W.; Fu, Y.; Halliday, G.M.; Sue, C.M. PARK Genes Link Mitochondrial Dysfunction and Alpha-Synuclein Pathology in Sporadic Parkinson's Disease. *Front. Cell Dev. Biol.* **2021**, *9*. [CrossRef]
40. Moujalled, D.; Strasser, A.; Liddell, J.R. Molecular mechanisms of cell death in neurological diseases. *Cell Death Differ.* **2021**, *28*, 2029–2044. [CrossRef]
41. Yang, X.Z.; Li, X.X.; Zhang, Y.J.; Rodriguez-Rodriguez, L.; Xiang, M.Q.; Wang, H.Y.; Zheng, X.F. Rab1 in cell signaling, cancer and other diseases. *Oncogene* **2016**, *35*, 5699–56704. [CrossRef]
42. Zhang, X.; Wang, X.; Yuan, Z.; Radford, S.J.; Liu, C.; Libutti, S.K.; Zheng, X.F.S. Amino acids-Rab1A-mTORC1 signaling controls whole-body glucose homeostasis. *Cell Rep.* **2021**, *34*, 108830. [CrossRef]
43. Du, J.; von Wrisberg, M.-K.; Gulen, B.; Stahl, M.; Pett, C.; Hedberg, C.; Lang, K.; Schneider, S.; Itzen, A. Rab1-AMPylation by Legionella DrrA is allosterically activated by Rab1. *Nat. Commun.* **2021**, *12*, 1–16. [CrossRef]
44. Kawabata, M.; Matsuo, H.; Koito, T.; Murata, M.; Kubori, T.; Nagai, H.; Tagaya, M.; Arasaki, K. Legionella hijacks the host Golgi-to-ER retrograde pathway for the association of Legionella-containing vacuole with the ER. *PLoS Pathog.* **2021**, *17*, e1009437. [CrossRef]
45. Rendón, W.O.; Martínez-Alonso, E.; Tomás, M.; Martínez-Martínez, N.; Martínez-Menárguez, J.A. Golgi fragmentation is Rab and SNARE dependent in cellular models of Parkinson's disease. *Histochem. Cell Biol.* **2012**, *139*, 671–684. [CrossRef]
46. Tomás, M.; Martínez-Alonso, E.; Martínez-Martínez, N.; Cara-Esteban, M.; Martínez-Menárguez, J.A. Fragmentation of the Golgi complex of dopaminergic neurons in human substantia nigra: New cytopathological findings in Parkinson's disease. *Histol. Histopathol.* **2021**, *36*, 47–60.
47. Soo, K.Y.; Halloran, M.; Sundaramoorthy, V.; Parakh, S.; Toth, R.P.; Southam, K.A.; McLean, C.A.; Lock, P.; King, A.; Farg, M.A.; et al. Rab1-dependent ER-Golgi transport dysfunction is a common pathogenic mechanism in SOD1, TDP-43 and FUS-associated ALS. *Acta Neuropathol.* **2015**, *130*, 679–697. [CrossRef]
48. Soo, K.Y.; Sultana, J.M.; King, A.; Atkinson, R.; Warraich, S.T.; Sundaramoorthy, V.; Blair, I.; Farg, M.A.; Atkin, J.D. ALS-associated mutant FUS inhibits macroautophagy which is restored by overexpression of Rab1. *Cell Death Discov.* **2015**, *1*, 15030. [CrossRef]
49. Dugan, J.M.; deWit, C.; McConlogue, L.; Maltese, W.A. The Ras-related GTP-binding protein, Rab1B, regulates early steps in exocytic transport and processing of beta-amyloid precursor protein. *J. Biol. Chem.* **1995**, *270*, 10982–10989. [CrossRef]
50. Tang, B.L. Rabs, Membrane Dynamics, and Parkinson's Disease. *J. Cell. Physiol.* **2017**, *232*, 1626–1633. [CrossRef]
51. Kiral, F.R.; Kohrs, F.E.; Jin, E.J.; Hiesinger, P.R. Rab GTPases and Membrane Trafficking in Neurodegeneration. *Curr. Biol.* **2018**, *28*, R471–R486. [CrossRef]
52. Lipatova, Z.; Hain, A.U.; Nazarko, V.Y.; Segev, N. Ypt/Rab GTPases: Principles learned from yeast. *Crit. Rev. Biochem. Mol. Biol.* **2015**, *50*, 203–211. [CrossRef]
53. Gao, Y.; Wilson, G.R.; Stephenson, S.; Bozaoglu, K.; Farrer, M.J.; Lockhart, P. The emerging role of Rab GTPases in the pathogenesis of Parkinson's disease. *Mov. Disord.* **2018**, *33*, 196–207. [CrossRef]
54. Aloisi, A.L.; Bucci, C. Rab GTPases-cargo direct interactions: Fine modulators of intracellular trafficking. *Histol. Histopathol.* **2013**, *28*, 839–849. [PubMed]
55. Voss, S.; Li, F.; Rätz, A.; Röger, M.; Wu, Y.W. Spatial Cycling of Rab GTPase, Driven by the GTPase Cycle, Controls Rab's Subcellular Distribution. *Biochemistry* **2019**, *58*, 276–285. [CrossRef] [PubMed]
56. Kirsten, M.L.; Baron, R.A.; Seabra, M.C.; Ces, O. Rab1a and Rab5a preferentially bind to binary lipid compositions with higher stored curvature elastic energy. *Mol. Membr. Biol.* **2013**, *30*, 303–314. [CrossRef] [PubMed]
57. Waschbüsch, D.; Khan, A.R. Phosphorylation of Rab GTPases in the regulation of membrane trafficking. *Traffic* **2020**, *21*, 712–719. [CrossRef]
58. Ng, E.L.; Tang, B.L. Rab GTPases and their roles in brain neurons and glia. *Brain Res. Rev.* **2008**, *58*, 236–246. [CrossRef]
59. Villarroel-Campos, D.; Bronfman, F.C.; Gonzalez-Billault, C. Rab GTPase signaling in neurite outgrowth and axon specification. *Cytoskeleton* **2016**, *73*, 498–507. [CrossRef]
60. Zheng, L.-Q.; Chi, S.-M.; Li, C.-X. Rab23's genetic structure, function and related diseases: A review. *Biosci. Rep.* **2017**, *37*. [CrossRef]
61. D'Adamo, P.; Masetti, M.; Bianchi, V.; Morè, L.; Mignogna, M.L.; Giannandrea, M.; Gatti, S. RAB GTPases and RAB-interacting proteins and their role in the control of cognitive functions. *Neurosci. Biobehav. Rev.* **2014**, *46*, 302–314. [CrossRef]
62. Sastre, A.A.; Montoro, M.L.; Lacerda, H.; Llavero, F.; Zugaza, J. Small GTPases of the Rab and Arf Families: Key Regulators of Intracellular Trafficking in Neurodegeneration. *Int. J. Mol. Sci.* **2021**, *22*, 4425. [CrossRef]
63. Gonçalves, S.A.; Macedo, D.; Raquel, H.; Simões, P.D.; Giorgini, F.; Ramalho, J.S.; Barral, D.C.; Ferreira-Moita, L.; Outeiro, T.F. shRNA-Based Screen Identifies Endocytic Recycling Pathway Components That Act as Genetic Modifiers of Alpha-Synuclein Aggregation, Secretion and Toxicity. *PLoS Genet.* **2016**, *12*, e1005995. [CrossRef]
64. Bae, E.-J.; Kim, D.-K.; Kim, C.; Mante, M.; Adame, A.; Rockenstein, E.; Ulusoy, A.; Klinkenberg, M.; Jeong, G.R.; Bae, J.R.; et al. LRRK2 kinase regulates α-synuclein propagation via RAB35 phosphorylation. *Nat. Commun.* **2018**, *9*, 1–16. [CrossRef]

65. Dinter, E.; Saridaki, T.; Nippold, M.; Plum, S.; Diederichs, L.; Komnig, D.; Fensky, L.; May, C.; Marcus, K.; Voigt, A.; et al. Rab7 induces clearance of α-synuclein aggregates. *J. Neurochem.* **2016**, *138*, 758–774. [CrossRef]
66. Chung, C.Y.; Khurana, V.; Yi, S.; Sahni, N.; Loh, K.H.; Auluck, P.K.; Baru, V.; Udeshi, N.D.; Freyzon, Y.; Carr, S.A.; et al. In Situ Peroxidase Labeling and Mass-Spectrometry Connects Alpha-Synuclein Directly to Endocytic Trafficking and mRNA Metabolism in Neurons. *Cell Syst.* **2017**, *4*, 242–250.e4. [CrossRef]
67. Dalfo, E.; Gómez-Isla, T.; Rosa, J.L.; Bodelón, M.N.; Cuadrado-Tejedor, M.; Barrachina, M.; Ambrosio, S.; Ferrer, I. Abnormal α-Synuclein Interactions with Rab Proteins in α-Synuclein A30P Transgenic Mice. *J. Neuropathol. Exp. Neurol.* **2004**, *63*, 302–313. [CrossRef]
68. Koss, D.J.; Campesan, S.; Giorgini, F.; Outeiro, T.F. Dysfunction of RAB39B- Mediated Vesicular Trafficking in Lewy Body Diseases. *Mov. Disord.* **2021**, *36*, 1744–1758. [CrossRef]
69. Touchot, N.; Zahraoui, A.; Vielh, E.; Tavitian, A. Biochemical properties of the YPT-related rab 1B protein. *FEBS Lett.* **1989**, *256*, 79–84. [CrossRef]
70. Zahraoui, A.; Touchot, N.; Chardin, P.; Tavitian, A. The human Rab genes encode a family of GTP-binding proteins related to yeast YPT1 and SEC4 products involved in secretion. *J. Biol. Chem.* **1989**, *264*, 12394–12401. [CrossRef]
71. Gurkan, C.; Lapp, H.; Alory, C.; Su, A.I.; HogenEsch, J.B.; Balch, W.E. Large-Scale Profiling of Rab GTPase Trafficking Networks: The Membrome. *Mol. Biol. Cell* **2005**, *16*, 3847–3864. [CrossRef]
72. Yang, J.; Zhou, X.; Zhang, R.; Sun, H.; You, F.; Jiang, Z. Differences in IFNβ secretion upon Rab1 inactivation in cells exposed to distinct innate immune stimuli. *Cell. Mol. Immunol.* **2021**, *18*, 1590–1592. [CrossRef]
73. Allan, B.B.; Moyer, B.D.; Balch, W.E. Rab1 Recruitment of p115 into a cis-SNARE Complex: Programming Budding COPII Vesicles for Fusion. *Science* **2000**, *289*, 444–448. [CrossRef]
74. Moyer, B.D.; Allan, B.B.; Balch, W.E. Rab1 interaction with a GM130 effector complex regulates COPII vesicle cis—Golgi tethering. *Traffic* **2001**, *2*, 268–276. [CrossRef]
75. Weide, T.; Bayer, M.; Köster, M.; Siebrasse, J.P.; Peters, R.; Barnekow, A. The Golgi matrix protein GM130: A specific interacting partner of the small GTPase rab1b. *EMBO Rep.* **2001**, *2*, 336–341. [CrossRef]
76. Satoh, A.; Wang, Y.; Malsam, J.; Beard, M.B.; Warren, G. Golgin-84 is a rab1 Binding Partner Involved in Golgi Structure. *Traffic* **2003**, *4*, 153–161. [CrossRef]
77. Beard, M.; Satoh, A.; Shorter, J.; Warren, G. A Cryptic Rab1-binding Site in the p115 Tethering Protein. *J. Biol. Chem.* **2005**, *280*, 25840–25848. [CrossRef]
78. Bayer, M.; Fischer, J.; Kremerskothen, J.; Ossendorf, E.; Matanis, T.; Konczal, M.; Weide, T.; Barnekow, A. Identification and characterization of Iporin as a novel interaction partner for rab1. *BMC Cell Biol.* **2005**, *6*, 15. [CrossRef]
79. Fischer, J.; Weide, T.; Barnekow, A. The MICAL proteins and rab1: A possible link to the cytoskeleton? *Biochem. Biophys. Res. Commun.* **2005**, *328*, 415–423. [CrossRef]
80. Cavieres, V.A.; Cerda-Troncoso, C.; Rivera-Dictter, A.; Castro, R.I.; Luchsinger, C.; Santibañez, N.; Burgos, P.V.; Mar-dones, G.A. Human Golgi phosphoprotein 3 is an effector of RAB1A and RAB1B. *PLoS ONE* **2020**, *15*, e0237514. [CrossRef] [PubMed]
81. Galindo, A.; Planelles-Herrero, V.J.; Degliesposti, G.; Munro, S. Cryo-EM structure of metazoan TRAPPIII, the multi-subunit complex that activates the GTPase Rab1. *EMBO J.* **2021**, *40*, e107608. [CrossRef] [PubMed]
82. Joiner, A.M.; Phillips, B.P.; Yugandhar, K.; Sanford, E.J.; Smolka, M.B.; Yu, H.; Miller, E.A.; Fromme, J.C. Structural basis of TRAPPIII-mediated Rab1 activation. *EMBO J.* **2021**, *40*, e107607. [CrossRef] [PubMed]
83. Cai, Y.; Chin, H.F.; Lazarova, D.; Menon, S.; Fu, C.; Cai, H.; Sclafani, A.; Rodgers, D.W.; De La Cruz, E.M.; Ferro-Novick, S.; et al. The Structural Basis for Activation of the Rab Ypt1p by the TRAPP Membrane-Tethering Complexes. *Cell* **2008**, *133*, 1202–1213. [CrossRef] [PubMed]
84. Cai, H.; Yu, S.; Menon, S.; Cai, Y.; Lazarova, D.L.; Fu, C.; Reinisch, K.M.; Hay, J.C.; Ferro-Novick, S. TRAPPI tethers COPII vesicles by binding the coat subunit Sec23. *Nature* **2007**, *445*, 941–944. [CrossRef]
85. Tan, D.; Cai, Y.; Wang, J.; Zhang, J.; Menon, S.; Chou, H.-T.; Ferro-Novick, S.; Reinisch, K.M.; Walz, T. The EM structure of the TRAPPIII complex leads to the identification of a requirement for COPII vesicles on the macroautophagy pathway. *Proc. Natl. Acad. Sci. USA* **2013**, *110*, 19432–19437. [CrossRef]
86. Lynch-Day, M.A.; Bhandari, D.; Menon, S.; Huang, J.; Cai, H.; Bartholomew, C.R.; Brumell, J.H.; Ferro-Novick, S.; Klionsky, D.J. Trs85 directs a Ypt1 GEF, TRAPPIII, to the phagophore to promote autophagy. *Proc. Natl. Acad. Sci. USA* **2010**, *107*, 7811–7816. [CrossRef]
87. Van Bergen, N.J.; Guo, Y.; Al-Deri, N.; Lipatova, Z.; Stanga, D.; Zhao, S.; Murtazina, R.; Gyurkovska, V.; Pehlivan, D.; Mitani, T.; et al. Deficiencies in vesicular transport mediated by TRAPPC4 are associated with severe syndromic intellectual disability. *Brain* **2019**, *143*, 112–130. [CrossRef]
88. Barr, F.; Lambright, D.G. Rab GEFs and GAPs. *Curr. Opin. Cell Biol.* **2010**, *22*, 461–470. [CrossRef]
89. Du, L.-L.; Novick, P. Yeast Rab GTPase-activating Protein Gyp1p Localizes to the Golgi Apparatus and Is a Negative Regulator of Ypt1p. *Mol. Biol. Cell* **2001**, *12*, 1215–1226. [CrossRef]
90. Thomas, L.L.; Highland, C.M.; Fromme, J.C. Arf1 orchestrates Rab GTPase conversion at the trans-Golgi network. *Mol. Biol. Cell* **2021**, *32*, 1104–1120. [CrossRef]
91. Kim, J.J.; Lipatova, Z.; Majumdar, U.; Segev, N. Regulation of Golgi Cisternal Progression by Ypt/Rab GTPases. *Dev. Cell* **2016**, *36*, 440–452. [CrossRef]

92. Novick, P. Regulation of membrane traffic by Rab GEF and GAP cascades. *Small GTPases* **2016**, *7*, 252–256. [CrossRef]
93. Mitter, A.L.; Schlotterhose, P.; Krick, R. Gyp1 has a dual function as Ypt1 GAP and interaction partner of Atg8 in selective autophagy. *Autophagy* **2019**, *15*, 1031–1050. [CrossRef]
94. Haas, A.K.; Yoshimura, S.-I.; Stephens, D.; Preisinger, C.; Fuchs, E.; Barr, F. Analysis of GTPase-activating proteins: Rab1 and Rab43 are key Rabs required to maintain a functional Golgi complex in human cells. *J. Cell Sci.* **2007**, *120*, 2997–3010. [CrossRef]
95. Sidjanin, D.J.; Park, A.K.; Ronchetti, A.; Martins, J.; Jackson, W.T. TBC1D20 mediates autophagy as a key regulator of autophagosome maturation. *Autophagy* **2016**, *12*, 1759–1775. [CrossRef]
96. Calero, M.; Collins, R.N. Saccharomyces cerevisiae Pra1p/Yip3p interacts with Yip1p and Rab proteins. *Biochem. Biophys. Res. Commun.* **2002**, *290*, 676–681. [CrossRef]
97. Lee, H.J.; Kang, S.J.; Lee, K.; Im, H. Human α-synuclein modulates vesicle trafficking through its interaction with prenylated Rab acceptor protein 1. *Biochem. Biophys. Res. Commun.* **2011**, *412*, 526–531. [CrossRef]
98. Wu, S.K.; Zeng, K.; Wilson, I.A.; Balch, W.E. Structural insights into the function of the Rab GDI superfamily. *Trends Biochem. Sci.* **1996**, *21*, 472–476. [CrossRef]
99. Saraste, J.; Lahtinen, U.; Goud, B. Localization of the small GTP-binding protein rab1p to early compartments of the secretory pathway. *J. Cell Sci.* **1995**, *108*, 1541–1552. [CrossRef]
100. Saraste, J. Spatial and Functional Aspects of ER-Golgi Rabs and Tethers. *Front. Cell Dev. Biol.* **2016**, *4*, 28. [CrossRef]
101. Sannerud, R.; Marie, M.; Nizak, C.; Dale, H.A.; Pernet-Gallay, K.; Perez, F.; Goud, B.; Saraste, J. Rab1 Defines a Novel Pathway Connecting the Pre-Golgi Intermediate Compartment with the Cell Periphery. *Mol. Biol. Cell* **2006**, *17*, 1514–1526. [CrossRef]
102. Monetta, P.; Slavin, I.; Romero, N.; Alvarez, C. Rab1b Interacts with GBF1 and Modulates both ARF1 Dynamics and COPI Association. *Mol. Biol. Cell* **2007**, *18*, 2400–2410. [CrossRef]
103. Cao, X.; Ballew, N.; Barlowe, C. Initial docking of ER-derived vesicles requires Uso1p and Ypt1p but is independent of SNARE proteins. *EMBO J.* **1998**, *17*, 2156–2165. [CrossRef]
104. Cao, X.; Barlowe, C. Asymmetric Requirements for a Rab Gtpase and Snare Proteins in Fusion of Copii Vesicles with Acceptor Membranes. *J. Cell Biol.* **2000**, *149*, 55–66. [CrossRef]
105. Morsomme, P.; Riezman, H. The Rab GTPase Ypt1p and tethering factors couple protein sorting at the ER to vesicle targeting to the Golgi apparatus. *Dev. Cell* **2002**, *2*, 307–317. [CrossRef]
106. Tisdale, E.J.; Bourne, J.R.; Khosravi-Far, R.; Der, C.J.; Balch, W.E. GTP-binding mutants of rab1 and rab2 are potent inhibitors of vesicular transport from the endoplasmic reticulum to the Golgi complex. *J. Cell Biol.* **1992**, *119*, 749–761. [CrossRef]
107. Slavin, I.; García, I.A.; Monetta, P.; Martinez, H.; Romero, N.; Alvarez, C. Role of Rab1b in COPII dynamics and function. *Eur. J. Cell Biol.* **2011**, *90*, 301–311. [CrossRef]
108. Westrate, L.M.; Hoyer, M.J.; Nash, M.J.; Voeltz, G.K. Vesicular and uncoated Rab1-dependent cargo carriers facilitate ER to Golgi transport. *J. Cell Sci.* **2020**, *133*. [CrossRef]
109. Witkos, T.; Lowe, M. The Golgin Family of Coiled-Coil Tethering Proteins. *Front. Cell Dev. Biol.* **2016**, *3*, 86. [CrossRef]
110. Diao, A.; Rahman, D.; Pappin, D.; Lucocq, J.; Lowe, M. The coiled-coil membrane protein golgin-84 is a novel rab effector required for Golgi ribbon formation. *J. Cell Biol.* **2003**, *160*, 201–212. [CrossRef]
111. Liu, X.; Wang, Z.; Yang, Y.; Li, Q.; Zeng, R.; Kang, J.; Wu, J. Rab1A mediates proinsulin to insulin conversion in beta-cells by maintaining Golgi stability through interactions with golgin-84. *Protein Cell* **2016**, *7*, 692–696. [CrossRef] [PubMed]
112. Kamena, F.; Diefenbacher, M.; Kilchert, C.; Schwarz, H.; Spang, A. Ypt1p is essential for retrograde Golgi-ER transport and for Golgi maintenance in S. cerevisiae. *J. Cell Sci.* **2008**, *121*, 1293–1302. [CrossRef] [PubMed]
113. Suvorova, E.S.; Duden, R.; Lupashin, V. The Sec34/Sec35p complex, a Ypt1p effector required for retrograde intra-Golgi trafficking, interacts with Golgi SNAREs and COPI vesicle coat proteins. *J. Cell Biol.* **2002**, *157*, 631–643. [CrossRef] [PubMed]
114. Willett, R.; Ungar, D.; Lupashin, V. The Golgi puppet master: COG complex at center stage of membrane trafficking interactions. *Histochem. Cell Biol.* **2013**, *140*, 271–283. [CrossRef] [PubMed]
115. Neuman, S.D.; Lee, A.R.; Selegue, J.E.; Cavanagh, A.T.; Bashirullah, A. A novel function for Rab1 and Rab11 during secretory granule maturation. *J. Cell Sci.* **2021**, *134*, 259037. [CrossRef] [PubMed]
116. Russo, A.J.; Mathiowetz, A.J.; Hong, S.; Welch, M.D.; Campellone, K.G. Rab1 recruits WHAMM during membrane remodeling but limits actin nucleation. *Mol. Biol. Cell* **2016**, *27*, 967–978. [CrossRef]
117. Campellone, K.G.; Webb, N.J.; Znameroski, E.A.; Welch, M.D. WHAMM is an Arp2/3 complex activator that binds microtubules and functions in ER to Golgi transport. *Cell* **2008**, *134*, 148–161. [CrossRef]
118. Mukhopadhyay, A.; Quiroz, J.A.; Wolkoff, A.W. Rab1a regulates sorting of early endocytic vesicles. *Am. J. Physiol. Liver Physiol.* **2014**, *306*, G412–G424. [CrossRef]
119. Rivero, S.; Cardenas, J.; Bornens, M.; Ríos, R.M. Microtubule nucleation at the cis-side of the Golgi apparatus requires AKAP450 and GM130. *EMBO J.* **2009**, *28*, 1016–1028. [CrossRef]
120. Mikhaylova, M.; Bera, S.; Kobler, O.; Frischknecht, R.; Kreutz, M.R. A Dendritic Golgi Satellite between ERGIC and Retromer. *Cell Rep.* **2015**, *14*, 189–199. [CrossRef]
121. Saraste, J.; Marie, M. Intermediate compartment (IC): From pre-Golgi vacuoles to a semi-autonomous membrane system. *Histochem. Cell Biol.* **2018**, *150*, 407–430. [CrossRef]
122. Solis, G.P.; Bilousov, O.; Koval, A.; Lüchtenborg, A.M.; Lin, C.; Katanaev, V.L. Golgi-Resident Galphao Promotes Protrusive Membrane Dynamics. *Cell* **2017**, *170*, 939–955. [CrossRef]

123. de Lima, N.C.R.; Melo, T.Q.; Sakugawa, A.Y.; Melo, K.P.; Ferrari, M.F. Restoration of Rab1 Levels Prevents Endoplasmic Reticulum Stress in Hippocampal Cells during Protein Aggregation Triggered by Rotenone. *Neuroscience* **2019**, *419*, 5–13. [CrossRef]
124. Tsvetanova, N.G. The secretory pathway in control of endoplasmic reticulum homeostasis. *Small GTPases* **2013**, *4*, 28–33. [CrossRef]
125. Kang, C.H.; Park, J.H.; Lee, E.S.; Paeng, S.K.; Chae, H.B.; Chi, Y.H.; Lee, S.Y. Exploring Novel Functions of the Small GTPase Ypt1p under Heat-Shock by Characterizing a Temperature-Sensitive Mutant Yeast Strain, ypt1-G80D. *Int. J. Mol. Sci.* **2019**, *20*, 132. [CrossRef]
126. Kaushik, S.; Cuervo, A.M. The coming of age of chaperone-mediated autophagy. *Nat. Rev. Mol. Cell Biol.* **2018**, *19*, 365–381. [CrossRef]
127. Yang, Q.; Wang, R.; Zhu, L. Chaperone-Mediated Autophagy. *Adv. Exp. Med. Biol.* **2019**, *1206*, 435–452.
128. Schuck, S. Microautophagy—Distinct molecular mechanisms handle cargoes of many sizes. *J. Cell Sci.* **2020**, *133*. [CrossRef]
129. Lamark, T.; Johansen, T. Mechanisms of Selective Autophagy. *Annu. Rev. Cell Dev. Biol.* **2021**, *37*, 143–169. [CrossRef]
130. Fernandez-Fernandez, M.R.; Ruiz-Garcia, D.; Martin-Solana, E.; Chichon, F.J.; Carrascosa, J.L.; Fernandez, J.-J. 3D electron tomography of brain tissue unveils distinct Golgi structures that sequester cytoplasmic contents in neurons. *J. Cell Sci.* **2016**, *130*, 83–89. [CrossRef]
131. Rana, T.; Behl, T.; Sehgal, A.; Mehta, V.; Singh, S.; Bhatia, S.; Al-Harrasi, A.; Bungau, S. Exploring the Role of Autophagy Dysfunction in Neurodegenerative Disorders. *Mol. Neurobiol.* **2021**, *58*, 4886–4905. [CrossRef]
132. Suzuki, K.; Kirisako, T.; Kamada, Y.; Mizushima, N.; Noda, T.; Ohsumi, Y. The pre-autophagosomal structure organized by concerted functions of APG genes is essential for autophagosome formation. *EMBO J.* **2001**, *20*, 5971–5981. [CrossRef]
133. Axe, E.L.; Walker, S.A.; Manifava, M.; Chandra, P.; Roderick, H.L.; Habermann, A.; Griffiths, G.; Ktistakis, N.T. Autophagosome formation from membrane compartments enriched in phosphatidylinositol 3-phosphate and dynamically connected to the endoplasmic reticulum. *J. Cell Biol.* **2008**, *182*, 685–701. [CrossRef]
134. Roberts, R.; Ktistakis, N.T. Omegasomes: PI3P platforms that manufacture autophagosomes. *Essays Biochem.* **2013**, *55*, 17–27. [CrossRef]
135. Nascimbeni, A.C.; Codogno, P.; Morel, E. Phosphatidylinositol-3-phosphate in the regulation of autophagy membrane dynamics. *FEBS J.* **2017**, *284*, 1267–1278. [CrossRef]
136. Ylä-Anttila, P.; Vihinen, H.; Jokitalo, E.; Eskelinen, E.-L. 3D tomography reveals connections between the phagophore and endoplasmic reticulum. *Autophagy* **2009**, *5*, 1180–1185. [CrossRef]
137. Uemura, T.; Yamamoto, M.; Kametaka, A.; Sou, Y.-S.; Yabashi, A.; Yamada, A.; Annoh, H.; Kametaka, S.; Komatsu, M.; Waguri, S. A Cluster of Thin Tubular Structures Mediates Transformation of the Endoplasmic Reticulum to Autophagic Isolation Membrane. *Mol. Cell. Biol.* **2014**, *34*, 1695–1706. [CrossRef]
138. Shen, H.-M.; Mizushima, N. At the end of the autophagic road: An emerging understanding of lysosomal functions in autophagy. *Trends Biochem. Sci.* **2014**, *39*, 61–71. [CrossRef]
139. Ganesan, D.; Cai, Q. Understanding amphisomes. *Biochem. J.* **2021**, *478*, 1959–1976. [CrossRef] [PubMed]
140. Mercer, T.J.; Gubas, A.; Tooze, S.A. A molecular perspective of mammalian autophagosome biogenesis. *J. Biol. Chem.* **2018**, *293*, 5386–5395. [CrossRef]
141. Nakatogawa, H. Mechanisms governing autophagosome biogenesis. *Nat. Rev. Mol. Cell Biol.* **2020**, *21*, 439–458. [CrossRef]
142. Matoba, K.; Noda, N.N. Structural catalog of core Atg proteins opens new era of autophagy research. *J. Biochem.* **2021**, *169*, 517–525. [CrossRef] [PubMed]
143. Nishimura, T.; Tamura, N.; Kono, N.; Shimanaka, Y.; Arai, H.; Yamamoto, H.; Mizushima, N. Autophagosome formation is initiated at phosphatidylinositol synthase-enriched ER subdomains. *EMBO J.* **2017**, *36*, 1719–1735. [CrossRef] [PubMed]
144. Zhou, C.; Ma, K.; Gao, R.; Mu, C.; Chen, L.; Liu, Q.; Luo, Q.; Feng, D.; Zhu, Y.; Chen, Q. Regulation of mATG9 trafficking by Src- and ULK1-mediated phosphorylation in basal and starvation-induced autophagy. *Cell Res.* **2016**, *27*, 184–201. [CrossRef]
145. De Tito, S.; Hervás, J.H.; van Vliet, A.R.; Tooze, S.A. The Golgi as an Assembly Line to the Autophagosome. *Trends Biochem. Sci.* **2020**, *45*, 484–496. [CrossRef]
146. Matoba, K.; Kotani, T.; Tsutsumi, A.; Tsuji, T.; Mori, T.; Noshiro, D.; Sugita, Y.; Nomura, N.; Iwata, S.; Ohsumi, Y.; et al. Atg9 is a lipid scramblase that mediates autophagosomal membrane expansion. *Nat. Struct. Mol. Biol.* **2020**, *27*, 1185–1193. [CrossRef]
147. Polson, H.E.; de Lartigue, J.; Rigden, D.J.; Reedijk, M.; Urbé, S.; Clague, M.J.; Tooze, S.A. Mammalian Atg18 (WIPI2) localizes to omegasome-anchored phagophores and positively regulates LC3 lipidation. *Autophagy* **2010**, *6*, 506–522. [CrossRef]
148. Mizushimam, N.; Kuma, A.; Kobayashi, Y.; Yamamoto, A.; Matsubae, M.; Takao, T.; Natsume, T.; Ohsumi, Y.; Yoshimori, T. Mouse Apg16L, a novel WD-repeat protein, targets to the autophagic isolation membrane with the Apg12-Apg5 con-jugate. *J. Cell Sci.* **2003**, *116*, 1679–1688. [CrossRef]
149. Slobodkin, M.R.; Elazar, Z. The Atg8 family: Multifunctional ubiquitin-like key regulators of autophagy. *Essays Biochem.* **2013**, *55*, 51–64. [CrossRef]
150. Noda, T. Autophagy in the context of the cellular membrane-trafficking system: The enigma of Atg9 vesicles. *Biochem. Soc. Trans.* **2017**, *45*, 1323–1331. [CrossRef]
151. Jedd, G.; Richardson, C.; Litt, R.; Segev, N. The Ypt1 GTPase is essential for the first two steps of the yeast secretory pathway. *J. Cell Biol.* **1995**, *131*, 583–590. [CrossRef]

152. Zoppino, F.C.M.; Militello, R.D.; Slavin, I.; Álvarez, C.; Colombo, M.I. Autophagosome Formation Depends on the Small GTPase Rab1 and Functional ER Exit Sites. *Traffic* **2010**, *11*, 1246–1261. [CrossRef]
153. Huang, J.; Birmingham, C.L.; Shahnazari, S.; Shiu, J.; Zheng, Y.T.; Smith, A.C.; Campellone, K.G.; Heo, W.D.; Gruenheid, S.; Meyer, T.; et al. Antibacterial autophagy occurs at PI(3)P-enriched domains of the endoplasmic reticulum and requires Rab1 GTPase. *Autophagy* **2011**, *7*, 17–26. [CrossRef]
154. Lipatova, Z.; Belogortseva, N.; Zhang, X.Q.; Kim, J.; Taussig, D.; Segev, N. Regulation of selective autophagy onset by a Ypt/Rab GTPase module. *Proc. Natl. Acad. Sci. USA* **2012**, *109*, 6981–6986. [CrossRef]
155. Mochizuki, Y.; Ohashi, R.; Kawamura, T.; Iwanari, H.; Kodama, T.; Naito, M.; Hamakubo, T. Phosphatidylinositol 3-Phosphatase Myotubularin-related Protein 6 (MTMR6) Is Regulated by Small GTPase Rab1B in the Early Secretory and Autophagic Pathways. *J. Biol. Chem.* **2013**, *288*, 1009–1021. [CrossRef]
156. Davis, S.; Ferro-Novick, S. Ypt1 and COPII vesicles act in autophagosome biogenesis and the early secretory pathway. *Biochem. Soc. Trans.* **2015**, *43*, 92–96. [CrossRef]
157. Ao, X.; Zou, L.; Wu, Y. Regulation of autophagy by the Rab GTPase network. *Cell Death Differ.* **2014**, *21*, 348–358. [CrossRef]
158. Barz, S.; Kriegenburg, F.; Sánchez-Martín, P.; Kraft, C. Small but mighty: Atg8s and Rabs in membrane dynamics during autophagy. *Biochim. Biophys. Acta Bioenerg.* **2021**, *1868*, 119064. [CrossRef]
159. Wang, J.; Menon, S.; Yamasaki, A.; Chou, H.-T.; Walz, T.; Jiang, Y.; Ferro-Novick, S. Ypt1 recruits the Atg1 kinase to the preautophagosomal structure. *Proc. Natl. Acad. Sci. USA* **2013**, *110*, 9800–9805. [CrossRef]
160. Webster, C.P.; Smith, E.F.; Bauer, C.S.; Moller, A.; Hautbergue, G.M.; Ferraiuolo, L.; Myszczynska, M.; Higginbottom, A.; Walsh, M.J.; Whitworth, A.J.; et al. The C9orf72 protein interacts with Rab1a and the ULK 1 complex to regulate initiation of autophagy. *EMBO J.* **2016**, *35*, 1656–1676. [CrossRef] [PubMed]
161. Tremel, S.; Ohashi, Y.; Morado, D.R.; Bertram, J.; Perisic, O.; Brandt, L.T.L.; von Wrisberg, M.K.; Chen, Z.A.; Maslen, S.L.; Kovtun, O.; et al. Structural basis for VPS34 kinase activation by Rab1 and Rab5 on membranes. *Nat. Commun.* **2021**, *12*, 1564. [CrossRef] [PubMed]
162. Backues, S.K.; Klionsky, D.J. Atg11: A Rab-dependent, coiled-coil membrane protein that acts as a tether for autophagy. *Autophagy* **2012**, *8*, 1275–1278. [CrossRef] [PubMed]
163. Kakuta, S.; Yamaguchi, J.; Suzuki, C.; Sasaki, M.; Kazuno, S.; Uchiyama, Y. Small GTPase Rab1B is associated with ATG9A vesicles and regulates autophagosome formation. *FASEB J.* **2017**, *31*, 3757–3773. [CrossRef]
164. Lamb, C.A.; Nühlen, S.; Judith, D.; Frith, D.; Snijders, B.; Behrends, C.; Tooze, S.A. TBC 1D14 regulates autophagy via the TRAPP complex and ATG 9 traffic. *EMBO J.* **2015**, *35*, 281–301. [CrossRef]
165. Winslow, A.R.; Chen, C.W.; Corrochano, S.; Acevedo-Arozena, A.; Gordon, D.E.; Peden, A.A.; Lichtenberg, M.; Menzies, F.M.; Ravikumar, B.; Imarisio, S.; et al. α-Synuclein impairs macroautophagy: Implications for Parkinson's disease. *J. Cell Biol.* **2010**, *190*, 1023–1037. [CrossRef]
166. Graef, M.; Friedman, J.; Graham, C.; Babu, M.; Nunnari, J. ER exit sites are physical and functional core autophagosome biogenesis components. *Mol. Biol. Cell* **2013**, *24*, 2918–2931. [CrossRef]
167. Shima, T.; Kirisako, H.; Nakatogawa, H. COPII vesicles contribute to autophagosomal membranes. *J. Cell Biol.* **2019**, *218*, 1503–1510. [CrossRef]
168. Li, Z.; Huang, W.; Wang, W. Multifaceted roles of COPII subunits in autophagy. *Biochim. Biophys. Acta Bioenerg.* **2019**, *1867*, 118627. [CrossRef]
169. Ge, L.; Melville, D.; Zhang, M.; Schekman, R. The ER–Golgi intermediate compartment is a key membrane source for the LC3 lipidation step of autophagosome biogenesis. *eLife* **2013**, *2*, e00947. [CrossRef]
170. Ge, L.; Zhang, M.; Schekman, R. Phosphatidylinositol 3-kinase and COPII generate LC3 lipidation vesicles from the ER-Golgi intermediate compartment. *eLife* **2014**, *3*, e04135. [CrossRef]
171. Ge, L.; Zhang, M.; Kenny, S.J.; Liu, D.; Maeda, M.; Saito, K.; Mathur, A.; Xu, K.; Schekman, R. Remodeling of ER -exit sites initiates a membrane supply pathway for autophagosome biogenesis. *EMBO Rep.* **2017**, *18*, 1586–1603. [CrossRef]
172. Jeong, Y.-T.; Simoneschi, D.; Keegan, S.; Melville, D.; Adler, N.S.; Saraf, A.; Florens, L.; Washburn, M.P.; Cavasotto, C.N.; Fenyö, D.; et al. The ULK1-FBXW5-SEC23B nexus controls autophagy. *eLife* **2018**, *7*, e42253. [CrossRef]
173. Gan, W.; Zhang, C.; Siu, K.Y.; Satoh, A.; Tanner, J.A.; Yu, S. ULK1 phosphorylates Sec23A and mediates autophagy-induced inhibition of ER-to-Golgi traffic. *BMC Cell Biol.* **2017**, *18*, 22. [CrossRef]
174. Wang, J.; Davis, S.; Menon, S.; Zhang, J.; Ding, J.; Cervantes, S.; Miller, E.; Jiang, Y.; Ferro-Novick, S. Ypt1/Rab1 regulates Hrr25/CK1delta kinase activity in ER-Golgi traffic and macroautophagy. *J. Cell Biol.* **2015**, *210*, 273–285. [CrossRef] [PubMed]
175. Li, Y.; Chen, X.; Xiong, Q.; Chen, Y.; Zhao, H.; Tahir, M.; Song, J.; Zhou, B.; Wang, J. Casein Kinase 1 Family Member CK1delta/Hrr25 Is Required for Autophagosome Completion. *Front. Cell Dev. Biol.* **2020**, *8*, 460.
176. Cooper, A.A.; Gitler, A.D.; Cashikar, A.; Haynes, C.M.; Hill, K.J.; Bhullar, B.; Liu, K.; Xu, K.; Strathearn, K.E.; Liu, F.; et al. Alpha-synuclein blocks ER-Golgi traffic and Rab1 rescues neuron loss in Parkinson's models. *Science* **2006**, *313*, 324–328. [CrossRef]
177. Gitler, A.D.; Bevis, B.J.; Shorter, J.; Strathearn, K.E.; Hamamichi, S.; Su, L.J.; Caldwell, K.A.; Caldwell, G.A.; Rochet, J.C.; McCaffery, J.M.; et al. The Parkinson's disease protein alpha-synuclein disrupts cellular Rab homeo-stasis. *Proc. Natl. Acad. Sci. USA* **2008**, *105*, 145–150. [CrossRef]

178. Thayanidhi, N.; Helm, J.R.; Nycz, D.C.; Bentley, M.; Liang, Y.; Hay, J.C. Alpha-synuclein delays endoplasmic reticulum (ER)-to-Golgi transport in mammalian cells by antagonizing ER/Golgi SNAREs. *Mol. Biol. Cell* **2010**, *21*, 1850–1863. [CrossRef]
179. Mazzulli, J.R.; Zunke, F.; Isacson, O.; Studer, L.; Krainc, D. α-Synuclein–induced lysosomal dysfunction occurs through disruptions in protein trafficking in human midbrain synucleinopathy models. *Proc. Natl. Acad. Sci. USA* **2016**, *113*, 1931–1936. [CrossRef]
180. Glick, B.S.; Nakano, A. Membrane Traffic within the Golgi Apparatus. *Annu. Rev. Cell Dev. Biol.* **2009**, *25*, 113–132. [CrossRef]
181. Klumperman, J. Architecture of the Mammalian Golgi. *Cold Spring Harb. Perspect. Biol.* **2011**, *3*, a005181. [CrossRef]
182. Li, J.; Ahat, E.; Wang, Y. Golgi Structure and Function in Health; Stress, and Diseases. *Results Probl. Cell Differ.* **2019**, *67*, 441–485.
183. Coune, P.G.; Bensadoun, J.C.; Aebischer, P.; Schneider, B.L. Rab1A over-expression prevents Golgi apparatus frag-men-tation and partially corrects motor deficits in an alpha-synuclein based rat model of Parkinson's disease. *J. Parkinsons Dis.* **2011**, *1*, 373–387. [CrossRef]
184. Gosavi, P.; Gleeson, P.A. The Function of the Golgi Ribbon Structure—An Enduring Mystery Unfolds! *Bioessays* **2017**, *39*, 1700063. [CrossRef]
185. Martínez-Menárguez, J.A.; Tomás, M.; Martínez-Martínez, N.; Martínez-Alonso, E. Golgi Fragmentation in Neuro-degenerative Diseases: Is There a Common Cause? *Cells* **2019**, *8*, 748. [CrossRef]
186. Lashuel, H.A.; Hirling, H. Rescuing Defective Vesicular Trafficking Protects against α-Synuclein Toxicity in Cellular and Animal Models of Parkinson's Disease. *ACS Chem. Biol.* **2006**, *1*, 420–424. [CrossRef]
187. Sevlever, D.; Jiang, P.; Yen, S.-H.C. Cathepsin D Is the Main Lysosomal Enzyme Involved in the Degradation of α-Synuclein and Generation of Its Carboxy-Terminally Truncated Species. *Biochemistry* **2008**, *47*, 9678–9687. [CrossRef]
188. Wilson, B.S.; Nuoffer, C.; Meinkoth, J.L.; McCaffery, M.; Feramisco, J.R.; Balch, W.E.; Farquhar, M.G. A Rab1 mutant affecting guanine nucleotide exchange promotes disassembly of the Golgi apparatus. *J. Cell Biol.* **1994**, *125*, 557–571. [CrossRef] [PubMed]
189. Aizawa, M.; Fukuda, M. Small GTPase Rab2B and Its Specific Binding Protein Golgi-associated Rab2B Interactor-like 4 (GARI-L4) Regulate Golgi Morphology. *J. Biol. Chem.* **2015**, *290*, 22250–22261. [CrossRef] [PubMed]
190. Galea, G.; Bexiga, M.G.; Panarella, A.; O'Neill, E.D.; Simpson, J.C. A high-content screening microscopy approach to dissect the role of Rab proteins in Golgi-to-ER retrograde trafficking. *J. Cell Sci.* **2015**, *128*, 2339–2349. [CrossRef]
191. Hara, T.; Nakamura, K.; Matsui, M.; Yamamoto, A.; Nakahara, Y.; Suzuki-Migishima, R.; Yokoyama, M.; Mishima, K.; Saito, I.; Okano, H.; et al. Suppression of basal autophagy in neural cells causes neurodegenerative disease in mice. *Nature* **2006**, *441*, 885–889. [CrossRef]
192. Komatsu, M.; Waguri, S.; Chiba, T.; Murata, S.; Iwata, J.-I.; Tanida, I.; Ueno, T.; Koike, M.; Uchiyama, Y.; Kominami, E.; et al. Loss of autophagy in the central nervous system causes neurodegeneration in mice. *Nature* **2006**, *441*, 880–884. [CrossRef]
193. Nixon, R.A.; Wegiel, J.; Kumar, A.; Yu, W.H.; Peterhoff, C.; Cataldo, A.; Cuervo, A.M. Extensive involvement of autophagy in Alzheimer disease: An immuno-electron microscopy study. *J. Neuropathol. Exp. Neurol.* **2005**, *64*, 113–122. [CrossRef]
194. Anglade, P.; Vyas, S.; Javoy-Agid, F.; Herrero, M.T.; Michel, P.P.; Marquez, J.; Mouatt-Prigent, A.; Ruberg, M.; Hirsch, E.C.; Agid, Y. Apoptosis and autophagy in nigral neurons of patients with Parkinson's disease. *Histol. Histopathol.* **1997**, *12*, 25–31. [PubMed]
195. Dehay, B.; Bové, J.; Rodríguez-Muela, N.; Perier, C.; Recasens, A.; Boya, P.; Vila, M. Pathogenic lysosomal depletion in Parkinson's disease. *J. Neurosci.* **2010**, *30*, 12535–12544. [CrossRef]
196. Tanji, K.; Mori, F.; Kakita, A.; Takahashi, H.; Wakabayashi, K. Alteration of autophagosomal proteins (LC3, GABARAP and GATE-16) in Lewy body disease. *Neurobiol. Dis.* **2011**, *43*, 690–697. [CrossRef]
197. Friedman, L.G.; Lachenmayer, M.L.; Wang, J.; He, L.; Poulose, S.M.; Komatsu, M.; Holstein, G.R.; Yue, Z. Disrupted autophagy leads to dopaminergic axon and dendrite degeneration and promotes presynaptic accumulation of α-synuclein and LRRK2 in the brain. *J. Neurosci.* **2012**, *32*, 7585–7593. [CrossRef] [PubMed]
198. Hou, X.; Watzlawik, J.O.; Fiesel, F.C.; Springer, W. Autophagy in Parkinson's Disease. *J. Mol. Biol.* **2020**, *432*, 2651–2672. [CrossRef]
199. Pantazopoulou, M.; Brembati, V.; Kanellidi, A.; Bousset, L.; Melki, R.; Stefanis, L. Distinct alpha-Synuclein species induced by seeding are selectively cleared by the Lysosome or the Proteasome in neuronally differentiated SH-SY5Y cells. *J. Neurochem.* **2020**, *156*, 880–896. [CrossRef]
200. Cuervo, A.M.; Stefanis, L.; Fredenburg, R.; Lansbury, P.T.; Sulzer, D. Impaired Degradation of Mutant α-Synuclein by Chaperone-Mediated Autophagy. *Science* **2004**, *305*, 1292–1295. [CrossRef]
201. Huang, Y.; Chegini, F.; Chua, G.; Murphy, K.; Gai, W.; Halliday, G.M. Macroautophagy in sporadic and the genetic form of Parkinson's disease with the A53T alpha-synuclein mutation. *Transl. Neurodegener.* **2012**, *1*, 2. [CrossRef]
202. Jiang, T.-F.; Zhang, Y.-J.; Zhou, H.-Y.; Wang, H.-M.; Tian, L.-P.; Liu, J.; Ding, J.-Q.; Chen, S.-D. Curcumin Ameliorates the Neurodegenerative Pathology in A53T α-synuclein Cell Model of Parkinson's Disease Through the Downregulation of mTOR/p70S6K Signaling and the Recovery of Macroautophagy. *J. Neuroimmune Pharmacol.* **2013**, *8*, 356–369. [CrossRef]
203. Sarkar, S.; Olsen, A.L.; Sygnecka, K.; Lohr, K.M.; Feany, M.B. α-synuclein impairs autophagosome maturation through abnormal actin stabilization. *PLoS Genet.* **2021**, *17*, e1009359. [CrossRef]
204. Orenstein, S.J.; Kuo, S.-H.; Tasset, I.; Arias, E.; Koga, H.; Carasa, I.F.; Cortes, E.; Honig, L.S.; Dauer, W.; Consiglio, A.; et al. Interplay of LRRK2 with chaperone-mediated autophagy. *Nat. Neurosci.* **2013**, *16*, 394–406. [CrossRef]
205. Alegre-Abarrategui, J.; Christian, H.; Lufino, M.M.; Mutihac, R.; Venda, L.L.; Ansorge, O.; Wade-Martins, R. LRRK2 regulates autophagic activity and localizes to specific membrane microdomains in a novel human genomic reporter cellular model. *Hum. Mol. Genet.* **2009**, *18*, 4022–4034. [CrossRef]

206. Plowey, E.D.; Cherra, S.J., 3rd; Liu, Y.J.; Chu, C.T. Role of autophagy in G2019S-LRRK2-associated neurite shortening in differentiated SH-SY5Y cells. *J. Neurochem.* **2008**, *105*, 1048–1056. [CrossRef]
207. Ramonet, D.; Daher, J.P.L.; Lin, B.M.; Stafa, K.; Kim, J.; Banerjee, R.; Westerlund, M.; Pletnikova, O.; Glauser, L.; Yang, L.; et al. Dopaminergic Neuronal Loss, Reduced Neurite Complexity and Autophagic Abnormalities in Transgenic Mice Expressing G2019S Mutant LRRK2. *PLoS ONE* **2011**, *6*, e18568. [CrossRef]
208. Boecker, C.A.; Goldsmith, J.; Dou, D.; Cajka, G.G.; Holzbaur, E.L.F. Increased LRRK2 kinase activity alters neuronal autophagy by disrupting the axonal transport of autophagosomes. *Curr. Biol.* **2021**, *31*, 2140–2154. [CrossRef] [PubMed]
209. Steger, M.; Tonelli, F.; Ito, G.; Davies, P.; Trost, M.; Vetter, M.; Wachter, S.; Lorentzen, E.; Duddy, G.; Wilson, S.; et al. Phosphoproteomics reveals that Parkinson's disease kinase LRRK2 regulates a subset of Rab GTPases. *eLife* **2016**, *5*, e12813. [CrossRef] [PubMed]
210. Bonet-Ponce, L.; Cookson, M.R. The role of Rab GTPases in the pathobiology of Parkinson' disease. *Curr. Opin. Cell Biol.* **2019**, *59*, 73–80. [CrossRef] [PubMed]
211. Nirujogi, R.S.; Tonelli, F.; Taylor, M.; Lis, P.; Zimprich, A.; Sammler, E.; Alessi, D.R. Development of a multiplexed targeted mass spectrometry assay for LRRK2-phosphorylated Rabs and Ser910/Ser935 biomarker sites. *Biochem. J.* **2021**, *478*, 299–326. [CrossRef]
212. Liu, K.; Shi, N.; Sun, Y.; Zhang, T.; Sun, X. Therapeutic effects of rapamycin on MPTP-induced Parkinsonism in mice. *Neurochem. Res.* **2013**, *38*, 201–207. [CrossRef]
213. Lizama, B.N.; Chu, C.T. Neuronal autophagy and mitophagy in Parkinson's disease. *Mol. Asp. Med.* **2021**, 100972, in press. [CrossRef]
214. Savitt, D.; Jankovic, J. Targeting α-Synuclein in Parkinson's Disease: Progress Towards the Development of Disease-Modifying Therapeutics. *Drugs* **2019**, *79*, 797–810. [CrossRef]
215. Lukman, S.; Nguyen, M.N.; Sim, K.; Teo, J.C. Discovery of Rab1 binding sites using an ensemble of clustering methods. *Proteins Struct. Funct. Bioinform.* **2017**, *85*, 859–871. [CrossRef]
216. Fleming, J.; Outeiro, T.F.; Slack, M.; Lindquist, S.L.; Bulawa, C.E. Detection of Compounds That Rescue Rab1-Synuclein. *Toxicity* **2008**, *439*, 339–351. [CrossRef]
217. Gray, J.L.; Von Delft, F.; Brennan, P.E. Targeting the Small GTPase Superfamily through Their Regulatory Proteins. *Angew. Chem. Int. Ed.* **2019**, *59*, 6342–6366. [CrossRef]
218. Gendaszewska-Darmach, E.; Garstka, M.A.; Błażewska, K.M. Targeting Small GTPases and Their Prenylation in Diabetes Mellitus. *J. Med. Chem.* **2021**, *64*, 9677–9710. [CrossRef]
219. Carroll, C.B.; Wyse, R.K.H. Simvastatin as a Potential Disease-Modifying Therapy for Patients with Parkinson's Disease: Rationale for Clinical Trial, and Current Progress. *J. Parkinsons Dis.* **2017**, *7*, 545–568. [CrossRef]

International Journal of
Molecular Sciences

MDPI

Review

Circulating miRNAs as Potential Biomarkers Distinguishing Relapsing–Remitting from Secondary Progressive Multiple Sclerosis. A Review

Sylwia Pietrasik [1], Angela Dziedzic [1,*], Elzbieta Miller [2], Michal Starosta [2] and Joanna Saluk-Bijak [1]

1 Department of General Biochemistry, Faculty of Biology and Environmental Protection, University of Lodz, Pomorska 141/143, 90-236 Lodz, Poland; sylwia.pietrasik@edu.uni.lodz.pl (S.P.); joanna.saluk@biol.uni.lodz.pl (J.S.-B.)

2 Department of Neurological Rehabilitation, Medical University of Lodz, Milionowa 14, 93-113 Lodz, Poland; elzbieta.dorota.miller@umed.lodz.pl (E.M.); michal.starosta@umed.lodz.pl (M.S.)

* Correspondence: angela.dziedzic@edu.uni.lodz.pl

Abstract: Multiple sclerosis (MS) is a debilitating neurodegenerative, highly heterogeneous disease with a variable course. The most common MS subtype is relapsing–remitting (RR), having interchanging periods of worsening and relative stabilization. After a decade, in most RR patients, it alters into the secondary progressive (SP) phase, the most debilitating one with no clear remissions, leading to progressive disability deterioration. Among the greatest challenges for clinicians is understanding disease progression molecular mechanisms, since RR is mainly characterized by inflammatory processes, while in SP, the neurodegeneration prevails. This is especially important because distinguishing RR from the SP subtype early will enable faster implementation of appropriate treatment. Currently, the MS course is not well-correlated with the biomarkers routinely used in clinical practice. Despite many studies, there are still no reliable indicators correlating with the disease stage and its activity degree. Circulating microRNAs (miRNAs) may be considered valuable molecules for the MS diagnosis and, presumably, helpful in predicting disease subtype. MiRNA expression dysregulation is commonly observed in the MS course. Moreover, knowledge of diverse miRNA panel expression between RRMS and SPMS may allow for deterring disability progression through successful treatment. Therefore, in this review, we address the current state of research on differences in miRNA panel expression between the phases.

Keywords: circulating microRNA; relapsing–remitting multiple sclerosis; secondary progressive multiple sclerosis; biomarker; neuroinflammation

Citation: Pietrasik, S.; Dziedzic, A.; Miller, E.; Starosta, M.; Saluk-Bijak, J. Circulating miRNAs as Potential Biomarkers Distinguishing Relapsing–Remitting from Secondary Progressive Multiple Sclerosis. A Review. *Int. J. Mol. Sci.* **2021**, *22*, 11887. https://doi.org/10.3390/ijms222111887

Academic Editors: Marcello Ciaccio and Luisa Agnello

Received: 21 September 2021
Accepted: 31 October 2021
Published: 2 November 2021

1. Introduction

Multiple sclerosis (MS) is an autoimmune disease characterized by chronic inflammation, demyelination, and neurodegeneration of the central nervous system (CNS) [1]. According to the latest data, the estimated number of people who suffer from MS in 2020 was 2.8 million worldwide (35.9 per 100,000 population) [2]. Just as in other autoimmune diseases, MS is found to be more prevalent in women than in men (female:men ratio 3:1) [3]. The disease is most often diagnosed in young adults between 20 and 40 years of age, who are in the period of the greatest activity in life; however, there are cases of the disease onset both in children (2–10% of patients being younger than 18 years) and older people (near 4% after age 50) [4,5]. MS is generally most prevalent in northern geographic latitudes, and it is suggested to be mainly due to the influence of environmental, genetic, and ethnicity, as well as behavioral differences [6,7].

The first neurological incident suggesting MS is known as a clinically isolated syndrome (CIS), which lasts at least 24 h and is characterized by symptoms and signs indicating either the presence of a single lesion (monofocal episode) or several lesions (multifocal episode) within the CNS. Patients who experience CIS may or may not develop MS [8]. Most people who suffer from MS (85–90%) are diagnosed with relapsing-remitting multiple sclerosis (RRMS), portrayed by active bouts of disease followed by a relative stabilization until the next relapse. Eventually, after 10–15 years from disease onset, more than 70% of RRMS patients transit to secondary progressive multiple sclerosis (SPMS), characterized by steadily increasing neurological disability, independent of relapses. The remaining 10–15% of MS patients develop primary progressive multiple sclerosis (PPMS), characterized by slowly worsening symptoms from the beginning, without prior or intermittent exacerbations and remissions [9].

This is the result of its very heterogeneous, multifactorial etiopathogenesis and unpredictable disease course, in which several clinical phenotypes with distinct underlying pathogenic mechanisms can be distinguished [9–12]. Another difficulty in the form of a varied response to applied therapies among patients additionally aggravated the challenges doctors face [13].

The pathophysiology of MS is a complex autoimmunological process characterized by the progressive loss of neurological function caused by the destruction of the axonal myelin sheath in several areas of the brain and the spinal cord [14]. MS is believed to be driven by systemic immune activation of autoimmunity against CNS components. In MS, the inflammatory state is mediated by the interaction between several immune cells, such as T and B cells, macrophages, CNS glial cells (microglia and astrocytes), as well as antigens reactive against myelin, including myelin basic protein (MBP) and myelin oligodendrocyte glycoprotein (MOG) [15]. The commonly implicated myelin-reactive CD4+ cells in the pathophysiology of MS are Th1 and Th17 lineage, defined on the basis of the production of interferon-γ (IFN-γ) and interleukin-17 (IL-17), respectively [16]. Autoreactive Th cells and activated monocytes secrete elevated amounts of proteolytic enzymes, such as matrix metalloproteinases (MMPs), which can degrade tight junction proteins and cause blood–brain barrier (BBB) disruption [17]. Moreover, Th1 and Th17 cells can cross the BBB and migrate to the CNS, followed by microglia activation and secretion of pro-inflammatory cytokines, such as tumor necrosis factor α (TNF-α), IL-1β, and IL-6 [18].

The interplay between relapses and progression during MS has led to the split of the disease into two leading phases, which are characterized by different but still mutually interacting pathological processes within the CNS [19]. In the early RRMS, the critical mechanism of the initiation of the disease process is inflammation and BBB damage. Patients with the RR subtype display stronger inflammatory features than progressive forms (PP and SP subtypes) [20]. In the advanced stages of the disease, the ongoing inflammation process gradually contributes to neurodegeneration, which seems to dominate in progressive MS [21]. SPMS is manifested by predominant neurodegeneration, brain atrophy, and steady clinical exacerbation, even when a patient is not experiencing a relapse [22]. Since therapies mainly exert a dampening effect on the immune system, this may be one explanation as to why therapeutic effects are poor in progressive MS. Moreover, many disease-modifying therapies (DMTs) used in RRMS are ineffective or even harmful for SPMS patients, thus highlighting the need to modify the therapeutic interventions used [23–26].

Currently, diagnosis of the conversion from RRMS to SPMS is based only on retrospective clinical and radiological assessment. However, the retrospective assessment presents many difficulties in terms of establishing the time point of disease progression [27]. Lublin et al. proposed a one-year retrospective clinical assessment of disability progression, as measured by the expanded disability status scale (EDSS), as diagnostic criteria for SPMS [28]. Nevertheless, disability progression in SPMS is not frequently clearly measurable by clinical scales. Moreover, EDSS is widely criticized for its lack of linearity, over-reliance on inferior limb function and ambulation, low sensitivity, and high interrater variability [29].

According to the 2017 McDonald's criteria, the primary tests for MS diagnosis are magnetic resonance imaging (MRI) findings and cerebrospinal fluid (CSF) examination. Indeed, oligoclonal band (OCB) analysis, despite being an invasive diagnostic method linked to the lumbar puncture, is a recommended analysis in CSF samples, while basic CSF biochemistry, as well as tests for intrathecal immunoglobulin G (IgG) index, are not recommended [30]. At the early MS stage, the volume of demyelinating lesions is superior in the white compared to the gray matter, whereas progressive MS manifests by widespread gray matter demyelination. It was shown that focal gray matter atrophy is an untimely indicator of progression, and the pace of gray matter atrophy correlated with MS development [9,31–33]. Nevertheless, standard MRI-based imaging does not fully reflect the ongoing disease mechanisms, such as demyelination/remyelination, microglial activation, astrogliosis, as well as neurodegeneration, which contribute to subclinical disease activity [34].

The available literature data report the existence of biologically active molecules that could be a potentially helpful tool for differentiating the RR phase from SP. Pasquali et al. reported that the plasmatic levels of proinflammatory cytokines, both IFN-γ and IL-17, are higher in RRMS compared to SPMS patients, while the level of transforming growth factor-β (TGF-β), a molecule with immunosuppressant activity, was much lower in RRMS in comparison to SPMS [35]. Another group of active molecules indicated in the literature are light neurofilaments (NFL) and glial fibrillary acid protein (GFAP), a marker of astrocyte damage and astrogliosis, in serum [36,37]. Högel et al. claimed that GFAP and NFL levels in serum are higher in patients with SPMS than RRMS, as well as correlate with a higher EDSS parameter [38]. Whereas Ayrignac et al. demonstrated higher levels of both NFL and GFAP in serum from PPMS compared to RRMS, indicating that they might be markers of the disease progression [37]. Based on the above studies, we consider NFL and GFAP as potential progressive MS and RRMS distinguishing biomarkers.

Most importantly, the majority of autoimmune diseases are accompanied by inflammation, which is why it is strongly recommended not to take inflammatory factors into account as proper MS markers.

Therefore, it is imperative to identify suitable diagnostic tools, for instance, in the form of sensitive, reliable, and stable biomarkers that can help distinguish the clinical phenotypes of MS, predict disease progression, and provide a correlation with disability [33]. It is firmly not recommended to consider cytokines/chemokines measured in serum/plasma as a reliable marker, especially because they are highly non-specific to concrete disease entity [39]. Therefore, the inflammatory markers mentioned above can only complement MRI and patients' clinical characteristics [9]. Recent studies have demonstrated that altered expression of some miRNAs may serve as valuable biomarkers to diagnose MS, and rapidly and effectively distinguish RR from the SP phase [40,41].

2. Biogenesis and Characteristics of miRNA

In the past few years, numerous studies have confirmed the essential role of small (19–25 nucleotides) non-coding RNA molecules, called microRNAs (miRNAs), as significant regulators of biological processes associated with the pathophysiology of various autoimmune and neurodegenerative disorders, including MS [42–45]. MiRNAs are remarkably stable, resistant to endogenous RNase activity, simple to obtain, and above all, highly sensitive to the processes taking place in the organism [46]. Furthermore, microRNAs have distinct expression level patterns, which could be characteristic of the specific disease [47]. Those features have made circulating miRNAs a potentially promising prognostic biomarker, being investigated for various human disorders, including neurodegenerative diseases and other neurological pathologies [48]. Despite those analyses, no diagnostic miRNA has been effectively applied in clinical examination until now. Still, more research regarding miRNA activity in MS, especially at the genetic level, needs further clarification. The analysis of the expression level of miRNAs potentially involved in neurodegeneration processes might provide new knowledge of MS etiopathogenesis and might be helpful while choosing the proper therapy used in different stages of the disease.

The formation of miRNA consists of several steps [49]. Gene-encoding miRNA is transcribed by RNA polymerase II, which leads to the formation of a long primary transcript, termed primary miRNA (pri-miRNA). In the nucleus, the pri-miRNA is cleaved by a microprocessor complex of the RNase III endonuclease Drosha with DiGeorge syndrome critical region 8 (DCGR8) protein into pre-miRNA. Next, pre-miRNA is transported to the cytoplasm by exportin-5, where it is cleaved by the RNase III endonuclease Dicer to produce an RNA duplex [50]. Finally, the mature miRNA is transferred to Argonaute family proteins (Ago) in the RNA-induced silencing complex (RISC) core. One strand of the duplex (-3p) is typically degraded by Ago. The other strand (-5p) is loaded into RISC (usually, this strand is the one with lower thermodynamical stability at the 5′end) [51]. However, both strands (-5p and -3p) may be loaded into RISC at similar frequencies for some miRNAs. In fact, while RISC loading may strongly favor the incorporation of one strand (-5p), recent research has demonstrated a small fraction of the -3p strand loaded into RISC for essentially all miRNA families [52].

Together with the RISC, single-stranded miRNA, using as few as 6–8 nucleotides (so-called 'seed region') which are located at the 5′end, hybridizes to the complementary sequences of the target messenger RNA (mRNA) in the 3′UTR (untranslated region), known as a miRNA response element (MREs). The hybridization results in specific post-transcriptional gene silencing—RNA interference (RNAi) [53]. Moreover, the interaction of miRNAs with other regions, including coding sequence and 5′UTR, has also been reported [54].

The effects of RNAi depend on the grade of the complementarity of miRNA with the target mRNA. Excellent complementarity leads to the destabilization and degradation of the mRNA transcript by the catalytic Ago2 protein. Whereas, if the complementarity is not full (this effect is more often), it results in translational repression [55]. In this way, miRNAs modulate more than 60% of all mammalian mRNAs, and therefore a wide range of cellular processes, including development, differentiation, proliferation, metabolism, and apoptosis [56]. Furthermore, there is a complex network of mutual interactions between miRNAs and mRNAs—one miRNA can bind to hundreds of target mRNAs, and a single mRNA can be targeted by various miRNAs [57]. This means that the combination of multiple miRNAs determines the expression of the same gene, and the lack of a single regulator can be compensated by other miRNA molecules complementary to the same target transcripts [58,59].

MiRNA expression profiling is conducted using a set of various methods, such as miRNA microarray platforms, quantitative real-time polymerase chain reaction (qRT-PCR), digital PCR (dPCR), in situ hybridization, and high-throughput sequencing (next-generation sequencing, NGS). Nowadays, the NGS technique is most frequently used to profile miRNA in distinct sets of diseases. It is noteworthy that each of these methods provides a large amount of information about miRNAs profile in different diseases; nevertheless, every method also has some disadvantages. Although NGS is challenging due to cost, labor, time consumption, and professional bioinformatics support for data analysis, it profiles both known and unknown miRNAs, which are beyond the capabilities of qRT-PCR and microarrays. Microarrays require a higher concentration of miRNAs and have lower specificity than qRT-PCR. On the other hand, sensitive and specific qRT-PCR, the standard technique to measure miRNA expression, provides medium-throughput concerning the number of samples processed per day [60–63]. Furthermore, quantification of samples with low concentrations of nucleic acids using this method can be challenging, while dPCR technology enables absolute quantification through partitioning the reaction. dPCR, highly sensitive and accurate in molecular detection, has demonstrated applications such as trace DNA detection, rare mutation detection, and copy number variation. Unfortunately, its disadvantages compared to qPCR are narrow dynamic range and high cost [64].

It was demonstrated that circulating miRNAs present in various bio-fluids, such as saliva, blood, plasma, serum, and CSF, can be released or produced due to different events, including (1) passive leakage from damaged cells due to chronic inflammation, apoptosis, or necrosis, (2) active secretion through cell-derived microparticles, exosomes, shedding vesicles, and apoptotic bodies, and (3) active transport by a complex with protein, such as Ago2 [65]. Furthermore, exosomes can provide an excellent source of miRNA biomarkers because of their easy extraction from body fluids [66]. Circulating miRNAs enclosed in membrane vesicles are extremely stable in the extracellular environment, with high RNAse activity [67]. High stability means that circulating miRNAs are long-lived in bio-fluids and are consequently proposed as attractive diagnostic [68] and prognostic biomarkers [69,70]. Some experimental studies also demonstrated strong resistance of endogenous circulating miRNAs for stressful conditions, such as repeated freeze–thaw cycles, temperature differences, and prolonged storage time, which are favorable traits for their analyses [71,72]. miRNAs circulating in the human body have the potential for being used as multi-marker models that could be non-invasive, less expensive, and less time-consuming methods in monitoring the disease course and its response to treatment than the classical protein markers used so far [73]. This is due to the simplicity of their obtaining, high sensitivity, and specificity for a given disease entity [74].

According to the miRNA repository, miRbase (www.mirbase.org; accessed date: 21 September 2021) currently lists 1917 precursors and 2654 mature miRNAs in *Homo sapiens*, which control the expression of hundreds of different genes [75]. The understanding of which miRNAs or panels are dysregulated in particular diseases makes miRNAs emerge as valuable therapeutic targets [76]. The human miRNA-associated disease database (HMDD) (www.cuilab.cn/hmdd; accessed date: 21 September 2021) operates as a rich resource for scientists, screening the permanently growing number of miRNA profiles for a broad range of diseases.

3. MiRNAs as Potential Biomarkers in MS

At the time of writing this review, the HMDD covered 148 different miRNAs associated with MS pathogenesis, predisposing to a disease marker [77].

Using a literature search for peer-reviewed articles, conducted through the PubMed and Sage Journals databases, we have retrieved and compared all significantly different circulating miRNAs between RRMS and SPMS. Google Scholar was also used to seek open-access articles. The data from the World Health Organization (WHO) and clinicaltrials.gov websites were also considered in the above review. The review consists of 176 literature positions, including 98 original research (case reports, clinical trials, cohort studies), 73 reviews (systemic reviews, literature reviews, and meta-analyses), 1 website, and 4 chapters in books. The cited works are in the 1982 to 2021 range of years. Identified relevant reviews were hand-searched for additional relevant references. The following search terms were used to locate articles specific to this study: relapsing–remitting multiple sclerosis, secondary progressive multiple sclerosis, biomarkers, circulating miRNAs, plasma, serum, differences, and expression. Variations of these terms were used to ensure exhaustive search results. After identifying all the keywords, synonyms, and phrases, the Boolean operators "AND" and "OR" were used. The PubMed search was performed using terms and database-appropriate syntax: "expression" AND "differences" AND "plasma" OR "serum" AND "circulating miRNAs" OR "miR-92a-1-3p" OR "miR-633-5p" OR "miR-485-3p" OR "miR-337-3p" OR "miR-326-5p" OR "miR-30b-5p" OR "miR-27a-3p" OR "miR-223-3p" OR "miR-181c-5p" OR "miR-145-5p" OR "let-7c" OR "let-7d" AND "biomarkers" AND "relapsing–remitting multiple sclerosis" OR "secondary progressive multiple sclerosis". Furthermore, the search terms relating to "multiple sclerosis" were not useful as the search facilities were not precise. Based on the PRISMA (Preferred Reporting Items for Systematic Reviews and Meta-Analyses) template, a diagram representing the literature strategy was created (Figure 1) [78].

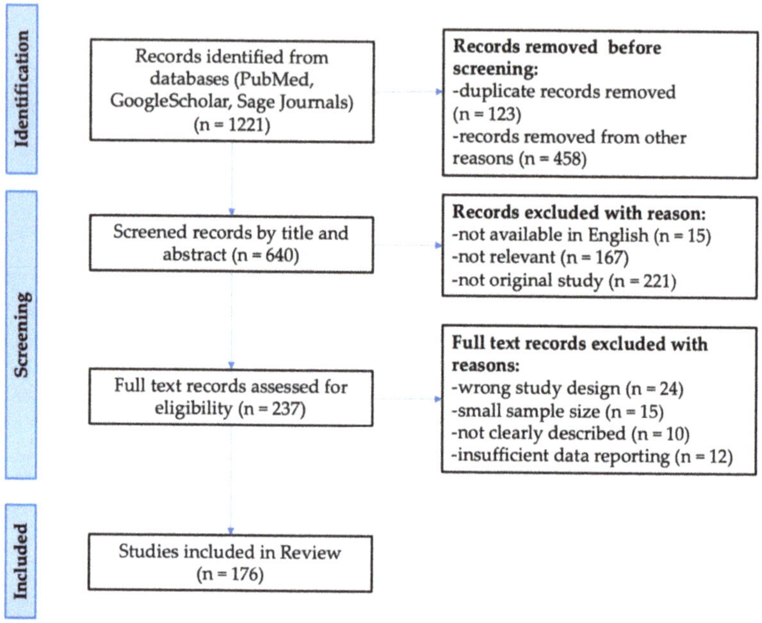

Figure 1. Flow diagram representing the study selection process based on the PRISMA template.

It is evident that miRNAs are recognized as essential modulators of inflammatory responses; however, against intuition, the activity of miRNAs is not only restricted to a simple inhibition of inflammation. The earlier discoveries of a handful of miRNAs involved in inflammatory responses have encouraged researchers to focus their studies on their systematic profiling in different cells and cell-free biological fluids to understand their function in regulating the inflammatory process [79]. Recent studies have uncovered a significant role for microRNAs as regulators of major cellular functions, such as development, differentiation, growth, and metabolism. Due to their abilities, microRNAs are believed to be implicated in many human pathologies, including inflammatory, autoimmune, and neurodegeneration, all characteristics of MS [80–83]. What is more, miRNAs modify the transcripts for proteins engaged in remyelination, neurogenesis, and gliogenesis [84–87].

In the literature are emerging several circulating miRNAs being either up- or down-regulated in MS patients vs. healthy controls (HCs) and differently expressed in various disease subtypes [88]. Dysregulated miRNAs in MS obtained from different tissues and cell types were analyzed, for instance, in a comprehensive review by Piket et al. or Gandhi [89,90]. Piket et al. provided a detailed overview of 61 studies that examined miRNAs in MS (most of them analyzed RRMS vs. HCs). They focused on the mechanisms of the most dysregulated miRNAs and also used predicted targets of the most dysregulated miRNAs to highlight affected pathways. From their research, it resulted that the prime affected pathway was TGF-β signaling, which is important in the differentiation and function of Th17 and Treg cells. Moreover, some of the miRNAs which can be found in their comprehensive overview overlap with miRNAs which we found dysregulated in RRMS vs. SPMS, including let-7d, 181c, or 145-5p [89]. Otaegui et al. indicated that cell-derived miRNAs are promising biomarkers in MS and selected those that may be indicators of the relapse and remission phases [91]. It was also shown that circulating miRNAs help distinguish MS subtypes [92]. Furthermore, other miRNA analyses have been performed between MS subtypes (Vistbakka et al. refers to PPMS vs. SPMS, Mandolesi et al. refers to CIS/Radiologically Isolated Syndrome vs. RRMS vs. progressive MS, or Perdaens et al. refers to relapsing and remitting MS vs. HCs) [88,93,94]. The panel of miRNAs with disturbed expression in the course of MS obtained from different bio-fluids, such as CSF, whole blood, plasma, as well as T- and B-cells, which includes, e.g., let-7, miR-320, miR-27a, miR-223, miR-155, and their target genes, has been found [95]. One of the more thoroughly studied miRNAs in the context of MS pathogenesis is miRNA-155, for which the pro-inflammatory effect is associated with activation of inflammatory cells, increasing permeability of the BBB, phagocytosis of myelin by macrophages, and neurodegenerative processes [96,97]. Circulating miR-155 contributes to microglial activation and polarization of astrocytes towards the neurotoxic A1 phenotype, thereby driving demyelination processes [98,99]. Additionally, miR-155, by having a suppressive effect on mothers against decapentaplegic homolog 2 (SMAD2) and SMAD4, leads to upregulation of the Nogo receptor (NgR), which is engaged in the suppression of neuronal growth. Hence, upregulated miR-155 may induce demyelination [100,101]. MiR-155 also has a huge impact on the development of neuropathic pain and indirectly influences a Treg cell differentiation involved in alleviating pain hypersensitivity [102]. Another documented miRNA engaged in neuroinflammatory processes, which could have an effect on MS, is let-7a. As proven by Kacperska et al., this miRNA revealed a statistically significant difference in the expression in patients' plasma (downregulated in RRMS vs. HCs ($p = 0.002$)), whereas Gandhi et al. showed a statistically significant difference in the expression of let-7a in the plasma obtained from SPMS and the HCs ($p = 0.002$); however, no significant differences were reported for the RRMS vs. HC comparison [103,104]. Let-7a serves as an essential modulator of neuroinflammatory processes, inhibits apoptosis, and promotes the neuroprotective M2 phenotype in microglia under active inflammatory conditions, as a result being involved in the anti-inflammatory process [105]. In response to inflammation, upregulated circulating let-7a represses the production of IL-6 and promotes the expression of brain-derived neurotrophic factor (BDNF) and anti-inflammatory cytokines, such as IL-4 and IL-10 in microglia [106]. IL-4′s role in regulating inflammation within the CNS was demonstrated in the experimental autoimmune encephalomyelitis (EAE), an animal model of MS, in which mice deficient in IL-4 exhibited more severe EAE clinical disease [107]. Another huge miRNA family involved in MS pathology is the miR-320 family (miR-320a, miR-320b, miR-320c, miR-320d, and miR-320e). Studies on the mice with EAE showed that overexpression of cell-free miR-320 isoforms plays a possible role in the autoimmune neuroinflammation and the pathogenesis of MS disease. Talebi et al. found that miR-320 isoforms could target SMAD2 and TGFBR2 (TGF-β Receptor 2), thereby diminishing the TGF-β signaling pathway [108]. This pathway contributes to the increasing differentiation of naive T cells into Treg cells and inhibition of Th1, leading to the decrease in IFN-γ production and lessening EAE [109]. As a result, overexpression of miR-320 isoforms might be involved in the neuroinflammation and pathogenesis of MS through reducing neuroprotective and immunomodulatory effects of TGF-β [108]. Moreover, in B cells of MS patients during a disease relapse, the expression of miR-320a is decreased, leading to increased MMP9 expression, being a marker of disease activity in patients with MS [110,111]. An elevated level of MMP9 can be implicated in the pathogenesis of MS, as this enzyme is involved in BBB degradation, thereby intensifying neuroinflammation and worsening the disease course [112].

In 2016, the clinical trial, which was based on a pilot study that seeks to characterize differences in miRNA profiles and cell products obtained from blood and CSF of patients in the early (CIS) and later (SP) stage of MS, as well as healthy participants (with no neurological or autoimmune illness), was completed. Although no results have been published, the aforementioned study is particularly promising, taking into account that obtained miRNA panels could then be correlated to clinical manifestations and subtypes of MS [113].

4. Potential Candidates for a Panel Distinguishing the RR from SP Phase

The use of miRNAs as biomarkers in MS is still being developed. Until now, in the case of differences between RRMS and SPMS, only a few reported studies have analyzed circulating miRNA in cell-free biological fluids.

Recently, it has been a trend to use a panel of multiple miRNAs, more favorable than a single miRNA, focusing on early diagnostic and/or prognostic biomarkers in many diseases [114–116]. Using a specific panel of several miRNAs increases the credibility of the obtained results and reduces the risk of a false-positive diagnosis to a minimum.

The study conducted by Haghikia et al. showed that miR-181c ($p = 0.02$) and miR-633 ($p = 0.0005$) are downregulated in the CSF of SPMS when compared to RRMS [117]. Kramer et al. confirmed that those miRNAs are dysregulated in the CSF in MS. However, their study claimed that miR-181c is upregulated in SPMS compared to RRMS ($p = 0.036$) and miR-633-5p does not reach the statistically significant differences ($p = 0.468$) [118]. It was reported that this miRNA could predict the conversion from CIS to RRMS, and the enhanced level of miRNA-181c in the CSF could be the early marker of the highly active phase of MS [119]. There is an experimental study on rat cortical neurons, which demonstrated the involvement of miR-181c-5p in the regulation of neuronal maturation and synaptogenesis in the cortex and the molecular responses of astrocytes under inflammatory conditions [120]. What is more, several target genes were identified for miR-181c-5p, including SMAD7 (a negative regulator of TGF-β signaling) engaged in Th17 differentiation, being a major driver of CNS autoimmunity in MS [121]. Nevertheless, no targets for miR-633-5p have been validated so far. However, the bioinformatics prediction tool, TargetScan, revealed a few potential binding sites, including macrophage scavenger receptor 1 (MSR1), which is also discussed in the context of TGF-β-induced microglia toxicity [118].

The study of Gandhi et al. explored the potential of miRNA profiling to distinguish subtypes of MS. Their research based on qRT-PCR analysis showed that expression of miR-92a-1-3p, let-7c, let-7d, and miR-145-5p is significantly upregulated in plasma from RRMS in comparison to SPMS ($p = 0.022$, $p = 0.04$, $p = 0.047$, and $p = 0.01$, respectively) [104]. In the multivariate model, miR-92a-1-3p showed a significant association with RRMS (adjusted OR = 1.35, $p = 0.022$), whereas it demonstrated no link with SPMS. Furthermore, they showed a positive correlation between miR-92a-1-3p expression and EDSS and disease duration, which may indicate that miR-92a-1-3p may be an early marker of disease progression [104]. In the same study, the additional biostatistical analysis showed that miR-92a-1-3p targets genes which regulate the CD40-CD40L pathway [104]. CD40-CD40L dyad was found to be essential in modulating the aberrant inflammatory response in MS. Although CD4+ and CD8+ T cells are extensively present during immune activation and can express CD40L, in MS, CD40L expression is only detected on CD4+ T cells, not on CD8+ T cells [122]. CD40L is neither identified in the CNS of healthy people nor patients with other neurodegenerative diseases [123]. CD40–CD40L interactions can promote the inflammatory response underlying MS, whereas it is confirmed that inhibition of the CD40-CD40L pathway reduces disease severity in the EAE [124]. Additionally, Balashov et al. demonstrated that CD40L-dependent Th1 differentiation and immune activation were only observed in the progressive MS but not in the RRMS, suggesting a link to disease pathogenesis and progression and providing a basis for immune response intervention in the disease [125]. The data considered above suggest that upregulated miR-92a-1-3p expression in RRMS via the CD40-CD40L pathway may be a genuine marker of early inflammation burst induced by CD4+ T cells, both in MS and EAE. MiR-92a-1-3p is involved in the pathogenesis of EAE, and it might function as a positive regulator of T cell differentiation towards pathogenic Th1 cells [126].

Subsequent miRNAs from the let-7 family, apart from let-7a, differently expressed in MS, are let-7c and let-7d. Banerjee et al. suggested an important role of let-7c in regulating macrophage polarization derived from bone marrow cells of mice. They found that overexpression of let-7c in macrophages promotes their polarization into the anti-inflammatory M2 phenotype, possibly by targeting the C/EBP-δ transcription factor, which plays a vital role in the inflammatory response [127]. Kim et al. conducted experiments on mice with a dendritic cell (DC)-specific deletion of the transcriptional repressor B lymphocyte-induced maturation protein1 (Blimp1), which showed that the increased level of let-7c that is present when Blimp1 expression is low results in a proinflammatory DC phenotype [128]. Besides, toll-like receptor 7 (TLR7), a noncatalytic receptor protein present in macrophages and DCs involved in innate immunity and inflammation, recognizes let-7c and plays a negative role in controlling neuronal growth in cultured neurons. Thereby, let-7c might suppress the dendritic growth of cortical neurons [129]. It is worth emphasizing that neuronal innate immune responses may influence neurodevelopment and neurodegeneration through the regulation of neuronal morphology [129]. Let-7d is predicted to target IL-10 since there is a strong positive correlation between the pro-inflammatory cytokine IL-1B and let-7d [130]. This is important, taking into consideration that IL-10 has significant regulatory effects on immunological and inflammatory responses, for instance, because of its capacity to inhibit the production of proinflammatory cytokines by monocytes [131].

In the next study, Regev et al. found that the miR-27a-3p was the only miRNA out of 652 analyzed miRNAs that differentiated progressive patients from relapsing ones. Expression of miR-27a-3p in serum was significantly upregulated in the RRMS compared to SPMS. AUC for miR-27a-3p was high and reached a value of 0.78 [92]. It was also demonstrated that miR-27a-3p expression shows a strong link to the neurotrophin signaling pathway [92]. In CNS, neurotrophins, which play a protective role for neuronal circuitry, facilitate synaptic transmission, and regulate brain plasticity, are essential for memory and regenerative processes [132]. It was shown that reduced production of neurotrophic factors, such as BDNF in patients with SPMS, can contribute to the progression of demyelinating disease and axonal loss [133]. Probably, miR-27a-3p, through regulation of the neurotrophin pathway, may be an early marker of neurodegeneration in patients with late RRMS and play a role as the prognostic marker of RR-to-SP transition. MiR-27a-3p targets many proteins of intracellular signaling networks, which regulate, for example, the activity of nuclear factor kappa-light-chain-enhancer of activated B cells (NF-κB) and mitogen-activated protein kinases (MAPKs), which control numerous cellular events associated with the inflammatory response, apoptosis, and cell proliferation [134,135]. Lu et al. revealed that downregulation of miR-27a-3p inhibits the inflammatory response (reduced the expression levels of pro-inflammatory cytokines, such as IL-1β, IL-6, and TNF-α) and hippocampal neuronal cell apoptosis by targeting mitogen-activated protein kinase 4 (MAP2K4) in epilepsy. Thus, increased expression of miR-27a-3p may induce inflammatory burst and hippocampal neuronal apoptosis by targeting MAP2K4 [136]. Ahmadian-Elmi et al. suggested that high expression of miR-27a-3p can suppress the TGF-β signaling pathway by directly targeting some key transcription factors, such as SMADs and Cul1 proteins [137]. It was shown that the SMAD2/3 complex and SMAD4 can regulate forkhead/winged-helix transcription factor 3 (FOXP3) gene expression [138]. FOXP3 is especially involved in the development and function of Treg cells, and it can inhibit the expression of the IL-17A gene by interacting with retinoic acid receptor-related orphan receptor-γt (RORγt), a transcription factor that directs the differentiation of inflammatory Th17 cells [139]. In RRMS patients, silencing SMADs and Cul1 by upregulated miR-27a-3p may block the TGF-β signaling pathway, leading to the suppression of naïve T cell differentiation into Treg cells. Finally, reducing the Treg number and enhancing the number of Th17 cells in the relapsing phase will promote inflammation in the RR stage of MS [137]. Additionally, miR-27a-3p is a regulator of oligodendrocytes' development and survival. Increased miR-27a-3p level leads to inhibition of oligodendrocytes differentiation/maturation in mice and human oligodendrocyte progenitor cells and myelination and remyelination in vivo in mice. Therefore, decreasing the levels of miR-27a-3p following demyelination is critical for facilitating remyelination [140].

In another study, Regev et al. reported that miR-337-3p in serum was significantly downregulated in SPMS compared to RRMS in one of the cohorts (p = 0.01), while no significant differences were found for the other cohorts [141]. Moreover, its increased expression negatively correlated with the EDSS in three independent MS cohort studies. Thus, it may be accepted that miR-337-3p might be a potential biomarker candidate for disability and disease progression [141]. Of interest, it was demonstrated that miR-337-3p targets Ras-related protein 1 (Rap1) A protein, which is a well-established major component of the integrin activation pathway, hence indicating a potential role of miR-337-3p to serve as a biomarker for predicting the therapy response to natalizumab (an $\alpha 4\beta 1$-integrin inhibitor) in MS patients [142]. Additionally, Rap1 signaling impacts upon autoimmune T cells at various levels and confirms the concept that sustained Rap1 activation diminishes T cell-mediated autoimmunity. Therefore, miR-337-3p via Rap1 signaling may initiate the pathogenic character of T cells in immune-mediated inflammatory diseases, such as MS [143].

There is also Sharaf-Eldin et al.'s study, which is promising, however, it requires future verification on a larger number of patients and detailed validation. Sharaf-Eldin et al. carried out a study on miR-145-5p, miR-223-3p, and miR-326-5p, and concluded that only miR-326-5p indicated a statistically significant difference ($p = 0.018$) between RRMS and SPMS patients (overexpression in RRMS vs. SPMS). Additionally, combinations of miR-145-5p + miR-326-5p, miR-223-3p + miR-326-5p, and miR-145-5p + miR-223-3p + miR-326-5p can differentiate RRMS from SPMS, with the area under the curve (AUC) and 95% confidence intervals (95% CI) values of (0.737 (0.57–0.904), $p = 0.014$), (0.713 (0.531–0.896), $p = 0.027$), and (0.772 (0.619–0.925), $p = 0.005$), respectively [144]. AUC is a parameter providing an estimate of the miRNA's ability to discriminate the groups compared, known as an area under the receiver operating characteristic curve [145]. Kornfeld et al. demonstrated that miR-145-5p targets myelin gene regulatory factor (MYRF), a transcriptional regulator required for CNS myelination and oligodendrocyte maturation. This was confirmed by the fact that mice lacking MYRF display severe neurological abnormalities and severe deficits in myelin gene expression [146]. Studies on transient middle cerebral artery occlusion in rats indicated that miR-145 plays a role in the brain's antioxidant defense because its lower expression led to increased protein expression of superoxide dismutase-2 (SOD2), one of the major antioxidants [147]. Moreover, miR-145-5p was identified as a putative regulator of nuclear receptor subfamily 4 group A member 2 (NR4A2), also known as Nurr1 [148]. The research performed on the secondary spinal cord injury in the rat model indicated that miR-145-5p inhibition decreases inflammation and oxidative stress, which, together with mitochondria dysfunction, feature prominently in MS [149], by targeting Nurr1 to block TNF-α signaling [150].

It was reported that miR-223-3p is involved in regulating hematopoiesis, myeloid progenitor proliferation, granulocyte differentiation, and thereby immune response [151]. Studies on the EAE model suggested that miR-223-3p has an important role in the development of CNS inflammation. MiR-223-3p regulates myeloid DC-induced activation of pathologic Th17 responses during autoimmune inflammation, controlling IL-1β, IL-6, and IL-23 cytokines [152]. Mice with miR-223–3p-knockout exhibited reduced numbers of myeloid DCs and Th17 cells in the CNS, thereby reducing EAE disease severity linked to decreased inflammation [152]. Besides, miR-223 is required for efficient macrophage M2 polarization, and mice lacking miR-223 display impaired CNS remyelination [153]. Depending on the type of regulated pathway, miR-233-3p exhibits anti- and pro-inflammatory properties. On the one hand, Li et al. demonstrated that miR-223-3p modulates the noncanonical NF-κB (nuclear factor-κB) pathway by targeting transcripts of the inhibitor of kappa B kinase alpha (IKKα) (engaged in activation of NF-κB), which is an anti-inflammatory factor that may prevent the spontaneous activation of macrophages, thus promoting controlled inflammation in the human myeloid leukemia cell line [154]. On the other hand, miR-223-3p, by targeting TNF receptor-associated factor 6 (TRAF6) and TGF-β activated kinase 1 binding protein 1, suppresses the canonical NF-κB pathway. Therefore, miR-223-3p expression was indicated to decrease neutrophil activation, suggesting anti-inflammatory effects [155]. Furthermore, Chen et al. demonstrated that miR-223-3p directly targets the transcription factor signal transducer and activator of transcription 3 (STAT3). It was shown that miR-223-3p overexpression was associated with a significant decrease in STAT3 levels and reduction in the production of IL-1β and IL-6, but not TNF-α, in macrophages. Thereby, miR-223-3p might regulate processes associated with the regulation of inflammatory responses in macrophages [156]. In addition, miR-223-3p could downregulate the nod-like receptor pyrin domain containing 3 (NLRP3), which is thought to be a critical and necessary factor in MS development. Decreasing the NLRP3 level inhibits inflammation through caspase-1 and IL-1β, thus reducing brain edema and improving neurological functions [157–159]. Besides, miR-223-3p appears to promote neuronal protection partly through the regulation of glutamate receptor signaling. Glutamate receptor 2 and N-methyl D-aspartate receptor 2B expressions are subunits for glutamate's ionotropic transmembrane receptors that mediate fast synaptic transmission in the CNS. It was demonstrated that miR-223-3p overexpression in the retina and optic nerve, by reducing the expression of the above subunits, had blocked the formation of EAE-driven pathological axonal swellings, which are attributed to excitotoxicity of glutamate [160,161].

It was investigated that miR-326-5p downregulates the expression of transcription factor Ets-1 [162], a negative regulator of Th17 differentiation, thereby promoting Th17 differentiation. MiR-326-5p expression is closely correlated with disease severity, both in patients with MS and mice with EAE [162]. Nevertheless, the exact role of Ets-1 in regulating the differentiation and function of Th17 cells still remains unknown [163]. In another study, Honardoost et al. also confirmed the potential of miR-326-5p, obtained from peripheral blood lymphocytes, as a diagnostic biomarker to discriminate between relapsing and remitting phases of MS disease [164]. Studies performed on phosphatase and tensin homolog-induced kinase 1 (PINK1)-deficient mice suggest that miR-326 may upregulate GFAP expression during neural stem cells' (NSCs) differentiation and brain development [165]. In addition, Junker et al. showed that miR-326, upregulated in active MS lesions, targets the 3′-UTR of CD47 in brain-resident cells. The decreasing level of CD47 releases macrophages from the inhibitory control, thereby increasing myelin phagocytosis [166].

Additionally, there is a study that compares RRMS patients with a group of progressive MS, including both SPMS and PPMS. Although the researchers did not split up those progressive phases, we would like to draw attention to their work because they detected miRNAs, whose combination improves discriminatory power between RRMS and progressive MS, and which might play a role in neuroinflammation and neurodegeneration in MS. Ebrahimkhani et al. profiled exosomal miRNAs and reported nine miRNAs (miR-15b-5p, miR-23a-3p, miR-223-3p, miR-374a-5p, miR-30b-5p, miR-433-3p, miR-485-3p, miR-342-3p, miR-432-5p), differently expressed in RRMS and progressive MS [167]. They showed that the combination of three miRNAs (miR-223-3p + miR-485-3p + miR-30b-5p) improves discriminatory power between RRMS and progressive MS with a 95% accuracy rate [167].

A recent study revealed that miR-485-3p might impact the differentiation and proliferation of NSCs to neuron and astrocytes cells, the activity of which is disrupted in neurodegenerative diseases. It was indicated that miR-485-3p targets thyroid receptor-interacting protein 6 (TRIP6) expression, which mediates signal transduction modulation during cell migration and adhesion, thereby diminishing proliferation and inducing differentiation of NSCs [168]. Yu et al. claimed that miR-485-3p might play a neurotoxic role while reducing neuronal viability and exacerbating neuroinflammation. Their experiments established that the knockdown of miR-485-3p promoted decreased cell proliferation and increased cell apoptosis induced by amyloid β-peptide in Alzheimer's disease [169]. Studies performed on the BV2 microglial cells after lipopolysaccharide treatment exhibited that the reduction of miR-485-3p could inhibit inflammatory responses, which suggested the negative regulatory effects of miR-485-3p on neuroinflammation [170].

It was reported that downregulation of miR-30b-5p is necessary to generate fully functional macrophages and DCs. Its upregulation inhibits the release of inflammatory cytokines, such as TNF-α, IL-6, and IL-12, in lipopolysaccharide-stimulated cells in cultured and transfected cells [171]. The examination of the Kyoto Encyclopedia of Genes and

Genomes (KEGG) database for miR-30b-5p, performed by Brennan et al., pointed out that this miRNA is able to target genes of pathways connected to neurodegeneration, including the Wnt pathway engaged in oligodendrocyte development and the remyelination process [172,173]. Zheng et al. demonstrated that miR-30b-5p is associated with a diminished expression level of the anti-apoptotic protein Bcl2 on the mouse pancreatic β-cell line [174]. It could be possible that this miRNA could also have an impact on Bcl2 expressed in peripheral lymphocytes whose dysregulated expression is a feature of clinically active MS [175]. Moreover, as proved by Zettl et al., chronic progressive MS patients exhibited a higher proportion of bcl-2-expressing T cells than patients with RRMS. In addition, active demyelinating lesions revealed a lower bcl-2-positive T cells number than remyelinating and demyelinated lesions. Therefore, this protein, expressed in MS plaques, might have important effects on the regulation of the persistence of the inflammatory cells in the CNS [176].

MiRNAs differentiating RR from SP described in this review are illustrated in Figure 2 and summarized in Table 1.

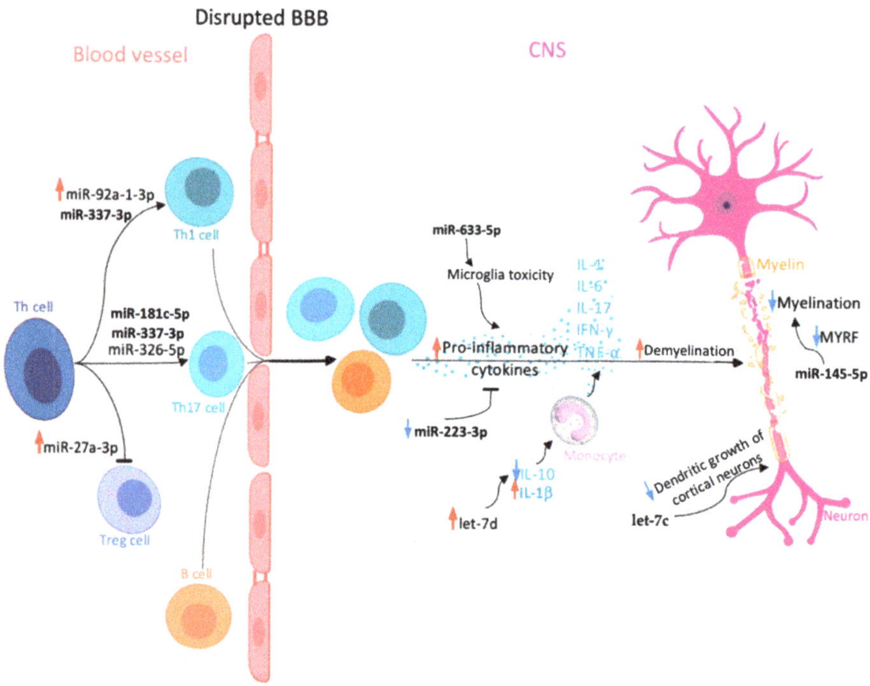

Figure 2. Potential role of circulating microRNAs (miRNAs) involved in multiple sclerosis (MS) pathogenesis, which could be useful in discriminating between relapsing-remitting multiple sclerosis (RRMS) and secondary progressive multiple sclerosis (SPMS) (miR-145-5p, miR-181c-5p, miR-223-3p, miR-27a-3p, miR-326-5p, miR-92a-1-3p, let-7c, let-7d, miR-337-3p, miR-633-5p). Upregulation and downregulation of specific miRNAs influence the Th cell differentiation into pro-inflammatory Th17 and Th1 cells, which together with B cells can cross the blood–brain barrier (BBB) and migrate to the central nervous system (CNS), where they proliferate and secrete pro-inflammatory cytokines (interleukin-1 (IL-1), IL-6, IL-17, interferon-γ (IFN-γ), and tumor necrosis factor α (TNF-α)). In the CNS, different miRNAs play various roles in cytokine secretion, monocyte stimulation, or neuronal maturation. As a consequence, dysregulation of these processes causes myelin sheath damage and, eventually, neuronal loss. Red arrow—upregulation of specific miRNA in MS compared to healthy controls; increased effect exerted by a given miRNA. Blue arrow—downregulation of specific miRNA in MS compared to healthy controls; decreased effect exerted by a given miRNA.

Table 1. Circulating miRNAs and combinations of miRNAs as potential tools in distinguishing RRMS from SPMS.

miRNA	Function	SPMS vs. RRMS	Biological Material	Patients Number	miRNAs Number Included in the Analysis	Ref.
miR-181c-5p	Regulates neuronal maturation and synaptogenesis in the cortex	Upregulated	CSF	81 RRMS 106 SPMS 0 HCs 211 OND	3	[118]
		Downregulated		17 RRMS 30 SPMS 0 HCs 39 OND	760	[117]
miR-633-5p	Targets mRNAs transcripts involved in neuroinflammatory pathways	Downregulated	CSF	17 RRMS 30 SPMS 0 HCs 39 OND	760	[117]
miR-27a-3p	Regulates the differentiation of Th1 and Th17 cells and the accumulation of Tregs Regulates oligodendrocytes differentiation	Downregulated	Serum	29 RRMS 19 SPMS 30 HCs	652	[92]
miR-337-3p	Initiates through Rap1 signaling pathogenic character of T cells	Downregulated	Serum	115 RRMS 51 SPMS 88 HCs	652	[141]
miR-92a-1-3p	Targets CD40-CD40L pathway Regulates cell cycle and cell signaling Promotes Th1 differentiation	Downregulated	Plasma	51 SPMS 50 RRMS 32 HCs	368	[104]
miR-145-5p	Regulates the expression of myelin gene regulatory factor					
let-7c	Plays a role in suppressing the dendritic growth of cortical neuron					
let-7d	Regulates immunological and inflammatory responses					
Combination of: miR-223-3p + miR-326-5p + miR-145-5p; miR-326-5p + miR-145-5p; miR-223-3p + miR-326-5p	miR-223-3p: Modulates the nuclear factor κB pathway Regulates inflammatory responses in macrophages	No statistical significance	Serum	18 RRMS 19 SPMS 23 HCs	3	[144]
	miR-326-5p: Regulates Th17 differentiation	Downregulated				
	miR-145-5p: Regulates the expression of myelin gene regulatory factor	No statistical significance				

All patients were untreated with disease-modifying therapies at the time of the sample collection, except the cohort of Kramer et al. (miR-181c-5p), which comprised both untreated patients and patients treated with azathioprine, interferons, glatiramer acetate, mitoxantrone, natalizumab, fingolimod, or fumaric acid. Detection technique: qRT-PCR. Abbreviations: other neurologic diseases—OND, relapsing–remitting multiple sclerosis—RRMS, secondary progressive multiple sclerosis—SPMS, healthy controls—HCs, cerebrospinal fluid—CSF.

5. Limitations in miRNA Biomarker Studies

To summarize, available published studies point out that circulating miRNAs may have an important role as biomarkers in the diagnosis and prognosis of MS in future clinical practice. However, there are a small number of studies concerning circulating miRNAs' expression in MS, especially in the context of miRNA profile difference between RRMS and SPMS subtypes. Furthermore, the research on circulating miRNAs as biomarkers is still in the early stage; thus, the findings usually lack reproducibility and are often even divergent. Studies evaluating circulating miRNAs have significant limitations, including the variability in biological sources and different techniques used to isolate and analyze (e.g., microarray, qRT-PCR, dPCR, and NGS) miRNAs. There are still no studies that analyze the differences in the level of miRNA expression measured in cell populations between RR and SP patients. Performing such an analysis would significantly narrow the pool of analyzed miRNA molecules to those which are cell-specific, and could constitute an important object in targeted therapies. Currently, there are studies that reported differences in miRNAs' expression in different types of cells in each subtype of MS separately, in comparison to the control group. However, there are still not enough studies that analyze the level of miRNA expression in cells as a potential marker distinguishing the RR from the SP subtype. The next limitation is the lack of a large sample size and validation of results in independent cohorts, which is why the works with limited sample size and those with poor study design should be carefully considered. Another limitation factor of many original works is the fact that the authors do not specify the time of the biological fluids' collection (at the diagnosis or at a different time point) and do not add the information on the treatment of patients. It would certainly allow us to take into consideration the potential influence of drugs on the level of miRNA and increase the reliability of the obtained results. Ultimately, one of the biggest hurdles facing researchers who want to determine the role of individual miRNAs in various disease entities is the complex network of relationships between the miRNAs and target transcripts. All those factors pose a huge problem as far as finding reliable and foolproof miRNAs, which might be characteristic for specific MS stage and effective prognosis of the RRMS-to-SPMS transition. Therefore, although some miRNAs might be useful during discrimination between RRMS and SPMS, there is a need to confirm their effectiveness via clinical trials.

6. Conclusions

Circulating miRNAs are particularly interesting molecules that may be able to identify differences in epigenetic regulation and pathophysiological mechanisms in different types of diseases. Research conducted on miRNAs in MS seems to be very promising. The findings suggest significant differences in the expression of many miRNAs between RRMS and SPMS subtypes. Currently, there is no sufficient data that would support the association of the expression level of specific circulating miRNA molecules with the development of pathological processes characteristic of particular stages of the MS course. While current immunomodulatory therapies have been shown to be efficacious in the early stages of MS, these therapies are less potent in the later phases of MS. Thus, distinguishing the conversion of RRMS to SPMS early is a great need. When the disease transits from the RR to the SP phase, the modification of the therapeutic protocol is required, ensuring a slower development of the disease by increasing the effectiveness of the treatment. Potential markers for RR-to-SP transition include: miR-92a-1-3p, let-7c, let-7d, miR-145-5p, miR-27a-3p, miR-337-3p, miR-633-5p, miR-181c-5p, and panels of miR-223-3p + miR-326-5p + miR-145-5p, miR-326-5p + miR-145-5p, and miR-223-3p + miR-326-5p. Nevertheless, the functions of most miRNAs in the pathophysiology of MS still remain largely unexplained. Currently, there are many miRNAs as potential markers of neuroinflammation and neurodegeneration, but further studies are desirable to verify those biomarkers as useful tools in clinical application, being novel therapeutic targets for treating MS. Each subtype of MS still requires its own specific miRNA profile.

Author Contributions: S.P., A.D., E.M., M.S. and J.S.-B. conceived the figures and wrote the manuscript. All authors have read and agreed to the published version of the manuscript.

Funding: This research received no external funding.

Institutional Review Board Statement: Not applicable.

Informed Consent Statement: Not applicable.

Data Availability Statement: No new data were created or analyzed in this study. Data sharing is not applicable to this article.

Conflicts of Interest: The authors declare no conflict of interest.

References

1. Thompson, A.J.; Baranzini, S.E.; Geurts, J.; Hemmer, B.; Ciccarelli, O. Multiple Sclerosis. *Lancet* **2018**, *391*, 1622–1636. [PubMed]
2. Walton, C.; King, R.; Rechtman, L.; Kaye, W.; Leray, E.; Marrie, R.A.; Robertson, N.; la Rocca, N.; Uitdehaag, B.; van der Mei, I.; et al. Rising Prevalence of Multiple Sclerosis Worldwide: Insights from the Atlas of MS, Third Edition. *Mult. Scler. J.* **2020**, *26*, 1816–1821. [CrossRef]
3. Miclea, A.; Salmen, A.; Zoehner, G.; Diem, L.; Kamm, C.P.; Chaloulos-Iakovidis, P.; Miclea, M.; Briner, M.; Kilidireas, K.; Stefanis, L.; et al. Age-dependent Variation of Female Preponderance across Different Phenotypes of Multiple Sclerosis: A Retrospective Cross-sectional Study. *CNS Neurosci. Ther.* **2019**, *25*, 527–531. [CrossRef] [PubMed]
4. Padilha, I.G.; Fonseca, A.P.A.; Pettengill, A.L.M.; Fragoso, D.C.; Pacheco, F.T.; Nunes, R.H.; Maia, A.C.M.; da Rocha, A.J. Pediatric Multiple Sclerosis: From Clinical Basis to Imaging Spectrum and Differential Diagnosis. *Pediatric Radiol.* **2020**, *50*, 776–792.
5. Roohani, P.; Emiru, T.; Carpenter, A.; Luzzio, C.; Freeman, J.; Scarberry, S.; Beaver, G.; Davidson, L.; Parry, G. Late Onset Multiple Sclerosis: Is It Really Late Onset? *Mult. Scler. Relat. Disord.* **2014**, *3*, 444–449. [CrossRef]
6. GBD 2016 Multiple Sclerosis Collaborators. Global, Regional, and National Burden of Multiple Sclerosis 1990-2016: A Systematic Analysis for the Global Burden of Disease Study 2016. *Lancet. Neurol.* **2019**, *18*, 269–285. [CrossRef]
7. Taylor, B.V.; Pearson, J.F.; Clarke, G.; Mason, D.F.; Abernethy, D.A.; Willoughby, E.; Sabel, C. MS Prevalence in New Zealand, an Ethnically and Latitudinally Diverse Country. *Mult. Scler.* **2010**, *16*, 1422–1431. [CrossRef] [PubMed]
8. Miller, D.; Chard, D.; Ciccarelli, O. Clinically Isolated Syndromes. *Lancet. Neurol.* **2012**, *11*, 157–169. [CrossRef] [PubMed]
9. Klineova, S.; Lublin, F.D. Clinical Course of Multiple Sclerosis. *Cold Spring Harb. Perspect. Med.* **2018**, *8*, a028928. [CrossRef]
10. Lassmann, H. Multiple Sclerosis Pathology. *Cold Spring Harb. Perspect. Med.* **2018**, *8*, a028936. [CrossRef]
11. Lucchinetti, C.; Brück, W.; Parisi, J.; Scheithauer, B.; Rodriguez, M.; Lassmann, H. Heterogeneity of Multiple Sclerosis Lesions: Implications for the Pathogenesis of Demyelination. *Ann. Neurol.* **2000**, *47*, 707–717. [CrossRef]
12. Kuhlmann, T.; Ludwin, S.; Prat, A.; Antel, J.; Brück, W.; Lassmann, H. An Updated Histological Classification System for Multiple Sclerosis Lesions. *Acta Neuropathol.* **2017**, *133*, 13–24. [CrossRef] [PubMed]
13. Bitsch, A.; Brück, W. Differentiation of Multiple Sclerosis Subtypes: Implications for Treatment. *CNS Drugs* **2002**, *16*, 405–418. [CrossRef] [PubMed]
14. Goverman, J. Autoimmune T Cell Responses in the Central Nervous System. *Nat. Rev. Immunol.* **2009**, *9*, 393–407. [CrossRef]
15. Kasper, L.H.; Shoemaker, J. Multiple Sclerosis Immunology: The Healthy Immune System vs. the MS Immune System. *Neurology* **2010**, *74*, S2. [CrossRef]
16. Kalra, S.; Lowndes, C.; Durant, L.; Strange, R.; Al-Araji, A.; Hawkins, C.P.; Curnow, S.J. Th17 Cells Increase in RRMS as Well as in SPMS, Whereas Various Other Phenotypes of Th17 Increase in RRMS Only. *Mult. Scler. J. Exp. Transl. Clin.* **2020**, *6*, 2055217319899695. [CrossRef]
17. Yang, Y.; Estrada, E.Y.; Thompson, J.F.; Liu, W.; Rosenberg, G.A. Matrix Metalloproteinase-Mediated Disruption of Tight Junction Proteins in Cerebral Vessels Is Reversed by Synthetic Matrix Metalloproteinase Inhibitor in Focal Ischemia in Rat. *J. Cereb. Blood Flow Metab.* **2007**, *27*, 697–709. [CrossRef] [PubMed]
18. Murphy, Á.C.; Lalor, S.J.; Lynch, M.A.; Mills, K.H.G. Infiltration of Th1 and Th17 Cells and Activation of Microglia in the CNS during the Course of Experimental Autoimmune Encephalomyelitis. *Brain Behav. Immun.* **2010**, *24*, 641–651. [CrossRef]
19. Confavreux, C.; Vukusic, S. The Clinical Course of Multiple Sclerosis. *Handb. Clin. Neurol.* **2014**, *122*, 343–369. [CrossRef]
20. Compston, A.; Coles, A. Multiple Sclerosis. *Lancet* **2008**, *372*, 1502–1517. [CrossRef]
21. Frischer, J.M.; Bramow, S.; Dal-Bianco, A.; Lucchinetti, C.F.; Rauschka, H.; Schmidbauer, M.; Laursen, H.; Sorensen, P.S.; Lassmann, H. The Relation between Inflammation and Neurodegeneration in Multiple Sclerosis Brains. *Brain* **2009**, *132*, 1175–1189. [CrossRef] [PubMed]
22. Ferrazzano, G.; Crisafulli, S.G.; Baione, V.; Tartaglia, M.; Cortese, A.; Frontoni, M.; Altieri, M.; Pauri, F.; Millefiorini, E.; Conte, A. Early Diagnosis of Secondary Progressive Multiple Sclerosis: Focus on Fluid and Neurophysiological Biomarkers. *J. Neurol.* **2021**, *268*, 3626–3645. [CrossRef] [PubMed]
23. Hollen, C.W.; Soldán, M.M.P.; Rinker, J.R., II; Spain, R.I. The Future of Progressive Multiple Sclerosis Therapies. *Fed. Pract.* **2020**, *37*, S43. [PubMed]
24. de Angelis, F.; Plantone, D.; Chataway, J. Pharmacotherapy in Secondary Progressive Multiple Sclerosis: An Overview. *CNS Drugs* **2018**, *32*, 499–526. [CrossRef]
25. Adamczyk-Sowa, M.; Adamczyk, B.; Kułakowska, A.; Rejdak, K.; Nowacki, P. Secondary Progressive Multiple Sclerosis—From Neuropathology to Definition and Effective Treatment. *Neurol. I Neurochir. Pol.* **2020**, *54*, 384–398. [CrossRef]
26. Gross, H.J.; Watson, C. Characteristics, Burden of Illness, and Physical Functioning of Patients with Relapsing-Remitting and Secondary Progressive Multiple Sclerosis: A Cross-Sectional US Survey. *Neuropsychiatr. Dis. Treat.* **2017**, *13*, 1349–1357. [CrossRef]
27. Inojosa, H.; Proschmann, U.; Akgün, K.; Ziemssen, T. A Focus on Secondary Progressive Multiple Sclerosis (SPMS): Challenges in Diagnosis and Definition. *J. Neurol.* **2021**, *268*, 1210–1221. [CrossRef] [PubMed]
28. Lublin, F.D.; Reingold, S.C.; Cohen, J.A.; Cutter, G.R.; Sørensen, P.S.; Thompson, A.J.; Wolinsky, J.S.; Balcer, L.J.; Banwell, B.; Barkhof, F.; et al. Defining the Clinical Course of Multiple Sclerosis: The 2013 Revisions. *Neurology* **2014**, *83*, 278–286. [CrossRef] [PubMed]
29. Ontaneda, D.; Cohen, J.; Amato, M. Clinical Outcome Measures for Progressive MS Trials. *Mult. Scler.* **2017**, *23*, 1627–1635. [CrossRef]

30. Thompson, A.J.; Banwell, B.L.; Barkhof, F.; Carroll, W.M.; Coetzee, T.; Comi, G.; Correale, J.; Fazekas, F.; Filippi, M.; Freedman, M.S.; et al. Diagnosis of Multiple Sclerosis: 2017 Revisions of the McDonald Criteria. *Lancet Neurol.* **2018**, *17*, 162–173. [CrossRef]
31. Calabrese, M.; Favaretto, A.; Martini, V.; Gallo, P. Grey Matter Lesions in MS: From Histology to Clinical Implications. *Prion* **2013**, *7*, 20–27. [CrossRef] [PubMed]
32. Trip, S.; Miller, D. Imaging in Multiple Sclerosis. *J. Neurol. Neurosurg. Psychiatry* **2005**, *76*, iii11. [CrossRef]
33. Housley, W.; Pitt, D.; Hafler, D. Biomarkers in Multiple Sclerosis. *Clin. Immunol.* **2015**, *161*, 51–58. [CrossRef] [PubMed]
34. Filippi, M.; Preziosa, P.; Rocca, M. Microstructural MR Imaging Techniques in Multiple Sclerosis. *Neuroimaging Clin. N. Am.* **2017**, *27*, 313–333. [CrossRef]
35. Pasquali, L.; Lucchesi, C.; Pecori, C.; Metelli, M.R.; Pellegrini, S.; Iudice, A.; Bonuccelli, U. A Clinical and Laboratory Study Evaluating the Profile of Cytokine Levels in Relapsing Remitting and Secondary Progressive Multiple Sclerosis. *J. Neuroimmunol.* **2015**, *278*, 53–59. [CrossRef] [PubMed]
36. Gresle, M.M.; Liu, Y.; Dagley, L.F.; Haartsen, J.; Pearson, F.; Purcell, A.W.; Laverick, L.; Petzold, A.; Lucas, R.M.; van der Walt, A.; et al. Serum Phosphorylated Neurofilament-Heavy Chain Levels in Multiple Sclerosis Patients. *J. Neurol. Neurosurg. Psychiatry* **2014**, *85*, 1209–1213. [CrossRef]
37. Ayrignac, X.; le Bars, E.; Duflos, C.; Hirtz, C.; Maceski, A.M.; Carra-Dallière, C.; Charif, M.; Pinna, F.; Prin, P.; de Champfleur, N.M.; et al. Serum GFAP in Multiple Sclerosis: Correlation with Disease Type and MRI Markers of Disease Severity. *Sci. Rep.* **2020**, *10*, 10923. [CrossRef] [PubMed]
38. Högel, H.; Rissanen, E.; Barro, C.; Matilainen, M.; Nylund, M.; Kuhle, J.; Airas, L. Serum Glial Fibrillary Acidic Protein Correlates with Multiple Sclerosis Disease Severity. *Mult. Scler.* **2018**, *26*, 210–219. [CrossRef] [PubMed]
39. Harris, V.K.; Sadiq, S.A. Disease Biomarkers in Multiple Sclerosis. *Mol. Diagn. Ther.* **2012**, *13*, 225–244. [CrossRef]
40. Zhou, Z.; Xiong, H.; Xie, F.; Wu, Z.; Feng, Y. A Meta-Analytic Review of the Value of MiRNA for Multiple Sclerosis Diagnosis. *Front. Neurol.* **2020**, *11*, 132. [CrossRef] [PubMed]
41. Gao, Y.; Han, D.; Feng, J. MicroRNA in Multiple Sclerosis. *Clin. Chim. Acta* **2021**, *516*, 92–99. [CrossRef]
42. Ferrante, M.; Conti, G.O. Environment and Neurodegenerative Diseases: An Update on MiRNA Role. *MicroRNA* **2017**, *6*. [CrossRef]
43. Martinez, B.; Peplow, P. MicroRNAs in Blood and Cerebrospinal Fluid as Diagnostic Biomarkers of Multiple Sclerosis and to Monitor Disease Progression. *Neural Regen. Res.* **2020**, *15*, 606–619. [CrossRef]
44. Qiu, L.; Tan, E.; Zeng, L. MicroRNAs and Neurodegenerative Diseases. *Adv. Exp. Med. Biol.* **2015**, *888*, 51–70. [CrossRef]
45. Dolati, S.; Babaloo, Z.; Jadidi-Niaragh, F.; Ayromlou, H.; Sadreddini, S.; Yousefi, M. Multiple Sclerosis: Therapeutic Applications of Advancing Drug Delivery Systems. *Biomed. Pharmacother. = Biomed. Pharmacother.* **2017**, *86*, 343–353. [CrossRef] [PubMed]
46. Ajit, S.K. Circulating MicroRNAs as Biomarkers, Therapeutic Targets, and Signaling Molecules. *Sensors* **2012**, *12*, 3359–3369. [CrossRef] [PubMed]
47. Mitchell, P.S.; Parkin, R.K.; Kroh, E.M.; Fritz, B.R.; Wyman, S.K.; Pogosova-Agadjanyan, E.L.; Peterson, A.; Noteboom, J.; O'Briant, K.C.; Allen, A.; et al. Circulating MicroRNAs as Stable Blood-Based Markers for Cancer Detection. *Proc. Natl. Acad. Sci. USA* **2008**, *105*, 10513–10518. [CrossRef]
48. Sheinerman, K.S.; Umansky, S.R. Circulating Cell-Free MicroRNA as Biomarkers for Screening, Diagnosis, and Monitoring of Neurodegenerative Diseases and Other Neurologic Pathologies. *Front. Cell. Neurosci.* **2013**, *7*, 150. [CrossRef] [PubMed]
49. Ghildiyal, M.; Xu, J.; Seitz, H.; Weng, Z.; Zamore, P.D. Sorting of Drosophila Small Silencing RNAs Partitions MicroRNA* Strands into the RNA Interference Pathway. *RNA* **2010**, *16*, 43–56. [CrossRef]
50. MacFarlane, L.-A.; Murphy, P.R. MicroRNA: Biogenesis, Function and Role in Cancer. *Curr. Genom.* **2010**, *11*, 537–561. [CrossRef] [PubMed]
51. Sheu-Gruttadauria, J.; MacRae, I.J. Structural Foundations of RNA Silencing by Argonaute. *J. Mol. Biol.* **2017**, *429*, 2619–2639. [CrossRef]
52. Desvignes, T.; Batzel, P.; Berezikov, E.; Eilbeck, K.; Eppig, J.T.; McAndrews, M.S.; Singer, A.; Postlethwait, J.H. MicroRNA Nomenclature: A View Incorporating Genetic Origins, Biosynthetic Pathways, and Sequence Variants. *Trends Genet. TIG* **2015**, *31*, 613–626. [CrossRef]
53. Catalanotto, C.; Cogoni, C.; Zardo, G. MicroRNA in Control of Gene Expression: An Overview of Nuclear Functions. *Int. J. Mol. Sci.* **2016**, *17*, 1712. [CrossRef] [PubMed]
54. Broughton, J.P.; Lovci, M.T.; Huang, J.L.; Yeo, G.W.; Pasquinelli, A.E. Pairing Beyond the Seed Supports MicroRNA Targeting Specificity. *Mol. Cell* **2016**, *64*, 320–333. [CrossRef]
55. Denzler, R.; McGeary, S.E.; Title, A.C.; Agarwal, V.; Bartel, D.P.; Stoffel, M. Impact of MicroRNA Levels, Target-Site Complementarity, and Cooperativity on Competing Endogenous RNA-Regulated Gene Expression. *Mol. Cell* **2016**, *64*, 565–579. [CrossRef]
56. Friedman, R.C.; Farh, K.K.-H.; Burge, C.B.; Bartel, D.P. Most Mammalian MRNAs Are Conserved Targets of MicroRNAs. *Genome Res.* **2009**, *19*, 92–105. [CrossRef]
57. Ni, W.-J.; Leng, X.-M. Dynamic MiRNA–MRNA Paradigms: New Faces of MiRNAs. *Biochem. Biophys. Rep.* **2015**, *4*, 337–341. [CrossRef]
58. Selbach, M.; Schwanhäusser, B.; Thierfelder, N.; Fang, Z.; Khanin, R.; Rajewsky, N. Widespread Changes in Protein Synthesis Induced by MicroRNAs. *Nature* **2008**, *455*, 58–63. [CrossRef] [PubMed]

59. Peter, M.E. Targeting of MRNAs by Multiple MiRNAs: The next Step. *Oncogene* **2010**, *29*, 2161–2164. [CrossRef]
60. Dedeoğlu, B.G. High-Throughput Approaches for MicroRNA Expression Analysis. *Methods Mol. Biol.* **2014**, *1107*, 91–103. [CrossRef] [PubMed]
61. Iorio, M.V.; Croce, C.M. MicroRNA Dysregulation in Cancer: Diagnostics, Monitoring and Therapeutics. A Comprehensive Review. *EMBO Mol. Med.* **2012**, *4*, 143–159. [CrossRef] [PubMed]
62. Chen, C.; Ridzon, D.A.; Broomer, A.J.; Zhou, Z.; Lee, D.H.; Nguyen, J.T.; Barbisin, M.; Xu, N.L.; Mahuvakar, V.R.; Andersen, M.R.; et al. Real-Time Quantification of MicroRNAs by Stem–Loop RT–PCR. *Nucleic Acids Res.* **2005**, *33*, e179. [CrossRef] [PubMed]
63. Wang, N.; Zheng, J.; Chen, Z.; Liu, Y.; Dura, B.; Kwak, M.; Xavier-Ferrucio, J.; Lu, Y.-C.; Zhang, M.; Roden, C.; et al. Single-Cell MicroRNA-MRNA Co-Sequencing Reveals Non-Genetic Heterogeneity and Mechanisms of MicroRNA Regulation. *Nat. Commun.* **2019**, *10*, 95. [CrossRef]
64. Mao, X.; Liu, C.; Tong, H.; Chen, Y.; Liu, K. Principles of Digital PCR and Its Applications in Current Obstetrical and Gynecological Diseases. *Am. J. Transl. Res.* **2019**, *11*, 7209–7222. [PubMed]
65. Redis, R.S.; Calin, S.; Yang, Y.; You, M.J.; Calin, G.A. Cell-to-Cell MiRNA Transfer: From Body Homeostasis to Therapy. *Pharmacol. Ther.* **2012**, *136*, 169–174. [CrossRef] [PubMed]
66. Gallo, A.; Tandon, M.; Alevizos, I.; Illei, G.G. The Majority of MicroRNAs Detectable in Serum and Saliva Is Concentrated in Exosomes. *PLoS ONE* **2012**, *7*, e30679. [CrossRef]
67. Sohel, M.H. Extracellular/Circulating MicroRNAs: Release Mechanisms, Functions and Challenges. *Achiev. Life Sci.* **2016**, *10*, 175–186. [CrossRef]
68. Pirola, C.J.; Gianotti, T.F.; Castaño, G.O.; Mallardi, P.; Martino, J.S.; Ledesma, M.M.G.L.; Flichman, D.; Mirshahi, F.; Sanyal, A.J.; Sookoian, S. Circulating MicroRNA Signature in Non-Alcoholic Fatty Liver Disease: From Serum Non-Coding RNAs to Liver Histology and Disease Pathogenesis. *Gut* **2015**, *64*, 800–812. [CrossRef]
69. Wu, J.; Shen, Z. Exosomal MiRNAs as Biomarkers for Diagnostic and Prognostic in Lung Cancer. *Cancer Med.* **2020**, *9*, 6909–6922. [CrossRef]
70. Manna, I.; Iaccino, E.; Dattilo, V.; Barone, S.; Vecchio, E.; Mimmi, S.; Filippelli, E.; Demonte, G.; Polidoro, S.; Granata, A.; et al. Exosome-Associated MiRNA Profile as a Prognostic Tool for Therapy Response Monitoring in Multiple Sclerosis Patients. *FASEB J.* **2018**, *32*, 4241–4246. [CrossRef]
71. Gilad, S.; Meiri, E.; Yogev, Y.; Benjamin, S.; Lebanony, D.; Yerushalmi, N.; Benjamin, H.; Kushnir, M.; Cholakh, H.; Melamed, N.; et al. Serum MicroRNAs Are Promising Novel Biomarkers. *PLoS ONE* **2008**, *3*, e3148. [CrossRef] [PubMed]
72. Xi, Y.; Nakajima, G.; Gavin, E.; Morris, C.G.; Kudo, K.; Hayashi, K.; Ju, J. Systematic Analysis of MicroRNA Expression of RNA Extracted from Fresh Frozen and Formalin-Fixed Paraffin-Embedded Samples. *RNA* **2007**, *13*, 1668–1674. [CrossRef]
73. Condrat, C.E.; Thompson, D.C.; Barbu, M.G.; Bugnar, O.L.; Boboc, A.; Cretoiu, D.; Suciu, N.; Cretoiu, S.M.; Voinea, S.C. MiRNAs as Biomarkers in Disease: Latest Findings Regarding Their Role in Diagnosis and Prognosis. *Cells* **2020**, *9*, 276. [CrossRef]
74. Correia, C.N.; Nalpas, N.C.; McLoughlin, K.E.; Browne, J.A.; Gordon, S.V.; MacHugh, D.E.; Shaughnessy, R.G. Circulating MicroRNAs as Potential Biomarkers of Infectious Disease. *Front. Immunol.* **2017**, *8*, 1. [CrossRef] [PubMed]
75. Kozomara, A.; Griffiths-Jones, S. MiRBase: Annotating High Confidence MicroRNAs Using Deep Sequencing Data. *Nucleic Acids Res.* **2014**, *42*, D68–D73. [CrossRef] [PubMed]
76. Baumann, V.; Winkler, J. MiRNA-Based Therapies: Strategies and Delivery Platforms for Oligonucleotide and Non-Oligonucleotide Agents. *Future Med. Chem.* **2014**, *6*, 1967. [CrossRef]
77. HMDD v3.2. Available online: https://www.cuilab.cn/hmdd (accessed on 6 September 2021).
78. Page, M.J.; McKenzie, J.E.; Bossuyt, P.M.; Boutron, I.; Hoffmann, T.C.; Mulrow, C.D.; Shamseer, L.; Tetzlaff, J.M.; Akl, E.A.; Brennan, S.E.; et al. The PRISMA 2020 Statement: An Updated Guideline for Reporting Systematic Reviews. *BMJ* **2021**, *372*, n71. [CrossRef]
79. Nejad, C.; Stunden, H.J.; Gantier, M.P. A Guide to MiRNAs in Inflammation and Innate Immune Responses. *FEBS J.* **2018**, *285*, 3695–3716. [CrossRef]
80. Mendell, J.T.; Olson, E.N. MicroRNAs in Stress Signaling and Human Disease. *Cell* **2012**, *148*, 1172–1187. [CrossRef] [PubMed]
81. Ardekani, A.M.; Naeini, M.M. The Role of MicroRNAs in Human Diseases. *Avicenna J. Med. Biotechnol.* **2010**, *2*, 161–179. [PubMed]
82. Li, Y.; Kowdley, K. MicroRNAs in Common Human Diseases. *Genom. Proteom. Bioinform.* **2012**, *10*, 246–253. [CrossRef]
83. Sheinerman, K.S.; Toledo, J.B.; Tsivinsky, V.G.; Irwin, D.; Grossman, M.; Weintraub, D.; Hurtig, H.I.; Chen-Plotkin, A.; Wolk, D.A.; McCluskey, L.F.; et al. Circulating Brain-Enriched MicroRNAs as Novel Biomarkers for Detection and Differentiation of Neurodegenerative Diseases. *Alzheimer's Res. Ther.* **2017**, *9*, 89. [CrossRef] [PubMed]
84. Ha, T.-Y. MicroRNAs in Human Diseases: From Autoimmune Diseases to Skin, Psychiatric and Neurodegenerative Diseases. *Immune Netw.* **2011**, *11*, 227–244. [CrossRef]
85. Patterson, M.; Gaeta, X.; Loo, K.; Edwards, M.; Smale, S.; Cinkornpumin, J.; Xie, Y.; Listgarten, J.; Azghadi, S.; Douglass, S.M.; et al. Let-7 MiRNAs Can Act through Notch to Regulate Human Gliogenesis. *Stem Cell Rep.* **2014**, *3*, 758–773. [CrossRef]
86. Emery, B. Regulation of Oligodendrocyte Differentiation and Myelination. *Science* **2010**, *330*, 779–782. [CrossRef]
87. Janowska, J.; Gargas, J.; Ziemka-Nalecz, M.; Zalewska, T.; Buzanska, L.; Sypecka, J. Directed Glial Differentiation and Transdifferentiation for Neural Tissue Regeneration. *Exp. Neurol.* **2019**, *319*, 112813. [CrossRef] [PubMed]

88. Vistbakka, J.; Elovaara, I.; Lehtimäki, T.; Hagman, S. Circulating MicroRNAs as Biomarkers in Progressive Multiple Sclerosis. *Mult. Scler.* **2016**, *23*, 403–412. [CrossRef] [PubMed]
89. Piket, E.; Zheleznyakova, G.; Kular, L.; Jagodic, M. Small Non-Coding RNAs as Important Players, Biomarkers and Therapeutic Targets in Multiple Sclerosis: A Comprehensive Overview. *J. Autoimmun.* **2019**, *101*, 17–25. [CrossRef]
90. Gandhi, R. MiRNA in Multiple Sclerosis: Search for Novel Biomarkers. *Mult. Scler. J.* **2015**, *21*, 1095–1103. [CrossRef]
91. Otaegui, D.; Baranzini, S.E.; Armañanzas, R.; Calvo, B.; Muñoz-Culla, M.; Khankhanian, P.; Inza, I.; Lozano, J.A.; Castillo-Triviño, T.; Asensio, A.; et al. Differential Micro RNA Expression in PBMC from Multiple Sclerosis Patients. *PLoS ONE* **2009**, *4*, e6309. [CrossRef] [PubMed]
92. Regev, K.; Paul, A.; Healy, B.; von Glenn, F.; Diaz-Cruz, C.; Gholipour, T.; Antonietta Mazzola, M.; Raheja, R.; Nejad, P.; Glanz, B.I.; et al. Comprehensive Evaluation of Serum MicroRNAs as Biomarkers in Multiple Sclerosis. *Neurol. Neuroimmunol. NeuroInflammation* **2016**, *3*. [CrossRef]
93. Mandolesi, G.; Rizzo, F.R.; Balletta, S.; Bassi, M.S.; Gilio, L.; Guadalupi, L.; Nencini, M.; Moscatelli, A.; Ryan, C.P.; Licursi, V.; et al. The MicroRNA Let-7b-5p Is Negatively Associated with Inflammation and Disease Severity in Multiple Sclerosis. *Cells* **2021**, *10*, 330. [CrossRef] [PubMed]
94. Perdaens, O.; Dang, H.A.; D'Auria, L.; van Pesch, V. CSF MicroRNAs Discriminate MS Activity and Share Similarity to Other Neuroinflammatory Disorders. *Neurol. Neuroimmunol. Neuroinflammation* **2020**, *7*, e673. [CrossRef] [PubMed]
95. Baulina, N.M.; Kulakova, O.G.; Favorova, O.O. MicroRNAs: The Role in Autoimmune Inflammation. *Acta Nat.* **2016**, *8*, 21–33. [CrossRef]
96. Maciak, K.; Dziedzic, A.; Miller, E.; Saluk-Bijak, J. MiR-155 as an Important Regulator of Multiple Sclerosis Pathogenesis. A Review. *Int. J. Mol. Sci.* **2021**, *22*, 4332. [CrossRef]
97. McCoy, C.E. MiR-155 Dysregulation and Therapeutic Intervention in Multiple Sclerosis. *Adv. Exp. Med. Biol.* **2017**, *1024*, 111–131. [CrossRef] [PubMed]
98. Guo, Y.; Hong, W.; Wang, X.; Zhang, P.; Körner, H.; Tu, J.; Wei, W. MicroRNAs in Microglia: How Do MicroRNAs Affect Activation, Inflammation, Polarization of Microglia and Mediate the Interaction Between Microglia and Glioma? *Front. Mol. Neurosci.* **2019**, *12*, 125. [CrossRef]
99. Tarassishin, L.; Loudig, O.; Bauman, A.; Shafit-Zagardo, B.; Suh, H.-S.; Lee, S.C. Interferon Regulatory Factor 3 Inhibits Astrocyte Inflammatory Gene Expression through Suppression of the Proinflammatory MiR-155 and MiR-155. *Glia* **2011**, *59*, 1911–1922. [CrossRef] [PubMed]
100. Han, S.R.; Kang, Y.H.; Jeon, H.; Lee, S.; Park, S.-J.; Song, D.-Y.; Min, S.S.; Yoo, S.-M.; Lee, M.-S.; Lee, S.-H. Differential Expression of MiRNAs and Behavioral Change in the Cuprizone-Induced Demyelination Mouse Model. *Int. J. Mol. Sci.* **2020**, *21*, 646. [CrossRef]
101. Sharma, K.; Zhang, G.; Li, S. Astrogliosis and Axonal Regeneration. *Neural Regen.* **2015**, 181–196. [CrossRef]
102. Liu, S.; Zhu, B.; Sun, Y.; Xie, X. MiR-155 Modulates the Progression of Neuropathic Pain through Targeting SGK3. *Int. J. Clin. Exp. Pathol.* **2015**, *8*, 14374.
103. Kacperska, M.J.; Jastrzebski, K.; Tomasik, B.; Walenczak, J.; Konarska-Krol, M.; Glabinski, A. Selected Extracellular MicroRNA as Potential Biomarkers of Multiple Sclerosis Activity—Preliminary Study. *J. Mol. Neurosci.* **2015**, *56*, 154–163. [CrossRef]
104. Gandhi, R.; Healy, B.; Gholipour, T.; Egorova, S.; Musallam, A.; Hussain, M.S.; Nejad, P.; Patel, B.; Hei, H.; Khoury, S.; et al. Circulating MicroRNAs as Biomarkers for Disease Staging in Multiple Sclerosis. *Ann. Neurol.* **2013**, *73*, 729–740. [CrossRef]
105. Bernstein, D.L.; Jiang, X.; Rom, S. Let-7 MicroRNAs: Their Role in Cerebral and Cardiovascular Diseases, Inflammation, Cancer, and Their Regulation. *Biomedicines* **2021**, *9*, 606. [CrossRef] [PubMed]
106. Cho, K.J.; Song, J.; Oh, Y.; Lee, J.E. MicroRNA-Let-7a Regulates the Function of Microglia in Inflammation. *Mol. Cell. Neurosci.* **2015**, *68*, 167–176. [CrossRef] [PubMed]
107. Falcone, M.; Rajan, A.; Bloom, B.; Brosnan, C. A Critical Role for IL-4 in Regulating Disease Severity in Experimental Allergic Encephalomyelitis as Demonstrated in IL-4-Deficient C57BL/6 Mice and BALB/c Mice. *J. Immunol.* **1998**, *160*, 4822–4830.
108. Talebi, F.; Ghourbani, S.; Vojgani, M.; Noorbakhsh, F. MiR-320 and Inflammation Regulation in Experimental Autoimmune Encephalomyelitis through Interference With Tumor Growth Factor-β Signaling Pathway. *Immunoregulation* **2020**, *2*, 111–120. [CrossRef]
109. Lee, P.W.; Severin, M.E.; Lovett-Racke, A.E. TGF-β Regulation of T Cells in Multiple Sclerosis. *Eur. J. Immunol.* **2017**, *47*, 446–453. [CrossRef]
110. Aung, L.L.; Mouradian, M.M.; Dhib-Jalbut, S.; Balashov, K.E. MMP-9 Expression Is Increased in B Lymphocytes during Multiple Sclerosis Exacerbation and Is Regulated by MicroRNA-320a. *J. Neuroimmunol.* **2015**, *278*, 185–189. [CrossRef]
111. Fainardi, E.; Castellazzi, M.; Bellini, T.; Manfrinato, M.C.; Baldi, E.; Casetta, I.; Paolino, E.; Granieri, E.; Dallocchio, F. Cerebrospinal Fluid and Serum Levels and Intrathecal Production of Active Matrix Metalloproteinase-9 (MMP-9) as Markers of Disease Activity in Patients with Multiple Sclerosis. *Mult. Scler.* **2016**, *12*, 294–301. [CrossRef]
112. Castellazzi, M.; Ligi, D.; Contaldi, E.; Quartana, D.; Fonderico, M.; Borgatti, L.; Bellini, T.; Trentini, A.; Granieri, E.; Fainardi, E.; et al. Multiplex Matrix Metalloproteinases Analysis in the Cerebrospinal Fluid Reveals Potential Specific Patterns in Multiple Sclerosis Patients. *Front. Neurol.* **2018**, *9*, 1080. [CrossRef] [PubMed]
113. A Pilot Study to Assess MicroRNA Biomarkers in Early and Later Stage Multiple Sclerosis—Full Text View—ClinicalTrials.Gov. Available online: https://clinicaltrials.gov/ct2/show/NCT01737372 (accessed on 6 September 2021).

114. Zou, X.; Wei, J.; Huang, Z.; Zhou, X.; Lu, Z.; Zhu, W.; Miao, Y. Identification of a Six-miRNA Panel in Serum Benefiting Pancreatic Cancer Diagnosis. *Cancer Med.* **2019**, *8*, 2810–2822. [CrossRef]
115. Yan, L.; Zhao, W.; Yu, H.; Wang, Y.; Liu, Y.; Xie, C. A Comprehensive Meta-Analysis of MicroRNAs for Predicting Colorectal Cancer. *Medicine* **2016**, *95*, e2738. [CrossRef] [PubMed]
116. Pogribny, I.P. MicroRNAs as Biomarkers for Clinical Studies. *Exp. Biol. Med.* **2018**, *243*, 283. [CrossRef] [PubMed]
117. Haghikia, A.; Haghikia, A.; Hellwig, K.; Baraniskin, A.; Holzmann, A.; Décard, B.F.; Thum, T.; Gold, R. Regulated MicroRNAs in the CSF of Patients with Multiple Sclerosis. *Neurology* **2012**, *79*, 2166–2170. [CrossRef]
118. Kramer, S.; Haghikia, A.; Bang, C.; Scherf, K.; Pfanne, A.; Duscha, A.; Kaisler, J.; Gisevius, B.; Gold, R.; Thum, T.; et al. Elevated Levels of MiR-181c and MiR-633 in the CSF of Patients with MS: A Validation Study. *Neurol. Neuroimmunol. Neuroinflammation* **2019**, *6*, e623. [CrossRef]
119. Ahlbrecht, J.; Martino, F.; Pul, R.; Skripuletz, T.; Sühs, K.-W.; Schauerte, C.; Yildiz, Ö.; Trebst, C.; Tasto, L.; Thum, S.; et al. Deregulation of MicroRNA-181c in Cerebrospinal Fluid of Patients with Clinically Isolated Syndrome Is Associated with Early Conversion to Relapsing–Remitting Multiple Sclerosis. *Mult. Scler.* **2015**, *22*, 1202–1214. [CrossRef]
120. Kos, A.; Loohuis, N.O.; Meinhardt, J.; van Bokhoven, H.; Kaplan, B.B.; Martens, G.; Aschrafi, A. MicroRNA-181 Promotes Synaptogenesis and Attenuates Axonal Outgrowth in Cortical Neurons. *Cell. Mol. Life Sci. CMLS* **2016**, *73*, 3555–3567. [CrossRef] [PubMed]
121. Zhang, Z.; Xue, Z.; Liu, Y.; Liu, H.; Guo, X.; Li, Y.; Yang, H.; Zhang, L.; Da, Y.; Yao, Z.; et al. MicroRNA-181c Promotes Th17 Cell Differentiation and Mediates Experimental Autoimmune Encephalomyelitis. *Brain Behav. Immun.* **2018**, *70*, 305–314. [CrossRef]
122. Mackey, M.F.; Barth, R.J.; Noelle, R.J. The Role of CD40/CD154 Interactions in the Priming, Differentiation, and Effector Function of Helper and Cytotoxic T Cells. *J. Leukoc. Biol.* **1998**, *63*, 418–428. [CrossRef]
123. Karnell, J.L.; Rieder, S.A.; Ettinger, R.; Kolbeck, R. Targeting the CD40-CD40L Pathway in Autoimmune Diseases: Humoral Immunity and Beyond. *Adv. Drug Deliv. Rev.* **2019**, *141*, 92–103. [CrossRef]
124. Aarts, S.A.B.M.; Seijkens, T.T.P.; van Dorst, K.J.F.; Dijkstra, C.D.; Kooij, G.; Lutgens, E. The CD40–CD40L Dyad in Experimental Autoimmune Encephalomyelitis and Multiple Sclerosis. *Front. Immunol.* **2017**, *8*, 1791. [CrossRef]
125. Balashov, K.E.; Smith, D.R.; Khoury, S.J.; Hafler, D.A.; Weiner, H.L. Increased Interleukin 12 Production in Progressive Multiple Sclerosis: Induction by Activated CD4+ T Cells via CD40 Ligand. *Proc. Natl. Acad. Sci. USA* **1997**, *94*, 599–603. [CrossRef] [PubMed]
126. Rezaei, N.; Talebi, F.; Ghorbani, S.; Rezaei, A.; Esmaeili, A.; Noorbakhsh, F.; Hakemi, M.G. MicroRNA-92a Drives Th1 Responses in the Experimental Autoimmune Encephalomyelitis. *Inflammation* **2018**, *42*, 235–245. [CrossRef] [PubMed]
127. Banerjee, S.; Xie, N.; Cui, H.; Tan, Z.; Yang, S.; Icyuz, M.; Abraham, E.; Liu, G. MicroRNA Let-7c Regulates Macrophage Polarization. *J. Immunol.* **2013**, *190*, 6542–6549. [CrossRef] [PubMed]
128. Kim, S.J.; Gregersen, P.K.; Diamond, B. Regulation of Dendritic Cell Activation by MicroRNA Let-7c and BLIMP1. *J. Clin. Investig.* **2013**, *123*, 823–833. [CrossRef]
129. Liu, H.Y.; Huang, C.M.; Hung, Y.F.; Hsueh, Y.P. The MicroRNAs Let7c and MiR21 Are Recognized by Neuronal Toll-like Receptor 7 to Restrict Dendritic Growth of Neurons. *Exp. Neurol.* **2015**, *269*, 202–212. [CrossRef]
130. Søndergaard, H.B.; Hesse, D.; Krakauer, M.; Sørensen, P.S.; Sellebjerg, F. Differential MicroRNA Expression in Blood in Multiple Sclerosis. *Mult. Scler.* **2013**, *19*, 1849–1857. [CrossRef] [PubMed]
131. De, R.; Malefyt, W.; Abrams, J.; Bennett, B.; Figdor, C.G.; de Vries, J.E. Interleukin 10(EL,.10) Inhibits Cytokine Synthesis by Human Monocytes: An Autoregulatory Role of IL-10 Produced by Monocytes. *J. Exp. Med.* **1991**, *174*, 1209–1220. [CrossRef]
132. Al-Faraj, S. Neurothrophins. Molecular Links between the Immune and Nervous Systems. *Neurosci. J.* **2001**, *6*, 23–25.
133. Sarchielli, P.; Greco, L.; Stipa, A.; Floridi, A.; Gallai, V. Brain-Derived Neurotrophic Factor in Patients with Multiple Sclerosis. *J. Neuroimmunol.* **2002**, *132*, 180–188. [CrossRef]
134. Saha, R.N.; Jana, M.; Pahan, K. MAPK P38 Regulates Transcriptional Activity of NF-KB in Primary Human Astrocytes via Acetylation of P65. *J. Immunol.* **2007**, *179*, 7101–7109. [CrossRef]
135. Torres, J.; Enríquez-de-Salamanca, A.; Fernández, I.; Rodríguez-Ares, M.T.; Quadrado, M.J.; Murta, J.; del Castillo, J.M.B.; Stern, M.E.; Calonge, M. Activation of MAPK Signaling Pathway and NF-KB Activation in Pterygium and Ipsilateral Pterygium-Free Conjunctival Specimens. *Investig. Ophthalmol. Vis. Sci.* **2011**, *52*, 5842–5852. [CrossRef] [PubMed]
136. Lu, J.; Zhou, N.; Yang, P.; Deng, L.; Liu, G. MicroRNA-27a-3p Downregulation Inhibits Inflammatory Response and Hippocampal Neuronal Cell Apoptosis by Upregulating Mitogen-Activated Protein Kinase 4 (MAP2K4) Expression in Epilepsy: In Vivo and In Vitro Studies. *Med. Sci. Monit. Int. Med. J. Exp. Clin. Res.* **2019**, *25*, 8499–8508. [CrossRef]
137. Ahmadian-Elmi, M.; Bidmeshki Pour, A.; Naghavian, R.; Ghaedi, K.; Tanhaei, S.; Izadi, T.; Nasr-Esfahani, M.H. MiR-27a and MiR-214 Exert Opposite Regulatory Roles in Th17 Differentiation via Mediating Different Signaling Pathways in Peripheral Blood CD4+ T Lymphocytes of Patients with Relapsing–Remitting Multiple Sclerosis. *Immunogenetics* **2016**, *68*, 43–54. [CrossRef] [PubMed]
138. Randall, R.A.; Howell, M.; Page, C.S.; Daly, A.; Bates, P.A.; Hill, C.S. Recognition of Phosphorylated-Smad2-Containing Complexes by a Novel Smad Interaction Motif. *Mol. Cell. Biol.* **2004**, *24*, 1106–1121. [CrossRef]
139. Zhou, L.; Lopes, J.E.; Chong, M.M.W.; Ivanov, I.I.; Min, R.; Victora, G.D.; Shen, Y.; Du, J.; Rubtsov, Y.P.; Rudensky, A.Y.; et al. TGF-β-Induced Foxp3 Inhibits Th17 Cell Differentiation by Antagonizing RORγt Function. *Nature* **2008**, *453*, 236–240. [CrossRef]

140. Tripathi, A.; Volsko, C.; Garcia, J.P.; Agirre, E.; Allan, K.C.; Tesar, P.J.; Trapp, B.D.; Castelo-Branco, G.; Sim, F.J.; Dutta, R. Oligodendrocyte Intrinsic MiR-27a Controls Myelination and Remyelination. *Cell Rep.* **2019**, *29*, 904–919. [CrossRef]
141. Regev, K.; Healy, B.C.; Paul, A.; Diaz-Cruz, C.; Mazzola, M.A.; Raheja, R.; Glanz, B.I.; Kivisäkk, P.; Chitnis, T.; Jagodic, M.; et al. Identification of MS-Specific Serum MiRNAs in an International Multicenter Study. *Neurol. Neuroimmunol. NeuroInflammation* **2018**, *5*, 491. [CrossRef]
142. Engelhardt, B.; Kappos, L. Natalizumab: Targeting A4-Integrins in Multiple Sclerosis. *Neurodegener. Dis.* **2008**, *5*, 16–22. [CrossRef]
143. Salinas, G.F.; Krausz, S.; Dontje, W.; Evavold, B.D.; Tak, P.P.; Baeten, D.L.; Reedquist, K.A. Sustained Rap1 Activation in Autoantigen-Specific T Lymphocytes Attenuates Experimental Autoimmune Encephalomyelitis. *J. Neuroimmunol.* **2012**, *250*, 35–43. [CrossRef]
144. Sharaf-Eldin, W.E.; Kishk, N.A.; Gad, Y.Z.; Hassan, H.; Ali, M.A.M.; Zaki, M.S.; Mohamed, M.R.; Essawi, M.L. Extracellular MiR-145, MiR-223 and MiR-326 Expression Signature Allow for Differential Diagnosis of Immune-Mediated Neuroinflammatory Diseases. *J. Neurol. Sci.* **2017**, *383*, 188–198. [CrossRef]
145. Hanley, J.A.; McNeil, B.J. The Meaning and Use of the Area under a Receiver Operating Characteristic (ROC) Curve. *Radiology* **1982**, *143*, 29–36. [CrossRef] [PubMed]
146. Kornfeld, S.F.; Cummings, S.E.; Fathi, S.; Bonin, S.R.; Kothary, R. MiRNA-145-5p Prevents Differentiation of Oligodendrocyte Progenitor Cells by Regulating Expression of Myelin Gene Regulatory Factor. *J. Cell. Physiol.* **2021**, *236*, 997–1012. [CrossRef] [PubMed]
147. Dharap, A.; Bowen, K.; Place, R.; Li, L.-C.; Vemuganti, R. Transient Focal Ischemia Induces Extensive Temporal Changes in Rat Cerebral MicroRNAome. *J. Cereb. Blood Flow Metab. Off. J. Int. Soc. Cereb. Blood Flow Metab.* **2009**, *29*, 675–687. [CrossRef] [PubMed]
148. Xie, X.; Peng, L.; Zhu, J.; Zhou, Y.; Li, L.; Chen, Y.; Yu, S.; Zhao, Y. MiR-145-5p/Nurr1/TNF-α Signaling-Induced Microglia Activation Regulates Neuron Injury of Acute Cerebral Ischemic/Reperfusion in Rats. *Front. Mol. Neurosci.* **2017**, *10*, 383. [CrossRef] [PubMed]
149. De Barcelos, I.P.; Troxell, R.M.; Graves, J.S. Mitochondrial Dysfunction and Multiple Sclerosis. *Biology* **2019**, *8*, 37. [CrossRef]
150. Jiang, L.; Wei, Z.-C.; Xu, L.-L.; Yu, S.-Y.; Li, C. Inhibition of MiR-145-5p Reduces Spinal Cord Injury-Induced Inflammatory and Oxidative Stress Responses via Affecting Nurr1-TNF-α Signaling Axis. *Cell Biochem. Biophys.* **2021**. [CrossRef]
151. Johnnidis, J.B.; Harris, M.H.; Wheeler, R.T.; Stehling-Sun, S.; Lam, M.H.; Kirak, O.; Brummelkamp, T.R.; Fleming, M.D.; Camargo, F.D. Regulation of Progenitor Cell Proliferation and Granulocyte Function by MicroRNA-223. *Nature* **2008**, *451*, 1125–1129. [CrossRef]
152. Ifergan, I.; Chen, S.; Zhang, B.; Miller, S.D. Cutting Edge: MicroRNA-223 Regulates Myeloid Dendritic Cell–Driven Th17 Responses in Experimental Autoimmune Encephalomyelitis. *J. Immunol.* **2016**, *196*, 1455–1459. [CrossRef]
153. Galloway, D.A.; Blandford, S.N.; Berry, T.; Williams, J.B.; Stefanelli, M.; Ploughman, M.; Moore, C.S. MiR-223 Promotes Regenerative Myeloid Cell Phenotype and Function in the Demyelinated Central Nervous System. *Glia* **2019**, *67*, 857–869. [CrossRef] [PubMed]
154. Li, T.; Morgan, M.J.; Choksi, S.; Zhang, Y.; Kim, Y.S.; Liu, Z.G. MicroRNAs Modulate the Noncanonical Transcription Factor NF-KB Pathway by Regulating Expression of the Kinase IKKα during Macrophage Differentiation. *Nat. Immunol.* **2010**, *11*, 799–805. [CrossRef]
155. Zhou, W.; Pal, A.S.; Hsu, A.Y.H.; Gurol, T.; Zhu, X.; Wirbisky-Hershberger, S.E.; Freeman, J.L.; Kasinski, A.L.; Deng, Q. MicroRNA-223 Suppresses the Canonical NF-KB Pathway in Basal Keratinocytes to Dampen Neutrophilic Inflammation. *Cell Rep.* **2018**, *22*, 1810–1823. [CrossRef]
156. Chen, Q.; Wang, H.; Liu, Y.; Song, Y.; Lai, L.; Han, Q.; Cao, X.; Wang, Q. Inducible MicroRNA-223 down-Regulation Promotes TLR-Triggered IL-6 and IL-1β Production in Macrophages by Targeting STAT3. *PLoS ONE* **2012**, *7*, e42971. [CrossRef]
157. Barclay, W.; Shinohara, M.L. Inflammasome Activation in Multiple Sclerosis and Experimental Autoimmune Encephalomyelitis (EAE). *Brain Pathol.* **2017**, *27*, 213–219. [CrossRef]
158. Bauernfeind, F.; Rieger, A.; Schildberg, F.A.; Knolle, P.A.; Schmid-Burgk, J.L.; Hornung, V. NLRP3 Inflammasome Activity Is Negatively Controlled by MiR-223. *J. Immunol.* **2012**, *189*, 4175–4181. [CrossRef] [PubMed]
159. Malhotra, S.; Costa, C.; Eixarch, H.; Keller, C.W.; Amman, L.; Martínez-Banaclocha, H.; Midaglia, L.; Sarró, E.; Machín-Díaz, I.; Villar, L.M.; et al. NLRP3 Inflammasome as Prognostic Factor and Therapeutic Target in Primary Progressive Multiple Sclerosis Patients. *Brain* **2020**, *143*, 1414–1430. [CrossRef] [PubMed]
160. Morquette, B.; Juźwik, C.A.; Drake, S.S.; Charabati, M.; Zhang, Y.; Lécuyer, M.A.; Galloway, D.A.; Dumas, A.; de Faria Junior, O.; Paradis-Isler, N.; et al. MicroRNA-223 Protects Neurons from Degeneration in Experimental Autoimmune Encephalomyelitis. *Brain* **2019**, *142*, 2979–2995. [CrossRef] [PubMed]
161. Wright, A.; Vissel, B. The Essential Role of AMPA Receptor GluA2 Subunit RNA Editing in the Normal and Diseased Brain. *Front. Mol. Neurosci.* **2012**, *5*, 34. [CrossRef]
162. Du, C.; Liu, C.; Kang, J.; Zhao, G.; Ye, Z.; Huang, S.; Li, Z.; Wu, Z.; Pei, G. MicroRNA MiR-326 Regulates TH-17 Differentiation and Is Associated with the Pathogenesis of Multiple Sclerosis. *Nat. Immunol.* **2009**, *10*, 1252–1259. [CrossRef]
163. Moisan, J.; Grenningloh, R.; Bettelli, E.; Oukka, M.; Ho, I.-C. Ets-1 Is a Negative Regulator of Th17 Differentiation. *J. Exp. Med.* **2007**, *204*, 2825–2835. [CrossRef] [PubMed]

164. Honardoost, M.A.; Kiani-Esfahani, A.; Ghaedi, K.; Etemadifar, M.; Salehi, M. MiR-326 and MiR-26a, Two Potential Markers for Diagnosis of Relapse and Remission Phases in Patient with Relapsing–Remitting Multiple Sclerosis. *Gene* **2014**, *544*, 128–133. [CrossRef] [PubMed]
165. Choi, I.; Woo, J.H.; Jou, I.; Joe, E. PINK1 Deficiency Decreases Expression Levels of Mir-326, Mir-330, and Mir-3099 during Brain Development and Neural Stem Cell Differentiation. *Exp. Neurobiol.* **2016**, *25*, 14–23. [CrossRef]
166. Junker, A.; Krumbholz, M.; Eisele, S.; Mohan, H.; Augstein, F.; Bittner, R.; Lassmann, H.; Wekerle, H.; Hohlfeld, R.; Meinl, E. MicroRNA Profiling of Multiple Sclerosis Lesions Identifies Modulators of the Regulatory Protein CD47. *Brain* **2009**, *132*, 3342–3352. [CrossRef]
167. Ebrahimkhani, S.; Vafaee, F.; Young, P.E.; Hur, S.S.J.; Hawke, S.; Devenney, E.; Beadnall, H.; Barnett, M.H.; Suter, C.M.; Buckland, M.E. Exosomal MicroRNA Signatures in Multiple Sclerosis Reflect Disease Status. *Sci. Rep.* **2017**, *7*, 14293. [CrossRef] [PubMed]
168. Gu, J.; Shao, R.; Li, M.; Yan, Q.; Hu, H. MiR-485-3p Modulates Neural Stem Cell Differentiation and Proliferation via Regulating TRIP6 Expression. *J. Cell. Mol. Med.* **2020**, *24*, 398–404. [CrossRef]
169. Yu, L.; Li, H.; Liu, W.; Zhang, L.; Tian, Q.; Li, H.; Li, M. MiR-485-3p Serves as a Biomarker and Therapeutic Target of Alzheimer's Disease via Regulating Neuronal Cell Viability and Neuroinflammation by Targeting AKT3. *Mol. Genet. Genom. Med.* **2021**, *9*, e1548. [CrossRef]
170. Lin, X.; Wang, R.; Li, R.; Tao, T.; Zhang, D.; Qi, Y. Diagnostic Performance of MiR-485-3p in Patients with Parkinson's Disease and Its Relationship with Neuroinflammation. *Neuromolecular Med.* **2021**. [CrossRef] [PubMed]
171. Fordham, J.B.; Naqvi, A.R.; Nares, S. Regulation of MiR-24, MiR-30b, and MiR-142-3p during Macrophage and Dendritic Cell Differentiation Potentiates Innate Immunity. *J. Leukoc. Biol.* **2015**, *98*, 195–207. [CrossRef]
172. Brennan, S.; Keon, M.; Liu, B.; Su, Z.; Saksena, N.K. Panoramic Visualization of Circulating MicroRNAs Across Neurodegenerative Diseases in Humans. *Mol. Neurobiol.* **2019**, *56*, 7380–7407. [CrossRef]
173. Xie, C.; Li, Z.; Zhang, G.-X.; Guan, Y. Wnt Signaling in Remyelination in Multiple Sclerosis: Friend or Foe? *Mol. Neurobiol.* **2013**, *49*, 1117–1125. [CrossRef] [PubMed]
174. Zheng, Y.; Wang, Z.; Tu, Y.; Shen, H.; Dai, Z.; Lin, J.; Zhou, Z. MiR-101a and MiR-30b Contribute to Inflammatory Cytokine-Mediated β-Cell Dysfunction. *Lab. Investig.* **2015**, *95*, 1387–1397. [CrossRef] [PubMed]
175. Sharief, M.K.; Matthews, H.; Noori, M.A. Expression Ratios of the Bcl-2 Family Proteins and Disease Activity in Multiple Sclerosis. *J. Neuroimmunol.* **2003**, *134*, 158–165. [CrossRef]
176. Zettl, U.K.; Kuhlmann, T.; Brück, W. Bcl-2 Expressing T Lymphocytes in Multiple Sclerosis Lesions. *Neuropathol. Appl. Neurobiol.* **1998**, *24*, 202–208. [CrossRef]

International Journal of *Molecular Sciences*

MDPI

Article

Memantine Modulates Oxidative Stress in the Rat Brain Following Experimental Autoimmune Encephalomyelitis

Beata Dąbrowska-Bouta †, Lidia Strużyńska †, Marta Sidoryk-Węgrzynowicz and Grzegorz Sulkowski *

Laboratory of Pathoneurochemistry, Department of Neurochemistry, Mossakowski Medical Research Institute, Polish Academy of Sciences, 5 Pawińskiego Str., 02-106 Warsaw, Poland; bbouta@imdik.pan.pl (B.D.-B.); lidkas@imdik.pan.pl (L.S.); msidoryk@imdik.pan.pl (M.S.-W.)
* Correspondence: gsulkowski@imdik.pan.pl; Tel.: +48-22-6086444; Fax: +48-22-6085423
† Equally contributing authors.

Abstract: Experimental autoimmune encephalomyelitis (EAE) is an animal model most commonly used in research on the pathomechanisms of multiple sclerosis (MS). The inflammatory processes, glutamate excitotoxicity, and oxidative stress have been proposed as determinants accompanying demyelination and neuronal degeneration during the course of MS/EAE. The aim of the current study was to characterize the role of NMDA receptors in the induction of oxidative stress during the course of EAE. The effect of memantine, the uncompetitive NMDA receptor antagonist, on modulation of neurological deficits and oxidative stress in EAE rats was analyzed using several experimental approaches. We demonstrated that the expression of antioxidative enzymes (superoxide dismutases SOD1 and SOD2) were elevated in EAE rat brains. Under the same experimental conditions, we observed alterations in oxidative stress markers such as increased levels of malondialdehyde (MDA) and decreased levels of sulfhydryl (-SH) groups, both protein and non-protein (indicating protein damage), and a decline in reduced glutathione. Importantly, pharmacological inhibition of ionotropic NMDA glutamate receptors by their antagonist memantine improved the physical activity of EAE rats, alleviated neurological deficits such as paralysis of tail and hind limbs, and modulated oxidative stress parameters (MDA, -SH groups, SOD's). Furthermore, the current therapy aiming to suppress NMDAR-induced oxidative stress was partially effective when NMDAR's antagonist was administered at an early (asymptomatic) stage of EAE.

Keywords: EAE; glutamate receptor antagonist; excitotoxicity; oxidative stress; -SH groups; superoxide dismutase

Citation: Dąbrowska-Bouta, B.; Strużyńska, L.; Sidoryk-Węgrzynowicz, M.; Sulkowski, G. Memantine Modulates Oxidative Stress in the Rat Brain Following Experimental Autoimmune Encephalomyelitis. *Int. J. Mol. Sci.* **2021**, *22*, 11330. https://doi.org/10.3390/ijms222111330

Academic Editors: Marcello Ciaccio and Luisa Agnello

Received: 3 September 2021
Accepted: 19 October 2021
Published: 20 October 2021

1. Introduction

Experimental autoimmune encephalomyelitis (EAE) is the most popular and well characterized animal model of multiple sclerosis (MS). MS is an immune-mediated demyelinating disease of the central nervous system (CNS) with inflammatory and neurodegenerative components that often affects young adults between the ages of 20 and 40, more frequently female than male [1]. The characteristic features of the disease are demyelinating areas in the white and grey matter of the spinal cord and brain. The etiology of MS is still not fully understood, although the role of both genetic and environmental factors has been demonstrated [1]. In demyelinating lesions, the presence of lymphocytes, macrophages and activated microglia has been observed in the proximity of the perivascular area [2,3], suggesting that these types of cells are involved in the process of demyelination [4]. Macrophages and reactive microglia activate the complement pathway, produce pro-inflammatory cytokines, release excitatory amino acids, and generate free radicals [2]. All of these factors can damage myelin and oligodendrocytes and, consequently, disrupt neurotransmission or induce injury and death of neurons. Inflammation, glutamate excitotoxicity and oxidative stress have all been proposed as the most important determinants associated with demyelination and degeneration of neurons during the course of MS [2,5,6].

Additionally, in the EAE rat brain, glutamate-induced cell death and activation of glial cells (microglia and astrocytes) has been reported, which consequently leads to the production and release of inflammatory mediators, such as cytokines and chemokines, and oxygen free radicals [7–9].

Glutamate is the main excitatory neurotransmitter in the mammalian brain and plays an important role in both physiological and pathological mechanisms operating in the CNS. The extracellular level of glutamate must be tightly controlled because an excess of this neurotransmitter leads to excitotoxic cell death [10]. In MS, excitotoxicity is caused by different mechanisms, but the ultimate consequence is neuronal damage due to overstimulation of glutamate receptors (GluRs), especially of NMDA subtype receptors [11]. Correlations between altered glutamate homeostasis, cell death, axonal damage, and disturbances in glutamatergic neurotransmission have been identified during both MS and EAE [2,12,13]. The exact mechanisms of glutamate-mediated disturbances are still not fully understood, but the overstimulation of glutamate receptors has been shown to be a main cause of the excessive production of reactive oxygen and nitrogen species (ROS/RNS) and the resulting oxidative stress (OS) [14]. It is widely accepted that acute glutamate-induced neuronal degeneration is mainly mediated by NMDA receptors, the activation of which leads to a massive influx of extracellular Ca^{2+} into the cells followed by an increase in intracellular Ca^{2+} concentration to pathological levels [10]. Increased intracellular Ca^{2+} levels may further lead to a series of downstream neurotoxic cascades, resulting in the increased formation of ROS and activation of both caspase-dependent and caspase-independent cell death, in which mitochondria play a key role [15]. Ca^{2+} overload depolarizes the mitochondrial membrane and initiates OS mediated by the mitochondrial pathway [16]. Overproduction of ROS causes deleterious effects on proteins, lipids, and nucleic acids, leading to the disruption of cellular functions [17]. High intracellular Ca^{2+} levels also activate a number of Ca^{2+}-dependent enzymes such as proteases, phospholipases, kinases, nitric oxide synthase (NOS), and endonucleases, which mediate proteolysis, free-radical production, or lipid peroxidation, thereby enhancing OS and subsequent oligodendrocyte and neuronal death. Ca^{2+}-related stimulation of mitogen-activated protein kinase p38 (MAPK p38) activates transcription factors that modify the nucleus and cause neuronal injury and apoptosis [8,10,18–20]. Several studies also suggest that overstimulation of GluRs may contribute to the pathogenesis of MS/EAE by altering the integrity of the blood–brain barrier (BBB). The mechanisms of GluRs-mediated changes in neurovascular integrity are unclear but have been shown to involve vasoactive molecules such as nitric oxide (NO) and superoxide radical ($O_2^{\cdot}$) that can combine to form damaging levels of peroxynitrite ($ONOO^-$) [21,22]. Therefore, it appears that glutamate has the potential to mediate both myelin and BBB breakdown in MS/EAE via the action of free radicals and up-regulation of various enzymes [23].

Under physiological conditions, ROS/RNS are neutralized by enzymatic (superoxide dismutases-SODs, catalase, and peroxidase) and non-enzymatic (glutathione, uric acid, and ascorbic acid) antioxidant defense systems. Under pathological conditions, elevated levels of ROS/RNS, which are produced as a result of inflammatory processes, mitochondrial respiratory chain dysfunction, or overstimulation of glutamate NMDA receptors, induce a tissue response that includes upregulation of various scavenger molecules, such as cytosolic SOD1 and mitochondrial SOD2 (the SOD family enzymes most commonly expressed in the CNS). Superoxide anion is neutralized by SOD, which converts the radical to the less toxic hydrogen peroxide, which is further neutralized by the antioxidant enzyme catalase and via the glutathione peroxidase pathway. The level of SODs has been reported to increase in astrocytes and macrophages of the brain of MS patients [5,24,25].

When ROS are not counterbalanced by cellular antioxidative defense systems, ROS metabolites cause OS, leading to protein and lipid peroxidation and DNA alkylation, all of which were observed in MS patients [26]. The induction of lipid peroxidation may be a major factor in free radical-mediated CNS damage. It is a complex process involving the interaction of oxygen-derived free radicals with polyunsaturated fatty acids, resulting in the formation of a variety of highly active electrophilic aldehydes with harmful potential [27].

In the current study, we assessed the contribution of NMDA glutamate receptors to oxidative/nitrosative stress during EAE. We investigated whether inhibition of NMDA receptors by their antagonist, memantine, affects markers of oxidative stress in the brain of rats subjected to EAE. An experimental approach included an analysis of the changes in the parameters indicating the oxidation of proteins (the level of protein sulfhydryl groups -SH) and lipids (the level of malondialdehyde, MDA). Furthermore, we evaluated the relevant antioxidative mechanisms focusing on the estimation of the enzymatic (SOD1 and SOD2) and non-enzymatic (glutathione-related non-protein -SH group) status during the different phases of the disease.

2. Results

2.1. The Effect of Memantine on the Course of EAE

The first symptom during the course of EAE is a change in body weight. In all experimental groups, except the control group, the rats achieved the highest body weight at about 8 d.p.i. At this time, body weight was in the same range for both the EAE untreated and EAE drug treated groups. From 8 d.p.i. to 14 d.p.i., EAE rats underwent a progressive 20–30% weight loss compared to their body weight at the beginning of the experiment, which corresponded to the acute phase of the disease and maximal neurological deficits. At 25 d.p.i., the mean body weight returned to the value observed at the beginning of the experiment. In the groups of EAE rats treated with NMDAR antagonist memantine, weight loss was significantly lower (by about 15%) relative to EAE rats (untreated) (Table 1).

Table 1. Changes in the body weight of experimental rats in different phases of EAE.

	Body Weight (g)		
Days Post Immunization (d.p.i.)	**Control**	**EAE**	**EAE + Memantine**
0	194.0 ± 8.7	193.6 ± 10.4	195.9 ± 11.8
4	198.4 ± 10.8	190.4 ± 10.1	191.6 ± 9.1
12	210.7 ± 4.3	157.9 ± 9.9 ***	166.0 ± 10.3 ***
20	212.0 ± 3.3	163.2 ± 7.7 ***	182.1 ± 5.1 ***,###
25	211.8 ± 2.6	172.0 ± 12.9 ***	188.4 ± 8.8 **,#

The values represent the means ± SD from $n = 64$ animals.** $p < 0.01$, *** $p < 0.001$ significantly different vs. control. # $p < 0.05$, ### $p < 0.01$ vs. experimental autoimmune encephalomyelitis (EAE) rats not subjected to therapy (one-way ANOVA with post hoc Tukey's test).

The neurological deficits observed during the course of EAE were classified daily. Neurological symptoms of the disease included developmental paralysis of tail and hind limbs and reduced physical activity of experimental rats. The neurological symptoms of EAE started to develop at 10–11 d.p.i. and peaked at 12–13 d.p.i. At 14 d.p.i., rats achieved partial recovery from neurological symptoms and full recovery was observed at 17 d.p.i. We did not observe any further neurological symptoms of the disease until the end of the experiment at 25 d.p.i. (Figure 1). Clinical parameters and the effects of memantine are presented in Table 2 and have been described in detail in our previous publications [28–30]. In the current study we observed a reduction in the severity and duration of neurological deficits after the administration of memantine. All rats in the memantine-treated EAE group showed better physiological conditions than the untreated EAE rats. Notably, the duration of the acute phase of the disease was also shortened by 1–2 days in the memantine-treated EAE group compared to the EAE rats. The lethality observed in the EAE rats after the administration of memantine, although not significant, was noticeable but was found to be lower than that of the EAE untreated rats.

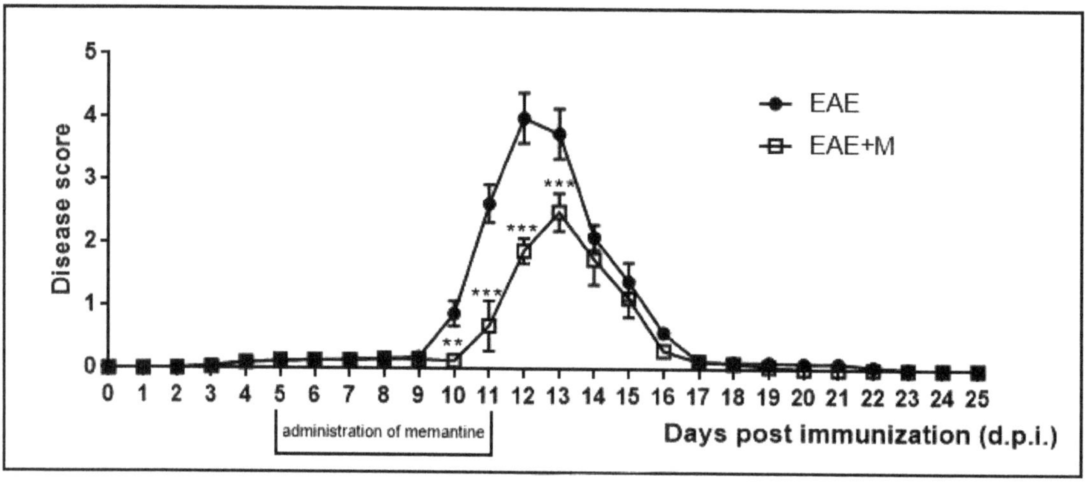

Figure 1. Scores of the neurological symptoms in experimental autoimmune encephalomyelitis (EAE) and memantine-treated EAE rats in different phases of the disease between 0 and 25 days post immunization (d.p.i.). Memantine was administered at a dose of 60 mg/kg b.w./day from 5 to 11 d.p.i. The values indicated neurological scores ± S.D. Results are combined data from 8 animals in each group. ** $p < 0.01$ *** $p < 0.001$ vs. untreated EAE rats (Student's *t*-test).

Table 2. Characteristics of the EAE model and the clinical parameters of EAE rats prior to and after treatment with memantine.

Characteristics/Clinical Parameters	EAE	EAE + Memantine
Animals with clinical symptoms (%)	100	98.44
Animals with severe EAE (%)	76.67	61.11
Lethality (%)	3.33	2.22
Inductive phase (days)	9.5 ± 1.4	11.3 ± 0.8 ***
Cumulative CI (score)	16.93 ± 1.16	9.57 ± 0.67 ***
Duration of disease (days)	17.5 ± 1.4	15.2 ± 1.0 ***
Number of animals	90	90

Administration of the NMDAR antagonist memantine reduced the neurological deficits and improved the condition of the experimental rats during the course of the disease. (CI, cumulative index). The values represent the means ±SD from $n = 64$ animals. *** $p < 0.001$ significantly different vs. experimental autoimmune encephalomyelitis (EAE) untreated rats (Student's *t*-test).

2.2. The Effect of NMDA Receptor Antagonists on the Level of Oxidative Stress Markers

To investigate the involvement of the NMDA receptor antagonist on the processes related to oxidative stress, we analyzed lipid peroxidation by measuring the concentration of a small end-product of oxidized fatty acid degradation, malondialdehyde (MDA). The obtained results showed that the level of MDA significantly increased between 4 and 25 d.p.i. in all experimental groups by about 40–60% compared to the control (Figure 2A). After administration of memantine, MDA levels decreased by about 50–100% between 12 and 25 d.p.i. relative to untreated EAE rats (Figure 2A).

The levels of total -SH groups (Figure 2C), protein -SH groups (Figure 2B), and non-protein -SH groups (Figure 2D) in brain homogenates obtained from untreated EAE rats decreased by about 20%, 30%, and 60%, respectively, compared to the control at 4 d.p.i., and these changes were stable until 25 d.p.i. (Figure 2B–D). The administration of memantine to EAE rats significantly prevented the decrease in non-protein-SH groups (Figure 2D) between 12 and 25 d.p.i. Moreover, it increased the levels of total protein (by about 20%) and protein–SH groups (by about 30%) at 12 d.p.i. (Figure 2B,C).

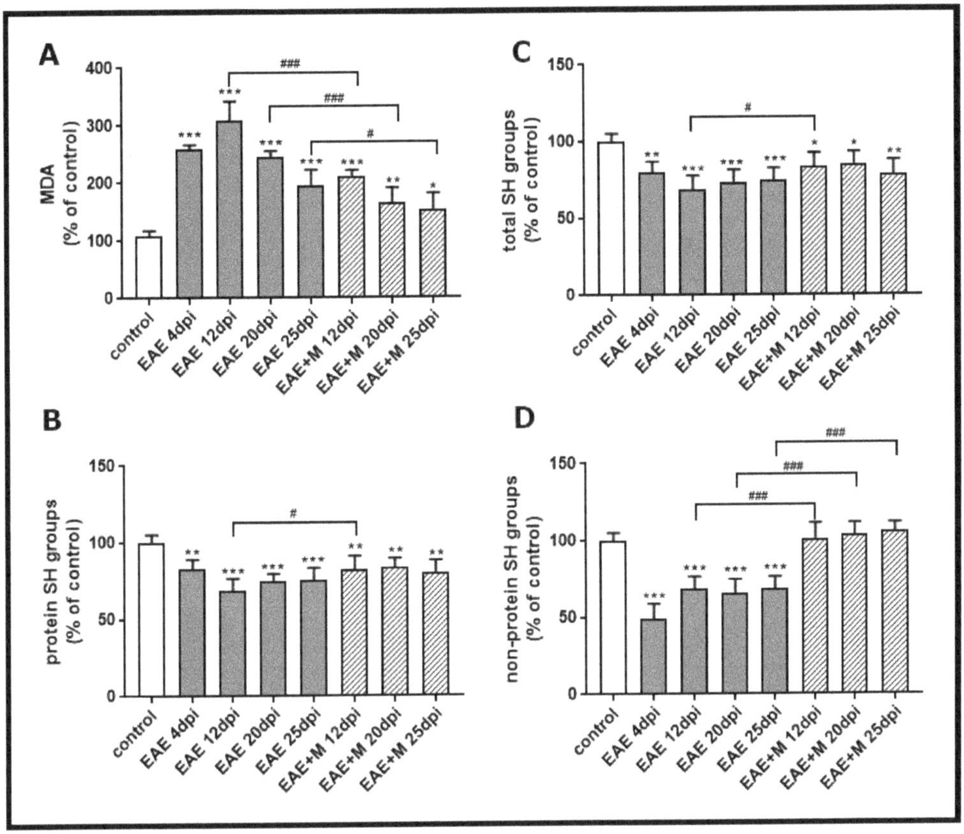

Figure 2. Thiobarbituric acid reactive substances (TBARS) content expressed as the level of malondialdehyde (MDA), a product of membrane lipid peroxidation (**A**), the level of total SH groups (**B**), protein SH groups (**C**), and non-protein SH groups (**D**) in brain homogenate obtained from the control, experimental autoimmune encephalomyelitis (EAE) rats and memantine-treated EAE rats in different phases of the disease. The results are the means from $n = 4$ animals and are expressed as percentages of the control. * $p < 0.05$, ** $p < 0.01$, *** $p < 0.001$ significantly different vs. control rats. # $p < 0.05$, ### $p < 0.01$ significantly different vs. untreated EAE rats (one-way ANOVA with post hoc Bonferroni's test).

2.3. Memantine-Induced Changes in the Expression of SOD Enzymes

To investigate the expression of mRNA coding for of SOD1 and SOD2 during the course of EAE and after treatment with memantine, we performed a qPCR test. Western blot was used to estimate immunoreactivity of SOD1 and SOD2 proteins in brain homogenate obtained from control rats, rats with EAE and memantine-treated rats. A strong positive immunoreaction was observed in a single band near 18 kDa and 25 kDa for SOD1 and SOD2, respectively.

The results revealed a statistically significant increase of around 50% in SOD1 mRNA in EAE rats between 4 and 20 d.p.i. compared to controls. At the end of the experiment (25 d.p.i.), the level of SOD1 mRNA was still above the control value (Figure 3B). After administration of memantine, it was found that EAE-developing animals still exhibited higher SOD1 mRNA expression levels than untreated EAE rats (Figure 3B), while significant upregulation of the enzyme by memantine (by about 20%) was only observed at 12 d.p.i. Identified changes in mRNA expression corresponded to the alterations observed in protein levels. The immunoreactivity of the SOD1 protein in EAE rats remained elevated above control values within the range of 90–110% at 4–20 d.p.i. (Figure 3A). Administration of memantine resulted in the increased expression of SOD1 protein by 10–20% relative to untreated EAE rats at 12–20 d.p.i. The memantine-treated rat model exhibited a 15% decrease in SOD1 protein compared to EAE untreated animals at 25 d.p.i. (Figure 3A).

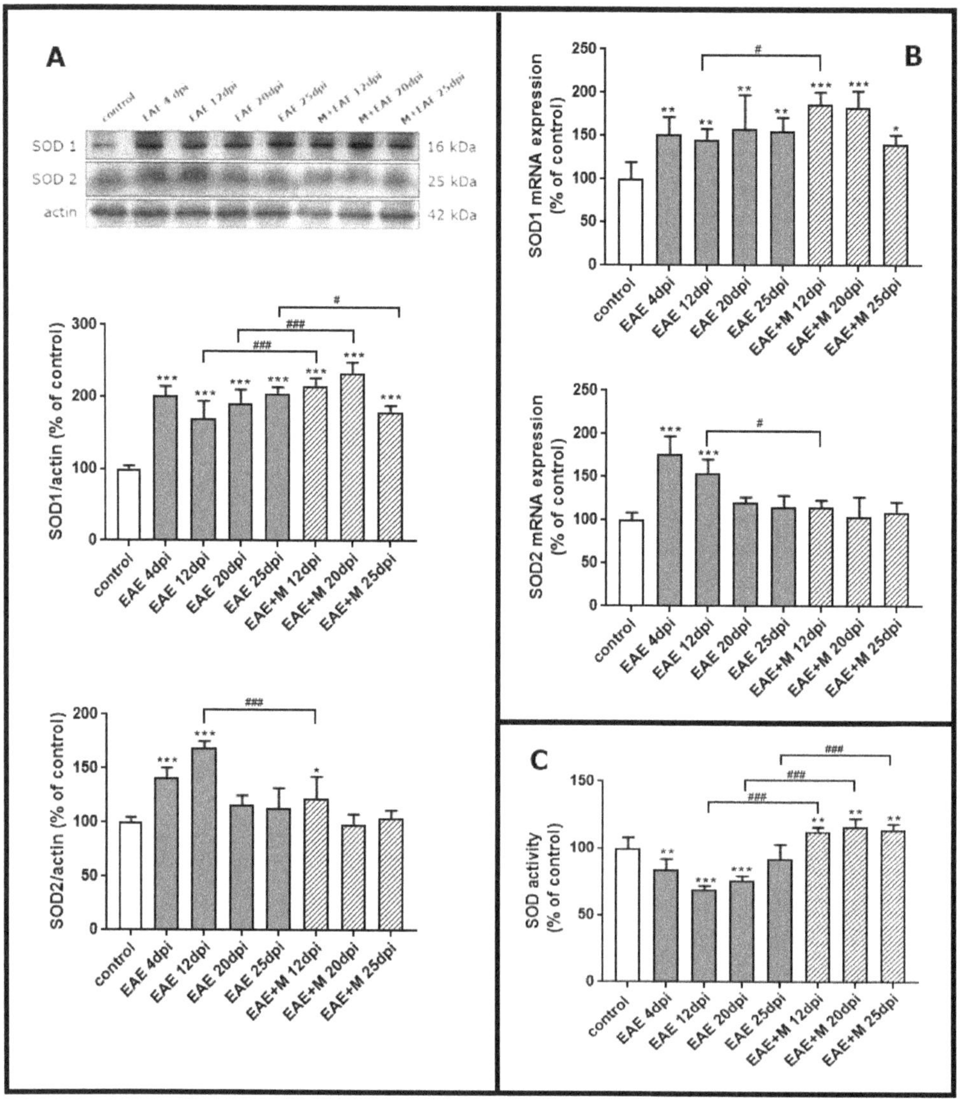

Figure 3. The expression of superoxide dismutases SOD1 and SOD2 in the brain of control and experimental autoimmune encephalomyelitis (EAE) rats at different times post-immunization and after therapeutic treatment with memantine. Representative immunoblots and bars showing the relative levels of enzymatic protein (**A**), bars showing the expression of enzymatic mRNA (**B**), and bars showing the total activity of SODs enzymes (**C**). The results are the means from $n = 4$ animals in each group and are expressed as a percentage of control. * $p < 0.05$; ** $p < 0.01$; *** $p < 0.001$ significantly different vs. control rats; # $p < 0.05$, ### $p < 0.01$ significantly different vs. EAE rats not subjected to therapy. (one-way ANOVA followed by Bonferroni's multiple comparison post-test).

Alterations in the SOD2 mRNA expressions and protein levels were found to be similar to those observed in SOD1 expression only between 4 and 12 d.p.i., reaching an elevation of about 70–80% compared to the control value (Figure 3D). In agreement with this observation, an elevated expression of SOD2 protein was also observed at 4–12 d.p.i., reaching 50–70% of control (Figure 3A). Interestingly, memantine administration significantly reduced both mRNA and SOD2 protein levels to control values at 12 d.p.i. (Figure 3A,B).

2.4. Memantine-Dependent Changes in SOD's Activity

We next investigated whether memantine administration affected specific enzyme activity. As shown in Figure 3C, total SOD activity markedly decreased in EAE rats compared to the control by about 20%, 40%, and 30% at 4, 12, and 20 d.p.i., respectively. In the group of animals treated with memantine, we observed a statistically significant increase in SOD activity of about 60%, 40%, and 20% compared to untreated EAE animals at 12, 20, and 25 d.p.i, respectively, and by about 10% relative to the control group (Figure 3C).

3. Discussion

In the current study, we analyzed the effect of the uncompetitive NMDA receptor antagonist, memantine, on the selective parameters of oxidative stress and neurological deficits in EAE rats. We demonstrated that administration of memantine: (i) significantly decreases the level of MDA relative to untreated EAE rats; (ii) significantly prevents the decrease in non-protein-SH groups and slightly increases the levels of total protein and protein–SH groups at a peak of the disease; (iii) increases expression of SOD1 protein relative to untreated EAE rats; (iv) significantly increases SOD activity compared to untreated EAE animals in all experimental groups; and (v) improves the physiological condition of immunized rats and partially ameliorates clinical symptoms.

Memantine, like other agents that specifically block the pathological stimulation of iGluRs (mainly of the NMDA class), might be expected to restore physiological function of synaptic nerve transmission and produce positive disease-modifying effects. Memantine is a weak, uncompetitive antagonist of the NMDA receptor that has been approved by the European Union and the U.S. Food and Drug Administration (FDA) for the treatment of dementia and Alzheimer's disease. In contrast to other NMDAR channel blockers such as dizocilpine, ketamine, and phencyclidine, which also inhibit NMDA receptors in a noncompetitive manner, memantine does not exhibit serious side effects [31]. Due to its uncompetitive antagonism and relatively fast off-rate, memantine blocks excessive NMDAR activation but allows for low (physiological) levels of NMDAR activity seen during normal neurotransmission. This may be due to the fact that it more effectively inhibits the extra-synaptic than the synaptic subpopulation of the NMDARs. Importantly, memantine binds at the 'intracellular' Mg^{2+} site in the channel pore and displays differential affinity for specific and non-specific binding sites of the NMDAR. These molecular interactions confer upon memantine favorable kinetic properties that contribute to the clinical tolerability of the drug, as well as its neuroprotective profile [32,33].

The potential use of NMDAR antagonists as neuroprotective agents has been established in preclinical studies [32]. Memantine has been found to reduce the NMDA-induced neuronal loss in culture [34] and to be effective in relieving symptoms of MS/EAE [28–30,35]. In our previous [28–30] and present studies, we investigated the neuroprotective effects of memantine in a rat model of MS. The method of administration and optimal therapeutic doses of the drug were selected based on the previously published data [12,35] and our own experience. The drug was administered starting on days 7 to 11 post-immunization, i.e., in an asymptomatic phase of the disease. We observed that memantine effectively reduced the development and duration of neurological deficits and modified all assessed parameters of the disease. The clinical status of treated rats was significantly improved and the severity of their developed neurological deficits was reduced compared to untreated EAE rats. After therapy with memantine, the disease score decreased to 2.2–2.6, while in untreated EAE rats, it was still 4.5. Additionally, the duration of the disease was reduced by about 2–3 days, whereas the inductive phase was prolonged by about 2 days relative to untreated animals.

The contribution of oxidative stress to the mechanisms operating during immune-mediated inflammatory disease has been previously demonstrated [2,5,6,23,27]. Tissue damage by ROS/RNS was an important cytotoxic mechanism of myeloid immune cells [25,36] and the presence of ROS/RNS has consistently been demonstrated in acute and chronic active MS lesions in correlation with the severity of the disease [36]. Within MS brains, active demyelinating lesions, in particular, show signs of ongoing severe OS as demonstrated by the extensive accumulation of oxidized phospholipids, proteins, and DNA, as well as the enhanced expression of antioxidant factors [37]. Our current study indicates that the applied rodent model of EAE replicates key aspects of the contribution of OS to MS/EAE pathology. The results of our study show changes in the level of both OS markers and antioxidant enzymes. The rate of lipid peroxidation (expressed as MDA level) in the brains of all experimental groups was elevated at all stages of the disease, showing a downward trend only at a late stage of EAE (25 d.p.i.), and was consistent with the data showing the high level of lipid peroxidation (LP) at 15 d.p.i. in the CNS of EAE rats [38,39]. In addition, the level of thiol groups, both protein-bound and non-protein-bound, decreased in the brain homogenates of EAE untreated rats from 4 to 25 d.p.i., suggesting the oxidative damage of proteins and a decrease in reduced forms of glutathione, a non-enzymatic antioxidant defense system. The results correspond with those observed in the early stages of EAE in mice, where the decline in GSH of the brain and spinal cord were also noticed [38]. The administration of memantine significantly prevented the oxidative damage of lipids and restored the levels of non-protein-SH groups at all the examined time points, while it significantly increased the concentration of protein-related and total -SH groups only at 12 d.p.i. Parallel to the markers indicating the presence of oxidative damage, we observed the upregulation of the enzymatic antioxidant defense system, represented by SOD enzymes. The expression of SOD1 was noticeably enhanced over the control during the whole course of EAE at both mRNA and protein levels. Interestingly, SOD2 protein and mRNA levels were elevated only in the early (4 d.p.i.) and acute (12 d.p.i.) phases of EAE, but not in the late phase of the disease (20–25 d.p.i.). Although the expression of both enzymes was enhanced, total SOD activity decreased below the control values in the early and late phases of the disease (12–25 d.p.i.), which is consistent with the previously published data [38]. It has been demonstrated that SOD2 plays a critical role in the mitochondrial signaling pathway and is particularly vulnerable to inactivation by ROS and RNS (RNS in particular), which nitrates its tyrosine residues [40]. The appearance of the SOD2 upregulation in our model may represent the compensatory mechanism of cellular systems to the diminished activity. Administration of memantine not only significantly prevents the loss of total SOD activity by reducing NMDA-mediated oxidative stress, but also upregulates the enzymes, particularly SOD1.Świadczy o tym również wzrost aktywności enzymatycznej SOD obserwowany u zwierząt z EAE między 4-20 d.p.i.

Taken together, our results suggest that blocking the NMDAR-mediated overproduction of ROS/RNS partially reduces the oxidative stress-related damage of lipids and proteins by stimulation of the SOD1-relevant antioxidant defense system. Notably, all these changes are accompanied by the improvement of the physiological condition of immunized rats and partial amelioration of EAE clinical symptoms. The current evidence described here highlights the positive effects of memantine that are important from the clinical point of view, suggesting a new therapeutic strategy for MS treatment.

4. Materials and Methods

4.1. Animal Model

Female Lewis rats weighing approximately 190–200 g were used throughout the study. Females are more often used in the EAE model because they are more susceptible to disease induction than males [41], reflecting the tendency observed in humans [1]. All procedures were carried out in accordance with ethical guidelines for the care and use of laboratory animals and were approved by the IV Local Care of Experimental Animal Committee in Warsaw (61/2009). To induce EAE, we immunized rats subcutaneously in both hind feet with an inoculum containing guinea pig spinal cord homogenate emulsified in Freund's complete adjuvant containing 5.5 mg/mL *Mycobacterium tuberculosis* H37Ra (Difco, Detroit, MI, USA).

During the experiment, the rats were housed in environmentally controlled conditions and were provided with free access to food and water. Body weight and neurological deficits were determined daily according to the following scale: 0 = no signs of neurological deficits, 1 = flaccid tail, 2 = impairment of fighting reflex and/or loss of muscle tone in hind limbs, 3 = complete paralysis of hind limbs, 4 = paraplegia, and 5 = a moribund state or death [42–44]. Sham-immunized rats (the control group) received subcutaneous injections of Freund's complete adjuvant containing exclusively *M. tuberculosis* (Difco, Detroit, MI, USA).

Glutamate receptor antagonist memantine was administered at a dose of 60 mg/kg b.w./day. Memantine was dissolved in PBS and administered intraperitoneally to the EAE rats once daily for 7 consecutive days, starting from day 5 post immunization (p.i.) until day 11 p.i. [12].

4.2. Experimental Groups and Tissue Processing

A total of 64 animals were used in the study. The rats were arranged into 8 groups (one control group and seven experimental groups sacrificed at different phases of EAE with different recovery periods after treatment with memantine): group I, control (healthy); group II, EAE 4 d.p.i.; group III, EAE 12 d.p.i.; group IV, EAE 20 d.p.i.; group V, EAE 25 d.p.i.; group VI, EAE 12 d.p.i. + memantine; group VII, EAE 20 d.p.i. + memantine; and group VIII—EAE 25 d.p.i. + memantine. Each group consisted of 8 animals. During the experiments, rats were monitored until days 4, 12, 20, and 25 after the initial injection of EAE-inducing inoculum.

At the respective time points, four rats from each experimental group were sacrificed for biochemical analysis and four rats were sacrificed for immunoblotting and real time PCR analysis. The determination of SOD activity was performed in fresh brain homogenates stored on ice and assayed the same day. The others' brains were quickly removed, frozen in liquid nitrogen and stored at −80 °C for further experiments that included extraction of RNA or preparation of tissue homogenates. To obtain homogenates for immunoblots, the forebrains were homogenized in 50 mM phosphate buffer (pH 7.4) containing 10 mM EGTA, 10 mM EDTA, 0.1 mM PMSF, and 100 mM NaCl in the presence of a protease inhibitor cocktail (1 μg/mL leupeptin, 0.1 μg/mL pepstatin and 1 μg/mL aprotinin).

4.3. Measurement of Lipid Peroxidation

Lipid peroxidation was measured in the brain homogenates using the thiobarbituric acid (TBAR) test, according to the methods reported by Wilbur and modified by Asakawa and Matsushita [45]. This method determines the concentration of malondialdehyde (MDA), which is the most important end-product of lipid peroxidation. Forebrains were homogenized and suspended in Krebs-Ringer buffer (pH = 4.0). The samples were preincubated with 25 μM Fe^{3+}, 800 μMadenosine diphosphate (ADP), and 200 μMascorbate at 30 °C in a water bath. After incubation, 1 mL of 30% TCA, 0.1 mL of 5 M HCl, and 1 mL of 0.75% TBAR were added. The mixture was heated at 100 °C for 15 min in boiling water and centrifuged. The optical density of the supernatant was determined at 535 nm. The molar extinction coefficient ($M = 1.56 \times 10^5\ M^{-1} \cdot cm^{-1}$) was used to calculate the amount of MDA. The results were expressed as percentage of control.

4.4. Measurement of the Level of Sulfhydryl Groups

The level of sulfhydryl (-SH) groups was determined by the method of Sedlak and Lindsay [46]. Briefly, samples of brain homogenates were mixed with 0.2 M Tris buffer, pH 8.2 and 0.1 M dithionitrobenzoic acid (DTNB) to determine total -SH groups. Non-protein SH groups that reflect the content of non-enzymatic antioxidant defense system (glutathione) were estimated after the addition of 50% TCA to each sample. The tubes were centrifuged at 3000× *g* for 10 min. The absorbance of the supernatants was read within 5 min at 412 nm after the addition of 0.4 M Tris buffer (pH = 8.9) and 0.1 M DTNB against a reagent blank. The content of protein-bound SH groups that reflect the status of protein oxidation during OS was calculated as a difference between total and non-protein SH groups.

4.5. Western Blot Analysis

Protein concentrations in the probes were determined according to the method of Lowry [47], using bovine albumin as a standard. Samples containing 50 μg of protein were subjected to SDS-polyacrylamide gel (10%) electrophoresis (Laemmli, 1970, [48]). Samples transferred onto nitrocellulose membranes were incubated overnight (4 °C) with primary monoclonal antibodies anti-SOD1 (1:1000) and anti-SOD2 (1:2000) (Santa Cruz Biotechnology, Dallas, TX, USA) and then with the secondary antibody conjugated with HRP (1:10,000) (Sigma-Aldrich, St. Louis, MO, USA). Polyclonal anti-β actin antibody (Sigma-Aldrich, St. Louis, MO, USA, 1:500) was used as an internal standard. Bands were visualized using the chemiluminescence ECL kit (Amersham, Buckinghamshire, UK), exposed to Hyperfilm ECL, and quantified by densitometric analysis (Image Scanner III (GE Healthcare, LabScan 6.0 Freiburg, Germany) with the Image Quant TL v. 2005 program).

4.6. Determination of SOD's Expression by Quantitative Real Time PCR

Total RNA was extracted from the brain cortex of all experimental groups of animals. Isolation was performed using TRI Reagent (Sigma, St. Louis, MO, USA) according to the method of Chomczyński [49]. RNA (2 μg) was reverse transcribed using random primers and AMV reverse transcriptase (Life Technologies, Forest City, CA, USA). The RT-PCR conditions included reverse transcription at 42 °C for 45 min followed by denaturation at 94 °C for 30 s. TaqMan assays were used for quantitative real time PCR analysis. Rat-specific primers for *SOD1 Rn00566938_m1* and *SOD2 Rn00690588_m1* were used. The probes were obtained from Life Technologies (Forest City, CA, USA). The levels of SODs and actin mRNAs were determined using the TaqMan assay reagents (Life Technologies, Forest City, CA, USA). Quantitative real time PCR (qPCR) analysis was conducted on a Roche LightCycler ® 96 system using 5 μL of RT product, TaqMan PCR Master Mix primers, and a TaqMan probe at a total volume of 20 μL. Cycle conditions for the qPCR were as follows: initial denaturation at 95 °C for 10 min, 45 cycles of 95 °C for 15 s, and 60 °C for 1 min. Each sample was analyzed in triplicate. The relative expression of SODs' mRNA was normalized to actin (Actb) as a reference gene and calculated on the basis of the $_{\Delta\Delta}$Ct method.

4.7. Measurement of SOD's Activity

To obtain homogenates for the measurement of SOD activity, the forebrains were homogenized in cold 20 mM HEPES buffer (pH 7.4) containing 1 mM EGTA and 70 mM sucrose. The homogenates were centrifuged at 10,000× *g* at 4 °C and the supernatants were stored on ice and assayed immediately for SOD activity. The activity of the enzyme was determined with a commercially available assay kit (catalog nr 19160, Sigma St. Louis, MO, USA) according to the manufacturer's instructions. The assay is based on a colorimetric method with a tetrazolium salt that is reduced by superoxide anions (O_2^-) to form a formazan dye. SOD activity is determined indirectly as an inhibition of the production rate of O_2^- by xanthine oxidase at 37 °C by measuring the decrease in absorbance of light at 440 nm. SOD activity was measured against standards, calculated using a standard curve and expressed as units (U)/mg protein.

4.8. Statistical Analysis

The results are expressed as percentages of the control or as the mean ±SD from 3–4 experiments as identified in the legends of the respective figures. Statistical significance was assessed by Student's t-test or a one-way-ANOVA with Bonferroni's or Tukey's multiple comparison tests to identify the changes that were significantly different from the values of the control or untreated EAE rats using GraphPad PRISM software, version 6.0 (San Diego, CA, USA).

Author Contributions: Conceptualization, G.S.; methodology, B.D.-B., M.S.-W. and G.S.; software, B.D.-B., M.S.-W.; validation, B.D.-B., L.S.; formal analysis, B.D.-B., L.S. and G.S.; investigation, B.D.-B., M.S.-W., L.S. and G.S.; resources, L.S.; data curation, G.S.; writing—original draft preparation, G.S.; writing—review and editing, G.S. and L.S.; visualization, B.D.-B.; supervision, G.S.; project administration, G.S.; funding acquisition, L.S. All authors have read and agreed to the published version of the manuscript.

Funding: This research is financed from the statutable funds provided by the Ministry of Education and Science for Mossakowski Medical Research Institute, Polish Academy of Sciences, Warsaw, Poland (9/2021).

Institutional Review Board Statement: All procedures were carried out in accordance with ethical guidelines for the care and use of laboratory animals and were approved by the IV Local Care of Experimental Animal Committee in Warsaw (61/2009).

Informed Consent Statement: Not applicable.

Data Availability Statement: The data presented in this study are available at the Laboratory of Pathoneurochemistry, Department of Neurochemistry; Mossakowski Medical Research Institute, Polish Academy of Sciences.

Acknowledgments: In this section, you can acknowledge any support given which is not covered by the author contribution or funding sections. This may include administrative and technical support, or donations in kind (e.g., materials used for experiments).

Conflicts of Interest: The authors declare no conflict of interest.

References

1. Orton, S.M.; Herrera, B.M.; Yee, I.M.; Valdar, W.; Ramagopalan, S.V.; Sadovnick, A.D.; Ebers, G.C.; Canadian Collaborative Study Group. Sex ratio of multiple sclerosis in Canada: A longitudinal study. *Lancet Neurol.* **2006**, *5*, 932–936. [CrossRef]
2. Groom, A.J.; Smith, T.; Turski, L. Multiple sclerosis and glutamate. *Ann. N. Y. Acad. Sci.* **2003**, *993*, 229–275. [CrossRef]
3. Ponomarev, E.D.; Shriver, L.P.; Maresz, K.; Dittel, B.N. Microglial cell activation and proliferation precedes the onset of CNS autoimmunity. *J. Neurosci. Res.* **2005**, *81*, 374–389. [CrossRef]
4. Cuzner, M.L.; Hayes, G.M.; Newcombe, J.; Wooroofe, M.N. The nature of inflammatory components during demyelination in multiple sclerosis. *J. Neuroimmunol.* **1988**, *20*, 203–209. [CrossRef]
5. Van Horssen, J.; Schreibelt, G.; Drexhage, J.; Hazes, T.; Dijkstra, C.D.; van der Valk, P.; de Vries, H.E. Severe oxidative damage in multiple sclerosis lesions coincides with enhanced antioxidant enzyme expression. *Free Radic. Biol. Med.* **2008**, *45*, 1729–1737. [CrossRef]
6. Gonsette, R.E. Neurodegeneration in multiple sclerosis: The role of oxidative stress and excitotoxicity. *J. Neurol. Sci.* **2008**, *274*, 48–53. [CrossRef] [PubMed]
7. Biber, K.; Laurie, D.J.; Berthele, A.; Sommer, B.; Tölle, T.R.; Gebicke-Härter, P.-J.; Van Calker, D.; Boddeke, H.W.G.M. Expression and Signaling of Group I Metabotropic Glutamate Receptors in Astrocytes and Microglia. *J. Neurochem.* **1999**, *72*, 1671–1680. [CrossRef]
8. Maiese, K.; Chong, Z.Z.; Li, F. Driving Cellular Plasticity and Survival Through the Signal Transduction Pathways of Metabotropic Glutamate Receptors. *Curr. Neurovasc. Res.* **2005**, *2*, 425–446. [CrossRef] [PubMed]
9. Sidoryk-Wegrzynowicz, M.; Struzynska, L. Astroglial and Microglial Purinergic P2X7 Receptor as a Major Contributor to Neuroinflammation during the Course of Multiple Sclerosis. *Int. J. Mol. Sci.* **2021**, *22*, 8404. [CrossRef]
10. Choi, D.W. Calcium and Excitotoxic Neuronal Injury. *Ann. N. Y. Acad. Sci.* **1994**, *747*, 162–171. [CrossRef]
11. Lipton, S.A. Paradigm shift in neuroprotection by NMDA receptor blockade: Memantine and beyond. *Nat. Rev. Drug Discov.* **2006**, *5*, 160–170. [CrossRef] [PubMed]
12. Paul, C.; Bolton, C. Modulation of Blood-Brain Barrier Dysfunction and Neurological Deficits during Acute Experimental Allergic Encephalomyelitis by theN-Methyl-d-aspartate Receptor Antagonist Memantine. *J. Pharmacol. Exp. Ther.* **2002**, *302*, 50–57. [CrossRef] [PubMed]

13. Pitt, D.; Werner, P.; Raine, C.S. Glutamate excitotoxicity in a model of multiple sclerosis. *Nat. Med.* **2000**, *6*, 67–70. [CrossRef]
14. Kumar, P.; Kalonia, H.; Kumar, A. Role of LOX/COX pathways in 3-nitropropionic acid-induced Huntington's Disease-like symptoms in rats: Protective effect of licofelone. *Br. J. Pharmacol.* **2011**, *164*, 644–654. [CrossRef]
15. Lipton, S.A. NMDA receptor activity regulates transcription of antioxidant pathways. *Nat. Neurosci.* **2008**, *11*, 381–382. [CrossRef] [PubMed]
16. Polster, B.M.; Fiskum, G. Mitochondrial mechanisms of neural cell apoptosis. *J. Neurochem.* **2004**, *90*, 1281–1289. [CrossRef]
17. Martindale, J.L.; Holbrook, N.J. Cellular response to oxidative stress: Signaling for suicide and survival. *J. Cell. Physiol.* **2002**, *192*, 1–15. [CrossRef] [PubMed]
18. Chen, H.S.; Pellergini, J.W.; Aggarwal, S.K.; Lei, S.Z.; Warach, S.; Jensen, F.E.; Lipton, S.A. Open-channel block of N-methyl-D-aspartate (NMDA) responses by memantine: Therapeutic advantage against NMDA receptor-mediated neurotoxicity. *J. Neurosci.* **1992**, *12*, 4427–4436. [CrossRef] [PubMed]
19. Joseph, E.K.; Levine, J.D. Caspase signalling in neuropathic and inflammatory pain in the rat. *Eur. J. Neurosci.* **2004**, *20*, 2896–2902. [CrossRef]
20. Marini, A.M.; Ueda, Y.; June, C.H. Intracellular Survival Pathways against Glutamate Receptor Agonist Excitotoxicity in Cultured Neurons: Intracellular Calcium Responses. *Ann. N. Y. Acad. Sci.* **1999**, *890*, 421–437. [CrossRef]
21. Kean, R.B.; Spitsin, S.V.; Mikheeva, T.; Scott, G.S.; Hooper, D.C. The Peroxynitrite Scavenger Uric Acid Prevents Inflammatory Cell Invasion into the Central Nervous System in Experimental Allergic Encephalomyelitis through Maintenance of Blood-Central Nervous System Barrier Integrity. *J. Immunol.* **2000**, *165*, 6511–6518. [CrossRef]
22. Hooper, D.C.; Scott, G.S.; Zborek, A.; Mikheeva, T.; Kean, R.B.; Koprowski, H.; Spitsin, S.V. Uric acid, a peroxynitrite scavenger, inhibits CNS inflammation, blood-CNS barrier permeability changes, and tissue damage in a mouse model of multiple sclerosis. *FASEB J.* **2000**, *14*, 691–698. [CrossRef] [PubMed]
23. Bolton, C.; Paul, C. Glutamate Receptors in Neuroinflammatory Demyelinating Disease. *Mediat. Inflamm.* **2006**, *2006*, 1–12. [CrossRef]
24. Fischer, M.T.; Sharma, R.; Lim, J.L.; Haider, L.; Frischer, J.M.; Drexhage, J.; Mahad, D.; Bradl, M.; Van Horssen, J.; Lassmann, H. NADPH oxidase expression in active multiple sclerosis lesions in relation to oxidative tissue damage and mitochondrial injury. *Brain* **2012**, *135*, 886–899. [CrossRef]
25. Liu, J.S.-H.; Zhao, M.-L.; Brosnan, C.F.; Lee, S.C. Expression of Inducible Nitric Oxide Synthase and Nitrotyrosine in Multiple Sclerosis Lesions. *Am. J. Pathol.* **2001**, *158*, 2057–2066. [CrossRef]
26. LeVine, S.M. The role of reactive oxygen species in the pathogenesis of multiple sclerosis. *Med. Hypotheses* **1992**, *39*, 271–274. [CrossRef]
27. Haider, L.; Fischer, M.T.; Frischer, J.M.; Bauer, J.; Höftberger, R.; Botond, G.; Esterbauer, H.; Binder, C.J.; Witztum, J.L.; Lassmann, H. Oxidative damage in multiple sclerosis lesions. *Brain* **2011**, *134*, 1914–1924. [CrossRef] [PubMed]
28. Sulkowski, G.; Dąbrowska-Bouta, B.; Strużyńska, L. Modulation of Neurological Deficits and Expression of Glutamate Receptors during Experimental Autoimmune Encephalomyelitis after Treatment with Selected Antagonists of Glutamate Receptors. *Biomed Res. Int.* **2013**, *2013*, 186068. [CrossRef] [PubMed]
29. Sulkowski, G.; Dabrowska-Bouta, B.; Chalimoniuk, M.; Struzyńska, L. Effects of antagonists of glutamate receptors on pro-inflammatory cytokines in the brain cortex of rats subjected to experimental autoimmune encephalomyelitis. *J. Neuroimmunol.* **2013**, *261*, 67–76. [CrossRef]
30. Sulkowski, G.; Dąbrowska-Bouta, B.; Salińska, E.; Strużyńska, L. Modulation of Glutamate Transport and Receptor Binding by Glutamate Receptor Antagonists in EAE Rat Brain. *PLoS ONE* **2014**, *9*, 113954. [CrossRef]
31. Czarnecka, K.; Chuchmacz, J.; Wójtowicz, P.; Szymański, P.J. Memantine in neurological disorders—Schizophrenia and depression. *J. Mol. Med.* **2021**, *99*, 327–334. [CrossRef]
32. Chen, H.S.; Lipton, S.A. The chemical biology of clinically tolerated NMDA receptor antagonists. *J. Neurochem.* **2006**, *97*, 1611–1626. [CrossRef] [PubMed]
33. Chen, H.-S.V.; Lipton, S.A. Pharmacological implications of two distinct mechanisms of interaction of memantine with N-methyl-D-aspartate-gated channels. *J. Pharmacol. Exp. Ther.* **2005**, *314*, 961–971. [CrossRef]
34. Volbracht, C.; Van Beek, J.; Zhu, C.; Blomgren, K.; Leist, M. Neuroprotective properties of memantine in different in vitro and in vivo models of excitotoxicity. *Eur. J. Neurosci.* **2006**, *23*, 2611–2622. [CrossRef] [PubMed]
35. Abdurasulova, I.N.; Serdyuk, S.E.; Gmiro, V.E. Comparative study of preventive and therapeutic effects of IEM-1966 and memantine in rats with experimental allergic encephalomyelitis. *Bull. Exp. Biol. Med.* **2007**, *144*, 217–220. [CrossRef]
36. Schuh, C.; Wimmer, I.; Hametner, S.; Haider, L.; Van Dam, A.-M.; Liblau, R.S.; Smith, K.J.; Probert, L.; Binder, C.J.; Bauer, J.; et al. Oxidative tissue injury in multiple sclerosis is only partly reflected in experimental disease models. *Acta Neuropathol.* **2014**, *128*, 247–266. [CrossRef] [PubMed]
37. Lassmann, H.; van Horssen, J.; Mahad, D. Progressive multiple sclerosis: Pathology and pathogenesis. *Nat. Rev. Neurol.* **2012**, *8*, 647–656. [CrossRef]
38. Ljubisavljevic, S.; Stojanovic, I.; Pavlovic, D.; Sokolovic, D.; Stevanovic, I. Aminoguanidine and N-acetyl-cysteine supress oxidative and nitrosative stress in EAE rat brains. *Redox Rep.* **2011**, *16*, 166–172. [CrossRef]

39. Ljubisavljevic, S.; Stojanovic, I.; Pavlovic, D.; Milojkovic, M.; Sokolovic, D.; Stevanovic, I.; Petrovic, A. Suppression of the lipid peroxidation process in the CNS reduces neurological expression of experimentally induced autoimmune encephalomyelitis. *Folia Neuropathol.* **2013**, *51*, 51–57. [CrossRef]
40. MacMillan-Crow, L.A.; Crow, J.P.; Thompson, J.A. Peroxynitrite-Mediated Inactivation of Manganese Superoxide Dismutase Involves Nitration and Oxidation of Critical Tyrosine Residues. *Biochemistry* **1998**, *37*, 1613–1622. [CrossRef] [PubMed]
41. Constantinescu, C.S.; Farooqi, N.; O'Brien, K.; Gran, B. Experimental autoimmune encephalomyelitis (EAE) as a model for multiple sclerosis (MS). *Br. J. Pharmacol.* **2011**, *164*, 1079–1106. [CrossRef]
42. Kerschensteiner, M.; Stadelmann, C.; Buddeberg, B.S.; Merkler, D.; Bareyre, F.M.; Anthony, D.C.; Linington, C.; Brück, W.; Schwab, M.E. Targeting Experimental Autoimmune Encephalomyelitis Lesions to a Predetermined Axonal Tract System Allows for Refined Behavioral Testing in an Animal Model of Multiple Sclerosis. *Am. J. Pathol.* **2004**, *164*, 1455–1469. [CrossRef]
43. Meyer, R.; Weissert, R.; Diem, R.; Storch, M.K.; de Graaf, K.L.; Kramer, B.; Bähr, M. Acute Neuronal Apoptosis in a Rat Model of Multiple Sclerosis. *J. Neurosci.* **2001**, *21*, 6214–6220. [CrossRef]
44. Ohgoh, M.; Hanada, T.; Smith, T.; Hashimoto, T.; Ueno, M.; Yamanishi, Y.; Watanabe, M.; Nishizawa, Y. Altered expression of glutamate transporters in experimental autoimmune encephalomyelitis. *J. Neuroimmunol.* **2002**, *125*, 170–178. [CrossRef]
45. Asakawa, T.; Matsushita, S. Coloring conditions of thiobarbituric acid test for detecting lipid hydroperoxides. *Lipids* **1980**, *15*, 137–140. [CrossRef]
46. Sedlak, J.; Lindsay, R.H. Estimation of total, protein-bound, and nonprotein sulfhydryl groups in tissue with Ellman's reagent. *Anal. Biochem.* **1968**, *25*, 192–205. [CrossRef]
47. Lowry, O.H.; Rosenbrough, N.J.; Farr, A.L.; Randal, R.J. Protein measurement with the Folin phenol reagent. *J. Biol. Chem.* **1951**, *193*, 265–275. [CrossRef]
48. Laemmli, U.K. Cleavage of structural proteins during the assembly of the head of bacteriophage T4. *Nature* **1970**, *227*, 680–685. [CrossRef]
49. Chomczynski, P.; Sacchi, N. Single-step method of RNA isolation by acid guanidinium thiocyanate-phenol-chloroform extraction. *Anal. Biochem.* **1987**, *162*, 156–159. [CrossRef]

International Journal of *Molecular Sciences*

MDPI

Review

Nuclear and Mitochondrial Genome, Epigenome and Gut Microbiome: Emerging Molecular Biomarkers for Parkinson's Disease

Gleyce Fonseca Cabral [1], Ana Paula Schaan [1], Giovanna C. Cavalcante [1], Camille Sena-dos-Santos [1], Tatiane Piedade de Souza [1], Natacha M. Souza Port's [2], Jhully Azevedo dos Santos Pinheiro [1], Ândrea Ribeiro-dos-Santos [1,3,4,*] and Amanda F. Vidal [1,4,5,*]

1 Laboratório de Genética Humana e Médica, Universidade Federal do Pará, R. Augusto Correa, Belém 66075-110, Brazil; cabralffg@gmail.com (G.F.C.); apschaan@gmail.com (A.P.S.); giovannaccavalcante@gmail.com (G.C.C.); camillebiologia@gmail.com (C.S.-d.-S.); tati_souz14@outlook.com (T.P.d.S.); jhully.asp@gmail.com (J.A.d.S.P.)

2 Laboratório de Neurofarmacologia Molecular, Universidade de São Paulo, São Paulo 05508-000, Brazil; natachamsports@gmail.com

3 Núcleo de Pesquisas em Oncologia, Universidade Federal do Pará–R. dos Mundurucus, Belém 66073-000, Brazil

4 Programa de Pós-Graduação em Genética e Biologia Molecular, Universidade Federal do Pará, R. Augusto Correa, Belém 66075-110, Brazil

5 ITVDS—Instituto Tecnológico Vale Desenvolvimento Sustentável–R. Boaventura da Silva, Belém 66055-090, Brazil

* Correspondence: akelyufpa@gmail.com (Â.R.-d.-S.); amandaferreiravidal@gmail.com (A.F.V.); Tel.: +55-(91)-3201-7843 (Â.R.-d.-S.)

Citation: Fonseca Cabral, G.; Schaan, A.P.; Cavalcante, G.C.; Sena-dos-Santos, C.; de Souza, T.P.; Souza Port's, N.M.; dos Santos Pinheiro, J.A.; Ribeiro-dos-Santos, Â.; Vidal, A.F. Nuclear and Mitochondrial Genome, Epigenome and Gut Microbiome: Emerging Molecular Biomarkers for Parkinson's Disease. *Int. J. Mol. Sci.* **2021**, *22*, 9839. https://doi.org/10.3390/ijms22189839

Academic Editors: Marcello Ciaccio and Luisa Agnello

Received: 2 June 2021
Accepted: 28 June 2021
Published: 11 September 2021

Abstract: Background: Parkinson's disease (PD) is currently the second most common neurodegenerative disorder, burdening about 10 million elderly individuals worldwide. The multifactorial nature of PD poses a difficult obstacle for understanding the mechanisms involved in its onset and progression. Currently, diagnosis depends on the appearance of clinical signs, some of which are shared among various neurologic disorders, hindering early diagnosis. There are no effective tools to prevent PD onset, detect the disease in early stages or accurately report the risk of disease progression. Hence, there is an increasing demand for biomarkers that may identify disease onset and progression, as treatment-based medicine may not be the best approach for PD. Over the last few decades, the search for molecular markers to predict susceptibility, aid in accurate diagnosis and evaluate the progress of PD have intensified, but strategies aimed to improve individualized patient care have not yet been established. Conclusions: Genomic variation, regulation by epigenomic mechanisms, as well as the influence of the host gut microbiome seem to have a crucial role in the onset and progress of PD, thus are considered potential biomarkers. As such, the human nuclear and mitochondrial genome, epigenome, and the host gut microbiome might be the key elements to the rise of personalized medicine for PD patients.

Keywords: Parkinson's disease; neurodegeneration; genetics; non-coding RNAs; microbiome; mitochondria; epigenetics; biomarkers; precision medicine

1. Introduction

As life expectancy rises as a result of technological advances, humanity faces an increased burden of aging diseases, such as cancer, diabetes, cardiovascular and neurodegenerative disorders. Degenerative diseases affecting the nervous system are recognized as major causes of death and disabilities among the elderly population worldwide [1]. However, the molecular mechanisms engaged in the onset and progression of neurodegenerative diseases remain elusive. A complete understanding of the molecular biology of neurodegeneration will benefit the search for biomarkers to be employed in strategies for disease detection and patient management, as seen with the efforts being made towards cancer research [2,3].

Among the most common neurodegenerative disorders, Parkinson's Disease (PD) has gained a leading position, preceded only by Alzheimer's Disease [1,4,5], affecting 1% of individuals above 60 years old and 3% of the elderly above 80 years old, and may also rarely affect individuals under 50 years old (early PD and juvenile PD) [4,6,7]. According to recent reports, there may be about 10 million individuals living with PD worldwide, a number that is predicted to multiply three-fold in the next few decades, as the elderly population grows.

Despite all the advances, the diagnosis of PD is based mainly on the observation of classic parkinsonism symptoms, such as muscle rigidness, dyskinesia, and tremor leading to postural imbalance and the investigation of family history of PD [8]. Individuals with PD also present other non-motor symptoms—most of which appear after 40–50% of neuronal loss, including the development of cognitive impairment and Parkinson-related dementia [9–11]. However, most PD patients are diagnosed in late stages, both because of the lack of tools for the evaluation of disease progress risk and the difficulty in differentiating PD from other neurological disorders, since many symptoms of PD overlap with clinical manifestations of other diseases, such as Essential Tremor, Multiple Sclerosis, and Alzheimer's Disease [12].

Currently, the lack of molecular markers to predict susceptibility, accurate diagnosis, and evaluate the progress of PD continues to hinder the establishment of precision medicine strategies. Moreover, it is essential that we consider the findings of multi-omics approaches, which reveal molecular aspects of PD from multiple perspectives and may lead to the establishment of genetic and epigenetic and other circulating markers, which are less invasive, to be used for accurate diagnosis and clinical management of the disease. Here, we discuss findings concerning the identification and validation of potential genetic, epigenetic and microbial biomarkers to enlighten the state-of-the-art in PD molecular biomarker research.

2. Parkinson's Disease

Since James Parkinson's first medical description in the 19th century, PD has been defined as a neurodegenerative disorder caused by progressive and irreversible degeneration of dopaminergic neurons of the substantia nigra (SN) pars compacta (SNpc) [9,13], although Parkinsonism descriptions can be found in earlier fragments [14,15]. PD is the principal cause of Parkinsonism, which describes a syndrome defined by muscular rigidity, resting tremor and bradykinesia [16]. Regarding the dopaminergic neuron degeneration in SNpc, the main feature that characterizes PD pathology is a progressive reduction in levels of dopamine (60–80%), a neurotransmitter involved in various brain functions, such as learning, memory, mood and sleep regulation [9,17,18].

This degenerative loss of neurons in substantia nigra occurs most profoundly in the lateral–ventral tier, which contains neurons that project to the dorsal putamen of the striatum. Therefore, progressive loss of neurons in this brain area explains the major clinical symptoms [19]. Another relevant hallmark in PD is Lewy pathology. The Lewy body is a neuronal inclusion not only found in substantia nigra but also in others brain regions in PD, mainly composed of altered neurofilaments that aggregation [20]. α-synuclein (α-syn) was identified as the main aberrantly folded protein that aggregates to form inclusions, called Lewy bodies, in PD [21].

PD is considered a heterogeneous disease that can progress slowly or quickly, depending on several factors, many of which are still not well understood [22]. It is commonly divided into (a) monogenic or familial PD, which the main cause is related to specific mutations present in key genes within members of the same family; and (b) idiopathic or sporadic PD, found in 85–90% of cases and whose etiology is still unknown [23]. In addition, α-syn is encoded by the *SNCA* gene, in which the first causal missense mutation was identified to be related to the monogenic form of this disease [24].

Therapeutic strategies for many of the disabling motor features are insufficient due to poor response rates to dopaminergic therapies or the development of long-term complications after its use, likely caused by the late appearance of clinical symptoms, in which there is severe damage in SNpc [25]. Thus, personalized and more efficient therapeutic strategies must be developed.

Currently, there are no effective tools to prevent PD onset, as there is no method of detecting the disease in its early stages or that can accurately inform the risk of disease progression. Consequently, further studies are needed to determine molecular factors that can be used as disease-specific biomarkers in preventive medicine (Figure 1). These can assist in the detection of PD before the onset of motor symptoms or in cases where the symptoms presented by the patient are insufficient for an accurate diagnosis. In addition to allowing an increase in the accuracy in the diagnosis process, these biomarkers could also differentiate PD from other forms of parkinsonism and other neurodegenerative diseases [6]. With this, many studies have explored the role of different genetic and epigenetic alterations, as well as alterations in the gut microbiome that may be involved in PD onset and progress, aiming to identify PD biomarkers and novel therapeutic targets.

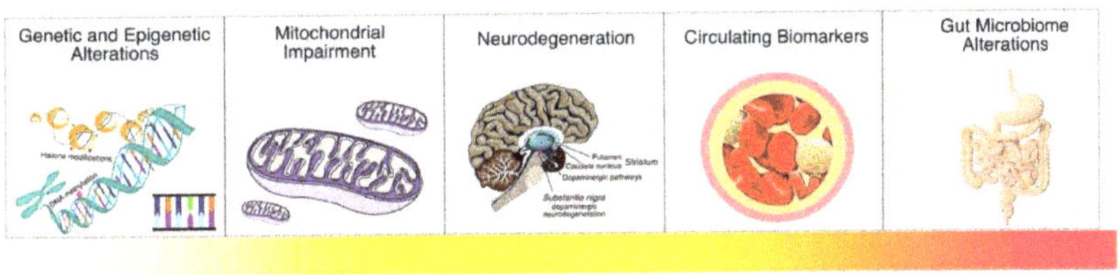

Figure 1. Molecular factors that may be involved in Parkinson's Disease onset and progression. There are various points of view in PD biomarker research, thus, there is a lot to be understood from the earliest molecular alterations until de development of the firsts symptoms. In this review, we decided to discuss these multi-omics factors to highlight how they can be applied for preventive measures.

3. Genetic Alterations in Parkinson's Disease

The characterization of the genetic architecture of PD is essential for understanding the cascade of events that leads to PD onset and to find genetic biomarkers to identify individuals or populations at risk of developing PD and patient prognosis. Hence, over the last few decades, various studies investigated the relationship between monogenic and/or idiopathic PD and gene variants (Table 1). Several genes have been associated with monogenic, sporadic, and both forms of PD. Among these, variants in the *SNCA* gene, which encodes α-syn, are the main genetic factor associated with PD in both etiologies [25,26]. Under pathological conditions, α-syn cannot fulfill its physiological functions, plaques of α-syn aggregates are formed, which may be toxic [25,27]. In addition, α-syn is an inhibitor of tyrosine hydroxylase (TH), an essential enzyme for dopamine production [28–30], and it may also interact with dopamine transporters (DAT), affecting the dopaminergic synapse [31]. Moreover, α-syn may undergo exocytosis and be absorbed by nearby astrocytes and microglia, forming more aggregates, which stimulates the release of pro-inflammatory cytokines and chemokines—such as interleukins, Tumor Necrosis Factor α (TNF-α), and interferon-γ (IFN-γ) [32].

Table 1. Parkinson's disease-associated genes.

Gene	Protein	PD Type	Main Function	Reference
SNCA	α-Synuclein	Monogenic and Sporadic PD	Increase local Ca^{2+} release to enhance ATP-induced exocytosis; Regulation of synaptic vesicle trafficking and neurotransmitter release; Modulation of Dopamine transporter (DAT)	[26] [33] [34] [31]
GBA	Glucocerebrosidase	Sporadic PD	Degradation of complex lipids; Cholesterol metabolism	[35] [36] [37] [38] [39]
LRRK2	Dardarin	Monogenic and Sporadic PD	Cellular response to Dopamine Mitochondrial organization, location Regulation of autophagy Mitochondrial depolarization	[40] [41] [42] [43] [44]
PARK2	Parkin	Familial and Sporadic PD	Mediates ubiquitination to remove and/or detox abnormal folded or damaged proteins	[45] [46]
PARK6	Pink1	Familial and Sporadic PD	Regulation of damaged mitochondrial clearance by mitophagy	[47] [45] [48]
PARK7	DJ-1	Sporadic PD	Mitophagy Response to ROS Regulation of neural apoptosis	[49] [50] [51] [52]
MAPT	TAU	Sporadic PD	Astrocyte activation; Axonal transport of mitochondria Microglial cellular activation Cellular response against ROS Regulation of Autophagy Regulation of Synaptic Plasticity	[53] [54] [55] [56]

Further, α-syn aggregation increases the production of reactive oxygen (ROS) and nitrogen species (RNS) [32,57,58]. Under healthy conditions, ROS and RNS molecules are readily neutralized. However, in neurons affected by advanced neurodegenerative processes, the decline in production and functionality of ROS and RNS neutralizing enzymes—such as Superoxide Dismutase (SOD), Catalase (CAT), Glutathione (GSH), and Glutathione Peroxidase (GPx), intensifies the damage caused by oxidative stress [59,60]. This leads to the development of microenvironmental alterations that affect the entire neuronal circuitry and stimulates immune response through the activation of Toll-like receptors (TLRs) present in the brain microglia [61]. In turn, this initiates a cascade of events that culminates in the activation of NFκb, which mediates the expression of genes related to the inflammatory response, like cytokines, IFN-γ, TNF-α, complement proteins, and other pro-inflammatory mediators that reactivate TLRs, creating a state of chronic inflammation [57,62,63].

Moreover, the literature also shows that α-syn is involved in several mitochondrial functions such as cell respiration, mitochondrial membrane potential (MMP) modulation, and mitochondrial integrity [64], hence, gene variants may also lead to mitochondrial impairment. Additionally, *SNCA* may induce ER stress and intracellular release of Ca^{2+}, leading to mitochondrial dysfunctions, which are the main cause of increased production of ROS and RNS, and activation of pro-apoptotic pathways [65–67].

In fact, several variants in nuclear genes essential for the pathogenesis of PD can lead to mitochondrial dysfunctions. For example, the *PARK2* and *PARK6* genes (which encode parkin and PINK1 proteins, respectively) are essential for mitophagy—a process in which damaged mitochondria are selectively degraded, crucial for functional mitochondria maintenance in senescent cells, such as neurons [68].

The *LRRK2* gene, the most frequent cause of late-onset autosomal dominant and sporadic PD [69], contributes to the recognition of damaged mitochondria by promoting the inhibition of mitochondrial motility [70], also being important for mitochondrial fusion/fission processes [71]. The DJ-1 protein, encoded by the *PARK7* gene, is associated with the recognition and neutralization of ROS and modulates the MMP, being important for the regulation of calcium levels and the stabilization of anti-apoptotic protein Bcl-X [64,72,73]. Variants in the glucocerebrosidase gene (*GBA*) were also found in both familial and idiopathic cases of PD and there is evidence that it may be involved in early-onset PD [39,74,75].

The activation of transposable elements (TE) in brain cells has also been a topic of interest in PD research, since retrotransposons such as *LINE-1*. These ancient viral particles (which were inserted in the human genome throughout the evolution) remain mostly inactive in our genome due to their potential to cause impairments in functional genes, however, they can be reactivated in the aging brain. In fact, it has been suggested that TE insertions are more common in brain cells [76–78]. Their activation leads to cellular mosaicism which is important for neuronal evolution [79]. Moreover, it is suggested that TE retrotransposition may also regulate gene expression in Neuronal Precursor Cells (NPCs) and differentiated neurons [78,80,81]. Despite that, TE insertions can also cause DNA damage and gene expression disruption, and their transcription may trigger microglial antiviral response, activating inflammatory pathways [76,82].

Furthermore, TE recombination events are associated with early-onset PD. Bravo et al., (2018) reported five structural variations in the *PRKN* gene, which is the most frequently mutated gene in this condition, including deletions and exon rearrangements, all of which were caused by recombinatory events triggered by TE reactivation [83]. It was also reported that pathogenic Tau protein leads to a decrease in piRNA-mediated regulation, which increases TE retrotransposition in PD neurons, triggering oxidative stress and cell death [84,85]. Although the mechanisms of TE reactivation in the brain remain elusive, the reported piRNA-mediated regulation of TEs in mammalian brain cells may be an important mechanism of neuroprotection and their deregulation may be involved in the development of neurodegenerative disorders as PD [86].

The investigation of pathogenic variants in PD-related genes is important to characterize the disease at a population level and better understand disease progression and clinical presentations in diverse patient groups, as what has been done for other diseases [87]. Nonetheless, most known variants were identified in populations of Europe, South and East Asia, and North America. For instance, recent studies reported pathogenic variants in *PARK2* and *LRRK2* in Spanish PD patients, and rare variants in *ATP13A2* and *GIGYF2* that may contribute to PD risk on a population scale [88]. Moreover, Wu et al., (2021) [89] showed an association between the p.V16A variant in superoxide dismutase gene (*SOD2*) and PD in Han Chinese individuals, while a large-scale investigation in the UK revealed copy number variants in the *SNCA* gene region in individuals without phenotypic PD manifestation [89,90]. Furthermore, a study performed by the *International Parkinson's Disease Genomics Consortium* showed 24 novel risk loci for PD in the European population [91]; and 17 risk loci for PD in *ITPKB* and *ZNF184* were identified in a study with East Asians [92]. On the other hand, the contribution of Amerindian and/or African Ancestry to the development of PD in populations of Central and South America remains unknown, mostly due to insufficient data. Thus, it is important to highlight the importance of these studies since they may enlighten individual and populational variations in the course of the disease.

Although it is not completely understood, the genetic aspects of PD have been widely explored, in both nuclear and mitochondrial genomics. Recent studies point to the influence of epigenetic aspects in PD onset, as well as how host–microbiome interactions may influence PD progression. These findings are a valuable source of new biomarkers and therapeutic targets that may not only help us understand the mechanisms involved in neurodegeneration, but also provide improvements in clinical practice and patient management through the comprehension of differences and particularities of each patient and population. As highlighted above, a lot of knowledge was elucidated using brain tissue biomarkers. Currently, though, it is important to convert this knowledge into less invasive biomarkers, as circulating molecules altered in tissue and biofluids, which could be useful in the patient management in precision medicine approaches.

4. Epigenetic Alterations in Parkinson's Disease

Advances in sequencing technologies allowed the exploration of the role of alterations not related to the DNA sequence, but associated with mechanisms of gene regulation. The main mechanisms involved in this process are DNA methylation, histone modifications and the regulation performed by non-coding RNAs (ncRNAs). Their modulation is involved in processes such as modulation of cell homeostasis and the onset of human diseases [93]. Among the various mechanisms of epigenetic regulation, the role of ncRNAs has been largely explored in different organs and tissues, including the brain. However, the epigenetic modulation of neurons remains elusive and a controversial issue.

We still do not fully understand how epigenetic modulators are involved in neuron activity, survival and death. Nonetheless, there is evidence that they may be important in healthy conditions and may contribute to the onset and progression of neurodegenerative disorders. These mechanisms may be an important source of PD biomarkers and therapeutic targets, since the epigenetic regulators can be modulated in patients to return to a healthy condition [94–96]. Below, we discuss some of the main findings regarding the role of ncRNAs in neurons and evidence that shows how these molecules may be involved in neurodegeneration and PD development, focusing mainly on microRNAs (miRNAs), Piwi-interacting RNAs (piRNAs) and Circular RNAs (circRNAs).

4.1. Chromatin Remodeling Mechanisms

The mechanisms of DNA methylation and histone modifications regulate the accessibility of DNA strands to transcription factors, enhancers and RNA polymerase, thus interfering directly in the transcription process (Figure 2). The involvement of DNA methylation and histone modifications in PD has been recently reported, suggesting a mechanism of genome reprogramming. Young et al., (2019) reported the alteration of DNA methylation profile in patients with PD, mainly located not only in dopaminergic neurons but also in regions such as the Cingulate gyrus and, especially, the dorsal motor nucleus of the vagus nerve, which is one of the first areas affected in early PD stages [97]. Thus, it is possible that the deregulation in DNA methylation patterns may explain the gastrointestinal symptoms related to the pre-diagnostic stages.

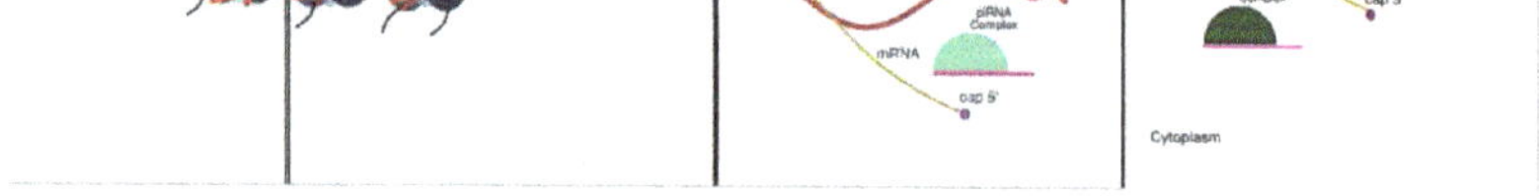

Figure 2. Mechanisms of gene expression regulation. Gene expression is regulated by a network which includes chromatin modulators, with the addition of molecules such as acetyl and/or methyl in histones and the methylation of DNA strands. Additionally, ncRNAs participated in several steps both in transcriptional and post-transcriptional levels. Within the nucleus, ncRNAs may play the role of enhancers, transcription factors or as gene silencing tools. In cytoplasm, they may inhibit RNA translation or modulate the action of other ncRNAs and be used as a sponge of RBP, inducing mRNA degradation.

Alterations in DNA methylation profile were also found in blood and saliva samples of patients living with PD, mainly affecting regions harboring genes related to the immune response, leading to an alteration in blood cells composition [98]. More recently, Wang et al., (2019) found 85 hypomethylated and hypermethylated genes in blood samples of patients with PD, suggesting PD-associated blood biomarkers [99]. It is important to emphasize studies investigating PD biomarkers present in blood and other biofluids, since the sample collection of these fluids is less invasive, therefore, less harmful to the patients, especially those in advanced stages and/or individuals living with dementia.

Södersten et al., (2018) performed an analysis of histone modification patterns in neural progenitor cells and in differentiated dopaminergic and serotoninergic neurons [100]. They observed how the repression of progenitor genes and genes involved in the differentiation of each neuronal subtype is correlated with the distribution of epigenetic landscapes of gene activation and silencing for both adult neuron subtypes. Furthermore, they revealed a stress-induced differential gene expression profile in murine models of PD, which is characterized by the repression of various genes with a dual pattern of activation marks (H3K4me3 and H3K27me3) in promoter regions, especially in promoter regions of genes involved in cell cycle regulation. Toker et al., (2020), also suggested genome-wide alterations in histone acetylation profiles in PD-affected brains, triggering the hyperacetylation of H3K27 in genes such as *SNCA, PARK7, PRKN* and *MAPT*, which are associated with the development of PD [101].

Considering these studies, it is possible to infer that, although not fully understood, differential gene expression caused by alterations in chromatin states may be directly associated with the events involved in the pathogenesis of neurodegenerative disorders, such as the α-syn aggregation observed in PD. Moreover, it means that might the comprehensive map of epigenetic alterations in PD might be a good source of disease-specific biomarkers.

4.2. *Non-Coding RNA Regulation*

4.2.1. MicroRNAs

MicroRNAs (miRNAs) are a class of 21-nucleotide long molecules, distributed in a wide range of organisms [102–104]. They are known for their role in post-translational gene silencing in which they are loaded into Argonaut proteins, forming RNA-induced silencing complexes (miRISC) to mediate the degradation of the target messenger RNA (mRNA). In human cells, most of the canonical miRNAs are transcribed from intronic sequences of coding and non-coding regions, although there are miRNAs encoded from exons [103]. Additionally, many miRNA loci are in close proximity, forming clusters, and miRNAs within the same cluster are usually co-transcribed, although some might be suppressed depending on cellular necessity [103].

Briefly, a mature miRNA binds to the argonaut 2 protein (AGO2), forming a complex called miRISC [102,103]. Here, the miRNA molecule works as a guide to recognition of the target mRNA, meanwhile the AGO2 works as an effector of the endonucleolytic cleavage of the mRy6 NA [102,105]. The recognition of the target mRNA depends on the complementarity between the miRNA binding site—located at the 3′ UTR region of the mRNA, and the seed region of the miRNA [102]. Depending on the strength of this complementarity, the mRNA can (i) be cleaved by the miRISC (in case of the perfect match); or (ii) the complex can mediate translational repression through the recruitment of additional proteins and enzymes that will degrade the mRNA by decapping and deadenylation [106].

This mechanism is part of a strictly regulated network and it was reported that, in neurons, the regulation of miRNAs was related to virtually all neuronal functions, including neurogenesis and neural development [107–109], synaptic plasticity [110,111], neural activities [112,113], and processes related to learning and memory [114]. Therefore, it would be expected that the deregulation of miRNAs would affect these processes, potentially leading to disease. In fact, age-related alterations in miRNA profile were reported in several studies [115,116].

Moreover, the deregulation of miRNAs was reported in several neurodevelopmental and neurodegenerative diseases, such as autism spectrum disorder [117], schizophrenia [118,119], and Huntington's disease [120]. It also has been linked to the pathogenesis and progression of Alzheimer's disease (AD) [121,122] and PD [123].

Several studies have demonstrated the potential of miRNAs for acting as PD biomarkers and therapeutic targets given their role in regulating genes classically associated with PD pathogenesis. Cardo et al., (2014) revealed a set of 11 miRNAs in the SNpc of PD patients [124]. Kim et al., (2007) reported that miR-133b was specifically expressed in midbrain dopaminergic neurons, associated with cell survival, and that PD patients presented hypoexpression of this miRNA [125]. MiR-7, miR-153 and miR-433 were shown to confer neuroprotective effects by regulating the levels of α-syn, and downregulation of these miRNAs was associated with *SNCA* overexpression and increased oxidative stress [126–128]. In PD models, Thome et al., (2016) [129] showed that miR-155 overexpression was associated with proinflammatory responses to α-syn aggregation. As α-syn aggregates are one of the various sources of oxidative stress in PD, these findings highlight the importance of further investigating the role of miRNAs in this disease.

Besides α-syn, it was demonstrated that miR-205 regulates the expression of *LRRK2* contributing to a pathogenic increase of this gene in PD patients [130] and Gehrke et al., (2010) demonstrated a bidirectional regulation between miRNAs and *LRRK2*, where pathogenic variants of the gene modulated the levels of miR-let-7 and miR-184 [131]. In addition, Xiong et al., (2014) [132] showed that miR-494 is directly involved in PD pathogenesis by targeting *DJ-1* and that the upregulation of the miRNA leads to hypoexpression of the gene. Furthermore, in a recent study, miR-221 levels were demonstrated to be modulated by the expression of *DJ-1*, a function lost in pathogenic mutants [133], and miR-221 showed a neuroprotective effect against the exposure to the dopaminergic neurotoxin 1-methyl-4-phenylpyridinium (MPP+) and the inhibition of the miRNA reduced the cell survival response against the MPP+-induced neurotoxicity.

It is important to emphasize that disease-induced deregulation of miRNAs (and some other ncRNAs) can be detected in patient blood circulation. Blood-based biomarkers are important for their applicability in clinical practice, for tracing of individuals at risk (especially among patient relatives), early diagnosis, and prognosis. Furthermore, in the case of neurodegenerative disorders, it is common to observe the overlapping of symptoms leading to additional complexity in identifying the disease [12,134]. Hence, circulating biomarkers may be helpful for differential diagnosis among neurological diseases, and the deregulation of ncRNAs such as miRNAs is a potential source of these biomarkers. For instance, mir-7 controls dopaminergic neuron and oligodendroglial cell fate by regulating genes of Wnt/β-catenin signaling [135]. Moreover, Zhang et al., (2019) observed that mir-let-7a can suppress neuroinflammation caused by microglia activation in PD patients [136]. Moreover, research of miRNA expression profiles identified apoptotic, inflammatory and axonal guidance pathways shared by PD and Alzheimer's disease [137]. Another study revealed a unique molecular signature based on the deregulation of miRNAs and piRNAs in Alzheimer's disease [121]. To our knowledge, this was the first study to investigate both small ncRNA classes in these diseases, and here we suggest that this approach could also be used to aid in PD diagnosis and patient management.

Recent studies have investigated the functional roles of miRNAs and other ncRNAs in neurodegeneration, exploring aspects such as the expression profile of miRNAs and potential target genes, thus indicating the biological processes affected by the deregulation of these molecules. One of the main points that remain to be elucidated is the identification of miRNA species that are shared among neurodegenerative disorders and disease-specific features miRNAs.

4.2.2. Piwi-Interacting RNAs

PIWI-interacting RNAs (piRNAs) are molecules with up to 31 nucleotides in length that were identified in 2006, whose sequences are, generally, not conserved [138–140].

Currently, there are at least 30,000 piRNAs described in the human genome and they are suggested to be the most diverse class of small ncRNAs [141]. PiRNAs bind to proteins of the Argonaut/PIWI subfamily—in humans called Piwi-like (Piwil) proteins, forming regulatory complexes (piRISC) that perform (i) post-transcriptional gene silencing by mediating the deadenylation or endonucleolytic cleavages of the mRNA; and (ii) transcriptional silencing by inducing chromatin remodeling by interacting with proteins involved in this process [142–144].

They are highly abundant and mostly described in germline cells—especially in models such as *Drosophila melanogaster*, *Mus musculus*, and *Caenorhabditis elegans*. The first and main function attributed to piRNAs was the silencing of TE, which is the reason why they are known as the guardians of genome integrity [143,145]. However, not all piRNAs target TE sequences, indicating that piRNA regulation might be more complex than what was first thought [146]. Further studies in germline and somatic cells revealed other functions, including the euchromatin remodeling and epigenetic reprogramming [147,148], modulation of mRNA stability [140], and regulation of the translation of protein-encoding genes, sequences of other ncRNAs (such as long non-coding RNA—lncRNAs), and pseudogenes [149–151].

Currently, there are several reports concerning their role in the progression of several diseases, especially cancer. The deregulation of piRNAs was reported in various tumors such as gastric [2,152], colorectal [153], and breast [154] cancers, being associated with biological processes that contribute to tumor progression and metastasis. Meanwhile, their functions in the central nervous system (CNS) and their roles in the development of neurodegenerative diseases remain poorly understood. Among somatic cells, piRNAs are slightly more abundant in the brain [155,156]. Rouget et al., (2010) observed that impaired piRNA regulation of the mRNA encoding the posterior morphogen Nanos (Nos) in *D. melanogaster* early embryo leads to head development malformations [157]. Studies using MILI-piRNA mutant mice showed that piRNA pathways may act in locomotory and exploratory drives and normal anxiety-like behavior [86].

Notably, the mechanisms of piRNAs have been investigated in different organisms. For instance, Lee et al., (2011) reported the presence of piRNAs in murine brain and hippocampus cells, where they were associated with the regulation of genes related to the dendritic spine development and the organization of postsynaptic density [158]. In addition, Yan et al., (2011) reported the presence of piRNAs in neurons of Rhesus macaque (*Macaca mulatta*) [159]. Rajasethupathy et al., (2012) suggested that, in cells of the sea worm of genus *Aplysia*, neuron-specific piRNAs were associated with neuronal plasticity, learning, and the establishment of long-term memories; the synaptic release of the serotonin neurotransmitter induces the neurotransmitter-dependent activation of piRNAs and, in response, serotonin-induced piRNAs modulate DNA methylation patterns in the promoter region of the gene encoding the activating transcription factor 4 (CREB2) [151]. The CREB protein family is composed of transcription factors mediating cAMP responses, associated with neuroprotective pathways, related to neuronal plasticity, neurogenesis and memory formation [160,161]. CREB2 is also an antagonist of CREB1, which is modulated by miRNAs, and the modulation of CREB2 and CREB1 neurons may signalize which neurons hold memory traces and which are susceptible to draw new memories [151].

The interplay among piRNA and other ncRNAs was demonstrated to be important for regulating the permeability of the blood–brain barrier (BBB) in patients diagnosed with glioma. As revealed by Shen et al., (2018), the deregulation of the piR-DQ593109/Piwil1 complex leads to the upregulation of lncRNA maternally expressed 3 (MEG3)—which sponges the miRNA miR-330-5p [162]. Therefore, MEG3 upregulation leads to an increased hijacking of this miRNA. Meanwhile, miR-330-5p targets the Runt-related transcription factor (*RUNX3*), which binds to the promoter region of genes encoding the proteins occludin, claudin-5, and Zonula Occludens-1 (ZO-1), associated with the enhancement BBB permeability. Hence, the downregulation of miR-330-5p contributes to increased BBB permeability by promoting the upregulation of *RUNX3*, and consequently causing the hyperexpression of ZO-1, occludin and claudin-5 [162].

There is evidence linking piRNAs to the pathological features of multiple CNS-related diseases. For instance, piRNA deregulation was reported in studies involving Ischemia, Rett, Alzheimer's disease, and PD [155,163–166]. Additionally, in neurons of *C. elegans*, the piRNA pathway was related to the inhibition of axonal regeneration via post-transcriptional regulation of the target genes, suggesting that understanding piRNA biology in the brain may be relevant to the treatment of neuronal injuries [167].

It is important to consider the roles of piRNAs in aging and in the progression of neurodegenerative diseases. Differentially expressed (DE) piRNAs play important roles in apoptotic cell death and neurodegeneration in Alzheimer's Disease [164]. Alterations in the expression profile of piRNAs were also reported among cell lines of healthy and sporadic PD patients, as revealed by Schulzer et al., (2018) [163]. In this study, piRNA expression profile was able to distinguish cell lines of (i) fibroblasts; (ii) fibroblast-derived induced pluripotent stem cells (iPSCs); and (iii) midbrain neurons, from healthy individuals and PD patients. The authors revealed a PD-specific piRNA signature in all three cell lines, as well as the presence of Piwil2 and Piwil4 proteins in midbrain neurons. Among the downregulated piRNA set, there was an enrichment of Line-derived piRNAs, indicating the activation of these TEs. They also observed that CREB pathways were impaired in PD patients [163]. As discussed above, CREB pathways are involved in neuroprotective processes and some elements of this pathway are regulated by piRNAs in response to neurotransmitter activities [151,160,161]. Therefore, piRNAs might play an important role in PD development, although more studies are needed to further comprehend this process.

Moreover, recent studies have reported the occurrence of polymorphisms lying within piRNA genes, although the effects of these variants are still not well understood, especially when related to neurodegenerative diseases, since most studies are focused on cancer-related polymorphisms [155,164]. Single-Nucleotide Polymorphisms (SNPs) in piRNA genes were associated with increased risk for breast cancer [168], melanoma [169], and glioma [170] development, suggesting a probable impact of SNPs in piRNA functions in the CNS.

Finally, as mitochondrial dysfunctions are believed to play key roles in PD progression, the discovery of mitochondrial piRNAs in cancer cells [171–173] derived from genes involved in the oxidative phosphorylation chain, indicate a more complex regulatory network involving mitochondrial DNA (mtDNA), and is one more reason we should consider investigating the functional roles of piRNAs in healthy CNS and in disease onset.

4.2.3. Long-Noncoding RNAs

Long non-coding RNAs (lncRNA) are the most studied class of ncRNAs. The term is commonly used to refer to linear RNA molecules with more than 200 nucleotides that may be involved in various processes such as DNA translation—acting as enhancers, protein scaffolds, guides for transcription factors, regulating pre-mRNA splicing and other processes [174,175]. Thus, it is suggested that lncRNAs can be involved in several mechanisms of disease onset, as reported in oncology studies. However, regarding PD, few studies associated the deregulation of lncRNAs with PD progression by regulating miRNA repression in neurons. Taurine Upregulated Gene 1 (*TUG1*) was associated with PD by regulating the miR-152-3p/PTEN pathway [176]. LncRNA GAS5 acts as a sponge of miR-223-3p, promoting microglial inflammatory response in PD by regulating inflammatory pathways [177].

LncRNAs also have a potential effect on PD onset and progression for being associated with damage and inflammatory responses in microglia and dopaminergic neurons. This is the case of lncRNA AK039862, for which upregulation was associated with the neuronal injury provoked by pesticides, inhibiting dopaminergic neuron proliferation and microglia migration [178]. NEAT1 is another lncRNA that is overexpressed in PD patients. Several studies demonstrated that its downregulation represses α-syn-induced activation of NLRP3 inflammation, apoptosis, and cytotoxicity by targeting miR-1301-3p, the miR-212-5p/RAB3IP axis, miR-124-3p and miR-212-3p [179–181]. The deregulation of NEAT1 downregulates microRNA-212-3p to accelerate the progression of PD. The knockdown of NEAT1 negatively affects the expression of *AXIN1*, a target of miR-212_3p reversing the suppression effect [179], meaning that the deregulation of this lncRNA is associated with PD development, characterizing it as a candidate therapeutic target. Furthermore, increased levels of NEAT1 were also detected in peripheral blood cells of patients with PD [182], revealing its potential as a PD progression biomarker.

Regarding lncRNAs in PD patient biofluids, the downregulation of lncRNAs MEG3 was reported as a potential PD biomarker in the plasma of PD patients when compared to that of healthy individuals, being associated with the aggravation of non-motor symptoms, cognitive decline, and PD stage [183]. Additionally, lncRNAs and α-syn were detected in plasma L1CAM exosomes and, combined with β-Glucocerebrosidase activity in plasma, they were correlated with the increase in motor/cognitive impairment [184].

In murine models of PD, lncRNAs decreased during PD development and were further decreased after the administration of L-DOPA therapy [185], suggesting that this set of lncRNAs were associated with both PD and Levodopa-induced dyskinesia (LID) pathogenesis, the latter is a common complication of the chronic dopamine replacement therapy. Gene Ontology analysis revealed the involvement of lncRNAs in processes such as oxidative stress response, inflammation, neurotransmission and apoptosis [185].

4.2.4. Circular RNAs

Circular RNAs (circRNAs) are a class of long ncRNAs formed by the covalent bonding of their 5′-3′ ends . The lack of free terminations makes them resistant to degradation by exonucleases and, therefore, highly stable [186,187]. The biogenesis of circRNAs occurs in parallel with mRNA transcription in a process called back-splicing, processed with the spliceosome machinery, cis sequences in flanking introns, and some specific regulatory proteins [188,189].

About 183.000 circRNAs have been identified in humans [188,190,191]. They function by regulating gene expression mainly at the transcriptional and post-transcriptional levels, can act as microRNA or protein scaffolds and some can be translated [187,192]. Of all the known circRNAs, about 20% are produced by genes of neural function, making circRNAs significantly more abundant in nervous tissue compared to other tissues [187]. This can be explained, in part, by the high number of exons and long introns present in neural genes, making them more likely to be back-spliced [193]. In addition, studies report that the levels of circRNAs, produced in the CNS and other tissues are higher in the brain, suggesting specific neuronal regulation [193,194]. In the CNS, most circRNAs are produced by synaptic function genes and are transported from the cell body to the dendrites, indicating a potential role in neuronal communication [187]. As they are more stable than linear RNAs and extremely abundant in the CNS, it is suggested that circRNAs may serve as potential biomarkers of neurological disorders [195].

Some studies have demonstrated a relationship between changes in circRNA expression and the development of PD [187,196,197]. Among these circRNAs, ciRS-7, which stands out as the most abundant and the most studied in the CNS, is highly expressed in neurons. It has 73 binding sites for miR-7, working as a sponge for this microRNA that regulates several genes, including *SNCA* [187,197]. Cells transfected with miR-7 impair α-syn production, which is restored with the overexpression of ciRS-7 [128,197]. Another circRNA involved in the regulation of miR-7 is circSNCA. Sang et al., (2018) [196] demonstrated that the increased expression of this circRNA leads to increased production of α-syn.

Another circRNA, CircDLGAP4 has shown a key role in the PD pathogenesis, as its expression was significantly reduced and in 1-methyl-4-phenyl-1,2,5,6-tetrahydropyridine (MPTP+) cell cultures and in animal models with MPTP-induced PD. Among its functions, CircDLGAP4 can reduce mitochondrial damage and apoptosis, attenuate the neurotoxic effect of MPP+, and increase autophagy and cell viability [198]. In silico investigations suggested the regulation of miR-124-5p by the circDLGAP4 [198]. This miRNA directly targets genes involved in the CREB pathway, which is essential for the expression of neuroprotective factors, such as BDNF, BCL-2, and PGC-1α. Feng et al., (2019) demonstrated in MPTP-induced PD cellular and animal models that a reduction in circDLGAP4 expression leads to an increased expression of miR-134-5p. This suggests that the circDLGAP4/miR-134-5p/CREB axis has a key role in PD onset and progression [198].

Hanan et al., (2020) [199] stated that in PD brain tissues, the SNpc had a lower number of expressed circRNAs when compared to other brain regions. In contrast, circSLC8A1 was significantly more expressed in PD. They also observed that the increase in oxidative stress leads to higher levels of this circRNA. Despite this, there was no change in the expression of host mRNA SLC8A1, previously associated with neurodegeneration [199]. Data suggest that alterations of circSLC8A1 expression occur due to either increased circularization or deregulation and consequent reduced degradation [199,200]. CircSLC8A1 has seven miR-128 binding sites, participating in the regulation of this microRNA. After circSLC8A1 knockdown, 24 miR-128 target mRNAs were differentially expressed, including neurodegenerative regulators, dopaminergic neuron protectors, regulators of mitochondrial function, and mRNAs involved in chronic inflammation [201,202]. Recent studies have focused on evaluating the roles of circRNAs in CNS disorders. Despite that, few studies focus on the role of circRNAs in PD [193,194].

5. Mitochondrial Genetics and Epigenetics in Parkinson's Disease

Mitochondria are cytoplasmic organelles that are involved in essential processes for proper cellular functioning, such as calcium (Ca^{2+}) buffering, regulation of cell death, lipid homeostasis, among other metabolic signals [203,204], but their most notable function is the generation of chemical energy (ATP, adenosine triphosphate) through the electron transport chain (ETC) in the process of oxidative phosphorylation (OXPHOS) [173,204,205].

The mtDNA encodes 13 proteins for OXPHOS complexes and 24 encode for RNA molecules (22 tRNAs and two rRNAs), besides containing non-coding regions such as the displacement loop (D-loop), where the sequences for the initiation of replication and transcription are located [206]. Mitochondria also need 1200–1400 nuclei-encoded proteins for their overall functioning, a process controlled by mitonuclear communication, which is crucial for the performance of both parties and the consequent cellular balance, including energy generation [207,208]. For instance, the ETC comprises five protein complexes (I-V), of which four (complexes I, III, IV and V) are encoded by both the mitochondrial genome (mtGenome) and the nuclear genome, and one (complex II) is encoded exclusively by the nuclear genome [205,209].

Remarkably, most ATP is produced during OXPHOS and, thus, mitochondrial dysfunctions would greatly affect the energy supply. These dysfunctions may be caused by genetic or epigenetic alterations. In this sense, it should be noted that reactive oxygen species (ROS) are also generated during OXPHOS; in excess, ROS may lead to cellular damages and oxidative stress, with a special impact on mitochondria. Therefore, mtDNA is particularly vulnerable to oxidative damage and mutations, which in turn may affect mitochondrial bioenergetics and increase ROS production [203,210]. Considering that neurons demand high ATP rates, mitochondrial malfunction might have a particular influence on the energy generation of these cells, impairing neural circuitry homeostasis, neurotransmission and neuroplasticity [211–213].

Mitochondria started to be associated with the etiology of PD from the observation of animal models developing parkinsonism after treatment with (MPTP) [214] and, in the following years, with other drugs such as rotenone, all potent neurotoxins that are now known to inhibit complex I activities in the SNpc [215]. Indeed, defects in complex I were identified in different tissues of PD patients, including the SN of postmortem brains [216], skeletal muscles [217] and platelets [218]. Interestingly, the transfection of platelet mtDNA from PD patients into normal cell lines caused the reduction of complex I and IV activity in the receiving cells [219]. While malfunction of Complex I seems to have an important role in the increase of neurotoxic vulnerability, oxidative stress and consequent loss of dopaminergic neurons, other mitochondrial mechanisms have also been pointed out as PD-related factors, such as calcium regulation, biogenesis, dynamics and mitophagy [220].

Currently, there is an increasing interest in the possibility of mtDNA variants predisposing to idiopathic PD [221–223]. Several studies have suggested an association of SNPs in mitochondrial genes with an enhanced risk for PD, such as 4216T>C in *MT-ND1* [224], 5460G>A in *MT-ND2*, as well as 4336T>C [225,226] and 4336A>G in *MT-TQ* [227]. Whole mtGenome sequencing of PD patients revealed an accumulation of mtDNA deletions in SN dopaminergic neurons [228] and heteroplasmic variants in genes of complex III (*CYTB*) and complex IV (*COXI* and *COXII*) in SN and frontal cortex tissues [229].

Although defects in complex I have been extensively shown as important etiological factors for idiopathic PD, the sequencing of whole mtGenome [228,229] or specific complex I genes [226] has failed to find variants that explain such defects and only a few studies have found mitochondrial SNPs associated with the risk of PD [224,230]. Interestingly, in different studies, the variants 10398A>G (*MT-ND3*) [231], 2158T>C (*MT-RNR2*) and 11251A>G (*MT-ND4*) [232], related to the classification of some European mitochondrial haplogroups, were associated with reduced risk to PD. Considering these studies together, haplogroups JT, T, J or K seem to present a decrease in the risk of PD in comparison to other European haplogroups (H or HV).

Dysfunctional mitochondria are also related to monogenic forms of PD: genetic methods have identified variants in mitophagy regulatory genes—e.g., *PINK1* (PTEN-induced putative protein kinase 1), *Parkin* (*PRKN*, parkin RBR E3 ubiquitin protein ligase) and *DJ-1* (protein deglycase DJ-1)—as hereditary factors of PD etiology, being associated with early-onset autosomal recessive PD (age less than 45 years) through alterations in mitophagy [233,234]. In order to maintain mitochondrial homeostasis in the nervous system, mitophagy (the process of autophagic degradation of damaged mitochondria) may take place, mainly through the PINK1/Parkin pathway [223,235]. Thus, considering mitophagy is part of mitochondrial quality control, variants in genes related to this mechanism may impair the elimination of dysfunctional mitochondria in the brain of PD patients [236].

Over 20 years ago, the *Parkin* gene was first described and associated with autosomal recessive juvenile parkinsonism in Japanese families (being named *PARK2*) [237]. A few years later, independent studies have mapped novel loci and associated them with early-onset PD: one in an Italian family [238] and the other in the Netherlands [239], being named *PARK6* and *PARK7*, which were later related to *PINK1* and *DJ-1* genes, respectively. Since then, many studies have suggested the association of genetic factors in mitophagy with PD. Recently, a review by Cai and Jeong (2020) [240] highlighted that defects in mitophagy have been widely associated with PD (especially in the PINK1/Parkin pathway but also in others like lipid-mediated mitophagy) and that increased rates of mtDNA deletions have been observed in PD patients, which could be related to dysfunctions in this process of mitochondrial quality control.

In addition to mtDNA variants, epigenetic mechanisms such as DNA methylation and the presence of ncRNAs have also been described in these organelles, giving birth to a research field that is now called mitochondrial epigenetics (mitoepigenetics) [173]. Since the 1970s, the occurrence of mtDNA methylation has been widely discussed and considered to be controversial; currently, not only is this type of epigenetic regulation recognized in mitochondria, but it is known that the metabolism of these organelles is involved in regulating the production of the universal methyl donor (S-adenosylmethionine, SAM) [241,242]. In addition, a study investigating mtDNA methylation in different brain tissues demonstrated that such patterns in the mitochondrial epigenome may vary among tissues [243]. Recently, a study reported genome-wide mtDNA methylation in different cell lines and tissue samples, providing patterns in both CpG and non-CpG contexts (but especially in non-CpG sites) and suggesting the involvement of varied DNA-methyltransferases (DNMT1, DNMT3A and DNMT3B) in this mitochondrial process [244].

As for ncRNAs, their discovery in mitochondria is more recent than mtDNA methylation. In 2011, a study reported the first human mitochondrial transcriptome and reinforced some ncRNAs shown in mitochondria a few years earlier, highlighting novel levels of complexity in mitochondrial regulation and encouraging the investigation of ncRNAs in these organelles [245]. Since then, different classes of ncRNAs have been described in human mitochondria (circRNAs, miRNAs, piRNAs, lncRNAs and sncRNAs), as recently reviewed by Cavalcante et al., (2020) [173]. These ncRNAs in mitochondria are known to participate in mitonuclear communication and may be either nuclear-encoded ncRNAs (nuclear-ncRNAs) or mitochondrial-encoded ncRNAs (mt-ncRNAs), involved in anterograde signaling (nucleus regulating mitochondria) or retrograde signaling (mitochondria regulating nucleus), respectively. Although many aspects of these mechanisms remain largely unknown, several mitochondrial-located lncRNAs and miRNAs have presented target genes involved in different mitochondrial functions [246].

In this context, epigenetic mechanisms in mitochondria have been associated with neurodegenerative diseases such as PD. For instance, reduced levels of mtDNA methylation (5-methylcytosine, 5mC) were observed in the D-loop region (but not in the other analyzed region, *MT-ND6*, which encodes a subunit of Complex I) in the SNpc of PD patients when compared to controls [247]. It should be noted that the differential methylation in nuclear genes related to mitochondrial apoptosis in association with cognitive and motor progression in PD was also reported, reinforcing mitochondrial influence on the evolution of this disease [248]. Similarly, a recent review by Lyu et al., (2019) [249] highlighted that mitochondrial-related lncRNAs might also play a role in PD through mitochondrial dysfunction or abnormalities in processes related to the management of oxidative stress, like mitophagy and apoptosis. Although there are currently few studies on this matter, it is increasingly clear that mitochondrial epigenetics may play an important role in the development and progression of PD and should be given more attention. Therefore, the investigation of different aspects in both mitochondrial genetics and epigenetics is presented as a promising strategy in the search for biomarkers of onset and treatment of PD, known to be a multifactorial disease.

6. Gut Microbiome in Parkinson's Disease

The gut microbiome (GM) consists of trillions of microorganisms that inhabit the human gastrointestinal tract and constantly interact with the host at multiple genetic and metabolic levels [250]. This community is known to have a strong effect on human health and its disruption has been associated with the development of various human diseases, such as inflammatory bowel disease, metabolic, immune, and neurodegenerative disorders [251,252]. The crosstalk between the microbiome, the gut and the nervous system is collectively referred to as the microbiota–gut–brain axis [253].

In Braak and colleagues' dual-hit hypothesis regarding the staging of LB accumulation in neuronal cells [254], the enteric nervous system (ENS) is one of the sites in which neurodegeneration begins. In this scenario, α-synuclein toxic forms would initially invade the ENS and eventually reach the CNS via the vagus nerve, triggering PD motor symptoms. Several lines of evidence show support of such a mechanism. First, results show that complete vagotomy is effective in decreasing the risk of developing PD [255]. Secondly, chronic constipation is one of the major and widespread early symptoms of PD, affecting approximately 80% of patients and with such signals being detected decades before diagnosis [10]. Additionally, Sampson et al., (2016) [256] showed worsening of motor impairments when α-syn overexpressing mice were inoculated with the GM of PD patients when compared to non-PD control GM. Recently, such findings were corroborated with evidence showing exacerbated motor impairments in a mouse model of PD with gut dysbiosis and intestinal inflammation [257–260]. Altogether, these findings establish convincing evidence for the role of the GM in regulating motor deficits in PD development.

To date, no single microbial species or taxa has been determined to have a causal role in PD development. However, structural differences between the GM of PD patients and healthy controls have been previously documented and accumulating evidence shows the abundance of certain fecal microbial taxa are differentially distributed among PD patients and healthy controls. *Lactobacillus*, *Bifidobacterium*, *Verrucomicrobiaceae*, and *Akkermansia* show an increase in PD patients while *Faecalibacterium*, *Roseburia*, *Coprococcus*, *Blautia*, *Prevotella*, and *Prevotellaceae* have lower abundances [261–264].

Scheperjans et al., (2015) [265] found that the relative abundance of *Prevotellaceae* taxa was reduced by nearly 80% in PD patients when compared to controls, and that the abundance of four microbial taxa was able to classify PD patients with over 90% specificity. Further, disease phenotypes such as postural instability were associated with the abundance of *Enterobacteriaceae* in PD patients from the same cohort. More recently, Aho et al., (2019) also found that *Prevotella* taxa are less abundant in faster-progressing patients [261]. Hegelmaier et al., (2020) tested dietary and enema interventions in PD patients and observed changes in the abundance of certain bacterial taxa, such as *Ruminococcaceae* associated with improved motor symptoms and decreased levodopa daily doses [266]. Hertel et al., (2019) [267] and Baldini et al., (2020) [268] proposed that *B. wadsworthia*, found in higher abundances in PD patients, is crucial for sulfite production in the gut, thus mediating brain mitochondrial energy balance, acting as a neurotoxin and promoting impaired metabolite secretion in PD [267,268].

In fact, communication between the host and the gut microbiome often consists of interactions involving short-chain fatty acids (SCFA), a major group of metabolites produced by the GM from dietary substrates [269]. SCFA production is crucial to metabolic homeostasis promotes overall systemic health. For instance, butyrate and propionate, two of the main SCFAs, have neuroprotective effects and aid in the rescue of motor capacity in PD [270], However, fecal SCFA levels have been found to be reduced in PD patients [263,271]. Cirstea et al., (2020) [272] demonstrated that butyrate synthesis is reduced in PD while deleterious aminoacid metabolites are increased, aggravating gut inflammation and constipation [273]. GM is also responsible for converting dietary flavonols into phenolic acids, another important metabolite [273]. Ho et al., (2019) demonstrated that individual GM metabolic repertoires of polyphenol production have the ability to protect against neurological disorders involving α-syn toxicity [274].

The mechanisms by which the GM affects neurodegenerative conditions are likely impaired production of neuroprotective factors, increased levels of pro-inflammatory cytokines, and unbalanced immune responses [275]. A reduction in the abundance of anti-inflammatory and neuroprotective metabolite-producing bacteria, such as members of the *Lachnospiraceae* family, has been documented for PD patients [276]. Further, taxa such as Bacteroides and *Verrucomicrobia* species were found in positive correlation with the abundance of pro-inflammatory cytokines TNF-α and IFN-γ in Asian PD patients [264]. Another inflammation-mediated GM effect is through small intestinal bacterial overgrowth (SIBO), present in around 25% of PD patients [277,278]. SIBO, which influences gastrointestinal dysfunction, seems to interfere in PD pathogenesis by increasing intestinal permeability, pro-inflammatory cytokine activation and, consequently, microglial activation. This leads to worsening of motor capacities and may also interfere in levodopa absorption [278].

Host–microbiome interactions in PD gain another level of complexity when considering host-derived epigenetic interference upon microbial metabolism. It has been shown that gut microbial DNA expression is affected by ncRNA molecules produced by gut epithelial cells, and may influence important bacterial pathways [279]. Such findings still require validation and more fecal ncRNA data for PD are needed. However, such associations attest to the intricate involvement of the GM in PD pathogenesis and encourage future studies to investigate this relationship further.

Current therapeutic interventions related to the GM are focused on probiotics, prebiotics, antibiotics and fecal microbiota transplantations (FMT) in a wide range of disorders, such as cancer and neurological diseases [280,281]. Probiotics could be employed to improve gut health, and antibiotics have been shown to ameliorate motor dysfunction by preventing dopamine neuron loss [282,283]. To date, one report has shown a successful case of FMT in improving constipation and leg tremors in a PD patient [284]. In Alzheimer's disease, FMT has also been shown to be useful for improving symptoms following

C. difficile infections [285]. It is important to note that GM-based therapeutic approaches do not necessarily involve proof of the causal role of microbial communities in PD pathogenesis, considering the GM modulates various metabolic systems and could improve quality of life in diverse aspects, such as gastrointestinal inflammation or constipation [286].

Gut microbiome data may be employed to determine the initial steps and progression of PD, considering not all patients experience enteric neurodegeneration at first. Additionally, profiling and measurement of GM metabolites could be applied to the clinical environment once we determine its predictive value to disease progression and/or prognosis. This could lead to personalized approaches in patient management as the GM could become a crucial part of PD treatment and diagnosis [286].

7. Future Perspectives in Parkinson's Disease

The multifactorial nature of PD poses a difficult obstacle for understanding the mechanisms involved in its onset and progression. However, it does provide a multitude of possibilities when it comes to the discovery of cellular pathways and biomarkers that can be employed for diagnostic or treatment purposes. Future efforts should focus on expanding population representation in PD databases as well as standardizing methods to avoid conflicting results due to different approaches.

Perspectives for PD biomarker discovery are promising as we invest in multi-omics techniques for characterizing patient genomic, epigenomic and gut microbial profiles. In this context, we might be moving towards personalized medicine approaches that employ a combination of different biomarkers. For instance, as GI tract alterations are one of the first symptoms related to PD onset, metagenomic data may be an important tool for early diagnosis.

Furthermore, the study of PD-related variants at a population level may also highlight individuals at risk, leading to early clinical interventions, and the role of ncRNAs may shed light on PD regulation mechanisms. Although there is no cure for PD yet, it is important to emphasize that the molecular factors involved in PD onset and progression may not only be applied for population screening and early diagnosis, they can also become new therapy targets, being involved in the development of new pharmacological approaches. Therefore, these potential biomarkers might help the development of innovative therapeutic approaches, elucidate neurodegeneration processes and improve patient care.

We would like to highlight that, considering the increasing amount of data from genomics, epigenomics, metagenomics, and other various omics, it is essential to analyze data in an integrative approach to understand the complexity of PD, find disease-specific biomarkers for population screening, early detection, as well as discovering new therapies to attenuate disease progress.

8. Conclusions

The post-genomic era has expanded our knowledge on several molecular processes, producing data that have been applied for early diagnosis and improvement of patient care in various diseases, such as cancer. However, the same is not a reality when it comes to neurodegenerative disorders, mainly due to scarce data and the obstacles in fully understanding the molecular mechanisms leading to neuronal death. The data discussed in this review highlight that genomic variation, regulation by epigenomic mechanisms, as well as the influence of host gut microbiome have a crucial role in the onset and progress of PD. The investigation of these molecular aspects in a collaborative and multidisciplinary manner is key to the rise of personalized medicine for PD patients.

Author Contributions: Conceptualization: G.F.C.; Writing–original draft preparation: G.F.C.; Writing–review and editing: G.F.C., A.P.S., G.C.C., C.S.-d.-S., T.P.d.S., N.M.S.P., J.A.d.S.P.; Supervision: A.F.V. and Â.R.-d.-S.; Project Administration and Funding Acquisition: Â.R.-d.-S. All authors have read and agreed to the published version of the manuscript.

Funding: We thank Conselho Nacional do Desenvolvimento Científico e Tecnológico (CNPq), Coordenação de Aperfeiçoamento de Pessoal de Nível Superior (CAPES), Pró-Reitoria de Pesquisa (PROPESP) of Universidade Federal do Pará (UFPA), and Fundação de Apoio à FIOCRUZ (FIOTEC/FIOCRUZ) for the received grants. We highlight that: A.R.-d.-S., G.F.C. and T.P.d.S. are supported by CNPq/Brazil; A.P.S. and N.M.S.P. are supported by CAPES/Brazil; G.C.C. is supported by Fundação de Amparo e Desenvolvimento à Pesquisa (FADESP); A.F.V. is supported by FIOTEC/FIOCRUZ. The funders had no role in the review design or preparation of the manuscript.

Acknowledgments: This work is part of the Research Project Genética das Doenças Neurodegenerativas and Rede de Pesquisa em Genômica Populacional Humana (Biocomputacional—Protocol no. 3381/2013/CAPES).

Conflicts of Interest: The authors declare no conflict of interest. The funders had no role in the design of the study; in the writing of the manuscript, or in the decision to publish the results.

References

1. GBD 2016 Neurology Collaborators. Global, regional, and national burden of neurological disorders, 1990-2016: A systematic analysis for the global burden of disease study 2016. *Lancet Neurol.* **2019**, *18*, 459–480. [CrossRef]
2. Vinasco-Sandoval, T.; Moreira, F.C.; Vidal, A.F.; Pinto, P.; Ribeiro-dos-Santos, A.M.; Cruz, R.L.S.; Cabral, G.F.; Anaissi, A.K.M.; Lopes, K.d.P.; Ribeiro-dos-Santos, A. Global analyses of expressed piwi-interacting RNAs in gastric cancer. *Int. J. Mol. Sci.* **2020**, *21*, 7656. [CrossRef]
3. Vidal, A.F.; Ribeiro-dos-Santos, A.M.; Vinasco-Sandoval, T.; Magalhães, L.; Pinto, P.; Anaissi, A.K.M.; Demachki, S.; de Assumpção, P.P.; dos Santos, S.E.B.; Ribeiro-dos-Santos, A. The comprehensive expression analysis of circular RNAs in gastric cancer and its association with field cancerization. *Sci. Rep.* **2017**, *7*, 14551. [CrossRef]
4. Reed, X.; Bandrés-Ciga, S.; Blauwendraat, C.; Cookson, M.R. The role of monogenic genes in idiopathic Parkinson's disease. *Neurobiol. Dis.* **2019**, *124*, 230–239. [CrossRef] [PubMed]
5. Cook Shukla, L.; Schulze, J.; Farlow, J.; Pankratz, N.D.; Wojcieszek, J.; Foroud, T. Parkinson Disease Overview. GeneReviews® 2004. Available online: https://www.ncbi.nlm.nih.gov/books/NBK1223/ (accessed on 25 July 2019).
6. Li, T.; Le, W. Biomarkers for Parkinson's disease: How good are they? *Neurosci. Bull.* **2020**, *36*, 183–194. [CrossRef] [PubMed]
7. Erkkinen, M.G.; Kim, M.-O.; Geschwind, M.D. Clinical neurology and epidemiology of the major neurodegenerative diseases. *Cold Spring Harb. Perspect. Biol.* **2018**, *10*. [CrossRef] [PubMed]
8. Radhakrishnan, D.; Goyal, V. Parkinson's disease: A review. *Neurol. India* **2018**, *66*, 26–35.
9. Draoui, A.; El Hiba, O.; Aimrane, A.; El Khiat, A.; Gamrani, H. Parkinson's disease: From bench to bedside. *Rev. Neurol.* **2020**, *176*, 543–559. [CrossRef]
10. Berg, D.; Postuma, R.B.; Adler, C.H.; Bloem, B.R.; Chan, P.; Dubois, B.; Gasser, T.; Goetz, C.G.; Halliday, G.; Joseph, L.; et al. MDS research criteria for prodromal Parkinson's disease. *Mov. Disord.* **2015**, *30*, 1600–1611. [CrossRef] [PubMed]
11. Postuma, R.B.; Berg, D.; Stern, M.; Poewe, W.; Olanow, C.W.; Oertel, W.; Obeso, J.; Marek, K.; Litvan, I.; Lang, A.E.; et al. MDS clinical diagnostic criteria for Parkinson's disease. *Mov. Disord.* **2015**, *30*, 1591–1601. [CrossRef] [PubMed]
12. Arnerić, S.P.; Kern, V.D.; Stephenson, D.T. Regulatory-accepted drug development tools are needed to accelerate innovative CNS disease treatments. *Biochem. Pharmacol.* **2018**, *151*, 291–306. [CrossRef]
13. Lang, A.E.; Obeso, J.A. Challenges in Parkinson's disease: Restoration of the nigrostriatal dopamine system is not enough. *Lancet Neurol.* **2004**, *3*, 309–316. [CrossRef]
14. de le Boë, S.F. *Opera Medica. (Editio Altera Correctior and Emendatior)*; Daniel Elsevir & Abraham Wolfgang: Amsterdam, The Netherlands, 1860.
15. de Sauvages, F.B. *Nosologia Methodica Sistens Morborum Classes: Juxtà Sydenhami Mentem & Botanicorum Ordinem. Ultima, Auctior, Emendatior*; Sumptibus fratrum de Tournes: Amsterdam, The Netherlands, 1768.
16. Samii, A.; Nutt, J.G.; Ransom, B.R. Parkinson's disease. *Lancet* **2004**, *363*, 1783–1793. [CrossRef]
17. Strafella, C.; Caputo, V.; Galota, M.R.; Zampatti, S.; Marella, G.; Mauriello, S.; Cascella, R.; Giardina, E. Application of precision medicine in neurodegenerative diseases. *Front. Neurol.* **2018**, *9*, 701. [CrossRef]
18. Ko, J.H.; Strafella, A.P. Dopaminergic neurotransmission in the human brain: New lessons from perturbation and imaging. *Neuroscientist* **2012**, *18*, 149–168. [CrossRef] [PubMed]
19. Goetz, C.G. The history of Parkinson's disease: Early clinical descriptions and neurological therapies. *Cold Spring Harb. Perspect. Med.* **2011**, *1*. [CrossRef] [PubMed]
20. Gibb, W.R.; Lees, A.J. The relevance of the Lewy body to the pathogenesis of idiopathic Parkinson's disease. *J. Neurol. Neurosurg. Psychiatry* **1988**, *51*, 745–752. [CrossRef]
21. Dugger, B.N.; Dickson, D.W. Pathology of neurodegenerative diseases. *Cold Spring Harb. Perspect. Biol.* **2017**, 9. [CrossRef] [PubMed]
22. Armstrong, M.J.; Okun, M.S. Diagnosis and treatment of Parkinson disease: A review. *JAMA* **2020**, *323*, 548. [CrossRef] [PubMed]
23. Emamzadeh, F.N.; Surguchov, A. Parkinson's disease: Biomarkers, treatment, and risk factors. *Front. Neurosci.* **2018**, *12*, 612. [CrossRef]

24. Polymeropoulos, M.H.; Lavedan, C.; Leroy, E.; Ide, S.E.; Dehejia, A.; Dutra, A.; Pike, B.; Root, H.; Rubenstein, J.; Boyer, R.; et al. Mutation in the alpha-synuclein gene identified in families with Parkinson's disease. *Science* **1997**, *276*, 2045–2047. [CrossRef]
25. Kalia, L.V.; Lang, A.E. Parkinson's disease. *Lancet* **2015**, *386*, 896–912. [CrossRef]
26. Bridi, J.C.; Hirth, F. Mechanisms of α-synuclein induced synaptopathy in Parkinson's disease. *Front. Neurosci.* **2018**, *12*, 80. [CrossRef] [PubMed]
27. Zolezzi, J.M.; Bastías-Candia, S.; Clnestrosa, N. Molecular basis of neurodegeneration: Lessons from Alzheimer's and Parkinson's diseases. In *Recent Advances in Neurodegeneration*; Borreca, A., Ed.; IntechOpen: London, UK, 2019.
28. Bellucci, A.; Zaltieri, M.; Navarria, L.; Grigoletto, J.; Missale, C.; Spano, P. From α-synuclein to synaptic dysfunctions: New insights into the pathophysiology of Parkinson's disease. *Brain Res.* **2012**, *1476*, 183–202. [CrossRef] [PubMed]
29. Perez, R.G.; Waymire, J.C.; Lin, E.; Liu, J.J.; Guo, F.; Zigmond, M.J. A role for α-synuclein in the regulation of dopamine biosynthesis. *J. Neurosci.* **2002**, *22*, 3090–3099. [CrossRef] [PubMed]
30. Yu, S.; Zuo, X.; Li, Y.; Zhang, C.; Zhou, M.; Zhang, Y.A.; Uéda, K.; Chan, P. Inhibition of tyrosine hydroxylase expression in α-synuclein-transfected dopaminergic neuronal cells. *Neurosci. Lett.* **2004**, *367*, 34–39. [CrossRef] [PubMed]
31. Butler, B.; Saha, K.; Rana, T.; Becker, J.P.; Sambo, D.; Davari, P.; Goodwin, S.J.; Khoshbouei, H. Dopamine transporter activity is modulated by α-synuclein. *J. Biol. Chem.* **2015**, *290*, 29542–29554. [CrossRef]
32. Valdinocci, D.; Radford, R.; Goulding, M.; Hayashi, J.; Chung, R.; Pountney, D. Extracellular interactions of alpha-synuclein in multiple system atrophy. *Int. J. Mol. Sci.* **2018**, *19*, 4129. [CrossRef]
33. Huang, C.-C.; Chiu, T.-Y.; Lee, T.-Y.; Hsieh, H.-J.; Lin, C.-C.; Kao, L.-S. Soluble α-synuclein facilitates priming and fusion by releasing Ca^{2+} from the thapsigargin-sensitive Ca^{2+} pool in PC12 cells. *J. Cell Sci.* **2018**, 131. [CrossRef] [PubMed]
34. Logan, T.; Bendor, J.; Toupin, C.; Thorn, K.; Edwards, R.H. α-Synuclein promotes dilation of the exocytotic fusion pore. *Nat. Neurosci.* **2017**, *20*, 681–689. [CrossRef]
35. Rieder, C.R.M. GBA mutations and Parkinson's disease in Brazil. *Arq. Neuro-Psiquiatr.* **2019**, *77*, 71–72. [CrossRef] [PubMed]
36. Magalhaes, J.; Gegg, M.E.; Migdalska-Richards, A.; Doherty, M.K.; Whitfield, P.D.; Schapira, A.H.V. Autophagic lysosome reformation dysfunction in glucocerebrosidase deficient cells: Relevance to Parkinson disease. *Hum. Mol. Genet.* **2016**, *25*, 3432–3445. [CrossRef] [PubMed]
37. Marques, A.R.A.; Mirzaian, M.; Akiyama, H.; Wisse, P.; Ferraz, M.J.; Gaspar, P.; Alfonso, P.; Irún, P.; Dahl, M.; Karlsson, S.; et al. Glucosylated cholesterol in mammalian cells and tissues: Formation and degradation by multiple cellular β-glucosidases. *J. Lipid Res.* **2016**, *57*, 451–463. [CrossRef] [PubMed]
38. Akiyama, H.; Kobayashi, S.; Hirabayashi, Y.; Murakami-Murofushi, K. Cholesterol glucosylation is catalyzed by transglucosylation reaction of β-glucosidase 1. *Biochem. Biophys. Res. Commun.* **2013**, *441*, 838–843. [CrossRef] [PubMed]
39. Sidransky, E.; Lopez, G. The link between the GBA gene and parkinsonism. *Lancet Neurol.* **2012**, *11*, 986–998. [CrossRef]
40. Manzoni, C. The LRRK2-macroautophagy axis and its relevance to Parkinson's disease. *Biochem. Soc. Trans.* **2017**, *45*, 155–162. [CrossRef] [PubMed]
41. Roth, J.A.; Eichhorn, M. Down-regulation of LRRK2 in control and DAT transfected HEK cells increases manganese-induced oxidative stress and cell toxicity. *Neurotoxicology* **2013**, *37*, 100–107. [CrossRef] [PubMed]
42. Cooper, O.; Seo, H.; Andrabi, S.; Guardia-Laguarta, C.; Graziotto, J.; Sundberg, M.; McLean, J.R.; Carrillo-Reid, L.; Xie, Z.; Osborn, T.; et al. Pharmacological rescue of mitochondrial deficits in iPSC-derived neural cells from patients with familial Parkinson's disease. *Sci. Transl. Med.* **2012**, *4*. [CrossRef]
43. Papkovskaia, T.D.; Chau, K.-Y.; Inesta-Vaquera, F.; Papkovsky, D.B.; Healy, D.G.; Nishio, K.; Staddon, J.; Duchen, M.R.; Hardy, J.; Schapira, A.H.V.; et al. G2019S leucine-rich repeat kinase 2 causes uncoupling protein-mediated mitochondrial depolarization. *Hum. Mol. Genet.* **2012**, *21*, 4201–4213. [CrossRef]
44. Melrose, H.L.; Dächsel, J.C.; Behrouz, B.; Lincoln, S.J.; Yue, M.; Hinkle, K.M. Impaired dopaminergic neurotransmission and microtubule-associated protein tau alterations in human LRRK2 transgenic mice. *Neurobiol. Dis.* **2010**, *40*, 503–517. [CrossRef]
45. Xiong, H.; Wang, D.; Chen, L.; Choo, Y.S.; Ma, H.; Tang, C. Parkin, PINK1, and DJ-1 form a ubiquitin E3 ligase complex promoting unfolded protein degradation. *J. Clin. Investig.* **2009**, *119*, 650–660. [CrossRef]
46. Chung, K.K.; Zhang, Y.; Lim, K.L.; Tanaka, Y.; Huang, H.; Gao, J. Parkin ubiquitinates the alpha-synuclein-interacting protein, synphilin-1: Implications for Lewy-body formation in Parkinson disease. *Nat. Med.* **2001**, *7*, 1144–1150. [CrossRef]
47. Wu, S.; Lei, L.; Song, Y.; Liu, M.; Lu, S.; Lou, D. Mutation of hop-1 and pink-1 attenuates vulnerability of neurotoxicity in C. elegans: The role of mitochondria-associated membrane proteins in Parkinsonism. *Exp. Neurol.* **2018**, *309*, 67–78. [CrossRef]
48. Valente, E.M.; Abou-Sleiman, P.M.; Caputo, V.; Muqit, M.M.K.; Harvey, K.; Gispert, S. Hereditary early-onset Parkinson's disease caused by mutations in PINK1. *Science* **2004**, *304*, 1158–1160. [CrossRef]
49. Fu, K.; Ren, H.; Wang, Y.; Fei, E.; Wang, H.; Wang, G. DJ-1 inhibits TRAIL-induced apoptosis by blocking pro-caspase-8 recruitment to FADD. *Oncogene* **2012**, *31*, 1311–1322. [CrossRef] [PubMed]
50. Haque, M.E.; Mount, M.P.; Safarpour, F.; Abdel-Messih, E.; Callaghan, S.; Mazerolle, C. Inactivation of pink1 gene in vivo sensitizes dopamine-producing neurons to 1-methyl-4-phenyl-1,2,3,6-tetrahydropyridine (MPTP) and can be rescued by autosomal recessive Parkinson disease genes, Parkin or DJ-1. *J. Biol. Chem.* **2012**, *287*, 23162–23170. [CrossRef]
51. Lee, J.; Song, J.; Kwon, K.; Jang, S.; Kim, C.; Baek, K. Human DJ-1 and its homologs are novel glyoxalases. *Hum. Mol. Genet.* **2012**, *21*, 3215–3225. [CrossRef]

52. Choi, D.-H.; Hwang, O.; Lee, K.-H.; Lee, J.; Beal, M.F.; Kim, Y.-S. DJ-1 cleavage by matrix metalloproteinase 3 mediates oxidative stress-induced dopaminergic cell death. *Antioxid. Redox Signal.* **2011**, *14*, 2137–2150. [CrossRef]
53. Guo, C.; Jeong, H.-H.; Hsieh, Y.-C.; Klein, H.-U.; Bennett, D.A.; De Jager, P.L. Tau activates transposable elements in Alzheimer's disease. *Cell Rep.* **2018**, *23*, 2874–2880. [CrossRef] [PubMed]
54. Pascale, E.; Di Battista, M.E.; Rubino, A.; Purcaro, C.; Valente, M.; Fattapposta, F. Genetic architecture of MAPT gene region in parkinson disease subtypes. *Front. Cell. Neurosci.* **2016**, *10*, 96. [CrossRef] [PubMed]
55. Zhang, C.-C.; Xing, A.; Tan, M.-S.; Tan, L.; Yu, J.-T. The role of MAPT in neurodegenerative diseases: Genetics, mechanisms and therapy. *Mol. Neurobiol.* **2016**, *53*, 4893–4904. [CrossRef] [PubMed]
56. Pacheco, C.D.; Elrick, M.J.; Lieberman, A.P. Tau deletion exacerbates the phenotype of Niemann-Pick type C mice and implicates autophagy in pathogenesis. *Hum. Mol. Genet.* **2009**, *18*, 956–965. [CrossRef] [PubMed]
57. Allen Reish, H.E.; Standaert, D.G. Role of α-synuclein in inducing innate and adaptive immunity in Parkinson disease. *J. Parkinson's Dis.* **2015**, *5*, 1–19. [CrossRef] [PubMed]
58. Lee, H.-J.; Kim, C.; Lee, S.-J. Alpha-synuclein stimulation of astrocytes: Potential role for neuroinflammation and neuroprotection. *Oxidative Med. Cell. Longev.* **2010**, *3*, 283–287. [CrossRef]
59. Meng, J.; Lv, Z.; Qiao, X.; Li, X.; Li, Y.; Zhang, Y. The decay of redox-stress response capacity is a substantive characteristic of aging: Revising the redox theory of aging. *Redox Biol.* **2017**, *11*, 365–374. [CrossRef]
60. Pham-Huy, L.A.; He, H.; Pham-Huy, C. Free radicals, antioxidants in disease and health. *Int. J. Biomed. Sci.* **2008**, *4*, 89–96.
61. Okun, E.; Griffioen, K.J.; Mattson, M.P. Toll-like receptor signaling in neural plasticity and disease. *Trends Neurosci.* **2011**, *34*, 269–281. [CrossRef]
62. Franceschi, C.; Garagnani, P.; Vitale, G.; Capri, M.; Salvioli, S. Inflammaging and 'garb-aging'. *Trends Endocrinol. Metab.* **2017**, *28*, 199–212. [CrossRef]
63. Dandekar, A.; Mendez, R.; Zhang, K. Cross talk between er stress, oxidative stress, and inflammation in health and disease. In *Stress Responses*; Oslowski, C.M., Ed.; Springer New York: New York, NY, USA, 2015; pp. 205–214.
64. Sironi, L.; Restelli, L.M.; Tolnay, M.; Neutzner, A.; Frank, S. Dysregulated interorganellar crosstalk of mitochondria in the pathogenesis of Parkinson's disease. *Cells* **2020**, *9*, 233. [CrossRef]
65. Puspita, L.; Chung, S.Y.; Shim, J. Oxidative stress and cellular pathologies in Parkinson's disease. *Mol. Brain* **2017**, *10*. [CrossRef]
66. Calì, T.; Ottolini, D.; Brini, M. Mitochondria, calcium, and endoplasmic reticulum stress in Parkinson's disease. *BioFactors* **2011**, *37*, 228–240. [CrossRef]
67. Calì, T.; Ottolini, D.; Brini, M. Mitochondrial Ca^{2+} and neurodegeneration. *Cell Calcium* **2012**, *52*, 73–85. [CrossRef]
68. Pickrell, A.M.; Youle, R.J. The roles of PINK1, parkin, and mitochondrial fidelity in Parkinson's disease. *Neuron* **2015**, *85*, 257–273. [CrossRef] [PubMed]
69. Klein, C.; Westenberger, A. Genetics of Parkinson's disease. *Cold Spring Harb. Perspect. Med.* **2012**, *2*. [CrossRef]
70. Wang, X.; Winter, D.; Ashrafi, G.; Schlehe, J.; Wong, Y.L.; Selkoe, D. PINK1 and parkin target miro for phosphorylation and degradation to arrest mitochondrial motility. *Cell* **2011**, *147*, 893–906. [CrossRef] [PubMed]
71. Stafa, K.; Tsika, E.; Moser, R.; Musso, A.; Glauser, L.; Jones, A. Functional interaction of Parkinson's disease-associated LRRK2 with members of the dynamin GTPase superfamily. *Hum. Mol. Genet.* **2014**, *23*, 2055–2077. [CrossRef] [PubMed]
72. Ren, H.; Fu, K.; Wang, D.; Mu, C.; Wang, G. Oxidized DJ-1 interacts with the mitochondrial protein BCL-X_L. *J. Biol. Chem.* **2011**, *286*, 35308–35317. [CrossRef]
73. Taira, T.; Saito, Y.; Niki, T.; Iguchi-Ariga, S.M.M.; Takahashi, K.; Ariga, H. DJ-1 has a role in antioxidative stress to prevent cell death. *EMBO Rep.* **2004**, *5*, 213–218. [CrossRef]
74. Do, J.; McKinney, C.; Sharma, P.; Sidransky, E. Glucocerebrosidase and its relevance to Parkinson disease. *Mol. Neurodegener.* **2019**, *14*, 1–16. [CrossRef]
75. Wu, Y.-R.; Chen, C.-M.; Chao, C.-Y.; Ro, L.-S.; Lyu, R.-K.; Chang, K.-H. Glucocerebrosidase gene mutation is a risk factor for early onset of Parkinson disease among taiwanese. *J. Neurol. Neurosurg. Psychiatry* **2007**, *78*, 977–979. [CrossRef]
76. Tam, O.H.; Ostrow, L.W.; Gale Hammell, M. Diseases of the nERVous system: Retrotransposon activity in neurodegenerative disease. *Mob. DNA* **2019**, *10*, 1–14. [CrossRef]
77. Jacob-Hirsch, J.; Eyal, E.; Knisbacher, B.A.; Roth, J.; Cesarkas, K.; Dor, C. Whole-genome sequencing reveals principles of brain retrotransposition in neurodevelopmental disorders. *Cell Res.* **2018**, *28*, 187–203. [CrossRef] [PubMed]
78. Li, W.; Prazak, L.; Chatterjee, N.; Grüninger, S.; Krug, L.; Theodorou, D. Activation of transposable elements during aging and neuronal decline in drosophila. *Nat. Neurosci.* **2013**, *16*, 529–531. [CrossRef] [PubMed]
79. Bodea, G.O.; McKelvey, E.G.Z.; Faulkner, G.J. oRetrotransposon-induced mosaicism in the neural genme. *Open Biol.* **2018**, *8*. [CrossRef] [PubMed]
80. Garcia-Perez, J.L.; Widmann, T.J.; Adams, I.R. The impact of transposable elements on mammalian development. *Development* **2016**, *143*, 4101–4114. [CrossRef] [PubMed]
81. Saleh, A.; Macia, A.; Muotri, A.R. Transposable elements, inflammation, and neurological disease. *Front. Neurol.* **2019**, *10*, 894. [CrossRef]
82. Kassiotis, G.; Stoye, J.P. Immune responses to endogenous retroelements: Taking the bad with the good. *Nat. Rev. Immunol.* **2016**, *16*, 207–219. [CrossRef]

83. Bravo, J.I.; Nozownik, S.; Danthi, P.S.; Benayoun, B.A. Transposable elements, circular RNAs and mitochondrial transcription in age-related genomic regulation. *Development* **2020**, *147*. [CrossRef]
84. Huang, X.; Wong, G. An old weapon with a new function: PIWI-interacting RNAs in neurodegenerative diseases. *Transl. Neurodegener.* **2021**, *10*, 9. [CrossRef]
85. Sun, W.; Samimi, H.; Gamez, M.; Zare, H.; Frost, B. Pathogenic tau-induced piRNA depletion promotes neuronal death through transposable element dysregulation in neurodegenerative tauopathies. *Nat. Neurosci.* **2018**, *21*, 1038–1048. [CrossRef]
86. Nandi, S.; Chandramohan, D.; Fioriti, L.; Melnick, A.M.; Hébert, J.M.; Mason, C.E. Roles for small noncoding RNAs in silencing of retrotransposons in the mammalian brain. *Proc. Natl. Acad. Sci. USA* **2016**, *113*, 12697–12702. [CrossRef]
87. Carlice-dos-Reis, T.; Viana, J.; Moreira, F.C.; De Cardoso, G.L.; Guerreiro, J.; Santos, S. Investigation of mutations in the HBB gene using the 1,000 genomes database. *PLoS ONE* **2017**, *12*, e0174637. [CrossRef]
88. Cristina, T.-P.; Pablo, M.; Teresa, P.M.; Lydia, V.-D.; Irene, A.-R.; Araceli, A.-C. A genetic analysis of a spanish population with early onset Parkinson's disease. *PLoS ONE* **2020**, *15*, e0238098. [CrossRef] [PubMed]
89. Blauwendraat, C.; Makarious, M.B.; Leonard, H.L.; Bandres-Ciga, S.; Iwaki, H.; Nalls, M.A. A population scale analysis of rare SNCA variation in the UK biobank. *Neurobiol. Dis.* **2021**, *148*, 105182. [CrossRef] [PubMed]
90. Wu, Y.-R.; Chang, K.-H.; Chao, C.-Y.; Lin, C.-H.; Chen, Y.-C.; Liu, T.-W. Association of SOD2 p.V16A polymorphism with Parkinson's disease: A meta-analysis in Han Chinese. *J. Formos. Med. Assoc.* **2021**, *120*, 501–507. [CrossRef]
91. Nalls, M.A.; Blauwendraat, C.; Vallerga, C.L.; Heilbron, K.; Bandres-Ciga, S.; Chang, D. Identification of novel risk loci, causal insights, and heritable risk for Parkinson's disease: A meta-analysis of genome-wide association studies. *Lancet Neurol.* **2019**, *18*, 1091–1102. [CrossRef]
92. Chew, E.G.Y.; Tan, L.C.S.; Au, W.-L.; Prakash, K.-M.; Liu, J.; Foo, J.N. ITPKB and ZNF184 are associated with Parkinson's disease risk in East Asians. *Neurobiol. Aging* **2020**, *86*, 201.e15–201.e17. [CrossRef]
93. Moosavi, A.; Ardekani, A.M. Role of epigenetics in biology and human diseases. *Iran. Biomed. J.* **2016**, *20*, 246–258. [PubMed]
94. Biswas, S.; Rao, C.M. Epigenetic tools (the writers, the readers and the erasers) and their implications in cancer therapy. *Eur. J. Pharmacol.* **2018**, *837*, 8–24. [CrossRef]
95. Pereira, A.L.; Magalhães, L.; Moreira, F.C.; Reis-das-Mercês, L.; Vidal, A.F.; Ribeiro-dos-Santos, A.M. Epigenetic field cancerization in gastric cancer: MicroRNAs as promising biomarkers. *J. Cancer* **2019**, *10*, 1560–1569. [CrossRef]
96. Magalhães, L.; Quintana, L.G.; Lopes, D.C.F.; Vidal, A.F.; Pereira, A.L.; Pinto, L.C.D.A.; Pinheiro, J.D.J.V.; Khayat, A.S.; Goulart, L.R.; Burbano, R.; et al. APC gene is modulated by hsa-miR-135b-5p in both diffuse and intestinal gastric cancer subtypes. *BMC Cancer* **2018**, *18*, 1055. [CrossRef]
97. Young, J.I.; Sivasankaran, S.K.; Wang, L.; Ali, A.; Mehta, A.; Davis, D.A. Genome-wide brain DNA methylation analysis suggests epigenetic reprogramming in Parkinson disease. *Neurol. Genet.* **2019**, *5*, e342. [CrossRef]
98. Chuang, Y.-H.; Paul, K.C.; Bronstein, J.M.; Bordelon, Y.; Horvath, S.; Ritz, B. Parkinson's disease is associated with DNA methylation levels in human blood and saliva. *Genome Med.* **2017**, *9*, 76. [CrossRef]
99. Wang, C.; Chen, L.; Yang, Y.; Zhang, M.; Wong, G. Identification of potential blood biomarkers for Parkinson's disease by gene expression and DNA methylation data integration analysis. *Clin. Epigenet.* **2019**, *11*, 1–15. [CrossRef]
100. Södersten, E.; Toskas, K.; Rraklli, V.; Tiklova, K.; Björklund, Å.K.; Ringnér, M. A comprehensive map coupling histone modifications with gene regulation in adult dopaminergic and serotonergic neurons. *Nat. Commun.* **2018**, *9*, 1226. [CrossRef] [PubMed]
101. Toker, L.; Tran, G.T.; Sundaresan, J.; Tysnes, O.-B.; Alves, G.; Haugarvoll, K. Genome-wide dysregulation of histone acetylation in the Parkinson's disease brain. *BioRxiv* **2019**. [CrossRef]
102. Ha, M.; Kim, V.N. Regulation of microRNA biogenesis. *Nat. Rev. Mol. Cell Biol.* **2014**, *15*, 509–524. [CrossRef] [PubMed]
103. Bartel, D.P. MicroRNAs: Genomics, biogenesis, mechanism, and function. *Cell* **2004**, *116*, 281–297. [CrossRef]
104. Pfeffer, S.; Sewer, A.; Lagos-Quintana, M.; Sheridan, R.; Sander, C.; Grässer, F.A. Identification of microRNAs of the herpesvirus family. *Nat. Methods* **2005**, *2*, 269–276. [CrossRef] [PubMed]
105. Dueck, A.; Ziegler, C.; Eichner, A.; Berezikov, E.; Meister, G. microRNAs associated with the different human argonaute proteins. *Nucleic Acids Res.* **2012**, *40*, 9850–9862. [CrossRef] [PubMed]
106. Huntzinger, E.; Izaurralde, E. Gene silencing by microRNAs: Contributions of translational repression and mRNA decay. *Nat. Rev. Genet.* **2011**, *12*, 99–110. [CrossRef] [PubMed]
107. Prodromidou, K.; Matsas, R. Species-specific miRNAs in human brain development and disease. *Front. Cell. Neurosci.* **2019**, *13*, 559. [CrossRef] [PubMed]
108. Volvert, M.-L.; Prévot, P.-P.; Close, P.; Laguesse, S.; Pirotte, S.; Hemphill, J. MicroRNA targeting of CoREST controls polarization of migrating cortical neurons. *Cell Rep.* **2014**, *7*, 1168–1183. [CrossRef]
109. Leucht, C.; Stigloher, C.; Wizenmann, A.; Klafke, R.; Folchert, A.; Bally-Cuif, L. MicroRNA-9 directs late organizer activity of the midbrain-hindbrain boundary. *Nat. Neurosci.* **2008**, *11*, 641–648. [CrossRef]
110. Vatsa, N.; Kumar, V.; Singh, B.K.; Kumar, S.S.; Sharma, A.; Jana, N.R. Down-regulation of miRNA-708 promotes aberrant calcium signaling by targeting neuronatin in a mouse model of angelman syndrome. *Front. Mol. Neurosci.* **2019**, *12*, 35. [CrossRef]
111. Wayman, G.A.; Davare, M.; Ando, H.; Fortin, D.; Varlamova, O.; Cheng, H.-Y.M. An activity-regulated microRNA controls dendritic plasticity by down-regulating p250GAP. *Proc. Natl. Acad. Sci. USA* **2008**, *105*, 9093–9098. [CrossRef]

112. Rajman, M.; Schratt, G. MicroRNAs in neural development: From master regulators to fine-tuners. *Development* **2017**, *144*, 2310–2322. [CrossRef]
113. Edbauer, D.; Neilson, J.R.; Foster, K.A.; Wang, C.-F.; Seeburg, D.P.; Batterton, M.N. Regulation of Synaptic structure and function by FMRP-associated MicroRNAs miR-125b and miR-132. *Neuron* **2010**, *65*, 373–384. [CrossRef]
114. Wang, W.; Kwon, E.J.; Tsai, L.-H. MicroRNAs in learning, memory, and neurological diseases. *Learn. Mem.* **2012**, *19*, 359–368. [CrossRef] [PubMed]
115. Jin Jung, H.; Suh, Y. MicroRNA in aging: From discovery to biology. *Curr. Genom.* **2012**, *13*, 548–557. [CrossRef] [PubMed]
116. John, A.; Kubosumi, A.; Reddy, P.H. Mitochondrial MicroRNAs in aging and neurodegenerative diseases. *Cells* **2020**, *9*, 1345. [CrossRef]
117. Ozkul, Y.; Taheri, S.; Bayram, K.K.; Sener, E.F.; Mehmetbeyoglu, E.; Öztop, D.B. A heritable profile of six miRNAs in autistic patients and mouse models. *Sci. Rep.* **2020**, *10*, 9011. [CrossRef]
118. He, K.; Guo, C.; He, L.; Shi, Y. MiRNAs of peripheral blood as the biomarker of schizophrenia. *Hereditas* **2018**, *155*, 9. [CrossRef]
119. Beveridge, N.J.; Cairns, M.J. MicroRNA dysregulation in schizophrenia. *Neurobiol. Dis.* **2012**, *46*, 263–271. [CrossRef] [PubMed]
120. Hoss, A.G.; Labadorf, A.; Latourelle, J.C.; Kartha, V.K.; Hadzi, T.C.; Gusella, J.F. miR-10b-5p expression in huntington's disease brain relates to age of onset and the extent of striatal involvement. *BMC Med. Genom.* **2015**, *8*, 10. [CrossRef] [PubMed]
121. Jain, G.; Stuendl, A.; Rao, P.; Berulava, T.; Pena Centeno, T.; Kaurani, L. A combined miRNA–piRNA signature to detect Alzheimer's disease. *Transl. Psychiatry* **2019**, *9*, 1–12. [CrossRef] [PubMed]
122. Sierksma, A.; Lu, A.; Salta, E.; Vanden Eynden, E.; Callaerts-Vegh, Z.; D'Hooge, R. Deregulation of neuronal miRNAs induced by amyloid-β or TAU pathology. *Mol. Neurodegener.* **2018**, *13*, 1–15. [CrossRef] [PubMed]
123. Juźwik, C.A.; SDrake, S.; Zhang, Y.; Paradis-Isler, N.; Sylvester, A.; Amar-Zifkin, A. microRNA dysregulation in neurodegenerative diseases: A systematic review. *Prog. Neurobiol.* **2019**, *182*, 101664. [CrossRef]
124. Cardo, L.F.; Coto, E.; Ribacoba, R.; Menéndez, M.; Moris, G.; Suárez, E. MiRNA profile in the substantia nigra of Parkinson's disease and healthy subjects. *J. Mol. Neurosci.* **2014**, *54*, 830–836. [CrossRef] [PubMed]
125. Kim, J.; Inoue, K.; Ishii, J.; Vanti, W.B.; Voronov, S.V.; Murchison, E. A MicroRNA feedback circuit in midbrain dopamine neurons. *Science* **2007**, *317*, 1220–1224. [CrossRef]
126. Wang, R.; Yang, Y.; Wang, H.; He, Y.; Li, C. MiR-29c protects against inflammation and apoptosis in Parkinson's disease model in vivo and in vitro by targeting SP1. *Clin. Exp. Pharmacol. Physiol.* **2020**, *47*, 372–382. [CrossRef] [PubMed]
127. Doxakis, E. Post-transcriptional regulation of α-synuclein expression by mir-7 and mir-153. *J. Biol. Chem.* **2010**, *285*, 12726–12734. [CrossRef] [PubMed]
128. Junn, E.; Lee, K.-W.; Jeong, B.S.; Chan, T.W.; Im, J.-Y.; Mouradian, M.M. Repression of synuclein expression and toxicity by microRNA-7. *Proc. Natl. Acad. Sci. USA* **2009**, *6*, 13052–13057. [CrossRef] [PubMed]
129. Thome, A.D.; Harms, A.S.; Volpicelli-Daley, L.A.; Standaert, D.G. microRNA-155 regulates alpha-synuclein-induced inflammatory responses in models of parkinson disease. *J. Neurosci.* **2016**, *36*, 2383–2390. [CrossRef]
130. Cho, H.J.; Liu, G.; Jin, S.M.; Parisiadou, L.; Xie, C.; Yu, J. MicroRNA-205 regulates the expression of Parkinson's disease-related leucine-rich repeat kinase 2 protein. *Hum. Mol. Genet.* **2013**, *22*, 608–620. [CrossRef] [PubMed]
131. Gehrke, S.; Imai, Y.; Sokol, N.; Lu, B. Pathogenic LRRK2 negatively regulates microRNA-mediated translational repression. *Nature* **2010**, *466*, 637–641. [CrossRef]
132. Xiong, R.; Wang, Z.; Zhao, Z.; Li, H.; Chen, W.; Zhang, B. MicroRNA-494 reduces DJ-1 expression and exacerbates neurodegeneration. *Neurobiol. Aging* **2014**, *35*, 705–714. [CrossRef]
133. Oh, S.E.; Park, H.-J.; He, L.; Skibiel, C.; Junn, E.; Mouradian, M.M. The Parkinson's disease gene product DJ-1 modulates miR-221 to promote neuronal survival against oxidative stress. *Redox Biol.* **2018**, *19*, 62–73. [CrossRef]
134. Carrete, H., Jr. Parkinson's disease and atypical parkinsonism: The importance of magnetic resonance imaging as a potential biomarker. *Radiol. Bras.* **2017**, *50*, 5–6. [CrossRef]
135. Adusumilli, L.; Facchinello, N.; Teh, C.; Busolin, G.; Le, M.T.; Yang, H. miR-7 controls the dopaminergic/oligodendroglial fate through Wnt/β-catenin signaling regulation. *Cells* **2020**, *9*, 711. [CrossRef]
136. Zhang, J.; Dongwei, Z.; Zhang, Z.; Xinhui, Q.; Kunwang, B.; Guohui, L. miR-let-7a suppresses α-synuclein-induced microglia inflammation through targeting STAT3 in Parkinson's disease. *Biochem. Biophys. Res. Commun.* **2019**, *519*, 740–746. [CrossRef] [PubMed]
137. Sadlon, A.; Takousis, P.; Alexopoulos, P.; Evangelou, E.; Prokopenko, I.; Perneczky, R. miRNAs identify shared pathways in Alzheimer's and Parkinson's diseases. *Trends Mol. Med.* **2019**, *25*, 662–672. [CrossRef] [PubMed]
138. Aravin, A.; Gaidatzis, D.; Pfeffer, S.; Lagos-Quintana, M.; Landgraf, P.; Iovino, N. A novel class of small RNAs bind to MILI protein in mouse testes. *Nature* **2006**, *442*, 203–207. [CrossRef]
139. Girard, A.; Sachidanandam, R.; Hannon, G.J.; Carmell, M.A. A germline-specific class of small RNAs binds mammalian Piwi proteins. *Nature* **2006**, *442*, 199–202. [CrossRef]
140. Grivna, S.T.; Beyret, E.; Wang, Z.; Lin, H. A novel class of small RNAs in mouse spermatogenic cells. *Genes Dev.* **2006**, *20*, 1709–1714. [CrossRef]
141. Chalbatani, G.M.; Dana, H.; Memari, F.; Gharagozlou, E.; Ashjaei, S.; Kheirandish, P. Biological function and molecular mechanism of piRNA in cancer. *Pract. Lab. Med.* **2019**, *13*, e00113. [CrossRef]

142. Fonseca, C.G.; dos Santos, J.A.P.; Vidal, A.F.; Santos, S.; Ribeiro-dos-Santos, Â. piRNAs in ancer: A new approach towards translational research. *Int. J. Mol. Sci.* **2020**, *21*, 2126. [CrossRef]
143. Czech, B.; Hannon, G.J. One loop to rule them all: The ping-pong cycle and piRNA-guided silencing. *Trends Biochem. Sci.* **2016**, *41*, 324–337. [CrossRef]
144. Brennecke, J.; Aravin, A.A.; Stark, A.; Dus, M.; Kellis, M.; Sachidanandam, R. Discrete small RNA-generating loci as master regulators of transposon activity in drosophila. *Cell* **2007**, *128*, 1089–1103. [CrossRef] [PubMed]
145. Czech, B.; Munafò, M.; Ciabrelli, F.; Eastwood, E.L.; Fabry, M.H.; Kneuss, E. piRNA-guided genome defense: From biogenesis to silencing. *Annu. Rev. Genet.* **2018**, *52*, 131–157. [CrossRef] [PubMed]
146. Rojas-Ríos, P.; Simonelig, M. piRNAs and PIWI proteins: Regulators of gene expression in development and stem cells. *Development* **2018**, *145*, 161786. [CrossRef]
147. Ross, R.J.; Weiner, M.M.; Lin, H. PIWI proteins and PIWI-interacting RNAs in the soma. *Nature* **2014**, *505*, 353–359. [CrossRef]
148. Weick, E.M.; Miska, E.A. piRNAs: From biogenesis to function. *Development* **2014**, *141*, 3458–3471. [CrossRef]
149. Sheng, Z.-H. The interplay of axonal energy homeostasis and mitochondrial trafficking and anchoring. *Trends Cell Biol.* **2017**, *27*, 403–416. [CrossRef] [PubMed]
150. Pantano, L.; Jodar, M.; Bak, M.; Ballescà, J.L.; Tommerup, N.; Oliva, R. The small RNA content of human sperm reveals pseudogene-derived piRNAs complementary to protein-coding genes. *RNA* **2015**, *21*, 1085–1095. [CrossRef] [PubMed]
151. Rajasethupathy, P.; Antonov, I.; Sheridan, R.; Frey, S.; Sander, C.; Tuschl, T. A role for neuronal piRNAs in the epigenetic control of memory-related synaptic plasticity. *Cell* **2012**, *149*, 693–707. [CrossRef]
152. Martinez, V.D.; Enfield, K.S.S.; Rowbotham, D.A.; Lam, W.L. An atlas of gastric PIWI-interacting RNA transcriptomes and their utility for identifying signatures of gastric cancer recurrence. *Gastric Cancer* **2016**, *19*, 660–665. [CrossRef] [PubMed]
153. Weng, W.; Liu, N.; Toiyama, Y.; Kusunoki, M.; Nagasaka, T.; Fujiwara, T. Novel evidence for a PIWI-interacting RNA (piRNA) as an oncogenic mediator of disease progression, and a potential prognostic biomarker in colorectal cancer. *Mol. Cancer* **2018**, *17*, 1–12. [CrossRef] [PubMed]
154. Tan, L.; Mai, D.; Zhang, B.; Jiang, X.; Zhang, J.; Bai, R. PIWI-interacting RNA-36712 restrains breast cancer progression and chemoresistance by interaction with SEPW1 pseudogene SEPW1P RNA. *Mol. Cancer* **2019**, *18*, 1–15. [CrossRef]
155. Qiu, W.; Guo, X.; Lin, X.; Yang, Q.; Zhang, W.; Zhang, Y. Transcriptome-wide piRNA profiling in human brains of Alzheimer's disease. *Neurobiol. Aging* **2017**, *57*, 170–177. [CrossRef]
156. Zuo, L.; Wang, Z.; Tan, Y.; Chen, X.; Luo, X. piRNAs and their functions in the brain. *Int. J. Hum. Genet.* **2016**, *16*, 53–60. [CrossRef]
157. Rouget, C.; Papin, C.; Boureux, A.; Meunier, A.-C.; Franco, B.; Robine, N. Maternal mRNA deadenylation and decay by the piRNA pathway in the early drosophila embryo. *Nature* **2010**, *467*, 1128–1132. [CrossRef]
158. Lee, E.J.; Banerjee, S.; Zhou, H.; Jammalamadaka, A.; Arcila, M.; Manjunath, B.S. Identification of piRNAs in the central nervous system. *RNA* **2011**, *17*, 1090–1099. [CrossRef] [PubMed]
159. Yan, Z.; Hu, H.Y.; Jiang, X.; Maierhofer, V.; Neb, E.; He, L. Widespread expression of piRNA-like molecules in somatic tissues. *Nucleic Acids Res.* **2011**, *39*, 6596–6607. [CrossRef] [PubMed]
160. Jensen, P.; Myhre, C.L.; Lassen, P.S.; Metaxas, A.; Khan, A.M.; Lambertsen, K.L. TNFα affects CREB-mediated neuroprotective signaling pathways of synaptic plasticity in neurons as revealed by proteomics and phospho-proteomics. *Oncotarget* **2017**, *8*, 60223–60242. [CrossRef] [PubMed]
161. Ortega-Martínez, S. A new perspective on the role of the CREB family of transcription factors in memory consolidation via adult hippocampal neurogenesis. *Front. Mol. Neurosci.* **2015**, *8*, 46. [CrossRef] [PubMed]
162. Shen, S.; Yu, H.; Liu, X.; Liu, Y.; Zheng, J.; Wang, P. PIWIL1/piRNA-DQ593109 regulates the permeability of the blood-tumor barrier via the MEG3/miR-330-5p/RUNX3 axis. *Mol. Ther. Nucleic Acids* **2018**, *10*, 412–425. [CrossRef] [PubMed]
163. Schulze, M.; Sommer, A.; Plötz, S.; Farrell, M.; Winner, B.; Grosch, J. Sporadic Parkinson's disease derived neuronal cells show disease-specific mRNA and small RNA signatures with abundant deregulation of piRNAs. *Acta Neuropathol. Commun.* **2018**, *6*, 58. [CrossRef]
164. Roy, J.; Sarkar, A.; Parida, S.; Ghosh, Z.; Mallick, B. Small RNA sequencing revealed dysregulated piRNAs in alzheimer's disease and their probable role in pathogenesis. *Mol. Biosyst.* **2017**, *13*, 565–576. [CrossRef]
165. Saxena, A.; Tang, D.; Carninci, P. piRNAs warrant investigation in rett syndrome: An omics perspective. *Dis. Markers* **2012**, *33*, 261–275. [CrossRef]
166. Dharap, A.; Nakka, V.P.; Vemuganti, R. Altered expression of PIWI RNA in the rat brain after transient focal ischemia. *Stroke* **2011**, *42*, 1105–1109. [CrossRef]
167. Kim, K.W.; Tang, N.H.; Andrusiak, M.G.; Wu, Z.; Chisholm, A.D.; Jin, Y. A neuronal piRNA pathway inhibits axon regeneration in C. elegans. *Neuron* **2018**, *97*, 511–519.e6. [CrossRef] [PubMed]
168. Fu, A.; Jacobs, D.I.; Hoffman, A.E.; Zheng, T.; Zhu, Y. PIWI-interacting RNA 021285 is involved in breast tumorigenesis possibly by remodeling the cancer epigenome. *Carcinogenesis* **2015**, *36*, 1094–1102. [CrossRef] [PubMed]
169. Zhang, W.; Liu, H.; Yin, J.; Wu, W.; Zhu, D.; Amos, C.I. Genetic variants in the PIWI-piRNA pathway gene DCP1A predict melanoma disease-specific survival. *Int. J. Cancer* **2016**, *139*, 2730–2737. [CrossRef] [PubMed]
170. Jacobs, D.I.; Qin, Q.; Lerro, M.C.; Fu, A.; Dubrow, R.; Claus, E.B. PIWI-interacting RNAs in gliomagenesis: Evidence from post-GWAS and functional analyses. *Cancer Epidemiol. Biomark. Prev.* **2016**, *25*, 1073–1080. [CrossRef] [PubMed]

171. Martinez, V.D.; Vucic, E.A.; Thu, K.L.; Hubaux, R.; Enfield, K.S.S.; Pikor, L.A. Unique somatic and malignant expression patterns implicate PIWI-interacting RNAs in cancer-type specific biology. *Sci. Rep.* **2015**, *5*. [CrossRef]
172. Kwon, C.; Tak, H.; Rho, M.; Chang, H.R.; Kim, Y.H.; Kim, K.T. Detection of PIWI and piRNAs in the mitochondria of mammalian cancer cells. *Biochem. Biophys. Res. Commun.* **2014**, *446*, 218–223. [CrossRef]
173. Cavalcante, G.C.; Magalhães, L.; Ribeiro-dos-Santos, Â.; Vidal, A.F. Mitochondrial epigenetics: Non-coding RNAs as a novel layer of complexity. *Int. J. Mol. Sci.* **2020**, *21*, 1838. [CrossRef] [PubMed]
174. Malik, B.; Feng, F.Y. Long noncoding RNAs in prostate cancer: Overview and clinical implications. *Asian J. Androl.* **2016**, *18*, 568–574.
175. Mercer, T.R.; Mattick, J.S. Structure and function of long noncoding RNAs in epigenetic regulation. *Nat. Struct. Mol. Biol.* **2013**, *20*, 300–307. [CrossRef]
176. Zhai, K.; Liu, B.; Gao, L. *Long-Noncoding RNA TUG1* promotes Parkinson's disease via modulating *MiR-152-3p/PTEN* pathway. *Hum. Gene Ther.* **2020**, *31*, 1274–1287. [CrossRef]
177. Xu, W.; Zhang, L.; Geng, Y.; Liu, Y.; Zhang, N. Long noncoding RNA GAS5 promotes microglial inflammatory response in Parkinson's disease by regulating NLRP3 pathway through sponging miR-223-3p. *Int. Immunopharmacol.* **2020**, *85*, 106614. [CrossRef]
178. Zhang, Y.; Shao, W.; Wu, J.; Huang, S.; Yang, H.; Luo, Z. Inflammatory lncRNA AK039862 regulates paraquat-inhibited proliferation and migration of microglial and neuronal cells through the Pafah1b1/Foxa1 pathway in co-culture environments. *Ecotoxicol. Environ. Saf.* **2021**, *208*, 111424. [CrossRef]
179. Chen, M.-Y.; Fan, K.; Zhao, L.-J.; Wei, J.-M.; Gao, J.-X.; Li, Z.-F. Long non-coding RNA nuclear enriched abundant transcript 1 (NEAT1) sponges microRNA-124-3p to up-regulate phosphodiesterase 4B (PDE4B) to accelerate the progression of Parkinson's disease. *Bioengineered* **2021**, *12*, 708–719. [CrossRef]
180. Sun, Q.; Zhang, Y.; Wang, S.; Yang, F.; Cai, H.; Xing, Y. NEAT1 decreasing suppresses parkinson's disease progression via acting as miR-1301-3p sponge. *J. Mol. Neurosci.* **2021**, *71*, 369–378. [CrossRef]
181. Liu, R.; Li, F.; Zhao, W. Long noncoding RNA NEAT1 knockdown inhibits MPP+-induced apoptosis, inflammation and cytotoxicity in SK-N-SH cells by regulating miR-212-5p/RAB3IP axis. *Neurosci. Lett.* **2020**, *731*, 135060. [CrossRef]
182. Boros, F.A.; Maszlag-Török, R.; Vécsei, L.; Klivényi, P. Increased level of NEAT1 long non-coding RNA is detectable in peripheral blood cells of patients with Parkinson's disease. *Brain Res.* **2020**, *1730*, 146672. [CrossRef]
183. Quan, Y.; Wang, J.; Wang, S.; Zhao, J. Association of the plasma long non-coding RNA MEG3 with Parkinson's disease. *Front. Neurol.* **2020**, *11*, 532891. [CrossRef] [PubMed]
184. Zou, J.; Guo, Y.; Wei, L.; Yu, F.; Yu, B.; Xu, A. Long noncoding RNA POU3F3 and α-synuclein in plasma L1CAM exosomes combined with β-glucocerebrosidase activity: Potential predictors of Parkinson's disease. *Neurotherapeutics* **2020**, *17*, 1104–1119. [CrossRef] [PubMed]
185. Han, C.-L.; Liu, Y.-P.; Sui, Y.-P.; Chen, N.; Du, T.-T.; Jiang, Y. Integrated transcriptome expression profiling reveals a novel lncRNA associated with L-DOPA-induced dyskinesia in a rat model of Parkinson's disease. *Aging* **2020**, *12*, 718–739. [CrossRef] [PubMed]
186. Jeck, W.R.; Sorrentino, J.A.; Wang, K.; Slevin, M.K.; Burd, C.E.; Liu, J. Circular RNAs are abundant, conserved, and associated with ALU repeats. *RNA* **2013**, *19*, 141–157. [CrossRef]
187. D'Ambra, E.; Capauto, D.; Morlando, M. Exploring the regulatory role of circular RNAs in neurodegenerative disorders. *Int. J. Mol. Sci.* **2019**, *20*, 5477. [CrossRef] [PubMed]
188. Chen, L.-L. The expanding regulatory mechanisms and cellular functions of circular RNAs. *Nat. Rev. Mol. Cell Biol.* **2020**, *21*, 475–490. [CrossRef] [PubMed]
189. Wilusz, J.E. A 360° view of circular RNAs: From biogenesis to functions. *WIREs RNA* **2018**, *9*, e1478. [CrossRef] [PubMed]
190. Vo, J.N.; Cieslik, M.; Zhang, Y.; Shukla, S.; Xiao, L.; Zhang, Y. The landscape of circular RNA in cancer. *Cell* **2019**, *176*, 869–881.e13. [CrossRef] [PubMed]
191. Ji, P.; Wu, W.; Chen, S.; Zheng, Y.; Zhou, L.; Zhang, J. Expanded expression landscape and prioritization of circular RNAs in mammals. *Cell Rep.* **2019**, *26*, 3444–3460.e5. [CrossRef] [PubMed]
192. Yang, Y.; Fan, X.; Mao, M.; Song, X.; Wu, P.; Zhang, Y. Extensive translation of circular RNAs driven by N6-methyladenosine. *Cell Res.* **2017**, *27*, 626–641. [CrossRef]
193. Lu, S.; Yang, X.; Wang, C.; Chen, S.; Lu, S.; Yan, W. Current status and potential role of circular RNAs in neurological disorders. *J. Neurochem.* **2019**, *150*, 237–248. [CrossRef]
194. You, X.; Vlatkovic, I.; Babic, A.; Will, T.; Epstein, I.; Tushev, G. Neural circular RNAs are derived from synaptic genes and regulated by development and plasticity. *Nat. Neurosci.* **2015**, *18*, 603–610. [CrossRef]
195. Qu, S.; Yang, X.; Li, X.; Wang, J.; Gao, Y.; Shang, R. Circular RNA: A new star of noncoding RNAs. *Cancer Lett.* **2015**, *365*, 141–148. [CrossRef]
196. Sang, Q.; Liu, X.; Wang, L.; Qi, L.; Sun, W.; Wang, W. CircSNCA downregulation by pramipexole treatment mediates cell apoptosis and autophagy in Parkinson's disease by targeting miR-7. *Aging* **2018**, *10*, 1281–1293. [CrossRef]
197. Hansen, T.B.; Jensen, T.I.; Clausen, B.H.; Bramsen, J.B.; Finsen, B.; Damgaard, C.K. Natural RNA circles function as efficient microRNA sponges. *Nature* **2013**, *495*, 384–388. [CrossRef]
198. Feng, Z.; Zhang, L.; Wang, S.; Hong, Q. Circular RNA circDLGAP4 exerts neuroprotective effects via modulating miR-134-5p/CREB pathway in Parkinson's disease. *Biochem. Biophys. Res. Commun.* **2020**, *522*, 388–394. [CrossRef] [PubMed]

199. Hanan, M.; Simchovitz, A.; Yayon, N.; Vaknine, S.; Cohen-Fultheim, R.; Karmon, M. A Parkinson's disease circ RNA s resource reveals a link between circ SLC 8A1 and oxidative stress. *EMBO Mol. Med.* **2020**, *12*. [CrossRef]
200. Sirabella, R.; Sisalli, M.J.; Costa, G.; Omura, K.; Ianniello, G.; Pinna, A. NCX1 and NCX3 as potential factors contributing to neurodegeneration and neuroinflammation in the A53T transgenic mouse model of Parkinson's Disease. *Cell Death Dis.* **2018**, *9*, 725. [CrossRef]
201. Zhou, L.; Yang, L.; Li, Y.; Mei, R.; Yu, H.; Gong, Y. MicroRNA-128 protects dopamine neurons from apoptosis and upregulates the expression of excitatory amino acid transporter 4 in parkinson's disease by binding to AXIN1. *Cell Physiol. Biochem.* **2018**, *51*, 2275–2289. [CrossRef] [PubMed]
202. Min, S.-W.; Sohn, P.D.; Cho, S.-H.; Swanson, R.A.; Gan, L. Sirtuins in neurodegenerative diseases: An update on potential mechanisms. *Front. Aging Neurosci.* **2013**, *5*, 53. [CrossRef]
203. Shaughnessy, D.T.; McAllister, K.; Worth, L.; Haugen, A.C.; Meyer, J.N.; Domann, F.E. Mitochondria, energetics, epigenetics, and cellular responses to stress. *Environ. Health Perspect.* **2014**, *122*, 1271–1278. [CrossRef]
204. Brooks, H.R. Mitochondria: Finding the power to change. *Cell* **2018**, *175*, 891–893. [CrossRef]
205. Burke, P.J. Mitochondria, bioenergetics and apoptosis in cancer. *Trends Cancer* **2017**, *3*, 857–870. [CrossRef] [PubMed]
206. Anderson, S.; Bankier, A.T.; Barrell, B.G.; de Bruijn, M.H.L.; Coulson, A.R.; Drouin, J. Sequence and organization of the human mitochondrial genome. *Nature* **1981**, *290*, 457–465. [CrossRef]
207. Gammage, P.A.; Frezza, C. Mitochondrial DNA: The overlooked oncogenome? *BMC Biol.* **2019**, *17*, 53. [CrossRef]
208. Karakaidos, P.; Rampias, T. Mitonuclear interactions in the maintenance of mitochondrial integrity. *Life* **2020**, *10*, 173. [CrossRef]
209. Taanman, J.-W. The mitochondrial genome: Structure, transcription, translation and replication. *Biochim. Biophys. Acta BBA Bioenerg.* **1999**, *1410*, 103–123. [CrossRef]
210. Alexeyev, M.; Shokolenko, I.; Wilson, G.; LeDoux, S. The maintenance of mitochondrial DNA integrity-critical analysis and update. *Cold Spring Harb. Perspect. Biol.* **2013**, *5*. [CrossRef] [PubMed]
211. Picard, M.; McEwen, B.S. Mitochondria impact brain function and cognition. *Proc. Natl. Acad. Sci. USA* **2014**, *111*, 7–8. [CrossRef] [PubMed]
212. Cabral-Costa, J.V.; Kowaltowski, A.J. Neurological disorders and mitochondria. *Mol. Asp. Med.* **2020**, *71*, 100826. [CrossRef]
213. Cunnane, S.C.; Trushina, E.; Morland, C.; Prigione, A.; Casadesus, G.; Andrews, Z.B. Brain energy rescue: An emerging therapeutic concept for neurodegenerative disorders of ageing. *Nat. Rev. Drug Discov.* **2020**, *19*, 609–633. [CrossRef] [PubMed]
214. Langston, J.W.; Langston, E.B.; Irwin, I. MPTP-induced parkinsonism in human and non-human primates-clinical and experimental aspects. *Acta Neurol. Scand. Suppl.* **1984**, *100*, 49–54.
215. Cannon, J.R.; Tapias, V.; Na, H.M.; Honick, A.S.; Drolet, R.E.; Greenamyre, J.T. A highly reproducible rotenone model of Parkinson's disease. *Neurobiol. Dis.* **2009**, *34*, 279–290. [CrossRef]
216. Schapira, A.H.V.; Mann, V.M.; Cooper, J.M.; Dexter, D.; Daniel, S.E.; Jenner, P. Anatomic and disease specificity of NADH CoQ_1 reductase (Complex I) deficiency in Parkinson's disease. *J. Neurochem.* **1990**, *55*, 2142–2145. [CrossRef] [PubMed]
217. Blin, O.; Desnuelle, C.; Rascol, O.; Borg, M.; Paul, H.P.S.; Azulay, J.P. Mitochondrial respiratory failure in skeletal muscle from patients with Parkinson's disease and multiple system atrophy. *J. Neurol. Sci.* **1994**, *125*, 95–101. [CrossRef]
218. Haas, R.H.; Nasirian, F.; Nakano, K.; Ward, D.; Pay, M.; Hill, R. Low platelet mitochondrial complex I and complex II/III activity in early untreated parkinson's disease: Abnormalities of electron transport complexes in PD. *Ann. Neurol.* **1995**, *37*, 714–722. [CrossRef] [PubMed]
219. Gu, M.; Cooper, J.M.; Taanman, J.W.; Schapira, A.H.V. Mitochondrial DNA transmission of the mitochondrial defect in Parkinson's disease. *Ann. Neurol.* **1998**, *44*, 177–186. [CrossRef]
220. Bose, A.; Beal, M.F. Mitochondrial dysfunction in Parkinson's disease. *J. Neurochem.* **2016**, *139*, 216–231. [CrossRef]
221. Taylor, R.W.; Turnbull, D.M. Mitochondrial DNA mutations in human disease. *Nat. Rev. Genet.* **2005**, *6*, 389–402. [CrossRef] [PubMed]
222. Schon, E.A.; DiMauro, S.; Hirano, M. Human mitochondrial DNA: Roles of inherited and somatic mutations. *Nat. Rev. Genet.* **2012**, *13*, 878–890. [CrossRef] [PubMed]
223. Antonyová, V.; Kejík, Z.; Brogyányi, T.; Kaplánek, R.; Pajková, M.; Talianová, V. Role of mtDNA disturbances in the pathogenesis of Alzheimer's and Parkinson's disease. *DNA Repair* **2020**, *91–92*, 102871. [CrossRef]
224. Richter, G.; Sonnenschein, A.; Grünewald, T.; Reichmann, H.; Janetzky, B. Novel mitochondrial DNA mutations in Parkinson's disease. *J. Neural. Transm.* **2002**, *109*, 721–729. [CrossRef]
225. Shoffner, J.M.; Brown, M.D.; Torroni, A.; Lott, M.T.; Cabell, M.F.; Mirra, S.S. Mitochondrial DNA variants observed in Alzheimer disease and Parkinson disease patients. *Genomics* **1993**, *17*, 171–184. [CrossRef]
226. Huerta, C.; Castro, M.G.; Coto, E.; Blázquez, M.; Ribacoba, R.; Guisasola, L.M. Mitochondrial DNA polymorphisms and risk of Parkinson's disease in Spanish population. *J. Neurol. Sci.* **2005**, *236*, 49–54. [CrossRef]
227. Egensperger, R.; Kösel, S.; Schnopp, N.M.; Mehraein, P.; Graeber, M.B. Association of the mitochondrial tRNAA4336G mutation with Alzheimer's and Parkinson's diseases. *Neuropathol. Appl. Neurobiol.* **1997**, *23*, 315–321. [CrossRef] [PubMed]
228. Nido, G.S.; Dölle, C.; Flønes, I.; Tuppen, H.A.; Alves, G.; Tysnes, O.-B. Ultradeep mapping of neuronal mitochondrial deletions in Parkinson's disease. *Neurobiol. Aging* **2018**, *63*, 120–127. [CrossRef] [PubMed]
229. Coxhead, J.; Kurzawa-Akanbi, M.; Hussain, R.; Pyle, A.; Chinnery, P.; Hudson, G. Somatic mtDNA variation is an important component of Parkinson's disease. *Neurobiol. Aging* **2016**, *38*, 217.e1–217.e6. [CrossRef]

230. Ross, O.A.; McCormack, R.; Maxwell, L.D.; Duguid, R.A.; Quinn, D.J.; Barnett, Y.A. mt4216C variant in linkage with the mtDNA TJ cluster may confer a susceptibility to mitochondrial dysfunction resulting in an increased risk of Parkinson's disease in the Irish. *Exp. Gerontol.* **2003**, *38*, 397–405. [CrossRef]
231. van der Walt, J.M.; Nicodemus, K.K.; Martin, E.R.; Scott, W.K.; Nance, M.A.; Watts, R.L. Mitochondrial Polymorphisms Significantly Reduce the Risk of Parkinson Disease. *Am. J. Hum. Genet.* **2003**, *72*, 804–811. [CrossRef]
232. Hudson, G.; Nalls, M.; Evans, J.R.; Breen, D.P.; Winder-Rhodes, S.; Morrison, K.E. Two-stage association study and meta-analysis of mitochondrial DNA variants in Parkinson disease. *Neurology* **2013**, *80*, 2042–2048. [CrossRef] [PubMed]
233. Rosner, S.; Giladi, N.; Orr-Urtreger, A. Advances in the genetics of Parkinson's disease. *Acta Pharmacol. Sin.* **2008**, *29*, 21–34. [CrossRef]
234. Singleton, A.; Hardy, J. The evolution of genetics: Alzheimer's and Parkinson's Diseases. *Neuron* **2016**, *90*, 1154–1163. [CrossRef]
235. Grünewald, A.; Kumar, K.R.; Sue, C.M. New insights into the complex role of mitochondria in Parkinson's disease. *Prog. Neurobiol.* **2019**, *177*, 73–93. [CrossRef]
236. Ashrafi, G.; Schwarz, T.L. The pathways of mitophagy for quality control and clearance of mitochondria. *Cell Death Differ.* **2013**, *20*, 31–42. [CrossRef]
237. Kitada, T.; Asakawa, S.; Hattori, N.; Matsumine, H.; Yamamura, Y.; Minoshima, S. Mutations in the parkin gene cause autosomal recessive juvenile parkinsonism. *Nature* **1998**, *392*, 605–608. [CrossRef]
238. Valente, E.M.; Bentivoglio, A.R.; Dixon, P.H.; Ferraris, A.; Ialongo, T.; Frontali, M. Localization of a novel locus for autosomal recessive early-onset parkinsonism, park6, on human chromosome 1p35-p36. *Am. J. Hum. Genet.* **2001**, *68*, 895–900. [CrossRef] [PubMed]
239. van Duijn, C.M.; Dekker, M.C.J.; Bonifati, V.; Galjaard, R.J.; Houwing-Duistermaat, J.J.; Snijders, P.J.L.M. PARK7, a novel locus for autosomal recessive early-onset parkinsonism, on chromosome 1p36. *Am. J. Hum. Genet.* **2001**, *69*, 629–634. [CrossRef] [PubMed]
240. Cai, Q.; Jeong, Y.Y. Mitophagy in Alzheimer's disease and other age-related neurodegenerative diseases. *Cells* **2020**, *9*, 150. [CrossRef]
241. Iacobazzi, V.; Castegna, A.; Infantino, V.; Andria, G. Mitochondrial DNA methylation as a next-generation biomarker and diagnostic tool. *Mol. Genet. Metab.* **2013**, *110*, 25–34. [CrossRef]
242. D'Aquila, P.; Bellizzi, D.; Passarino, G. Mitochondria in health, aging and diseases: The epigenetic perspective. *Biogerontology* **2015**, *16*, 569–585. [CrossRef] [PubMed]
243. Devall, M.; Smith, R.G.; Jeffries, A.; Hannon, E.; Davies, M.N.; Schalkwyk, L. Regional differences in mitochondrial DNA methylation in human post-mortem brain tissue. *Clin. Epigenet.* **2017**, *9*, 47. [CrossRef]
244. Patil, V.; Cuenin, C.; Chung, F.; Aguilera, J.R.R.; Fernandez-Jimenez, N.; Romero-Garmendia, I. Human mitochondrial DNA is extensively methylated in a non-CpG context. *Nucleic Acids Res.* **2019**, *47*, 10072–10085. [CrossRef]
245. Mercer, T.R.; Neph, S.; Dinger, M.E.; Crawford, J.; Smith, M.A.; Shearwood, A.-M.J. The human mitochondrial transcriptome. *Cell* **2011**, *146*, 645–658. [CrossRef]
246. Gusic, M.; Prokisch, H. ncRNAs: New players in mitochondrial health and disease? *Front. Genet.* **2020**, *11*, 95. [CrossRef]
247. Blanch, M.; Mosquera, J.L.; Ansoleaga, B.; Ferrer, I.; Barrachina, M. Altered mitochondrial DNA methylation pattern in Alzheimer disease–related pathology and in Parkinson disease. *Am. J. Pathol.* **2016**, *186*, 385–397. [CrossRef]
248. Chuang, Y.-H.; Lu, A.T.; Paul, K.C.; Folle, A.D.; Bronstein, J.M.; Bordelon, Y. Longitudinal epigenome-wide methylation study of cognitive decline and motor progression in Parkinson's disease. *JPD* **2019**, *9*, 389–400. [CrossRef]
249. Lyu, Y.; Bai, L.; Qin, C. Long noncoding RNAs in neurodevelopment and Parkinson's disease. *Anim. Models. Exp. Med.* **2019**, *2*, 239–251. [CrossRef]
250. Cho, I.; Blaser, M.J. The human microbiome: At the interface of health and disease. *Nat. Rev. Genet.* **2012**, *13*, 260–270. [CrossRef]
251. Lloyd-Price, J.; Abu-Ali, G.; Huttenhower, C. The healthy human microbiome. *Genome Med.* **2016**, *8*, 51. [CrossRef]
252. Sasmita, A.O. Modification of the gut microbiome to combat neurodegeneration. *Rev. Neurosci.* **2019**, *30*, 795–805. [CrossRef]
253. Rhee, S.H.; Pothoulakis, C.; Mayer, E.A. Principles and clinical implications of the brain–gut–enteric microbiota axis. *Nat. Rev. Gastroenterol. Hepatol.* **2009**, *6*, 306–314. [CrossRef]
254. Braak, H.; Tredici, K.D.; Rüb, U.; de Vos, R.A.I.; Jansen Steur, E.N.H.; Braak, E. Staging of brain pathology related to sporadic Parkinson's disease. *Neurobiol. Aging* **2003**, *24*, 197–211. [CrossRef]
255. Svensson, E.; Horváth-Puhó, E.; Thomsen, R.W.; Djurhuus, J.C.; Pedersen, L.; Borghammer, P. Vagotomy and subsequent risk of Parkinson's disease: Vagotomy and risk of PD. *Ann. Neurol.* **2015**, *78*, 522–529. [CrossRef]
256. Sampson, T.R.; Debelius, J.W.; Thron, T.; Janssen, S.; Shastri, G.G.; Ilhan, Z.E. Gut microbiota regulate motor deficits and neuroinflammation in a model of Parkinson's disease. *Cell* **2016**, *167*, 1469–1480.e12. [CrossRef] [PubMed]
257. Bhattarai, Y.; Si, J.; Pu, M.; Ross, O.A.; McLean, P.J.; Till, L. Role of gut microbiota in regulating gastrointestinal dysfunction and motor symptoms in a mouse model of Parkinson's disease. *Gut Microbes* **2021**, *13*. [CrossRef] [PubMed]
258. Dodiya, H.B.; Forsyth, C.B.; Voigt, R.M.; Engen, P.A.; Patel, J.; Shaikh, M. Chronic stress-induced gut dysfunction exacerbates Parkinson's disease phenotype and pathology in a rotenone-induced mouse model of Parkinson's disease. *Neurobiol. Dis.* **2020**, *135*, 104352. [CrossRef] [PubMed]
259. Gorecki, A.M.; Preskey, L.; Bakeberg, M.C.; Kenna, J.E.; Gildenhuys, C.; MacDougall, G. Altered gut microbiome in Parkinson's disease and the influence of lipopolysaccharide in a human α-synuclein over-expressing mouse model. *Front. Neurosci.* **2019**, *13*, 839. [CrossRef]

260. Aho, V.T.E.; Pereira, P.A.B.; Voutilainen, S.; Paulin, L.; Pekkonen, E.; Auvinen, P. Gut microbiota in Parkinson's disease: Temporal stability and relations to disease progression. *EBioMedicine* **2019**, *44*, 691–707. [CrossRef]
261. Cilia, R.; Piatti, M.; Cereda, E.; Bolliri, C.; Caronni, S.; Ferri, V. Does gut microbiota influence the course of Parkinson's disease? A 3-year prospective exploratory study in de novo patients. *JPD* **2021**, *11*, 159–170. [CrossRef]
262. Unger, M.M.; Spiegel, J.; Dillmann, K.-U.; Grundmann, D.; Philippeit, H.; Bürmann, J. Short chain fatty acids and gut microbiota differ between patients with Parkinson's disease and age-matched controls. *Parkinsonism Relat. Disord.* **2016**, *32*, 66–72. [CrossRef]
263. Lin, C.-H.; Chen, C.-C.; Chiang, H.-L.; Liou, J.-M.; Chang, C.-M.; Lu, T.-P. Altered gut microbiota and inflammatory cytokine responses in patients with Parkinson's disease. *J. Neuroinflamm.* **2019**, *16*, 129. [CrossRef]
264. Scheperjans, F.; Aho, V.; Pereira, P.A.B.; Koskinen, K.; Paulin, L.; Pekkonen, E. Gut microbiota are related to Parkinson's disease and clinical phenotype. *Mov. Disord.* **2015**, *30*, 350–358. [CrossRef]
265. Hegelmaier, T.; Lebbing, M.; Duscha, A.; Tomaske, L.; Tönges, L.; Holm, J.B. Interventional influence of the intestinal microbiome through dietary intervention and bowel cleansing might improve motor symptoms in Parkinson's disease. *Cells* **2020**, *9*, 376. [CrossRef]
266. Hertel, J.; Harms, A.C.; Heinken, A.; Baldini, F.; Thinnes, C.C.; Glaab, E. Integrated analyses of microbiome and longitudinal metabolome data reveal microbial-host interactions on sulfur metabolism in Parkinson's disease. *Cell Rep.* **2019**, *29*, 1767–1777.e8. [CrossRef] [PubMed]
267. Baldini, F.; Hertel, J.; Sandt, E.; Thinnes, C.C.; Neuberger-Castillo, L.; Pavelka, L. Parkinson's disease-associated alterations of the gut microbiome predict disease-relevant changes in metabolic functions. *BMC Biol.* **2020**, *18*, 62. [CrossRef] [PubMed]
268. Morrison, D.J.; Preston, T. Formation of short chain fatty acids by the gut microbiota and their impact on human metabolism. *Gut Microbes* **2016**, *7*, 189–200. [CrossRef] [PubMed]
269. Hou, Y.; Shan, C.; Zhuang, S.; Zhuang, Q.; Ghosh, A.; Zhu, K. Gut microbiota-derived propionate mediates the neuroprotective effect of osteocalcin in a mouse model of Parkinson's disease. *Microbiome* **2021**, *9*, 34. [CrossRef]
270. Nishiwaki, H.; Hamaguchi, T.; Ito, M.; Ishida, T.; Maeda, T.; Kashihara, K. Short-chain fatty acid-producing gut microbiota is decreased in Parkinson's disease but not in rapid-eye-movement sleep behavior disorder. *MSystems* **2020**, *5*, e00797-20. [CrossRef]
271. Cirstea, M.S.; Yu, A.C.; Golz, E.; Sundvick, K.; Kliger, D.; Radisavljevic, N. Microbiota composition and metabolism are associated with gut function in Parkinson's disease. *Mov. Disord.* **2020**, *35*, 1208–1217. [CrossRef]
272. Rowland, I.; Gibson, G.; Heinken, A.; Scott, K.; Swann, J.; Thiele, I. Gut microbiota functions: Metabolism of nutrients and other food components. *Eur. J. Nutr.* **2018**, *57*, 1–24. [CrossRef]
273. Ho, L.; Zhao, D.; Ono, K.; Ruan, K.; Mogno, I.; Tsuji, M. Heterogeneity in gut microbiota drive polyphenol metabolism that influences α-synuclein misfolding and toxicity. *J. Nutr. Biochem.* **2019**, *64*, 170–181. [CrossRef]
274. Haikal, C.; Chen, Q.-Q.; Li, J.-Y. Microbiome changes: An indicator of Parkinson's disease? *Transl. Neurodegener.* **2019**, *8*, 38. [CrossRef]
275. Vascellari, S.; Palmas, V.; Melis, M.; Pisanu, S.; Cusano, R.; Uva, P. Gut microbiota and metabolome alterations associated with Parkinson's disease. *MSystems* **2020**, *5*, e00561-20. [CrossRef]
276. Losurdo, G.; D'Abramo, F.S.; Indellicati, G.; Lillo, C.; Ierardi, E.; Di Leo, A. The influence of small intestinal bacterial overgrowth in digestive and extra-intestinal disorders. *Int. J. Mol. Sci.* **2020**, *21*, 3531. [CrossRef]
277. Tan, A.H.; Mahadeva, S.; Thalha, A.M.; Gibson, P.R.; Kiew, C.K.; Yeat, C.M. Small intestinal bacterial overgrowth in Parkinson's disease. *Park. Relat. Disord.* **2014**, *20*, 535–540. [CrossRef]
278. Hewel, C.; Kaiser, J.; Wierczeiko, A.; Linke, J.; Reinhardt, C.; Endres, K. Common miRNA patterns of alzheimer's disease and Parkinson's disease and their putative impact on commensal gut microbiota. *Front. Neurosci.* **2019**, *13*, 113. [CrossRef]
279. Fong, W.; Li, Q.; Yu, J. Gut microbiota modulation: A novel strategy for prevention and treatment of colorectal cancer. *Oncogene* **2020**, *39*, 4925–4943. [CrossRef]
280. Vendrik, K.E.W.; Ooijevaar, R.E.; de Jong, P.R.C.; Laman, J.D.; van Oosten, B.W.; van Hilten, J.J. Fecal microbiota transplantation in neurological disorders. *Front. Cell Infect. Microbiol.* **2020**, *10*, 98. [CrossRef]
281. Koutzoumis, D.N.; Vergara, M.; Pino, J.; Buddendorff, J.; Khoshbouei, H.; Mandel, R.J. Alterations of the gut microbiota with antibiotics protects dopamine neuron loss and improve motor deficits in a pharmacological rodent model of Parkinson's disease. *Exp. Neurol.* **2020**, *325*, 113159. [CrossRef] [PubMed]
282. Uyar, G.Ö.; Yildiran, H. A nutritional approach to microbiota in Parkinson's disease. *Biosci. Microbiota Food Health* **2019**, *38*, 115–127. [CrossRef] [PubMed]
283. Huang, H.; Xu, H.; Luo, Q.; He, J.; Li, M.; Chen, H.; Tang, W.; Nie, Y.; Zhou, Y. Fecal microbiota transplantation to treat Parkinson's disease with constipation: A case report. *Medicine* **2019**, *98*, e16163. [CrossRef] [PubMed]
284. Hazan, S. Rapid improvement in Alzheimer's disease symptoms following fecal microbiota transplantation: A case report. *J. Int. Med. Res.* **2020**, *48*. [CrossRef] [PubMed]
285. Scheperjans, F.; Derkinderen, P.; Borghammer, P. The gut and Parkinson's disease: Hype or hope? *JPD* **2018**, *8*, S31–S39. [CrossRef] [PubMed]
286. Borghammer, P.; Van Den Berge, N. Brain-first versus gut-first parkinson's disease: A hypothesis. *JPD* **2019**, *9*, S281–S295. [CrossRef]

International Journal of
Molecular Sciences

MDPI

Review

From Menopause to Neurodegeneration—Molecular Basis and Potential Therapy

Yu-Jung Cheng [1,2], Chieh-Hsin Lin [3,4,5,6,*] and Hsien-Yuan Lane [3,4,7,8,*]

1 Department of Physical Therapy and Graduate Institute of Rehabilitation Science, China Medical University, Taichung 40402, Taiwan; chengyu@mail.cmu.edu.tw
2 Department of Rehabilitation, China Medical University Hospital, Taichung 40402, Taiwan
3 Institute of Clinical Medical Science, China Medical University, Taichung 40402, Taiwan
4 Graduate Institute of Biomedical Sciences, China Medical University, Taichung 40402, Taiwan
5 Kaohsiung Chang Gung Memorial Hospital, Chang Gung University College of Medicine, Kaohsiung 83301, Taiwan
6 School of Medicine, Chang Gung University, Taoyuan 33302, Taiwan
7 Department of Psychiatry & Brain Disease Research Center, China Medical University Hospital, Taichung 40402, Taiwan
8 Department of Psychology, College of Medical and Health Sciences, Asia University, Taichung 41354, Taiwan
* Correspondence: simone36@cgmh.org.tw (C.-H.L.); hylane@mail.cmuh.org.tw (H.-Y.L.); Tel.: +886-7-7317123 (ext. 8753) (C.-H.L.); Fax: +886-7-7326817 (C.-H.L.)

Abstract: The impacts of menopause on neurodegenerative diseases, especially the changes in steroid hormones, have been well described in cell models, animal models, and humans. However, the therapeutic effects of hormone replacement therapy on postmenopausal women with neurodegenerative diseases remain controversial. The steroid hormones, steroid hormone receptors, and downstream signal pathways in the brain change with aging and contribute to disease progression. Estrogen and progesterone are two steroid hormones which decline in circulation and the brain during menopause. Insulin-like growth factor 1 (IGF-1), which plays an import role in neuroprotection, is rapidly decreased in serum after menopause. Here, we summarize the actions of estrogen, progesterone, and IGF-1 and their signaling pathways in the brain. Since the incidence of Alzheimer's disease (AD) is higher in women than in men, the associations of steroid hormone changes and AD are emphasized. The signaling pathways and cellular mechanisms for how steroid hormones and IGF-1 provide neuroprotection are also addressed. Finally, the molecular mechanisms of potential estrogen modulation on N-methyl-d-aspartic acid receptors (NMDARs) are also addressed. We provide the viewpoint of why hormone therapy has inconclusive results based on signaling pathways considering their complex response to aging and hormone treatments. Nonetheless, while diagnosable AD may not be treatable by hormone therapy, its preceding stage of mild cognitive impairment may very well be treatable by hormone therapy.

Keywords: menopause; estrogen; neurodegenerative disease; Alzheimer's disease; IGF-1; NMDAR

Citation: Cheng, Y.-J.; Lin, C.-H.; Lane, H.-Y. From Menopause to Neurodegeneration—Molecular Basis and Potential Therapy. *Int. J. Mol. Sci.* **2021**, *22*, 8654. https://doi.org/10.3390/ijms22168654

Academic Editors: Marcello Ciaccio and Luisa Agnello

Received: 26 July 2021
Accepted: 9 August 2021
Published: 11 August 2021

1. Introduction

Neurodegenerative diseases include movement, cognitive, and behavioral disorders which are caused along the process of neurodegeneration. The most common neurodegenerative diseases are Alzheimer's disease (AD), Parkinson's disease (PD), and dementia with Lewy bodies (DLB) [1–3]. Slow progression, lack of definitive diagnosis tools, and complex pathophysiology make these neurodegenerative diseases, especially AD, difficult to treat. In these neurodegenerative diseases, increases in neuronal loss are found with disease progression [4–6]. The causes of neuronal loss can be protein misfolding, overloaded proteostasis networks, oxidative stress, programmed cell death, and neuroinflammation [7]. In addition, endocrine dyscrasia by menopause and andropause is also correlated to neuronal dysfunction, cell death, and cognitive decline [8].

Many neurodegenerative diseases display sex-specific features. Pieces of evidence show that aging and female gender are more commonly related with incidence and prevalence of AD but not parkinsonism [9,10]. In addition to incidence and prevalence, the disease progression and treatment responses are also different according to gender [11]. Estrogen and progesterone show protective activity on brain functions, and thus loss of these steroid hormones at menopause is an important risk factor for AD progression [12,13]. In rats, ovariectomy causes a decrease in dendritic spines in hippocampus pyramidal cells, and this reduction can be reversed by 17β-estradiol treatment [14]. The same situation has also been observed in non-human primates. Compared to premenopausal rhesus macaques, rhesus macaques with natural menopause showed worse recognition memory and lower synapse density in the hippocampus [15]. In addition, the frequency of multisynaptic boutons decreased by over 50% after ovariectomy in aged monkeys compared to normal aging monkeys [16].

In this review, we first present evidence showing the sex differences, especially the impacts of menopause, in major neurodegenerative diseases. Then, we review the literature on the possible mechanisms of estrogen, progesterone, and other neurotrophic factors in the progression of neurodegenerative diseases. Finally, we present details on what is known about the molecular pathways through which menopause contributes to neurodegenerative diseases.

2. Sex Differences in Major Neurodegenerative Diseases

2.1. Alzheimer's Disease (AD)

Sex-related differences in AD have been well identified in clinical manifestations [17]. Women are more susceptible to AD, and incidence rates are higher in women than in men in Europe and the U.S. [18–22]. Beam and colleagues analyzed 16,926 women and men with any dementia, AD alone, and non-AD dementia alone and found that the incidence rates of all were greater in women [20]. Compared to men, women with AD dementia reach partial loss of autonomy faster, which means they live with more disability longer [23]. Amyloid-beta (Aβ) and hyperphosphorylated tau accumulation are the two major pathological changes in AD progression. Several studies show there is no sex difference in diffuse Aβ plaque deposit in brain or in cerebrospinal fluid (CSF) concentrations [24,25]. Unlike Aβ, higher numbers of neurofibrillary tangles in brains are found in women with AD than in men [24]. PET imaging studies have revealed no gender associated with tau accumulation in aging and early AD [26,27]. Studies which evaluated total tau and phosphorylated tau levels in CSF showed that sex had no impact on patients with AD or mild cognitive impairment (MCI) [28,29].

2.2. Parkinson's Disease (PD)

The prevalence of PD, the second most common neurodegenerative disease, is about 315 per 100,000 persons of all ages [30]. PD is attributed to progressive degeneration of midbrain dopaminergic neurons and motor symptoms. The loss of dopaminergic neurons is sex-related, which has been found in PD animal models and PD patients. Epidemiological studies show that the male sex is one of the risk factors for PD at all ages [22,31–36]. Comparing women of similar ages, the incidence and prevalence of PD are higher in postmenopausal than in premenopausal women [37–39], which implicates the benefit of estrogen and progesterone on PD onset. However, hormone replacement therapy in women has inconsistent results. A clinical trial on women with Parkinson's disease with or without dementia showed that estrogen replacement therapy has protective effects for the development of dementia [40]. Another multicenter case–control study showed that oral contraceptives could increase the risk of PD [41]. A meta-analysis showed that estrogen replacement therapy can significantly decrease the risk of onset and/or development of PD [42].

2.3. Dementia with Lewy Bodies (DLB)

The gender difference in prevalence of DLB is still inconclusive. A systematic review from 2014 summarized eight studies of DLB and reported that five studies showed disproportionately more females with the disease [43]. Later studies also showed inconsistency in gender distribution. Some studies showed that male sex is more prevalent [44–47], some favored women [48], and others showed no significance [49]. Mouton et al. investigated gender differences in DLB, AD, PD, and Parkinson's disease dementia (PDD) from the French national database spanning from 2010 to 2015. The result from a total of 237,695 patients showed that the sex ratio (female percent/male percent) is 1.21 for DLB, 2.34 for AD, 0.76 for PD, and 0.83 for PDD. This large-scale study agrees with the balanced sex distribution in DLB compared with AD and PD–PDD. Regarding biomarkers in the CSF of patients with DLB, women had lower alpha-synuclein (α-syn) CSF Aβ42 levels compared to men, and no sex differences for phospho-tau concentrations were found [50].

3. Role of Hormones and Trophic Factors in Neurodegenerative Diseases

There are several possible factors that contribute to sex differences in neurodegenerative diseases. Estrogen is the most frequently discussed factor due to its neuroprotective effects. Loss of estrogen after menopause increases the risk of neurodegenerative diseases, which indicates that estrogen plays an important role in disease progression and onset. During menopause transition, the follicle-stimulating hormone (FSH) to estrogen ratio increases due to increased FSH levels and decreased estrogen levels. The FSH/estrogen ratio can be applied as a screening method for MCI in postmenopausal women [51]. In AD patients under sodium benzoate treatment, reversal of the FSH/estrogen ratio is associated with better cognition improvement [11]. FSH and the FSH receptor are found in hippocampal neurons [52], and FSH-mediated neuroestrogen generation is associated with the regulation of neuroplasticity [53].

In addition to estrogen, other neurosteroids—i.e., the steroids produced within the nervous system—are reported to have neuroprotective effects [54,55]. Circulating progesterone is decreased during menopausal transition and drops dramatically following menopause [56,57]. The neuroprotective effects of progesterone are well documented, and progesterone and its metabolites, 5α-dihydroprogesterone (5α-DHPROG) and 3α,5α-tetrahydroprogesterone (3α,5α-THPROG), decrease with aging [58]. Levels of insulin-like growth factor 1 (IGF-1), which can be regulated by estrogen, are also reported to decline in serum with aging, especially after the age of 40 years [59,60]. Interestingly, production and secretion of estrogen, progesterone, and IGF-1 by the Central Nervous System (CNS) have been also detected [61–63]. In this section, we will discuss the effects of estrogen and other neurosteroids on neuroprotection and the impacts of menopause on neurodegeneration.

3.1. Effects of Estrogen on Neuroplasticity

The association of estrogen with AD progression has been implicated in morphological changes of neuronal cells, the cholinergic system, and amyloid processing. Through cell culture systems and ovariectomies in rodents, the effects of estrogen or menopause on nerve degeneration have been well characterized. 17β-estradiol, an estrogen steroid hormone, can rapidly induce long-term potentiation (LTP) and increase the spine density of the neurons of hippocampal slices. These synaptic plasticity effects are mediated through mitogen-activated protein kinase (ERK/MAPK) pathways [64]. Through binding to estrogen receptor (ER), 17β-estradiol increases the phosphorylation of AKT and Tropomyosin receptor kinase B (TrkB), the receptor of brain-derived neurotrophic factor (BDNF), in the hippocampus [65]. As AKT and BDNF are molecules which play well-known roles in synaptic plasticity, this result indicates that 17β-estradiol and ER have the ability to regulate hippocampal synaptic plasticity and improve memory. There are several studies showing that activation ERs can increase dendritic spine density and/or cognitive performance [66]. Using selective ERβ agonists, it has been shown that estrogen increased key synaptic proteins in the hippocampus, which might affect synaptic plasticity and memory through Erβ [67]. Moreover, infusion of ERα or ERβ agonists to the hippocampus enhanced memory in ovariectomized female mice in an ERK-dependent manner [68]. The expression levels of membrane ERα in synapse are higher in female mice, and ERα is essential in 17β-estradiol-induced LTP in females, as well as ERβ in males [69]. In addition, 17β-estradiol and the G-protein-coupled estrogen receptor 1 (GPER1)-specific agonist G1 can induce BDNF release and trigger LTP [70].

The impacts of menopause on the central nervous system are mostly due to the dramatic loss of estrogen. The hypothalamic–pituitary–gonadal axis, in which endocrine and neural systems interact with each other, is disturbed during pre- and postmenopause, leading to neurodegeneration [71]. Lower 17β-estradiol concentration is related to hippocampal dysfunction and poorer performance on initial learning and memory retrieval [72,73]. Moreover, shorter 17β-estradiol exposure due to delayed age at menarche, younger age at menopause, shorter reproductive span, and hysterectomies are correlated to higher dementia risk [74]. The average density of N-methyl-d-aspartic acid receptors (NMDARs) current in dorsal root ganglia is much higher in female rats than male rats [75]. Interestingly, 17β-estradiol-increased LTP depends on NMDA receptors in females and AMPA receptors in males [76,77]. The positive effects of 17β-estradiol on the hippocampus and memory are enacted through different signal pathways in males and females. In females, 17β-estradiol enhances memory consolidation through ERK, but not in males [78], and metabotropic (mGluR1a) and ionotropic (NMDA) glutamate receptors are essential for 17β-estradiol-mediated ERK activation and memory consolidation [68,79]. However, the effects of hormone replacement therapy on preventing AD or dementia in postmenopausal women are inconsistent [80,81]. In addition, although epidemiology studies indicate that estrogen loss after menopause might trigger severe symptoms in AD, the results from hormone replacement therapy in postmenopausal AD patients are varied [82,83]. The latest meta-analysis study showed that hormone replacement therapy is correlated with reduced risk of AD. Moreover, longer duration and natural steroid formulations are linked to better protection [84]. A recent review suggested that during the time frame of MCI, typically preceding fully developed AD, hormone replacement treatment could postpone and possibly even prevent the onset of diagnosable AD [85].

3.2. *Immunomodulatory Effects of Estrogen*

In addition to reproductive and sexual functions, estrogen has been shown to be able to modulate the immune system. Most immune cells and epithelial cells express ERs and progesterone receptors, and steroid hormones regulate innate immunity against bacteria in epithelial tissues [86]. 17β-estradiol protects the epithelium through modulating the expression of inflammatory cytokines and antimicrobial proteins [87]. Treatment with 17β-estradiol in epithelial cells can reduce inflammation but increase antimicrobial molecules, which maintain the balance between inflammation-induced tissue damage and antimicrobial ability [88]. Moreover, loss of ERα on epithelial cells also impairs the vaginal structural integrity and increases leukocyte infiltration [89]. These effects on the epithelium protect the vagina, urinary tract, and respiratory tract from infectious agents [90–92]. In addition to antimicrobial ability, estrogen can modulate gastrointestinal epithelial secretions [93]. For instance, 17β-estradiol inhibits Cl- secretion in rat distal colonic epithelium caused by endotoxins, which may prevent diarrhea [94,95]. Thus, the gastrointestinal tract is also a sex-steroid-targeted organ.

Since estrogen can affect the GI tract and the gut–brain axis participates in the pathogenesis of neurodegenerative diseases, the impact of menopause on the gut microbiota needs to be discussed. The gut microbiota is essential in maintaining homeostasis, and changes in the gut microbiota can also lead to alteration of behavior and cognition [96]. During menopause transition and the early menopausal period, abdominal irritation was found and negatively related to estrone glucuronide levels [97]. In postmenopausal women, the urinary estrogen level is associated with fecal microbiota [98]. Soy isoflavones, which exhibit estrogen-like properties, can significantly change the fecal bacterial community in postmenopausal women [99]. In ovariectomized mice, ovariectomy was associated with a shift in the beta diversity of the gut microbiota analyzed by 16S rRNA gene sequence analysis [100]. A recent study showed that *Verrucomicrobia* and *Actinobacteria* were significantly increased in ovariectomized mice compared with a control group [101]. In addition, the reduction in microbiota diversity in ovariectomy was changed by a probiotic and soy isoflavone diet treatment [102]. Another study on ovariectomized mice demonstrated that long-term conjugated estrogens plus bazedoxifene did not change the overall cecal or fecal microbiome community but modulated gut microbial β-glucuronidase activity [103]. These results show that changes in sex steroid hormones did affect the intestinal microecological environment, which might contribute to neurodegenerative diseases.

3.3. The Role of Estrogen in AD

Neurotoxic Aβ is one of the major pathogenic agents involved in the onset and progression of AD. The amyloid precursor protein (APP), which is highly expressed in nerve cells, plays a crucial role in AD pathogenesis. Aβ is generated by APP processing. At first, APP is cleaved by β-secretase BACE-1, leading to soluble APPβ (sAPPβ) and the membrane-bound fragment β-CTF. Further processing of β-CTF by γ-secretase creates Aβ, and the depositions of Aβ induce neurotoxicity [104]. Another processing pathway of APP is mediated by α-secretase, which can generate sAPPα [105]. Interestingly, sAPPα can protect neurons against Aβ-dependent neuron toxicity [106]. Incubation with 17β-estradiol induces the release of sAPPα by the ERK pathway in mouse hippocampal HT22 cells and human neuroblastoma SK-N-MC cells [107]. A study on primary rat cortical neurons also showed that 17β-estradiol increased the secretion of sAPPα through protein kinase C [108]. These data imply that estrogen might decrease Aβ neurotoxicity by increasing sAPPα secretion (Figure 1). It is becoming more and more appreciated that viral, bacterial, and fungal brain infections can be causative for AD [109]. Similarly, it is becoming understood that generation of Aβ, including Aβ plaques, is part of the response process against such brain infections [85,110]. Interestingly, hormone replacement appears to be effective against at least several very serious infections, i.e., against the infectious agents [92,111]. Thus, the immunoenhancing effects of estrogen might also be involved in reducing abnormal Aβ accumulation.

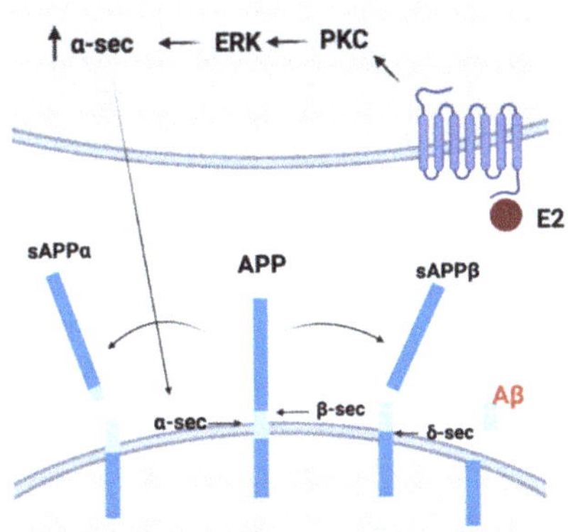

Figure 1. Proposed model of the mechanism of 17β-estradiol (E2) regulating amyloid precursor protein (APP) processing. E2 increases α-secretase (α-sec) via the PKC-ERK pathway and enhances sAPPα secretion (created with BioRender.com).

Estrogen can regulate Aβ expression, accumulation, and degradation. A brain-selective 17β-estradiol estrogen prodrug could decrease amyloid precursor and Aβ protein levels in an APPswe/PS1dE9 double-transgenic mouse model [112]. In the same transgenic model, 17β-estradiol and its prodrug could decrease Aβ (1–40) and Aβ (1–42) levels in the brain in both intact and ovariectomized animals [113]. 17β-estradiol regulates βAPP trafficking within the late secretory pathway in neuroblastoma cells and primary neurons [114]. 17β-estradiol and progesterone increased the expression level of insulin-degrading enzyme (IDE), which is related to Aβ clearance, and this overexpression of IDE is inversely associated with soluble Aβ levels [115]. Administration of 17β-estradiol reduced Aβ accumulation and plaque formation and increased hippocampal IDE expression in 3xTg-AD mice with ovariectomy [116]. In addition to IDE, 17β-estradiol-induced matrix metalloproteinases and neprilysin are also associated to Aβ degradation [117,118]. These Aβ clearance pathways are one of the treatment strategies which might be used to remove amyloids and improve cell survival in AD [119].

Another benefit of estrogen for AD is its neuroprotective effects, and this cellular effect might delay the progression of AD. In Aβ-indued neuronal death, estrogen can protect cells against death by enhancing BcL-2, BCLxL, and heat shock protein 70 and decreasing Bax proteins [120–122]. In the mouse cholinergic cell line SN56, 17β-estradiol could prevent Aβ-induced cell death through estrogen-receptor-mediated pathways [123]. Treatment with estren, 4-estren-3α, and 17β-diol could activate non-genomic estradiol pathways, restored loss of cholinergic neurons, and attenuated learning deficits which were induced by Aβ (1–42) injection to mouse basal forebrain [124]. In addition, estrogen promoted the degradation of Cav1.2 through ERα in neuronal cells and in an AD mouse model, and Cav1.2 is the subunit of the L-type calcium channel, which is important in calcium overload and cell death in AD [125]. 17β-estradiol protects against oxidative stress-induced cell death in hippocampal neurons through TRPV1 [126]. In a combined bioinformatics analysis of female AD patients and AD mouse models, thioredoxin-interacting protein (TXNIP) was found specifically expressed in the hippocampus in AD. Pan et al. found that estrogen treatment protects SH-SY5Y neuroblastoma from Aβ42-induced apoptosis by increasing thioredoxin and reducing TXNIP, which is mediated by the AMPK signal pathway [127]. Since TXNIP reduces thioredoxin and increases oxidative stress, the protective effects of estrogen in AD can be linked to oxidative stress and inflammation [128].

The antioxidative properties of estrogen in neuroprotection against AD have been well discussed. Aging or ovariectomy in rodents increases mitochondrial dysfunction, which can lead to reactive oxygen species (ROS) production. This phenomenon is observed in AD animal models and patients with AD [129,130]. In aged or AD brains, not only excessive ROS but also decreased antioxidant ability contribute to neurodegeneration [131,132]. Aβ induces calcium influx, leading to mitochondrial dysfunction and oxidative stress in astrocytes, and this overproduced oxidative stress causes antioxidant molecular depletion and cell death in neurons [133]. This evidence indicates that oxidant/antioxidant imbalance can be a therapeutic target in AD [134]. 17β-estradiol induces SH-SY5Y neuroblastoma cells to express neuronal nitric oxide synthase (NOS) and manganese superoxide dismutase (MnSOD), which are linked to antioxidative and neuroprotective ability [135]. Although estrogen shows inhibitory effects on lipid peroxidation, DNA damage, and intracellular peroxide accumulation [136,137], its protective effect on Aβ-induced oxidative stress is inconclusive. Chronic incubation with Aβ increases the lipid peroxidation level, and estrogen treatment can decrease peroxidation in human SK-N-SH neuroblastoma cells by almost half [138]. In another study of human neuroblastoma SH-SY5Y cells, 17β-estradiol had little effect on Aβ-induced ROS generation [139]. Along with Aβ-induced oxidative stress in the brain, increased biomarkers of lipid peroxidation in blood were found in mild cognitive impairment (MCI) or AD patients [140,141]. Plenty of evidence shows that estrogen displays a strong antioxidant ability, and estrogen treatment providing neuroprotection by reducing lipid peroxidation is a possible therapeutic approach to slow down or prevent progression of AD [142].

Inappropriate immune response in the CNS contributes to the severity of the pathology in AD. During AD progression, resting microglia, the immune cells in the brain, are activated by Aβ deposition and tau protein [143]. A longitudinal study on microglial activation in MCI and AD subjects showed that there are two peaks of microglial activation in the trajectory [144]. The pro-inflammatory cytokines and ROS released from activated microglia contribute to neuronal damage and neurodegeneration [145]. On microglia, the presence of estrogen receptors, namely ERα, ERβ, and GPER1, has been detected [146]. 17β-estradiol inhibits inflammatory genes' expression by reducing the transcription factor NF-κB [147]. 17β-estradiol can downregulate lipopolysaccharide-induced inducible nitric oxide synthase (iNOS) expression, and this anti-inflammatory effect is mediated by the MPAK pathway [148]. In addition to NO, estrogen inhibits cerebral ischemia-induced inflammatory cytokines such as IL-1β and TNF-α from releasing via GPER1on forebrain microglia [149]. ERα phosphorylated at Ser216 is only observed on microglia, not astrocytes, in mixed astrocyte and microglia culture or mouse brain. Stimulation of mouse brain or microglia with lipopolysaccharide induces inflammatory cytokines, and these pro-inflammatory cytokines are upregulated in the brain and in microglia bearing a phosphorylation-blocked ERα S216A mutation. These data validate that estrogen conducts anti-inflammatory action via ERα in microglia [150]. Moreover, patients with higher ERα promoter methylation rates showed decreased ERα mRNA expression and impaired cognition [151], which implies the importance of ERα in AD. As mentioned, it is becoming more and more appreciated that infection may be one causative factor for AD.

Apolipoprotein E (APOE), the strongest genetic risk factor for late-onset AD, is another key factor contributing to sex-specific differences in AD. In the brain, APOE is highly expressed on astrocytes, and APOE modulates lipid homeostasis [152]. Ratnakumar et al. compared neuronal genes regulated by estrogen in ovariectomized rhesus macaques and exome sequencing data from female AD patients and found that estrogen upregulates APOE gene expression [153]. In cultured hippocampal neurons and rat hippocampus, 17β-estradiol increased APOE mRNA and protein expression via ERα but not ERβ [154]. However, unlike rodents, which only have one APOE isoform, there are three different APOE isoforms, namely APOE2, APOE3, and APOE4, in humans. APOE4 is associated with a higher risk of AD, and people with APOE2 homozygotes present an exceptionally low likelihood of AD [155]. In transgenic mice with human APOE2, APOE3, or APOE4, Nathan et al. found that 17β-estradiol-induced APOE2 and APOE3, but not APOE4, can enhance neuritic outgrowth [156]. Thus, although estrogen can induce APOE expression, the APOE genotype may decide the outcomes of estrogen exposure.

Mitochondrial dysfunction has been associated with AD progression in pre- and postmenopausal patients. Pieces of evidence indicate that Aβ-mediated toxicity is the key factor that triggers mitochondrial dysfunction [157]. Aβ overload increases membranes' permeabilization of calcium, and excessive calcium leads to increased mitochondrial membrane permeability, which causes apoptosis and necrosis [158]. The voltage-dependent anion channel (VDAC), which is expressed on cell membranes and mitochondria, regulates calcium homeostasis and interacts with ERs in neuronal membranes [159]. Administration of estrogen modulates VDAC phosphorylation within a few minutes and leads to a reduction in VDAC opening. This phenomenon might be another mechanism of estrogen-provided neuroprotection [160]. In addition to VDAC, ERβ on neuronal mitochondria (mtERβ) can also affect mitochondrial function. In female AD patients, mtERβ expression and mitochondrial cytochrome C oxidase activity were significantly reduced, which implies that the deficiency of mtERβ might be associated with the dysfunction of mitochondria in AD [161]. A recent study showed that mitochondrial damage, including abnormal mitochondrial function and biogenesis, occurs prior to cognitive decline in the hippocampus of ovariectomized mice [162]. Compared to a premenopausal control, post- and perimenopausal females displayed reduced platelet mitochondrial activity, which was highly correlated to cerebral glucose metabolism [163]. The same group also analyzed in vivo brain mitochondrial ATP production of female participants across the pre-, peri-, and postmenopausal phases using multi-modality neuroimaging. Their results showed that postmenopausal brains have higher mitochondrial ATP production, which was correlated with preserved cognition [164], which implies that there might be a compensatory mechanism.

3.4. The Role of Estrogen in PD

Women have a lower risk of PD; however, PD symptoms increase in women after menopause due to the drop in endogenous estrogen [165]. Since estrogen can improve object recognition memory [166], which depends on dopamine neurons, it is not surprising that estrogen can modulate dopamine-dependent cognition. Administration of estrogen can alter dopamine transmission in the striatum, nucleus accumbens, and prefrontal cortex [167]. However, ERα- and GPER are found in cholinergic neurons and glia, not in axons and terminals of dopamine neurons [168]. Moreover, sERα/β and GPER express mainly on pyramidal cells of the hippocampus [169,170]; thus, the improvement of memory and cognition by estrogen might not be directly related to dopamine neurons. 17β-estradiol treatment stimulates the expression of glial cell-line-derived neurotrophic factor (GDNF) and protects dopaminergic neurons from 6-hydroxydopamine toxicity [171]. Moreover, estrogen can increase circulating IGF-1 and protect postmenopausal women from PD via IGF-1 [172]. Results from a meta-analysis also showed that hormone replacement therapy in menopausal women may reduce the risk of PD [42,173]. Altogether, lack of estrogen after menopause appears to play an important role in PD development.

3.5. The Effects of Progesterone on Neurodegenerative Diseases

Although the neuroprotective effects of progesterone are well known, the association of progesterone and neurodegenerative diseases, especially AD, is not very clear [174]. Endogenous progesterone and 3α,5α-THPROG are decreased by Aβ administration in vivo and in vitro. Exposure to progesterone of primary neuron culture and rat brain can induce Aβ-clearance-related factor expression, especially insulin-degrading enzyme [115]. In an AD mouse model, progesterone could prevent tau hyperphosphorylation but not Aβ accumulation [175]. However, some hormone replacement therapies including estrogen plus progesterone increased risk of AD [176]. A meta-analysis of hormone replacement therapy showed that synthetic progestin reduced the protective effects of estrogen on reducing risk of AD, but natural progesterone did not [84]. 3α,5α-THPROG, a metabolite of progesterone, showed superior effects on pathophysiology, cognition, and memory in a 3xTg AD mouse model [177,178], and the underlying mechanism involved 3α,5α-THPROG regulating glucose metabolism, mitochondrial bioenergetics, and cholesterol homeostasis in the brain [179]. In contrast to the inconclusive results of progesterone treatment for PD [58], 3α,5α-THPROG showed better improved cognitive and motor functions in MPTP- or 6-OHDA-induced PD mouse models [180,181].

3.6. The Role of IGF-1 in Neurodegenerative Diseases

Although the IGF-1 level decreases with aging, the role of IGF-1 in menopause-related neurodegenerative diseases is still unclear. Several studies have shown decreased IGF-1 levels in blood are associated with cognitive decline or risk of AD [182–189], and some studies found increasing IGF-1 in the serum of AD patients [190–192]. A meta-analysis did not show a clear correlation between IGF-1 level and AD in human subjects [193]. Compared to age-matched controls, a lower CSF/plasma ratio of IGF-1 was found in AD patients [192], which implies that IGF-1 transportation from blood to brain might be important in AD. Moreover, decreased mRNA expression of IGF-1, IGF-1R, and insulin was detected in late AD cases compared to controls [194]. In early PD patients, decreased serum IGF-1 levels are correlated to poor cognition, attention, and verbal memory [195]. Using the cyclic glycine-proline (cGP)/IGF-1 ratio to represent bioactive IGF-1, it has been shown that higher IGF-1 function is correlated to better cognition in normal aged people and the PD group [196]. However, this positive correlation of IGF-1 and cognitive function is sex-dependent. A cohort study showed that IGF-1 is positively related to cognitive function measured by the

Mini-Mental State Examination and Verbal Fluency test in males but not females [197]. A cross-sectional study in Ashkenazi Jews with exceptional longevity (age $\geq$ 95 years) showed that lower serum IGF-1 is associated with better cognitive function in females but not males [198]. These results imply that the role of IGF-1 in neurodegenerative diseases is not only sex-related but also age-dependent.

4. Menopause and Estrogen, Progesterone, and IGF-1 Signals

Numerous studies aimed to identify the role of estrogen, progesterone, and IGF-1 in neurodegenerative diseases. The following section will summarize the signaling pathways involved in menopause-related neurodegenerative diseases.

4.1. Genomic Action of Estrogen

Estrogens can mediate their neuroprotective effects in the brain through a number of mechanisms. In nature, there are three main estrogens: estriol, estrone, and 17β-estradiol. Frederiksen et al. analyzed these three estrogens' levels throughout life and found that the 17β-estradiol level is higher than that of estrone in premenopausal women but lower than that of estrone in postmenopausal women [199]. Regarding neuroprotection in AD, all these estrogens can inhibit Aβ oligomer formation, and estriol had the strongest in vitro activity [200]. 17β-estradiol is the most potent estrogen, and the decrease in 17β-estradiol in blood during menopause contributes to the increasing risk of cardiovascular disease due to loss of cholesterol and triglyceride metabolism regulation [201]. Furthermore, cardiovascular diseases also contribute to cognitive decline [202]. The typical mechanism is via the genomic pathway, for which estrogens bind to their receptors, ERα and ERβ. ERα and ERβ are steroid nuclear hormone receptors, and they also belong to ligand-activated transcription factors [203]. After estrogens bind to ERs, the conformational change of inactive ERs induces homodimerization of the estrogen receptors and removal of the regulatory receptor-associated protein Hsp90 [204,205]. The activated ERs translocate to the nucleus and bind to estrogen response elements (EREs) on specific genes [206]. This ERE binding facilitates the recruitment of other transcription factors, leading to an increase or decrease in target genes' expression, such as brain-derived neurotrophic factor (BDNF) and GDNF [171,207], which are important in neural protection. The responses described above are genomic responses which involve ERs and downstream signal cascades activating/repressing gene transcription. Interestingly, the response of BDNF transcription to estrogen during hippocampus development is gender- and region-dependent [208], which indicates cell-type- and gender-specific effects of estrogen on neuronal cells.

4.2. Non-Genomic Pathway Signals of Estrogen

The non-genomic response, also called the non-classical pathway, which does not involve ERE transcription factors, acts much faster than the genomic response. It usually responds within seconds because signals are mediated by the G-protein-coupled receptor and kinases [209]. There are four estrogen-binding membrane receptors, include G-protein-coupled estrogen receptor (GPER), Gq-coupled membrane estrogen receptor (Gq-mER), membrane subpopulation of ERs (mERα/β), and ER-X [210]. Gq-mER mainly distributes in hypothalamic proopiomelanocortin (POMC) neurons, and Gq-mER activation is related to homeostasis [211]. GPER, also named GPER1 or GPR30, is a transmembrane intracellular receptor and is found highly expressed on some breast cancer tissue and cell lines [212]. GPFER is expressed not only on cancer cells but also in the hippocampus. Immunohistochemistry staining results showed that GPER is strongly expressed in CA1 and the dentate gyrus in the hippocampus [213]. In addition, infusion of GPER agonists to the hippocampus could enhance memory in ovariectomized female mice [214]. Besides GPER, mERα/β, Erα, and ERβ localize to the plasma membrane and are found in neurons and glia [79]. Activation of mERα/β rapidly regulates neuron function in a sex-specific manner. 17β-estradiol treatment increases spontaneous excitatory postsynaptic current amplitude in female hippocampus, but not in that of males [215]. After estrogen binds to membrane estrogen receptors, it modulates neuroplasticity by activating second signals [216]. The most frequently mentioned downstream signal of GPER is extracellular signal-regulated kinase (ERK1/2) [217]. The other molecular signals which are involved in GPER pathways are phosphatidylinositol 3-kinase (PI3-K)/Akt and cAMP [218,219]. However, in an embryonic mouse hippocampal cell line, 17β-estradiol dose-dependently increased forskolin-stimulated cyclic AMP levels without activating the ERK1/2 pathway [220]. In a PD mouse model, GPER was required for ERα-induced neuroprotection on dopamine neurons [221]. In a global cerebral ischemia model, estrogen, as well as the GPER agonist G1, promoted neuronal survival through GPER activation of this pathway [213]. Similar protective effects were also found in a spinal cord injury mouse model and traumatic brain injury [222,223]. In these two studies, activation of GPER by estrogen or G1 could increase the phosphorylation of ERK and AKT, as well as increasing BDNF expression and decreasing cell apoptosis. Estrogen can increase Trkb receptor, BDNF receptor, phosphorylation, and ERK signals via GPER and can also stimulate RhoA and Rac signals via mERα/β. These signals regulate protein synthesis and actin polymerization in hippocampal neurons [224]. In addition to GPER-induced signals, GPER participates in the IGF-1-mediated neuroprotective effect [225,226]. IGF-1 receptor (IGF-1R) and GPER regulate each other through the PI3-K and ERK/MAPK signaling pathways. Interestingly, insulin-induced hypoglycemia significantly decreases GPER protein expression in A2 noradrenergic nerve cells in females but not males [227], which indicates a sex difference in GPER gene expression regulation. In addition, ERα and ERβ are also expressed in mitochondria and against mitochondrial-mediated oxidative stress via a non-genomic pathway [228], and estrogen deficiency causes mitochondrial dysfunction prior to cognitive decline in ovariectomized mice [162].

4.3. IGF-1 and IGF-1R Signals in AD

Among the neuroprotective factors that are affected by menopause, IGF-1 cannot be excluded. IGF-1 is involved in synaptic plasticity, neural regeneration, and cognitive function [229]. In aged rats, administration of IGF-1 can increase NMDAR2A and NMDAR2B protein expression, which is decreased in the hippocampus of aged rats [230]. Restoration of IGF-1 levels in an adult rat brain significantly restores the age-dependent reduction in neurogenesis in the dentate subgranular proliferative zone [231]. In peri- and postmenopausal women, serum levels of IGF-1 were significantly reduced compared with the premenopausal group [60]. In an APPSwe/PS1ΔE9 AD mouse model, decreased serum IGF-1 levels were associated with early Aβ deposition in the brain [232]. Systemic slow-release formulation of IGF-1 significantly rescued Aβ accumulation and memory decline in AD rodent models [233,234]. On the contrary, the role of IGF-1R in AD is different to that of IGF-1. Through IGF1RKO conditional mutant mice and an APP/PS1 AD mouse model, George et al. showed that reduction in neuronal IGF-1R prevents Aβ-induced cognitive deficits and inflammation, which indicates that suppression of IGF-1R signaling can improve AD progression [235]. The same group also showed that IGF-1R knockout in adult APP/PS1 AD mice results in improved spatial memory and lower accumulation of Aβ-containing autophagic vacuoles [236]. Treatment with picropodophyllin, an IGF-1R inhibitor, attenuated insoluble Aβ and microgliosis in the hippocampi of AβPP/PS1 mice [237]. IGF-1R can form hybrid receptors with insulin receptor (IR), and insulin resistance is one of risk factors of neurodegenerative diseases. Thus, the interactions between IGF-1, IGF-1R, and IR form a more complex signaling pathway. The controversial results from therapeutic strategies targeting modulation of IGF-1 or IGF-1R alone indicate that this may not be an appropriate therapy for AD patients with diabetes or insulin resistance. On the other hand, one-month administration of IGF-1 by subcutaneous minipump had no effects on Aβ protein level, plaque pathology, or phospho-tau expression in a Tg2576 AD mouse model [238]. A recent study showed that adenovirus-mediated transduction of IGF-1 protected mice from Aβ-induced memory impairment [239]. Although some studies in AD animal models show that IGF-1 treatment reduces Aβ accumulation and improves behavioral and pathological AD, the therapeutic strategy targeting increasing circulating IGF-1 failed to delay AD progression in a clinical trial [240]. In agreement with the clinical trial, increasing IGF-1 by MK-677 in 5xFAD AD mice failed to prevent hippocampal Aβ deposition and other AD pathogenesis [241].

4.4. Progesterone Signals and Neuroprotection

Similar to estrogen, progesterone can act through classical and non-classical pathways. In the classical pathway, progesterone binds to nuclear progesterone receptors (PGRs) and stimulates gene expression. In the non-classical pathway, progesterone can bind progesterone membrane receptors (mPRs) or the membrane-associated protein progesterone receptor-membrane component 1 (PGRMC1), leading to faster responses. In addition, progesterone can be converted to 3α,5α-THPROG, which can modulate GABA receptors [242]. Progesterone attenuates Aβ-induced primary cortical neuron apoptosis by inhibiting the mitochondria-associated apoptotic pathway via PGRMC1-mediated JNK inactivation [243]. 3α,5α-THPROG can reduce Aβ-induced ERK phosphorylation in SH-SY5Y human female neuroblastoma cells in a GABA-receptor-independent manner [55]. The neuroprotective effects of progesterone and 3α,5α-THPROG are also related to anti-inflammation and myelin repair. 3α,5α-THPROG can change microglia morphology and inhibit oligodendrocyte phagocytosis via microglia [244]. These results explain the protective effects of progesterone on demyelination and oligodendrocyte degeneration in the cuprizone-induced demyelination mouse model [245]. Progesterone and its synthetic derivative nestorone were also shown to promote myelin regeneration in chronic demyelinating lesions [246,247]. Since alterations in myelination are correlated to AD and MCI [248], the AD preventive ability of progesterone, 3α,5α-THPROG, and nestorone might be due to their protective effects on myelination. Progesterone and 3α,5α-THPROG also exhibit an anti-inflammatory ability for neuroprotection [249]. Progesterone treatment induced microglia phenotype change from pro-inflammatory M1 to anti-inflammatory M2-type in a demyelination mouse model [250]. Progesterone can also reduce lipopolysaccharide-induced microglia activation in brain [251]. In MPTP-induced PD mouse models, five-day administration of progesterone could reverse MPTP-induced striatal glial fibrillary acidic protein (GFAP) overexpression, which is the marker of activated astrocytes [252]. Results from the same paper also show that progesterone increased the BDNF level. In human Schwann cell-like cells which are differentiated from adipose stem cells, progesterone and mPRα agonist activate mPRα and increase the BDNF level, in which the Src and PI3K-Akt signaling pathways are involved [253]. Although the neuroprotective effects of progesterone and its associate neurosteroids are clear, continued treatment with progesterone or medroxyprogesterone acetate had no benefit, or even a negative impact, on memory in surgically menopausal rats [254]. This indicates that simply activating progesterone signals is not sufficient to treat AD. In addition, the age-associated changes in sensitivity of the brain to progesterone create "therapeutic time windows" for hormone replacement therapy. This issue has been discussed thoroughly in several reviews [254,255].

4.5. Estrogen, IGF-1, and IGF-1R Signals

Through in vitro studies, the neuroprotective effects of the estrogen/IGF-1/IGF-1R axis have been well characterized. Estrogen enhances IGF-1 transcription via the genomic pathway, which induces ER nucleus translocation and binds to ERE on the IGF-1 promoter [256]. Increased IGF-1 levels expand IGF-1R signals. Estrogen can also increase IGF-1R activity through non-genomic pathways. IGF-1-mediated cell proliferation requires GPER, and GPER transcription is also enhanced by the IGF-IR/PKCδ/ERK pathway [257]. In addition, ERα can also affect IGF-1R function [258,259]. Downstream of IGF-1R, the PI3K/Akt and Ras/MEK/ERK pathways are most frequently mentioned. These signals stimulate cell survival and proliferation. Moreover, IGF-1R activation can increase BDNF expression, which plays an important role in cognition [260]. Increased IGF-1 or IGF-1R activity can promote hippocampal progenitor cell proliferation [261], and lack of IGF-1 leads to a smaller hippocampal volume [262]. In addition, IGF-1R signals seem to interfere with Aβ clearance. Reduced IGF signaling in AD mice can protect mice from AD pathogenesis, which is associated with less hyperaggregation of Aβ in the brain [236,263]. As the disease progression is difficult to reverse, including Aβ accumulation formed in the late AD phase, the ineffectiveness of potent inducers of IGF-1 on AD patients might be due to the negative effects of IGF-1R on Aβ clearance. Considering that oxidative stress and glia-mediated inflammation also contribute to neurodegenerative diseases, the anti-inflammatory effects of IGF-1/IGF-1R still cannot be ruled out. Although IGF-1 suppresses MPTP/MPP +-induced astrocyte activation via GCER [225,264], the IGF-1R inhibitor prevented neuroinflammation in the hippocampus in an AD mouse model [237]. Since MPTP/MPP induced dopaminergic neurotoxicity in a PD model, these controversial results imply that IGF-1/IGF-1R signals might play different roles in PD compared to AD. Figure 2 illustrates the interaction of estrogen, estrogen receptors, IGF-1, and IGF-1R signals.

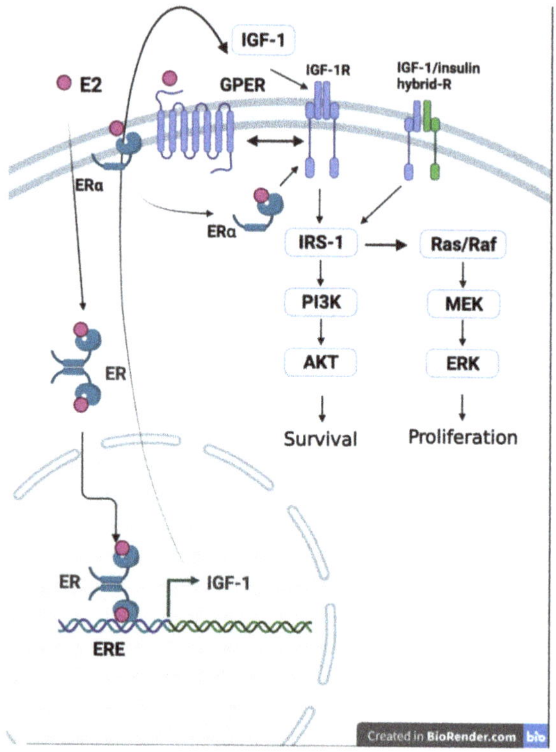

Figure 2. The correlation of estrogen, estrogen receptors, IGF-1, and IGF-1R. Estrogen (E2) increases IGF-1 gene expression through genomic pathways and enhances IGF-1 receptor activity through non-genomic pathways, including mER and GPER (created with BioRender.com).

5. Estrogen and NMDA Receptor Signals in Neurodegenerative Diseases

Estrogen is well known to influence cognition by regulating synapse structure and function. On CA1 pyramidal cells in the hippocampus of adult female rats, 17β-estradiol can increase dendritic spine density and excitatory synapses [14]. In addition, 17β-estradiol treatment increased the LTP magnitude in CA1, which involved NMDA receptors [265]. In CA1 pyramidal neurons, which are innervated by the temporoammonic pathway, 17β-estradiol induces LTP via ERα and PI 3-kinase signaling, and this ERα-induced LTP is NMDAR-dependent [266]. By using NMDA receptor antagonists and cAMP inhibitors, a study in mice showed that estrogen improves object memory through NMDA receptors and PKA activation within the dorsal hippocampus [267]. In ovariectomized rats, 17β-estradiol enhanced hippocampal LTP and novel object recognition via NMDA receptors that contained NR2B, and these enhancements could be reduced by the NR2B subunit antagonist Ro25-6981 [268]. However, 17β-estradiol treatment did not alter the protein expression level of NMDA receptor subunits, including NR1, NR2A, and NR2B [269]. The possible mechanism by which 17β-estradiol improves memory and cognition might include phosphorylation and recruitment of NR2B-containing NMDARs to synapses. In addition, a study in the dentate gyrus of juvenile male rats showed that low and high doses of 17β-estradiol had opposite effects on NMDA receptors. The bidirectional effect on NMDA receptors due to the different doses of 17β-estradiol implies activating ERα or ERβ as their downstream signaling [270]. Thus, dosage of 17β-estradiol should be considered, as high and low dosages result in vastly different effects.

Estrogen may also influence NMDA receptor signals through interaction with metabotropic glutamate receptors (mGluRs). In CA3-CA1 hippocampal pyramidal neurons, 17β-estradiol activates ERα and regulates CREB phosphorylation via activation of mGluRs [271]. Treating the dorsal hippocampus with ERα and ERβ agonists also enhanced novel object recognition and object placement memory via ERK signals in ovariectomized mice, and the 17β-estradiol-improved novel object recognition could be reduced by the mGluR1 antagonist [68]. Since ERα, ERβ, mGluR1, and ERK gather at specialized membrane microdomains, the 17β-estradiol-induced enhancement might be the consequence of ER/mGluR complex formation and downstream signaling. Moreover, activation of mGluR1 reinforces NMDA current through Src kinases [69,272], and 17β-estradiol is reported to be able to increase NR2B phosphorylation through Src signaling [273]. The depressive-like phenotype and reduced neurogenesis of male mice lacking sigma-1 receptor can be rescued by 17β-estradiol treatment. These neuroprotective and antidepressant effects of 17β-estradiol are mediated by Src and NR2B [274]. These results indicate that estrogen might enhance NMDA signals via the mGluR1 and Src pathways, and these signals participate in estrogen-induced neurogenesis [273] (Figure 3).

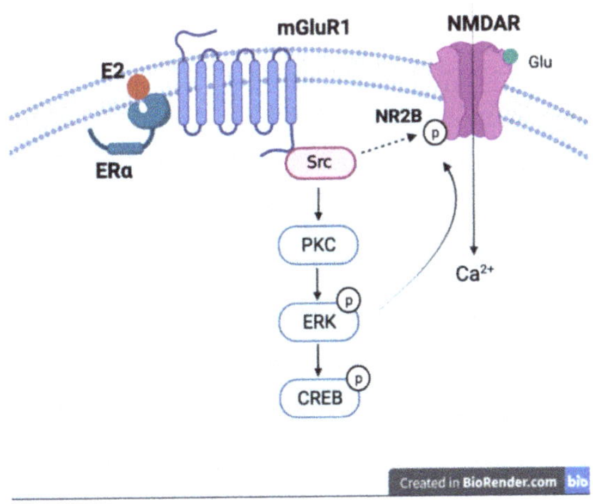

Figure 3. Interaction of estrogen and NMDA receptor with mGluR1 and signaling molecules. General description of the proposed mechanism of how estrogen activates mGluR1 through mERα and leads to activation of NMDA receptor by Src and other second messenger signaling cascades (created with BioRender.com).

Although cell and animal studies all show estrogen can provide neuroprotective effects via NMDARs [275], randomized clinical trials of hormone replace therapy on female AD patients show no improvements in cognitive symptoms [276]. One possibility is the imbalance of NMDARs and signal molecular expression at the postmenopause stage and under estrogen stimulation. The NMDA receptor type 2D (NR2D) gene, which contains estrogen response elements (EREs), is a target for estrogen signals in the brain [277]. Indeed, NR2D mRNA expression is decreased in ovariectomized mice and can be restored by estrogen [278]. In AD patients, as well as in aging mice, a decreased expression level of NR2A and NR2B, but not NR2D, has been noted [279–281]. Unlike NR2A, more related to neuroprotection, NR2D overexpression induces neuronal death [282]. Dedicator of Cytokinesis 3 (DOCK3), which is decreased in the brains of AD patients, can protect neurons from NR2D- and NR2B-induced excitotoxicity [283–285]. Although estrogen can prevent NR2B-induced neural cytotoxicity via the GPR30/ERK signal pathway [286], it might also increase neuronal death by increasing NR2D expression. Moreover, there is no evidence showing that estrogen can modulate DOCK3 expression. Thus, the failure of hormone replacement therapy in AD patients might be due to the unwanted excessive NR2D expression but no equally inhibitory signal from DOCK3 (Figure 4).

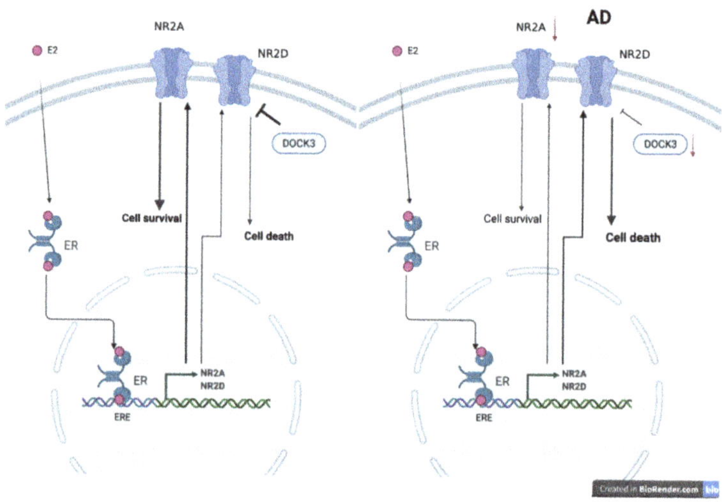

Figure 4. Proposed mechanism of why estrogen treatments fail in promoting neuron survival via NMDA receptors in AD patients (created with BioRender.com).

6. Other Potential Therapeutic Strategies

6.1. Effects of Phytoestrogens on AD

In addition to classic hormone replacement therapy, the use of phytoestrogens in women with menopause can help reduce symptoms. Phytoestrogens, having a similar structure to 17β-estradiol, are plant-based compounds, including isoflavones, lignans, cumestan, and lactones [287]. Among these phytoestrogens, isoflavones have the strongest estrogenicity [288] and have been proposed as synthetic selective estrogen modulators [289]. A meta-analysis study showed that isoflavone supplementation may improve cognition in postmenopausal women regardless of methodological flaws [290]. Isoflavones have shown their neuroprotective, antioxidant, and anti-inflammatory effects in several animal models [291–294]. Moreover, isoflavones can improve cognition and reduce Aβ and tau levels in cell and rodent AD models [295–298]. Genistein, an isoflavone, is able to reduce Aβ-induced neurotoxicity and oxidative stress through various intracellular signals, including the ERK/MAPK pathway [299]. A recent study showed that polyhydroxyisoflavones can serve as a scaffold that prevents Aβ and tau aggregation [300]. Compared to estrogen, the binding ability of phytoestrogens to ERs is weak, and some phytoestrogens have a higher binding affinity for Erβ, which can inhibit proliferation in breast cancer [301]. Thus, phytoestrogens seem to have similar neuroprotective effects as estrogen but less carcinogenic potential. However, more studies are needed for evaluating the risks and therapeutic effects of phytoestrogens on preventing or treating AD.

6.2. Genomic and Non-Genomic Aspects of AD

The exact cause of Alzheimer's disease is not clear, but both genetic and non-genetic factors contribute to the etiopathogenesis of AD. The genetic factors of early-onset AD are well studied, and mutations of the APP, PSEN1, and PSEN2 genes are the main cause of early-onset AD [302]. These genes mutation result in the Aβ1-42 peptide overproduction, which is connected to γ-secretase activity [303].

Genetic factors play important roles not only in early-onset AD but also in late-onset AD. One or two APOE ε4 alleles increase the risk of late-onset AD [304,305]. Carrying the APOE ε4 allele is associated with increasing Aβ accumulation and enhanced neuroinflammation [306]. In addition to APOE ε4, several gene variants, including BIN1, ABC7, PICALM, MS4A4E/MS4A6A, CD2Ap, CD33, EPHA1, CLU, CR1, and SORL1, were also found to be associated with Aβ accumulation and linked to late-onset AD [307]. The variants of estrogen metabolic-related genes, such as estrogen receptor β gene (ESR2), cytochrome P450 19A1 gene (CYP19A1), and cytochrome P450 11A1 gene (CYP11A1), are also associated with AD risk [308]. Through meta-analysis, polymorphisms of the CYP19A1 gene were found to be significantly correlated with increasing AD susceptibility [309]. To identify more genomic variants contributing to AD, several genome-wide association studies (GWASs) have been conducted to identify multiple susceptibility loci [310], and deep learning models can be used for detecting AD through the findings of GWASs [311]. These results suggest that combing GWAS on AD and menopause may identify novel pathways for therapeutic targets.

Although genomic factors are inherited and fixed, non-genomic factors associated with the onset and progression of AD may offer chances for treatment. The non-genomic factors contributing to AD risk include head injury in males, age, diabetes mellitus, conjugated equine estrogen use with medroxyprogesterone acetate, current smoking, and lower social engagement [312]. Estrogen and progesterone have neuroprotective effects on head injury [223], and the changes in sex hormones at menopause may increase diabetes risk as well [313]. Moreover, menopause and diabetes mellitus are well-known risk factors of cardiovascular diseases, and several cardiovascular diseases are related to dementia [314]. Thus, interventions targeted to menopause may also reduce non-genomic factors and enable prevention and treatment of AD.

7. Conclusions

There are growing lines of evidence correlating menopause with risk and progression of neurodegenerative diseases. In this review, we summarized the association of estrogen, progesterone, IGF-1, and their signal pathways with AD and PD. Based on the current literature, decreases in estrogen, progesterone, and IGF-1 increase inflammation, impair Aβ clearance, and hinder their neuroprotective effects. Not all clinical trials of hormone replacement or IGF-1R potentiator showed convincing benefits in AD patients. The modulation of estrogen, progesterone, or IGF-1 levels alone in aging populations may not be sufficient to reverse disease progression due to the complex signal pathways in neurodegenerative diseases. While diagnosed AD may not be consistently treatable by hormone replacement, its preceding stage entailing MCI may be. The combination of estrogens, IGF-1, or other neurotrophic factors in specific time windows, such as in the phase of MCI preceding fully developed AD, may provide a therapeutic strategy in postmenopausal AD patients.

Author Contributions: Conceptualization, C.-H.L. and H.-Y.L.; writing, Y.-J.C. All authors have read and agreed to the published version of the manuscript.

Funding: This work was supported by the National Health Research Institutes, Taiwan (NHRI-EX110-10816NC); the Ministry of Science and Technology in Taiwan (MOST 109-2628-B-182A-002; 110-2314-B-182A-048-; 108-2410-H-039-009-MY2); the Chang Gung Memorial Hospital, Taiwan (CMRPG8G1391, CMRPG8K1161, CMRPG8K1461), and the China Medical University (CMU104-S-14-04).

Acknowledgments: All figures were created with BioRender.com.

Conflicts of Interest: The authors declare no conflict of interest. The sponsors were not involved in the design of the study; the collection, analysis, and interpretation of the data; the writing of the report; and the decision to submit the article for publication.

References

1. Garre-Olmo, J. Epidemiology of Alzheimer's disease and other dementias. *Rev. Neurol.* **2018**, *66*, 377–386.
2. GBD 2016 Parkinson's Disease Collaborators. Global, regional, and national burden of Parkinson's disease, 1990–2016: A systematic analysis for the Global Burden of Disease Study 2016. *Lancet Neurol.* **2018**, *17*, 939–953. [CrossRef]
3. Kane, J.P.M.; Surendranathan, A.; Bentley, A.; Barker, S.A.H.; Taylor, J.P.; Thomas, A.J.; Allan, L.M.; McNally, R.J.; James, P.W.; McKeith, I.G.; et al. Clinical prevalence of Lewy body dementia. *Alzheimers Res. Ther.* **2018**, *10*, 19. [CrossRef]
4. West, M.J.; Coleman, P.D.; Flood, D.G.; Troncoso, J.C. Differences in the pattern of hippocampal neuronal loss in normal ageing and Alzheimer's disease. *Lancet* **1994**, *344*, 769–772. [CrossRef]
5. Schulz, J.B. Neuronal pathology in Parkinson's disease. *Cell Tissue Res.* **2005**, *320*, 211. [CrossRef] [PubMed]
6. Erskine, D.; Thomas, A.J.; Attems, J.; Taylor, J.P.; McKeith, I.G.; Morris, C.M.; Khundakar, A.A. Specific patterns of neuronal loss in the pulvinar nucleus in dementia with lewy bodies. *Mov. Disord.* **2017**, *32*, 414–422. [CrossRef] [PubMed]
7. Kurtishi, A.; Rosen, B.; Patil, K.S.; Alves, G.W.; Moller, S.G. Cellular Proteostasis in Neurodegeneration. *Mol. Neurobiol.* **2019**, *56*, 3676–3689. [CrossRef]
8. Atwood, C.S.; Bowen, R.L. The endocrine dyscrasia that accompanies menopause and andropause induces aberrant cell cycle signaling that triggers re-entry of post-mitotic neurons into the cell cycle, neurodysfunction, neurodegeneration and cognitive disease. *Horm. Behav.* **2015**, *76*, 63–80. [CrossRef]
9. Mielke, M.M. Sex and Gender Differences in Alzheimer's Disease Dementia. *Psychiatr. Times* **2018**, *35*, 14–17.
10. Jurado-Coronel, J.C.; Cabezas, R.; Avila Rodriguez, M.F.; Echeverria, V.; Garcia-Segura, L.M.; Barreto, G.E. Sex differences in Parkinson's disease: Features on clinical symptoms, treatment outcome, sexual hormones and genetics. *Front. Neuroendocrinol.* **2018**, *50*, 18–30. [CrossRef]
11. Lin, C.H.; Chen, P.K.; Wang, S.H.; Lane, H.Y. Effect of Sodium Benzoate on Cognitive Function Among Patients With Behavioral and Psychological Symptoms of Dementia: Secondary Analysis of a Randomized Clinical Trial. *JAMA Netw. Open* **2021**, *4*, e216156. [CrossRef] [PubMed]
12. Tang, M.X.; Jacobs, D.; Stern, Y.; Marder, K.; Schofield, P.; Gurland, B.; Andrews, H.; Mayeux, R. Effect of oestrogen during menopause on risk and age at onset of Alzheimer's disease. *Lancet* **1996**, *348*, 429–432. [CrossRef]
13. Kawas, C.; Resnick, S.; Morrison, A.; Brookmeyer, R.; Corrada, M.; Zonderman, A.; Bacal, C.; Lingle, D.D.; Metter, E. A prospective study of estrogen replacement therapy and the risk of developing Alzheimer's disease: The Baltimore Longitudinal Study of Aging. *Neurology* **1997**, *48*, 1517–1521. [CrossRef]
14. Gould, E.; Woolley, C.S.; Frankfurt, M.; McEwen, B.S. Gonadal steroids regulate dendritic spine density in hippocampal pyramidal cells in adulthood. *J. Neurosci.* **1990**, *10*, 1286–1291. [CrossRef]
15. Hara, Y.; Park, C.S.; Janssen, W.G.; Roberts, M.T.; Morrison, J.H.; Rapp, P.R. Synaptic correlates of memory and menopause in the hippocampal dentate gyrus in rhesus monkeys. *Neurobiol. Aging* **2012**, *33*, 421.e17–421.e28. [CrossRef] [PubMed]
16. Hara, Y.; Yuk, F.; Puri, R.; Janssen, W.G.; Rapp, P.R.; Morrison, J.H. Estrogen Restores Multisynaptic Boutons in the Dorsolateral Prefrontal Cortex while Promoting Working Memory in Aged Rhesus Monkeys. *J. Neurosci.* **2016**, *36*, 901–910. [CrossRef]
17. Zagni, E.; Simoni, L.; Colombo, D. Sex and Gender Differences in Central Nervous System-Related Disorders. *Neurosci. J.* **2016**, *2016*, 2827090. [CrossRef]
18. Ott, A.; Stolk, R.P.; van Harskamp, F.; Pols, H.A.; Hofman, A.; Breteler, M.M. Diabetes mellitus and the risk of dementia: The Rotterdam Study. *Neurology* **1999**, *53*, 1937–1942. [CrossRef]
19. Paykel, E.S.; Brayne, C.; Huppert, F.A.; Gill, C.; Barkley, C.; Gehlhaar, E.; Beardsall, L.; Girling, D.M.; Pollitt, P.; O'Connor, D. Incidence of dementia in a population older than 75 years in the United Kingdom. *Arch. Gen. Psychiatry* **1994**, *51*, 325–332. [CrossRef] [PubMed]
20. Beam, C.R.; Kaneshiro, C.; Jang, J.Y.; Reynolds, C.A.; Pedersen, N.L.; Gatz, M. Differences Between Women and Men in Incidence Rates of Dementia and Alzheimer's Disease. *J. Alzheimers Dis.* **2018**, *64*, 1077–1083. [CrossRef]
21. Corrada, M.M.; Brookmeyer, R.; Paganini-Hill, A.; Berlau, D.; Kawas, C.H. Dementia incidence continues to increase with age in the oldest old: The 90+ study. *Ann. Neurol.* **2010**, *67*, 114–121. [CrossRef]
22. Mouton, A.; Blanc, F.; Gros, A.; Manera, V.; Fabre, R.; Sauleau, E.; Gomez-Luporsi, I.; Tifratene, K.; Friedman, L.; Thummler, S.; et al. Sex ratio in dementia with Lewy bodies balanced between Alzheimer's disease and Parkinson's disease dementia: A cross-sectional study. *Alzheimers Res. Ther.* **2018**, *10*, 92. [CrossRef] [PubMed]

23. Sinforiani, E.; Citterio, A.; Zucchella, C.; Bono, G.; Corbetta, S.; Merlo, P.; Mauri, M. Impact of gender differences on the outcome of Alzheimer's disease. *Dement. Geriatr. Cogn. Disord.* **2010**, *30*, 147–154. [CrossRef] [PubMed]
24. Barnes, L.L.; Wilson, R.S.; Bienias, J.L.; Schneider, J.A.; Evans, D.A.; Bennett, D.A. Sex differences in the clinical manifestations of Alzheimer disease pathology. *Arch. Gen. Psychiatry* **2005**, *62*, 685–691. [CrossRef] [PubMed]
25. Mattsson, N.; Lonneborg, A.; Boccardi, M.; Blennow, K.; Hansson, O.; Geneva Task Force for the Roadmap of Alzheimer's Biomarkers. Clinical validity of cerebrospinal fluid Abeta42, tau, and phospho-tau as biomarkers for Alzheimer's disease in the context of a structured 5-phase development framework. *Neurobiol. Aging* **2017**, *52*, 196–213. [CrossRef]
26. Johnson, K.A.; Schultz, A.; Betensky, R.A.; Becker, J.A.; Sepulcre, J.; Rentz, D.; Mormino, E.; Chhatwal, J.; Amariglio, R.; Papp, K.; et al. Tau positron emission tomographic imaging in aging and early Alzheimer disease. *Ann. Neurol.* **2016**, *79*, 110–119. [CrossRef]
27. Jack, C.R., Jr.; Wiste, H.J.; Weigand, S.D.; Therneau, T.M.; Knopman, D.S.; Lowe, V.; Vemuri, P.; Mielke, M.M.; Roberts, R.O.; Machulda, M.M.; et al. Age-specific and sex-specific prevalence of cerebral beta-amyloidosis, tauopathy, and neurodegeneration in cognitively unimpaired individuals aged 50–95 years: A cross-sectional study. *Lancet Neurol.* **2017**, *16*, 435–444. [CrossRef]
28. Buckley, R.F.; Mormino, E.C.; Chhatwal, J.; Schultz, A.P.; Rabin, J.S.; Rentz, D.M.; Acar, D.; Properzi, M.J.; Dumurgier, J.; Jacobs, H.; et al. Associations between baseline amyloid, sex, and APOE on subsequent tau accumulation in cerebrospinal fluid. *Neurobiol. Aging* **2019**, *78*, 178–185. [CrossRef]
29. Knapskog, A.B.; Eldholm, R.S.; Braekhus, A.; Engedal, K.; Saltvedt, I. Factors that influence the levels of cerebrospinal fluid biomarkers in memory clinic patients. *BMC Geriatr.* **2017**, *17*, 210. [CrossRef]
30. Pringsheim, T.; Jette, N.; Frolkis, A.; Steeves, T.D. The prevalence of Parkinson's disease: A systematic review and meta-analysis. *Mov. Disord.* **2014**, *29*, 1583–1590. [CrossRef]
31. Baldereschi, M.; Di Carlo, A.; Rocca, W.A.; Vanni, P.; Maggi, S.; Perissinotto, E.; Grigoletto, F.; Amaducci, L.; Inzitari, D. Parkinson's disease and parkinsonism in a longitudinal study: Two-fold higher incidence in men. ILSA Working Group. Italian Longitudinal Study on Aging. *Neurology* **2000**, *55*, 1358–1363. [CrossRef]
32. Swerdlow, R.H.; Parker, W.D.; Currie, L.J.; Bennett, J.P.; Harrison, M.B.; Trugman, J.M.; Wooten, G.F. Gender ratio differences between Parkinson's disease patients and their affected relatives. *Parkinsonism Relat. Disord.* **2001**, *7*, 129–133. [CrossRef]
33. Van Den Eeden, S.K.; Tanner, C.M.; Bernstein, A.L.; Fross, R.D.; Leimpeter, A.; Bloch, D.A.; Nelson, L.M. Incidence of Parkinson's disease: Variation by age, gender, and race/ethnicity. *Am. J. Epidemiol.* **2003**, *157*, 1015–1022. [CrossRef]
34. Taylor, K.S.; Cook, J.A.; Counsell, C.E. Heterogeneity in male to female risk for Parkinson's disease. *J. Neurol. Neurosurg. Psychiatry* **2007**, *78*, 905–906. [CrossRef] [PubMed]
35. Wooten, G.F.; Currie, L.J.; Bovbjerg, V.E.; Lee, J.K.; Patrie, J. Are men at greater risk for Parkinson's disease than women? *J. Neurol. Neurosurg. Psychiatry* **2004**, *75*, 637–639. [CrossRef]
36. Meoni, S.; Macerollo, A.; Moro, E. Sex differences in movement disorders. *Nat. Rev. Neurol.* **2020**, *16*, 84–96. [CrossRef] [PubMed]
37. Currie, L.J.; Harrison, M.B.; Trugman, J.M.; Bennett, J.P.; Wooten, G.F. Postmenopausal estrogen use affects risk for Parkinson disease. *Arch. Neurol.* **2004**, *61*, 886–888. [CrossRef] [PubMed]
38. Ragonese, P.; D'Amelio, M.; Callari, G.; Salemi, G.; Morgante, L.; Savettieri, G. Age at menopause predicts age at onset of Parkinson's disease. *Mov. Disord.* **2006**, *21*, 2211–2214. [CrossRef]
39. Ragonese, P.; D'Amelio, M.; Savettieri, G. Implications for estrogens in Parkinson's disease: An epidemiological approach. *Ann. N. Y. Acad. Sci.* **2006**, *1089*, 373–382. [CrossRef]
40. Marder, K.; Tang, M.X.; Alfaro, B.; Mejia, H.; Cote, L.; Jacobs, D.; Stern, Y.; Sano, M.; Mayeux, R. Postmenopausal estrogen use and Parkinson's disease with and without dementia. *Neurology* **1998**, *50*, 1141–1143. [CrossRef]
41. Nicoletti, A.; Nicoletti, G.; Arabia, G.; Annesi, G.; De Mari, M.; Lamberti, P.; Grasso, L.; Marconi, R.; Epifanio, A.; Morgante, L.; et al. Reproductive factors and Parkinson's disease: A multicenter case-control study. *Mov. Disord.* **2011**, *26*, 2563–2566. [CrossRef]
42. Song, Y.J.; Li, S.R.; Li, X.W.; Chen, X.; Wei, Z.X.; Liu, Q.S.; Cheng, Y. The Effect of Estrogen Replacement Therapy on Alzheimer's Disease and Parkinson's Disease in Postmenopausal Women: A Meta-Analysis. *Front. Neurosci.* **2020**, *14*, 157. [CrossRef]
43. Vann Jones, S.A.; O'Brien, J.T. The prevalence and incidence of dementia with Lewy bodies: A systematic review of population and clinical studies. *Psychol. Med.* **2014**, *44*, 673–683. [CrossRef]
44. Fereshtehnejad, S.M.; Religa, D.; Westman, E.; Aarsland, D.; Lokk, J.; Eriksdotter, M. Demography, diagnostics, and medication in dementia with Lewy bodies and Parkinson's disease with dementia: Data from the Swedish Dementia Quality Registry (SveDem). *Neuropsychiatr. Dis. Treat.* **2013**, *9*, 927–935. [CrossRef]
45. Yue, W.; Wang, X.D.; Shi, Z.; Wang, Y.; Liu, S.; Liu, S.; Zhang, Y.; Zhang, Y.; Lu, H.; Su, W.; et al. The prevalence of dementia with Lewy bodies in a rural area of China. *Parkinsonism Relat. Disord.* **2016**, *29*, 72–77. [CrossRef]
46. Goodman, R.A.; Lochner, K.A.; Thambisetty, M.; Wingo, T.S.; Posner, S.F.; Ling, S.M. Prevalence of dementia subtypes in United States Medicare fee-for-service beneficiaries, 2011–2013. *Alzheimers Dement.* **2017**, *13*, 28–37. [CrossRef]
47. Savica, R.; Grossardt, B.R.; Bower, J.H.; Boeve, B.F.; Ahlskog, J.E.; Rocca, W.A. Incidence of dementia with Lewy bodies and Parkinson disease dementia. *JAMA Neurol.* **2013**, *70*, 1396–1402. [CrossRef]
48. Breitve, M.H.; Hynninen, M.J.; Bronnick, K.; Chwiszczuk, L.J.; Auestad, B.H.; Aarsland, D.; Rongve, A. A longitudinal study of anxiety and cognitive decline in dementia with Lewy bodies and Alzheimer's disease. *Alzheimers Res. Ther.* **2016**, *8*, 3. [CrossRef] [PubMed]

49. Farina, E.; Baglio, F.; Caffarra, P.; Magnani, G.; Scarpini, E.; Appollonio, I.; Bascelli, C.; Cheldi, A.; Nemni, R.; Franceschi, M.; et al. Frequency and clinical features of Lewy body dementia in Italian memory clinics. *Acta Biomed.* **2009**, *80*, 57–64. [PubMed]
50. Van de Beek, M.; Babapour Mofrad, R.; van Steenoven, I.; Vanderstichele, H.; Scheltens, P.; Teunissen, C.E.; Lemstra, A.W.; van der Flier, W.M. Sex-specific associations with cerebrospinal fluid biomarkers in dementia with Lewy bodies. *Alzheimers Res. Ther.* **2020**, *12*, 44. [CrossRef] [PubMed]
51. Hestiantoro, A.; Wiwie, M.; Shadrina, A.; Ibrahim, N.; Purba, J.S. FSH to estradiol ratio can be used as screening method for mild cognitive impairment in postmenopausal women. *Climacteric* **2017**, *20*, 577–582. [CrossRef] [PubMed]
52. Chu, C.; Gao, G.; Huang, W. A study on co-localization of FSH and its receptor in rat hippocampus. *J. Mol. Histol.* **2008**, *39*, 49–55. [CrossRef]
53. Kandasamy, M.; Radhakrishnan, R.K.; Poornimai Abirami, G.P.; Roshan, S.A.; Yesudhas, A.; Balamuthu, K.; Prahalathan, C.; Shanmugaapriya, S.; Moorthy, A.; Essa, M.M.; et al. Possible Existence of the Hypothalamic-Pituitary-Hippocampal (HPH) Axis: A Reciprocal Relationship Between Hippocampal Specific Neuroestradiol Synthesis and Neuroblastosis in Ageing Brains with Special Reference to Menopause and Neurocognitive Disorders. *Neurochem. Res.* **2019**, *44*, 1781–1795. [CrossRef] [PubMed]
54. Grimm, A.; Schmitt, K.; Lang, U.E.; Mensah-Nyagan, A.G.; Eckert, A. Improvement of neuronal bioenergetics by neurosteroids: Implications for age-related neurodegenerative disorders. *Biochim. Biophys. Acta* **2014**, *1842*, 2427–2438. [CrossRef] [PubMed]
55. Mendell, A.L.; MacLusky, N.J. Neurosteroid Metabolites of Gonadal Steroid Hormones in Neuroprotection: Implications for Sex Differences in Neurodegenerative Disease. *Front. Mol. Neurosci.* **2018**, *11*, 359. [CrossRef]
56. Santoro, N. The menopausal transition. *Am. J. Med.* **2005**, *118* (Suppl. 12B), 8–13. [CrossRef] [PubMed]
57. Santoro, N.; Lasley, B.; McConnell, D.; Allsworth, J.; Crawford, S.; Gold, E.B.; Finkelstein, J.S.; Greendale, G.A.; Kelsey, J.; Korenman, S.; et al. Body size and ethnicity are associated with menstrual cycle alterations in women in the early menopausal transition: The Study of Women's Health across the Nation (SWAN) Daily Hormone Study. *J. Clin. Endocrinol. Metab.* **2004**, *89*, 2622–2631. [CrossRef]
58. Guennoun, R. Progesterone in the Brain: Hormone, Neurosteroid and Neuroprotectant. *Int. J. Mol. Sci.* **2020**, *21*, 5271. [CrossRef] [PubMed]
59. Liu, J.M.; Zhao, H.Y.; Ning, G.; Chen, Y.; Zhang, L.Z.; Sun, L.H.; Zhao, Y.J.; Xu, M.Y.; Chen, J.L. IGF-1 as an early marker for low bone mass or osteoporosis in premenopausal and postmenopausal women. *J. Bone Miner. Metab.* **2008**, *26*, 159–164. [CrossRef]
60. Nasu, M.; Sugimoto, T.; Chihara, M.; Hiraumi, M.; Kurimoto, F.; Chihara, K. Effect of natural menopause on serum levels of IGF-I and IGF-binding proteins: Relationship with bone mineral density and lipid metabolism in perimenopausal women. *Eur. J. Endocrinol.* **1997**, *136*, 608–616. [CrossRef] [PubMed]
61. Cao, P.; Maximov, A.; Sudhof, T.C. Activity-dependent IGF-1 exocytosis is controlled by the Ca(2+)-sensor synaptotagmin-10. *Cell* **2011**, *145*, 300–311. [CrossRef]
62. Lu, Y.; Sareddy, G.R.; Wang, J.; Zhang, Q.; Tang, F.L.; Pratap, U.P.; Tekmal, R.R.; Vadlamudi, R.K.; Brann, D.W. Neuron-Derived Estrogen Is Critical for Astrocyte Activation and Neuroprotection of the Ischemic Brain. *J. Neurosci.* **2020**, *40*, 7355–7374. [CrossRef] [PubMed]
63. Kuo, J.; Hamid, N.; Bondar, G.; Prossnitz, E.R.; Micevych, P. Membrane estrogen receptors stimulate intracellular calcium release and progesterone synthesis in hypothalamic astrocytes. *J. Neurosci.* **2010**, *30*, 12950–12957. [CrossRef] [PubMed]
64. Hasegawa, Y.; Hojo, Y.; Kojima, H.; Ikeda, M.; Hotta, K.; Sato, R.; Ooishi, Y.; Yoshiya, M.; Chung, B.C.; Yamazaki, T.; et al. Estradiol rapidly modulates synaptic plasticity of hippocampal neurons: Involvement of kinase networks. *Brain Res.* **2015**, *1621*, 147–161. [CrossRef]
65. Spencer-Segal, J.L.; Tsuda, M.C.; Mattei, L.; Waters, E.M.; Romeo, R.D.; Milner, T.A.; McEwen, B.S.; Ogawa, S. Estradiol acts via estrogen receptors alpha and beta on pathways important for synaptic plasticity in the mouse hippocampal formation. *Neuroscience* **2012**, *202*, 131–146. [CrossRef] [PubMed]
66. Phan, A.; Lancaster, K.E.; Armstrong, J.N.; MacLusky, N.J.; Choleris, E. Rapid effects of estrogen receptor alpha and beta selective agonists on learning and dendritic spines in female mice. *Endocrinology* **2011**, *152*, 1492–1502. [CrossRef]
67. Liu, F.; Day, M.; Muniz, L.C.; Bitran, D.; Arias, R.; Revilla-Sanchez, R.; Grauer, S.; Zhang, G.; Kelley, C.; Pulito, V.; et al. Activation of estrogen receptor-beta regulates hippocampal synaptic plasticity and improves memory. *Nat. Neurosci.* **2008**, *11*, 334–343. [CrossRef]
68. Boulware, M.I.; Heisler, J.D.; Frick, K.M. The memory-enhancing effects of hippocampal estrogen receptor activation involve metabotropic glutamate receptor signaling. *J. Neurosci.* **2013**, *33*, 15184–15194. [CrossRef]
69. Wang, W.; Le, A.A.; Hou, B.; Lauterborn, J.C.; Cox, C.D.; Levin, E.R.; Lynch, G.; Gall, C.M. Memory-Related Synaptic Plasticity Is Sexually Dimorphic in Rodent Hippocampus. *J. Neurosci.* **2018**, *38*, 7935–7951. [CrossRef]
70. Briz, V.; Liu, Y.; Zhu, G.; Bi, X.; Baudry, M. A novel form of synaptic plasticity in field CA3 of hippocampus requires GPER1 activation and BDNF release. *J. Cell Biol.* **2015**, *210*, 1225–1237. [CrossRef]
71. Khadilkar, S.V.; Patil, V.A. Sex Hormones and Cognition: Where Do We Stand? *J. Obstet. Gynaecol. India* **2019**, *69*, 303–312. [CrossRef]
72. Jacobs, E.G.; Weiss, B.K.; Makris, N.; Whitfield-Gabrieli, S.; Buka, S.L.; Klibanski, A.; Goldstein, J.M. Impact of Sex and Menopausal Status on Episodic Memory Circuitry in Early Midlife. *J. Neurosci.* **2016**, *36*, 10163–10173. [CrossRef]
73. Rentz, D.M.; Weiss, B.K.; Jacobs, E.G.; Cherkerzian, S.; Klibanski, A.; Remington, A.; Aizley, H.; Goldstein, J.M. Sex differences in episodic memory in early midlife: Impact of reproductive aging. *Menopause* **2017**, *24*, 400–408. [CrossRef]

74. Gilsanz, P.; Lee, C.; Corrada, M.M.; Kawas, C.H.; Quesenberry, C.P., Jr.; Whitmer, R.A. Reproductive period and risk of dementia in a diverse cohort of health care members. *Neurology* **2019**, *92*, e2005–e2014. [CrossRef] [PubMed]
75. McRoberts, J.A.; Li, J.; Ennes, H.S.; Mayer, E.A. Sex-dependent differences in the activity and modulation of N-methyl-d-aspartic acid receptors in rat dorsal root ganglia neurons. *Neuroscience* **2007**, *148*, 1015–1020. [CrossRef] [PubMed]
76. Smith, C.C.; McMahon, L.L. Estradiol-induced increase in the magnitude of long-term potentiation is prevented by blocking NR2B-containing receptors. *J. Neurosci.* **2006**, *26*, 8517–8522. [CrossRef]
77. Kramar, E.A.; Chen, L.Y.; Brandon, N.J.; Rex, C.S.; Liu, F.; Gall, C.M.; Lynch, G. Cytoskeletal changes underlie estrogen's acute effects on synaptic transmission and plasticity. *J. Neurosci.* **2009**, *29*, 12982–12993. [CrossRef]
78. Koss, W.A.; Haertel, J.M.; Philippi, S.M.; Frick, K.M. Sex Differences in the Rapid Cell Signaling Mechanisms Underlying the Memory-Enhancing Effects of 17beta-Estradiol. *eNeuro* **2018**, *5*. [CrossRef]
79. Nicholson, K.; MacLusky, N.J.; Leranth, C. Synaptic effects of estrogen. *Vitam. Horm.* **2020**, *114*, 167–210. [PubMed]
80. Imtiaz, B.; Tuppurainen, M.; Rikkonen, T.; Kivipelto, M.; Soininen, H.; Kroger, H.; Tolppanen, A.M. Postmenopausal hormone therapy and Alzheimer disease: A prospective cohort study. *Neurology* **2017**, *88*, 1062–1068. [CrossRef]
81. Waring, S.C.; Rocca, W.A.; Petersen, R.C.; O'Brien, P.C.; Tangalos, E.G.; Kokmen, E. Postmenopausal estrogen replacement therapy and risk of AD: A population-based study. *Neurology* **1999**, *52*, 965–970. [CrossRef]
82. Manson, J.E.; Aragaki, A.K.; Rossouw, J.E.; Anderson, G.L.; Prentice, R.L.; LaCroix, A.Z.; Chlebowski, R.T.; Howard, B.V.; Thomson, C.A.; Margolis, K.L.; et al. Menopausal Hormone Therapy and Long-term All-Cause and Cause-Specific Mortality: The Women's Health Initiative Randomized Trials. *JAMA* **2017**, *318*, 927–938. [CrossRef]
83. Marongiu, R. Accelerated Ovarian Failure as a Unique Model to Study Peri-Menopause Influence on Alzheimer's Disease. *Front. Aging Neurosci.* **2019**, *11*, 242. [CrossRef] [PubMed]
84. Kim, Y.J.; Soto, M.; Branigan, G.L.; Rodgers, K.; Brinton, R.D. Association between menopausal hormone therapy and risk of neurodegenerative diseases: Implications for precision hormone therapy. *Alzheimers Dement.* **2021**, *7*, e12174.
85. Veenman, L. Raloxifene as Treatment for Various Types of Brain Injuries and Neurodegenerative Diseases: A Good Start. *Int. J. Mol. Sci.* **2020**, *21*, 7586. [CrossRef] [PubMed]
86. Blanquart, E.; Laffont, S.; Guery, J.C. Sex hormone regulation of innate lymphoid cells. *Biomed. J.* **2021**, *44*, 144–156. [CrossRef] [PubMed]
87. Amjadi, F.; Salehi, E.; Mehdizadeh, M.; Aflatoonian, R. Role of the innate immunity in female reproductive tract. *Adv. Biomed. Res.* **2014**, *3*, 1.
88. Medina-Estrada, I.; Alva-Murillo, N.; Lopez-Meza, J.E.; Ochoa-Zarzosa, A. Immunomodulatory Effects of 17beta-Estradiol on Epithelial Cells during Bacterial Infections. *J. Immunol. Res.* **2018**, *2018*, 6098961. [CrossRef]
89. Li, S.; Herrera, G.G.; Tam, K.K.; Lizarraga, J.S.; Beedle, M.T.; Winuthayanon, W. Estrogen Action in the Epithelial Cells of the Mouse Vagina Regulates Neutrophil Infiltration and Vaginal Tissue Integrity. *Sci. Rep.* **2018**, *8*, 11247. [CrossRef]
90. Chen, Y.Y.; Su, T.H.; Lau, H.H. Estrogen for the prevention of recurrent urinary tract infections in postmenopausal women: A meta-analysis of randomized controlled trials. *Int. Urogynecol. J.* **2021**, *32*, 17–25. [CrossRef] [PubMed]
91. Stanton, A.; Mowbray, C.; Lanz, M.; Brown, K.; Hilton, P.; Tyson-Capper, A.; Pickard, R.S.; Ali, A.S.M.; Hall, J. Topical Estrogen Treatment Augments the Vaginal Response to Escherichia coli Flagellin. *Sci. Rep.* **2020**, *10*, 8473. [CrossRef]
92. Abramenko, N.; Vellieux, F.; Tesarova, P.; Kejik, Z.; Kaplanek, R.; Lacina, L.; Dvorankova, B.; Rosel, D.; Brabek, J.; Tesar, A.; et al. Estrogen Receptor Modulators in Viral Infections Such as SARS-CoV-2: Therapeutic Consequences. *Int. J. Mol. Sci.* **2021**, *22*, 6551. [CrossRef]
93. Yang, X.; Guo, Y.; He, J.; Zhang, F.; Sun, X.; Yang, S.; Dong, H. Estrogen and estrogen receptors in the modulation of gastrointestinal epithelial secretion. *Oncotarget* **2017**, *8*, 97683–97692. [CrossRef] [PubMed]
94. Condliffe, S.B.; Doolan, C.M.; Harvey, B.J. 17beta-oestradiol acutely regulates Cl- secretion in rat distal colonic epithelium. *J. Physiol* **2001**, *530*, 47–54. [CrossRef]
95. Alzamora, R.; O'Mahony, F.; Harvey, B.J. Estrogen inhibits chloride secretion caused by cholera and Escherichia coli enterotoxins in female rat distal colon. *Steroids* **2011**, *76*, 867–876. [CrossRef] [PubMed]
96. Sampson, T.R.; Mazmanian, S.K. Control of brain development, function, and behavior by the microbiome. *Cell Host Microbe* **2015**, *17*, 565–576. [CrossRef]
97. Callan, N.G.L.; Mitchell, E.S.; Heitkemper, M.M.; Woods, N.F. Abdominal pain during the menopause transition and early postmenopause: Observations from the Seattle Midlife Women's Health Study. *Womens Midlife Health* **2019**, *5*, 2. [CrossRef]
98. Flores, R.; Shi, J.; Fuhrman, B.; Xu, X.; Veenstra, T.D.; Gail, M.H.; Gajer, P.; Ravel, J.; Goedert, J.J. Fecal microbial determinants of fecal and systemic estrogens and estrogen metabolites: A cross-sectional study. *J. Transl. Med.* **2012**, *10*, 253. [CrossRef] [PubMed]
99. Nakatsu, C.H.; Armstrong, A.; Clavijo, A.P.; Martin, B.R.; Barnes, S.; Weaver, C.M. Fecal bacterial community changes associated with isoflavone metabolites in postmenopausal women after soy bar consumption. *PLoS ONE* **2014**, *9*, e108924. [CrossRef]
100. Sau, L.; Olmstead, C.M.; Cui, L.J.; Chen, A.; Shah, R.S.; Kelley, S.T.; Thackray, V.G. Alterations in Gut Microbiota Do Not Play a Causal Role in Diet-independent Weight Gain Caused by Ovariectomy. *J. Endocr. Soc.* **2021**, *5*, bvaa173. [CrossRef] [PubMed]
101. Lei, Z.; Wu, H.; Yang, Y.; Hu, Q.; Lei, Y.; Liu, W.; Nie, Y.; Yang, L.; Zhang, X.; Yang, C.; et al. Ovariectomy Impaired Hepatic Glucose and Lipid Homeostasis and Altered the Gut Microbiota in Mice With Different Diets. *Front. Endocrinol.* **2021**, *12*, 708838. [CrossRef] [PubMed]

102. Chen, Q.; Wang, B.; Wang, S.; Qian, X.; Li, X.; Zhao, J.; Zhang, H.; Chen, W.; Wang, G. Modulation of the Gut Microbiota Structure with Probiotics and Isoflavone Alleviates Metabolic Disorder in Ovariectomized Mice. *Nutrients* **2021**, *13*, 1793. [CrossRef]
103. Chen, K.L.A.; Liu, X.; Zhao, Y.C.; Hieronymi, K.; Rossi, G.; Auvil, L.S.; Welge, M.; Bushell, C.; Smith, R.L.; Carlson, K.E.; et al. Long-Term Administration of Conjugated Estrogen and Bazedoxifene Decreased Murine Fecal beta-Glucuronidase Activity Without Impacting Overall Microbiome Community. *Sci. Rep.* **2018**, *8*, 8166. [CrossRef]
104. Urban, A.S.; Pavlov, K.V.; Kamynina, A.V.; Okhrimenko, I.S.; Arseniev, A.S.; Bocharov, E.V. Structural Studies Providing Insights into Production and Conformational Behavior of Amyloid-beta Peptide Associated with Alzheimer's Disease Development. *Molecules* **2021**, *26*, 2897. [CrossRef]
105. Guyon, A.; Rousseau, J.; Lamothe, G.; Tremblay, J.P. The protective mutation A673T in amyloid precursor protein gene decreases Abeta peptides production for 14 forms of Familial Alzheimer's Disease in SH-SY5Y cells. *PLoS ONE* **2020**, *15*, e0237122. [CrossRef]
106. Tackenberg, C.; Nitsch, R.M. The secreted APP ectodomain sAPPalpha, but not sAPPbeta, protects neurons against Abeta oligomer-induced dendritic spine loss and increased tau phosphorylation. *Mol. Brain* **2019**, *12*, 27. [CrossRef]
107. Manthey, D.; Heck, S.; Engert, S.; Behl, C. Estrogen induces a rapid secretion of amyloid beta precursor protein via the mitogen-activated protein kinase pathway. *Eur. J. Biochem.* **2001**, *268*, 4285–4291. [CrossRef] [PubMed]
108. Zhang, S.; Huang, Y.; Zhu, Y.C.; Yao, T. Estrogen stimulates release of secreted amyloid precursor protein from primary rat cortical neurons via protein kinase C pathway. *Acta Pharmacol. Sin.* **2005**, *26*, 171–176. [CrossRef] [PubMed]
109. Sochocka, M.; Zwolinska, K.; Leszek, J. The Infectious Etiology of Alzheimer's Disease. *Curr. Neuropharmacol.* **2017**, *15*, 996–1009. [CrossRef] [PubMed]
110. Shah, A.F.; Morris, J.A.; Wray, M. Pathogenesis of Alzheimer's disease: Multiple interacting causes against which amyloid precursor protein protects. *Med. Hypotheses* **2020**, *143*, 110035. [CrossRef] [PubMed]
111. Felberbaum, R.; Kupker, W. COVID-19 from the perspective of a gynecological endocrinologist. *Gynakol. Endokrinol.* **2021**, 1–4. [CrossRef]
112. Tschiffely, A.E.; Schuh, R.A.; Prokai-Tatrai, K.; Ottinger, M.A.; Prokai, L. An exploratory investigation of brain-selective estrogen treatment in males using a mouse model of Alzheimer's disease. *Horm. Behav.* **2018**, *98*, 16–21. [CrossRef]
113. Tschiffely, A.E.; Schuh, R.A.; Prokai-Tatrai, K.; Prokai, L.; Ottinger, M.A. A comparative evaluation of treatments with 17beta-estradiol and its brain-selective prodrug in a double-transgenic mouse model of Alzheimer's disease. *Horm. Behav.* **2016**, *83*, 39–44. [CrossRef] [PubMed]
114. Greenfield, J.P.; Leung, L.W.; Cai, D.; Kaasik, K.; Gross, R.S.; Rodriguez-Boulan, E.; Greengard, P.; Xu, H. Estrogen lowers Alzheimer beta-amyloid generation by stimulating trans-Golgi network vesicle biogenesis. *J. Biol. Chem* **2002**, *277*, 12128–12136. [CrossRef]
115. Jayaraman, A.; Carroll, J.C.; Morgan, T.E.; Lin, S.; Zhao, L.; Arimoto, J.M.; Murphy, M.P.; Beckett, T.L.; Finch, C.E.; Brinton, R.D.; et al. 17beta-estradiol and progesterone regulate expression of beta-amyloid clearance factors in primary neuron cultures and female rat brain. *Endocrinology* **2012**, *153*, 5467–5479. [CrossRef] [PubMed]
116. Zhao, L.; Yao, J.; Mao, Z.; Chen, S.; Wang, Y.; Brinton, R.D. 17beta-Estradiol regulates insulin-degrading enzyme expression via an ERbeta/PI3-K pathway in hippocampus: Relevance to Alzheimer's prevention. *Neurobiol. Aging* **2011**, *32*, 1949–1963. [CrossRef]
117. Merlo, S.; Sortino, M.A. Estrogen activates matrix metalloproteinases-2 and -9 to increase beta amyloid degradation. *Mol. Cell Neurosci.* **2012**, *49*, 423–429. [CrossRef]
118. Liang, K.; Yang, L.; Yin, C.; Xiao, Z.; Zhang, J.; Liu, Y.; Huang, J. Estrogen stimulates degradation of beta-amyloid peptide by up-regulating neprilysin. *J. Biol. Chem.* **2010**, *285*, 935–942. [CrossRef] [PubMed]
119. Mputhia, Z.; Hone, E.; Tripathi, T.; Sargeant, T.; Martins, R.; Bharadwaj, P. Autophagy Modulation as a Treatment of Amyloid Diseases. *Molecules* **2019**, *24*, 3372. [CrossRef]
120. Pike, C.J. Estrogen modulates neuronal Bcl-xL expression and beta-amyloid-induced apoptosis: Relevance to Alzheimer's disease. *J. Neurochem.* **1999**, *72*, 1552–1563. [CrossRef]
121. Zhang, Y.; Champagne, N.; Beitel, L.K.; Goodyer, C.G.; Trifiro, M.; LeBlanc, A. Estrogen and androgen protection of human neurons against intracellular amyloid beta1-42 toxicity through heat shock protein 70. *J. Neurosci.* **2004**, *24*, 5315–5321. [CrossRef] [PubMed]
122. Nilsen, J.; Chen, S.; Irwin, R.W.; Iwamoto, S.; Brinton, R.D. Estrogen protects neuronal cells from amyloid beta-induced apoptosis via regulation of mitochondrial proteins and function. *BMC Neurosci.* **2006**, *7*, 74. [CrossRef]
123. Marin, R.; Guerra, B.; Hernandez-Jimenez, J.G.; Kang, X.L.; Fraser, J.D.; Lopez, F.J.; Alonso, R. Estradiol prevents amyloid-beta peptide-induced cell death in a cholinergic cell line via modulation of a classical estrogen receptor. *Neuroscience* **2003**, *121*, 917–926. [CrossRef]
124. Kwakowsky, A.; Potapov, K.; Kim, S.; Peppercorn, K.; Tate, W.P.; Abraham, I.M. Treatment of beta amyloid 1-42 (Abeta(1-42))-induced basal forebrain cholinergic damage by a non-classical estrogen signaling activator in vivo. *Sci. Rep.* **2016**, *6*, 21101. [CrossRef]
125. Lai, Y.J.; Zhu, B.L.; Sun, F.; Luo, D.; Ma, Y.L.; Luo, B.; Tang, J.; Xiong, M.J.; Liu, L.; Long, Y.; et al. Estrogen receptor alpha promotes Cav1.2 ubiquitination and degradation in neuronal cells and in APP/PS1 mice. *Aging Cell* **2019**, *18*, e12961. [CrossRef]

126. Ramirez-Barrantes, R.; Carvajal-Zamorano, K.; Rodriguez, B.; Cordova, C.; Lozano, C.; Simon, F.; Diaz, P.; Munoz, P.; Marchant, I.; Latorre, R.; et al. TRPV1-Estradiol Stereospecific Relationship Underlies Cell Survival in Oxidative Cell Death. *Front. Physiol.* **2020**, *11*, 444. [CrossRef]
127. Pan, Q.; Guo, K.; Xue, M.; Tu, Q. Estrogen protects neuroblastoma cell from amyloid-beta 42 (Abeta42)-induced apoptosis via TXNIP/TRX axis and AMPK signaling. *Neurochem. Int.* **2020**, *135*, 104685. [CrossRef] [PubMed]
128. Tsubaki, H.; Tooyama, I.; Walker, D.G. Thioredoxin-Interacting Protein (TXNIP) with Focus on Brain and Neurodegenerative Diseases. *Int. J. Mol. Sci.* **2020**, *21*, 9357. [CrossRef] [PubMed]
129. Manoharan, S.; Guillemin, G.J.; Abiramasundari, R.S.; Essa, M.M.; Akbar, M.; Akbar, M.D. The Role of Reactive Oxygen Species in the Pathogenesis of Alzheimer's Disease, Parkinson's Disease, and Huntington's Disease: A Mini Review. *Oxid. Med. Cell Longev.* **2016**, *2016*, 8590578. [CrossRef] [PubMed]
130. Lejri, I.; Agapouda, A.; Grimm, A.; Eckert, A. Mitochondria- and Oxidative Stress-Targeting Substances in Cognitive Decline-Related Disorders: From Molecular Mechanisms to Clinical Evidence. *Oxid. Med. Cell Longev.* **2019**, *2019*, 9695412. [CrossRef]
131. Massaad, C.A. Neuronal and vascular oxidative stress in Alzheimer's disease. *Curr. Neuropharmacol.* **2011**, *9*, 662–673. [CrossRef] [PubMed]
132. Cervellati, C.; Wood, P.L.; Romani, A.; Valacchi, G.; Squerzanti, M.; Sanz, J.M.; Ortolani, B.; Zuliani, G. Oxidative challenge in Alzheimer's disease: State of knowledge and future needs. *J. Investig. Med.* **2016**, *64*, 21–32. [CrossRef]
133. Abramov, A.Y.; Canevari, L.; Duchen, M.R. Beta-amyloid peptides induce mitochondrial dysfunction and oxidative stress in astrocytes and death of neurons through activation of NADPH oxidase. *J. Neurosci.* **2004**, *24*, 565–575. [CrossRef] [PubMed]
134. Wojsiat, J.; Zoltowska, K.M.; Laskowska-Kaszub, K.; Wojda, U. Oxidant/Antioxidant Imbalance in Alzheimer's Disease: Therapeutic and Diagnostic Prospects. *Oxid. Med. Cell Longev.* **2018**, *2018*, 6435861. [CrossRef] [PubMed]
135. Lee, S.Y.; Andoh, T.; Murphy, D.L.; Chiueh, C.C. 17beta-estradiol activates ICI 182,780-sensitive estrogen receptors and cyclic GMP-dependent thioredoxin expression for neuroprotection. *FASEB J.* **2003**, *17*, 947–948. [CrossRef]
136. Ayres, S.; Abplanalp, W.; Liu, J.H.; Subbiah, M.T. Mechanisms involved in the protective effect of estradiol-17beta on lipid peroxidation and DNA damage. *Am. J. Physiol.* **1998**, *274*, E1002–E1008.
137. Behl, C.; Skutella, T.; Lezoualc'h, F.; Post, A.; Widmann, M.; Newton, C.J.; Holsboer, F. Neuroprotection against oxidative stress by estrogens: Structure-activity relationship. *Mol. Pharmacol.* **1997**, *51*, 535–541. [CrossRef]
138. Gridley, K.E.; Green, P.S.; Simpkins, J.W. Low concentrations of estradiol reduce beta-amyloid (25-35)-induced toxicity, lipid peroxidation and glucose utilization in human SK-N-SH neuroblastoma cells. *Brain Res.* **1997**, *778*, 158–165. [CrossRef]
139. Celsi, F.; Ferri, A.; Casciati, A.; D'Ambrosi, N.; Rotilio, G.; Costa, A.; Volonte, C.; Carri, M.T. Overexpression of superoxide dismutase 1 protects against beta-amyloid peptide toxicity: Effect of estrogen and copper chelators. *Neurochem. Int.* **2004**, *44*, 25–33. [CrossRef]
140. Pena-Bautista, C.; Baquero, M.; Vento, M.; Chafer-Pericas, C. Free radicals in Alzheimer's disease: Lipid peroxidation biomarkers. *Clin. Chim. Acta* **2019**, *491*, 85–90. [CrossRef]
141. Chang, Y.T.; Chang, W.N.; Tsai, N.W.; Huang, C.C.; Kung, C.T.; Su, Y.J.; Lin, W.C.; Cheng, B.C.; Su, C.M.; Chiang, Y.F.; et al. The roles of biomarkers of oxidative stress and antioxidant in Alzheimer's disease: A systematic review. *Biomed. Res. Int.* **2014**, *2014*, 182303. [CrossRef]
142. Schonfeld, P.; Reiser, G. How the brain fights fatty acids' toxicity. *Neurochem. Int.* **2021**, *148*, 105050. [CrossRef] [PubMed]
143. Dani, M.; Wood, M.; Mizoguchi, R.; Fan, Z.; Walker, Z.; Morgan, R.; Hinz, R.; Biju, M.; Kuruvilla, T.; Brooks, D.J.; et al. Microglial activation correlates in vivo with both tau and amyloid in Alzheimer's disease. *Brain* **2018**, *141*, 2740–2754. [CrossRef]
144. Fan, Z.; Brooks, D.J.; Okello, A.; Edison, P. An early and late peak in microglial activation in Alzheimer's disease trajectory. *Brain* **2017**, *140*, 792–803. [CrossRef] [PubMed]
145. Varnum, M.M.; Ikezu, T. The classification of microglial activation phenotypes on neurodegeneration and regeneration in Alzheimer's disease brain. *Arch. Immunol. Ther. Exp.* **2012**, *60*, 251–266. [CrossRef]
146. Sierra, A.; Gottfried-Blackmore, A.; Milner, T.A.; McEwen, B.S.; Bulloch, K. Steroid hormone receptor expression and function in microglia. *Glia* **2008**, *56*, 659–674. [CrossRef]
147. Ghisletti, S.; Meda, C.; Maggi, A.; Vegeto, E. 17beta-estradiol inhibits inflammatory gene expression by controlling NF-kappaB intracellular localization. *Mol. Cell Biol.* **2005**, *25*, 2957–2968. [CrossRef] [PubMed]
148. Bruce-Keller, A.J.; Keeling, J.L.; Keller, J.N.; Huang, F.F.; Camondola, S.; Mattson, M.P. Antiinflammatory effects of estrogen on microglial activation. *Endocrinology* **2000**, *141*, 3646–3656. [CrossRef]
149. Zhao, T.Z.; Ding, Q.; Hu, J.; He, S.M.; Shi, F.; Ma, L.T. GPER expressed on microglia mediates the anti-inflammatory effect of estradiol in ischemic stroke. *Brain Behav.* **2016**, *6*, e00449. [CrossRef]
150. Shindo, S.; Chen, S.H.; Gotoh, S.; Yokobori, K.; Hu, H.; Ray, M.; Moore, R.; Nagata, K.; Martinez, J.; Hong, J.S.; et al. Estrogen receptor alpha phosphorylated at Ser216 confers inflammatory function to mouse microglia. *Cell Commun. Signal.* **2020**, *18*, 117. [CrossRef]
151. Li, K.X.; Sun, Q.; Wei, L.L.; Du, G.H.; Huang, X.; Wang, J.K. ERalpha Gene Promoter Methylation in Cognitive Function and Quality of Life of Patients With Alzheimer Disease. *J. Geriatr. Psychiatry Neurol.* **2019**, *32*, 221–228. [CrossRef]
152. Gamache, J.; Yun, Y.; Chiba-Falek, O. Sex-dependent effect of APOE on Alzheimer's disease and other age-related neurodegenerative disorders. *Dis. Model. Mech.* **2020**, *13*, dmm045211. [CrossRef]

153. Ratnakumar, A.; Zimmerman, S.E.; Jordan, B.A.; Mar, J.C. Estrogen activates Alzheimer's disease genes. *Alzheimers Dement.* **2019**, *5*, 906–917. [CrossRef] [PubMed]
154. Wang, J.M.; Irwin, R.W.; Brinton, R.D. Activation of estrogen receptor alpha increases and estrogen receptor beta decreases apolipoprotein E expression in hippocampus in vitro and in vivo. *Proc. Natl. Acad. Sci. USA* **2006**, *103*, 16983–16988. [CrossRef]
155. Reiman, E.M.; Arboleda-Velasquez, J.F.; Quiroz, Y.T.; Huentelman, M.J.; Beach, T.G.; Caselli, R.J.; Chen, Y.; Su, Y.; Myers, A.J.; Hardy, J.; et al. Exceptionally low likelihood of Alzheimer's dementia in APOE2 homozygotes from a 5000-person neuropathological study. *Nat. Commun.* **2020**, *11*, 667. [CrossRef] [PubMed]
156. Nathan, B.P.; Barsukova, A.G.; Shen, F.; McAsey, M.; Struble, R.G. Estrogen facilitates neurite extension via apolipoprotein E in cultured adult mouse cortical neurons. *Endocrinology* **2004**, *145*, 3065–3073. [CrossRef]
157. Zaretsky, D.V.; Zaretskaia, M.V. Mini-review: Amyloid degradation toxicity hypothesis of Alzheimer's disease. *Neurosci. Lett.* **2021**, *756*, 135959. [CrossRef]
158. Shoshan-Barmatz, V.; Nahon-Crystal, E.; Shteinfer-Kuzmine, A.; Gupta, R. VDAC1, mitochondrial dysfunction, and Alzheimer's disease. *Pharmacol. Res.* **2018**, *131*, 87–101. [CrossRef] [PubMed]
159. Ramirez, C.M.; Gonzalez, M.; Diaz, M.; Alonso, R.; Ferrer, I.; Santpere, G.; Puig, B.; Meyer, G.; Marin, R. VDAC and ERalpha interaction in caveolae from human cortex is altered in Alzheimer's disease. *Mol. Cell Neurosci.* **2009**, *42*, 172–183. [CrossRef]
160. Herrera, J.L.; Diaz, M.; Hernandez-Fernaud, J.R.; Salido, E.; Alonso, R.; Fernandez, C.; Morales, A.; Marin, R. Voltage-dependent anion channel as a resident protein of lipid rafts: Post-transductional regulation by estrogens and involvement in neuronal preservation against Alzheimer's disease. *J. Neurochem.* **2011**, *116*, 820–827. [CrossRef]
161. Long, J.; He, P.; Shen, Y.; Li, R. New evidence of mitochondria dysfunction in the female Alzheimer's disease brain: Deficiency of estrogen receptor-beta. *J. Alzheimers Dis.* **2012**, *30*, 545–558. [CrossRef] [PubMed]
162. Zhao, W.; Hou, Y.; Song, X.; Wang, L.; Zhang, F.; Zhang, H.; Yu, H.; Zhou, Y. Estrogen Deficiency Induces Mitochondrial Damage Prior to Emergence of Cognitive Deficits in a Postmenopausal Mouse Model. *Front. Aging Neurosci.* **2021**, *13*, 713819. [CrossRef]
163. Mosconi, L.; Berti, V.; Quinn, C.; McHugh, P.; Petrongolo, G.; Osorio, R.S.; Connaughty, C.; Pupi, A.; Vallabhajosula, S.; Isaacson, R.S.; et al. Perimenopause and emergence of an Alzheimer's bioenergetic phenotype in brain and periphery. *PLoS ONE* **2017**, *12*, e0185926. [CrossRef] [PubMed]
164. Mosconi, L.; Berti, V.; Dyke, J.; Schelbaum, E.; Jett, S.; Loughlin, L.; Jang, G.; Rahman, A.; Hristov, H.; Pahlajani, S.; et al. Menopause impacts human brain structure, connectivity, energy metabolism, and amyloid-beta deposition. *Sci. Rep.* **2021**, *11*, 10867. [CrossRef] [PubMed]
165. Ragonese, P.; D'Amelio, M.; Salemi, G.; Aridon, P.; Gammino, M.; Epifanio, A.; Morgante, L.; Savettieri, G. Risk of Parkinson disease in women: Effect of reproductive characteristics. *Neurology* **2004**, *62*, 2010–2014. [CrossRef]
166. Walf, A.A.; Koonce, C.J.; Frye, C.A. Progestogens' effects and mechanisms for object recognition memory across the lifespan. *Behav. Brain Res.* **2015**, *294*, 50–61. [CrossRef] [PubMed]
167. Almey, A.; Milner, T.A.; Brake, W.G. Estrogen receptors in the central nervous system and their implication for dopamine-dependent cognition in females. *Horm. Behav.* **2015**, *74*, 125–138. [CrossRef]
168. Almey, A.; Filardo, E.J.; Milner, T.A.; Brake, W.G. Estrogen receptors are found in glia and at extranuclear neuronal sites in the dorsal striatum of female rats: Evidence for cholinergic but not dopaminergic colocalization. *Endocrinology* **2012**, *153*, 5373–5383. [CrossRef] [PubMed]
169. Mitterling, K.L.; Spencer, J.L.; Dziedzic, N.; Shenoy, S.; McCarthy, K.; Waters, E.M.; McEwen, B.S.; Milner, T.A. Cellular and subcellular localization of estrogen and progestin receptor immunoreactivities in the mouse hippocampus. *J. Comp. Neurol.* **2010**, *518*, 2729–2743. [CrossRef] [PubMed]
170. Funakoshi, T.; Yanai, A.; Shinoda, K.; Kawano, M.M.; Mizukami, Y. G protein-coupled receptor 30 is an estrogen receptor in the plasma membrane. *Biochem. Biophys. Res. Commun.* **2006**, *346*, 904–910. [CrossRef] [PubMed]
171. Campos, F.L.; Cristovao, A.C.; Rocha, S.M.; Fonseca, C.P.; Baltazar, G. GDNF contributes to oestrogen-mediated protection of midbrain dopaminergic neurones. *J. Neuroendocrinol.* **2012**, *24*, 1386–1397. [CrossRef] [PubMed]
172. Castilla-Cortazar, I.; Aguirre, G.A.; Femat-Roldan, G.; Martin-Estal, I.; Espinosa, L. Is insulin-like growth factor-1 involved in Parkinson's disease development? *J. Transl. Med.* **2020**, *18*, 70. [CrossRef] [PubMed]
173. Wang, P.; Li, J.; Qiu, S.; Wen, H.; Du, J. Hormone replacement therapy and Parkinson's disease risk in women: A meta-analysis of 14 observational studies. *Neuropsychiatr. Dis. Treat.* **2015**, *11*, 59–66. [PubMed]
174. Barron, A.M.; Pike, C.J. Sex hormones, aging, and Alzheimer's disease. *Front. Biosci.* **2012**, *4*, 976–997.
175. Carroll, J.C.; Rosario, E.R.; Chang, L.; Stanczyk, F.Z.; Oddo, S.; LaFerla, F.M.; Pike, C.J. Progesterone and estrogen regulate Alzheimer-like neuropathology in female 3xTg-AD mice. *J. Neurosci.* **2007**, *27*, 13357–13365. [CrossRef]
176. Savolainen-Peltonen, H.; Rahkola-Soisalo, P.; Hoti, F.; Vattulainen, P.; Gissler, M.; Ylikorkala, O.; Mikkola, T.S. Use of post-menopausal hormone therapy and risk of Alzheimer's disease in Finland: Nationwide case-control study. *BMJ* **2019**, *364*, l665. [CrossRef]
177. Chen, S.; Wang, J.M.; Irwin, R.W.; Yao, J.; Liu, L.; Brinton, R.D. Allopregnanolone promotes regeneration and reduces beta-amyloid burden in a preclinical model of Alzheimer's disease. *PLoS ONE* **2011**, *6*, e24293.
178. Irwin, R.W.; Brinton, R.D. Allopregnanolone as regenerative therapeutic for Alzheimer's disease: Translational development and clinical promise. *Prog. Neurobiol.* **2014**, *113*, 40–55. [CrossRef]

179. Wang, T.; Yao, J.; Chen, S.; Mao, Z.; Brinton, R.D. Allopregnanolone Reverses Bioenergetic Deficits in Female Triple Transgenic Alzheimer's Mouse Model. *Neurotherapeutics* **2020**, *17*, 178–188. [CrossRef]
180. Adeosun, S.O.; Hou, X.; Jiao, Y.; Zheng, B.; Henry, S.; Hill, R.; He, Z.; Pani, A.; Kyle, P.; Ou, X.; et al. Allopregnanolone reinstates tyrosine hydroxylase immunoreactive neurons and motor performance in an MPTP-lesioned mouse model of Parkinson's disease. *PLoS ONE* **2012**, *7*, e50040. [CrossRef]
181. Chen, Z.C.; Wang, T.T.; Bian, W.; Ye, X.; Li, M.Y.; Du, J.J.; Zhou, P.; Cui, H.R.; Ding, Y.Q.; Ren, Y.H.; et al. Allopregnanolone restores the tyrosine hydroxylase-positive neurons and motor performance in a 6-OHDA-injected mouse model. *CNS Neurosci. Ther.* **2020**, *26*, 1069–1082. [CrossRef]
182. Okereke, O.; Kang, J.H.; Ma, J.; Hankinson, S.E.; Pollak, M.N.; Grodstein, F. Plasma IGF-I levels and cognitive performance in older women. *Neurobiol. Aging* **2007**, *28*, 135–142. [CrossRef] [PubMed]
183. Salzmann, A.; James, S.N.; Williams, D.M.; Richards, M.; Cadar, D.; Schott, J.M.; Coath, W.; Sudre, C.H.; Chaturvedi, N.; Garfield, V. Investigating the Relationship Between IGF-I, IGF-II, and IGFBP-3 Concentrations and Later-Life Cognition and Brain Volume. *J. Clin. Endocrinol. Metab.* **2021**, *106*, 1617–1629. [CrossRef]
184. Doi, T.; Shimada, H.; Makizako, H.; Tsutsumimoto, K.; Hotta, R.; Nakakubo, S.; Suzuki, T. Association of insulin-like growth factor-1 with mild cognitive impairment and slow gait speed. *Neurobiol. Aging* **2015**, *36*, 942–947. [CrossRef]
185. Alvarez, A.; Cacabelos, R.; Sanpedro, C.; Garcia-Fantini, M.; Aleixandre, M. Serum TNF-alpha levels are increased and correlate negatively with free IGF-I in Alzheimer disease. *Neurobiol. Aging* **2007**, *28*, 533–536. [CrossRef]
186. Mustafa, A.; Lannfelt, L.; Lilius, L.; Islam, A.; Winblad, B.; Adem, A. Decreased plasma insulin-like growth factor-I level in familial Alzheimer's disease patients carrying the Swedish APP 670/671 mutation. *Dement. Geriatr. Cogn. Disord.* **1999**, *10*, 446–451. [CrossRef] [PubMed]
187. Watanabe, T.; Miyazaki, A.; Katagiri, T.; Yamamoto, H.; Idei, T.; Iguchi, T. Relationship between serum insulin-like growth factor-1 levels and Alzheimer's disease and vascular dementia. *J. Am. Geriatr. Soc.* **2005**, *53*, 1748–1753. [CrossRef]
188. Duron, E.; Funalot, B.; Brunel, N.; Coste, J.; Quinquis, L.; Viollet, C.; Belmin, J.; Jouanny, P.; Pasquier, F.; Treluyer, J.M.; et al. Insulin-like growth factor-I and insulin-like growth factor binding protein-3 in Alzheimer's disease. *J. Clin. Endocrinol. Metab.* **2012**, *97*, 4673–4681. [CrossRef]
189. Westwood, A.J.; Beiser, A.; Decarli, C.; Harris, T.B.; Chen, T.C.; He, X.M.; Roubenoff, R.; Pikula, A.; Au, R.; Braverman, L.E.; et al. Insulin-like growth factor-1 and risk of Alzheimer dementia and brain atrophy. *Neurology* **2014**, *82*, 1613–1619. [CrossRef] [PubMed]
190. Vardy, E.R.; Rice, P.J.; Bowie, P.C.; Holmes, J.D.; Grant, P.J.; Hooper, N.M. Increased circulating insulin-like growth factor-1 in late-onset Alzheimer's disease. *J. Alzheimers Dis.* **2007**, *12*, 285–290. [CrossRef] [PubMed]
191. Vargas, T.; Martinez-Garcia, A.; Antequera, D.; Vilella, E.; Clarimon, J.; Mateo, I.; Sanchez-Juan, P.; Rodriguez-Rodriguez, E.; Frank, A.; Rosich-Estrago, M.; et al. IGF-I gene variability is associated with an increased risk for AD. *Neurobiol. Aging* **2011**, *32*, 556.e3–556.e11. [CrossRef] [PubMed]
192. Trueba-Saiz, A.; Cavada, C.; Fernandez, A.M.; Leon, T.; Gonzalez, D.A.; Fortea Ormaechea, J.; Lleo, A.; Del Ser, T.; Nunez, A.; Torres-Aleman, I. Loss of serum IGF-I input to the brain as an early biomarker of disease onset in Alzheimer mice. *Transl. Psychiatry* **2013**, *3*, e330. [CrossRef] [PubMed]
193. Ostrowski, P.P.; Barszczyk, A.; Forstenpointner, J.; Zheng, W.; Feng, Z.P. Meta-Analysis of Serum Insulin-Like Growth Factor 1 in Alzheimer's Disease. *PLoS ONE* **2016**, *11*, e0155733. [CrossRef]
194. De la Monte, S.M.; Wands, J.R. Alzheimer's disease is type 3 diabetes-evidence reviewed. *J. Diabetes Sci. Technol.* **2008**, *2*, 1101–1113. [CrossRef] [PubMed]
195. Picillo, M.; Pivonello, R.; Santangelo, G.; Pivonello, C.; Savastano, R.; Auriemma, R.; Amboni, M.; Scannapieco, S.; Pierro, A.; Colao, A.; et al. Serum IGF-1 is associated with cognitive functions in early, drug-naive Parkinson's disease. *PLoS ONE* **2017**, *12*, e0186508. [CrossRef]
196. Fan, D.; Pitcher, T.; Dalrymple-Alford, J.; MacAskill, M.; Anderson, T.; Guan, J. Changes of plasma cGP/IGF-1 molar ratio with age is associated with cognitive status of Parkinson disease. *Alzheimers Dement.* **2020**, *12*, e12025. [CrossRef]
197. Al-Delaimy, W.K.; von Muhlen, D.; Barrett-Connor, E. Insulinlike growth factor-1, insulinlike growth factor binding protein-1, and cognitive function in older men and women. *J. Am. Geriatr. Soc.* **2009**, *57*, 1441–1446. [CrossRef]
198. Perice, L.; Barzilai, N.; Verghese, J.; Weiss, E.F.; Holtzer, R.; Cohen, P.; Milman, S. Lower circulating insulin-like growth factor-I is associated with better cognition in females with exceptional longevity without compromise to muscle mass and function. *Aging* **2016**, *8*, 2414–2424. [CrossRef]
199. Frederiksen, H.; Johannsen, T.H.; Andersen, S.E.; Albrethsen, J.; Landersoe, S.K.; Petersen, J.H.; Andersen, A.N.; Vestergaard, E.T.; Schorring, M.E.; Linneberg, A.; et al. Sex-specific Estrogen Levels and Reference Intervals from Infancy to Late Adulthood Determined by LC-MS/MS. *J. Clin. Endocrinol. Metab.* **2020**, *105*, 754–768. [CrossRef]
200. Morinaga, A.; Ono, K.; Takasaki, J.; Ikeda, T.; Hirohata, M.; Yamada, M. Effects of sex hormones on Alzheimer's disease-associated beta-amyloid oligomer formation in vitro. *Exp. Neurol.* **2011**, *228*, 298–302. [CrossRef]
201. Palmisano, B.T.; Zhu, L.; Stafford, J.M. Role of Estrogens in the Regulation of Liver Lipid Metabolism. *Adv. Exp. Med. Biol.* **2017**, *1043*, 227–256.
202. Dimache, A.M.; Salaru, D.L.; Sascau, R.; Statescu, C. The Role of High Triglycerides Level in Predicting Cognitive Impairment: A Review of Current Evidence. *Nutrients* **2021**, *13*, 2118. [CrossRef] [PubMed]

203. Beekman, J.M.; Allan, G.F.; Tsai, S.Y.; Tsai, M.J.; O'Malley, B.W. Transcriptional activation by the estrogen receptor requires a conformational change in the ligand binding domain. *Mol. Endocrinol.* **1993**, *7*, 1266–1274. [PubMed]
204. Fliss, A.E.; Benzeno, S.; Rao, J.; Caplan, A.J. Control of estrogen receptor ligand binding by Hsp90. *J. Steroid Biochem. Mol. Biol.* **2000**, *72*, 223–230. [CrossRef]
205. Bustamante-Barrientos, F.A.; Mendez-Ruette, M.; Ortloff, A.; Luz-Crawford, P.; Rivera, F.J.; Figueroa, C.D.; Molina, L.; Batiz, L.F. The Impact of Estrogen and Estrogen-Like Molecules in Neurogenesis and Neurodegeneration: Beneficial or Harmful? *Front. Cell Neurosci.* **2021**, *15*, 636176. [CrossRef]
206. Klinge, C.M. Estrogen receptor interaction with estrogen response elements. *Nucleic Acids Res.* **2001**, *29*, 2905–2919. [CrossRef]
207. Sohrabji, F.; Miranda, R.C.; Toran-Allerand, C.D. Identification of a putative estrogen response element in the gene encoding brain-derived neurotrophic factor. *Proc. Natl. Acad. Sci. USA* **1995**, *92*, 11110–11114. [CrossRef]
208. Kight, K.E.; McCarthy, M.M. Sex differences and estrogen regulation of BDNF gene expression, but not propeptide content, in the developing hippocampus. *J. Neurosci. Res.* **2017**, *95*, 345–354. [CrossRef]
209. Kwakowsky, A.; Milne, M.R.; Waldvogel, H.J.; Faull, R.L. Effect of Estradiol on Neurotrophin Receptors in Basal Forebrain Cholinergic Neurons: Relevance for Alzheimer's Disease. *Int. J. Mol. Sci.* **2016**, *17*, 2122. [CrossRef]
210. Krolick, K.N.; Zhu, Q.; Shi, H. Effects of Estrogens on Central Nervous System Neurotransmission: Implications for Sex Differences in Mental Disorders. *Prog. Mol. Biol. Transl. Sci.* **2018**, *160*, 105–171.
211. Vail, G.; Roepke, T.A. Membrane-initiated estrogen signaling via Gq-coupled GPCR in the central nervous system. *Steroids* **2019**, *142*, 77–83. [CrossRef]
212. Revankar, C.M.; Cimino, D.F.; Sklar, L.A.; Arterburn, J.B.; Prossnitz, E.R. A transmembrane intracellular estrogen receptor mediates rapid cell signaling. *Science* **2005**, *307*, 1625–1630. [CrossRef] [PubMed]
213. Tang, H.; Zhang, Q.; Yang, L.; Dong, Y.; Khan, M.; Yang, F.; Brann, D.W.; Wang, R. Reprint of "GPR30 mediates estrogen rapid signaling and neuroprotection". *Mol. Cell Endocrinol.* **2014**, *389*, 92–98. [CrossRef]
214. Kim, J.; Szinte, J.S.; Boulware, M.I.; Frick, K.M. 17beta-Estradiol and Agonism of G-protein-Coupled Estrogen Receptor Enhance Hippocampal Memory via Different Cell-Signaling Mechanisms. *J. Neurosci.* **2016**, *36*, 3309–3321. [CrossRef] [PubMed]
215. Zhang, M.; Weiland, H.; Schofbanker, M.; Zhang, W. Estrogen Receptors Alpha and Beta Mediate Synaptic Transmission in the PFC and Hippocampus of Mice. *Int. J. Mol. Sci.* **2021**, *22*, 1485. [CrossRef]
216. Frick, K.M. Molecular mechanisms underlying the memory-enhancing effects of estradiol. *Horm. Behav.* **2015**, *74*, 4–18. [CrossRef]
217. Filardo, E.J.; Quinn, J.A.; Bland, K.I.; Frackelton, A.R., Jr. Estrogen-induced activation of Erk-1 and Erk-2 requires the G protein-coupled receptor homolog, GPR30, and occurs via trans-activation of the epidermal growth factor receptor through release of HB-EGF. *Mol. Endocrinol.* **2000**, *14*, 1649–1660. [CrossRef]
218. Filardo, E.J.; Quinn, J.A.; Frackelton, A.R., Jr.; Bland, K.I. Estrogen action via the G protein-coupled receptor, GPR30: Stimulation of adenylyl cyclase and cAMP-mediated attenuation of the epidermal growth factor receptor-to-MAPK signaling axis. *Mol. Endocrinol.* **2002**, *16*, 70–84. [CrossRef] [PubMed]
219. Shi, D.; Zhao, P.; Cui, L.; Li, H.; Sun, L.; Niu, J.; Chen, M. Inhibition of PI3K/AKT molecular pathway mediated by membrane estrogen receptor GPER accounts for cryptotanshinone induced antiproliferative effect on breast cancer SKBR-3 cells. *BMC Pharmacol. Toxicol.* **2020**, *21*, 32. [CrossRef] [PubMed]
220. Evans, N.J.; Bayliss, A.L.; Reale, V.; Evans, P.D. Characterisation of Signalling by the Endogenous GPER1 (GPR30) Receptor in an Embryonic Mouse Hippocampal Cell Line (mHippoE-18). *PLoS ONE* **2016**, *11*, e0152138. [CrossRef]
221. Bourque, M.; Morissette, M.; Di Paolo, T. Neuroprotection in Parkinsonian-treated mice via estrogen receptor alpha activation requires G protein-coupled estrogen receptor 1. *Neuropharmacology* **2015**, *95*, 343–352. [CrossRef]
222. Cheng, Q.; Meng, J.; Wang, X.S.; Kang, W.B.; Tian, Z.; Zhang, K.; Liu, G.; Zhao, J.N. G-1 exerts neuroprotective effects through G protein-coupled estrogen receptor 1 following spinal cord injury in mice. *Biosci. Rep.* **2016**, *36*, 36. [CrossRef]
223. Brotfain, E.; Gruenbaum, S.E.; Boyko, M.; Kutz, R.; Zlotnik, A.; Klein, M. Neuroprotection by Estrogen and Progesterone in Traumatic Brain Injury and Spinal Cord Injury. *Curr. Neuropharmacol.* **2016**, *14*, 641–653. [CrossRef] [PubMed]
224. Briz, V.; Baudry, M. Estrogen Regulates Protein Synthesis and Actin Polymerization in Hippocampal Neurons through Different Molecular Mechanisms. *Front. Endocrinol.* **2014**, *5*, 22. [CrossRef]
225. Yuan, L.J.; Wang, X.W.; Wang, H.T.; Zhang, M.; Sun, J.W.; Chen, W.F. G protein-coupled estrogen receptor is involved in the neuroprotective effect of IGF-1 against MPTP/MPP(+)-induced dopaminergic neuronal injury. *J. Steroid Biochem. Mol. Biol.* **2019**, *192*, 105384. [CrossRef]
226. Wang, X.W.; Yuan, L.J.; Yang, Y.; Zhang, M.; Chen, W.F. IGF-1 inhibits MPTP/MPP(+)-induced autophagy on dopaminergic neurons through the IGF-1R/PI3K-Akt-mTOR pathway and GPER. *Am. J. Physiol. Endocrinol. Metab.* **2020**, *319*, E734–E743. [CrossRef]
227. Briski, K.P.; Ali, M.H.; Napit, P.R. Sex-specific acclimation of A2 noradrenergic neuron dopamine-beta-hydroxylase and estrogen receptor variant protein and 5′-AMP-Activated protein kinase reactivity to recurring hypoglycemia in rat. *J. Chem. Neuroanat.* **2020**, *109*, 101845. [CrossRef] [PubMed]
228. Velarde, M.C. Mitochondrial and sex steroid hormone crosstalk during aging. *Longev. Healthspan.* **2014**, *3*, 2. [CrossRef]
229. Dobolyi, A.; Leko, A.H. The insulin-like growth factor-1 system in the adult mammalian brain and its implications in central maternal adaptation. *Front. Neuroendocrinol.* **2019**, *52*, 181–194. [CrossRef] [PubMed]

230. Sonntag, W.E.; Bennett, S.A.; Khan, A.S.; Thornton, P.L.; Xu, X.; Ingram, R.L.; Brunso-Bechtold, J.K. Age and insulin-like growth factor-1 modulate N-methyl-D-aspartate receptor subtype expression in rats. *Brain Res. Bull.* **2000**, *51*, 331–338. [CrossRef]
231. Lichtenwalner, R.J.; Forbes, M.E.; Bennett, S.A.; Lynch, C.D.; Sonntag, W.E.; Riddle, D.R. Intracerebroventricular infusion of insulin-like growth factor-I ameliorates the age-related decline in hippocampal neurogenesis. *Neuroscience* **2001**, *107*, 603–613. [CrossRef]
232. Poirier, R.; Fernandez, A.M.; Torres-Aleman, I.; Metzger, F. Early brain amyloidosis in APP/PS1 mice with serum insulin-like growth factor-I deficiency. *Neurosci. Lett.* **2012**, *509*, 101–104. [CrossRef] [PubMed]
233. Carro, E.; Trejo, J.L.; Gerber, A.; Loetscher, H.; Torrado, J.; Metzger, F.; Torres-Aleman, I. Therapeutic actions of insulin-like growth factor I on APP/PS2 mice with severe brain amyloidosis. *Neurobiol. Aging* **2006**, *27*, 1250–1257. [CrossRef]
234. Carro, E.; Trejo, J.L.; Gomez-Isla, T.; LeRoith, D.; Torres-Aleman, I. Serum insulin-like growth factor I regulates brain amyloid-beta levels. *Nat. Med.* **2002**, *8*, 1390–1397. [CrossRef] [PubMed]
235. George, C.; Gontier, G.; Lacube, P.; Francois, J.C.; Holzenberger, M.; Aid, S. The Alzheimer's disease transcriptome mimics the neuroprotective signature of IGF-1 receptor-deficient neurons. *Brain* **2017**, *140*, 2012–2027. [CrossRef]
236. Gontier, G.; George, C.; Chaker, Z.; Holzenberger, M.; Aid, S. Blocking IGF Signaling in Adult Neurons Alleviates Alzheimer's Disease Pathology through Amyloid-beta Clearance. *J. Neurosci.* **2015**, *35*, 11500–11513. [CrossRef] [PubMed]
237. Sohrabi, M.; Floden, A.M.; Manocha, G.D.; Klug, M.G.; Combs, C.K. IGF-1R Inhibitor Ameliorates Neuroinflammation in an Alzheimer's Disease Transgenic Mouse Model. *Front. Cell Neurosci.* **2020**, *14*, 200. [CrossRef] [PubMed]
238. Lanz, T.A.; Salatto, C.T.; Semproni, A.R.; Marconi, M.; Brown, T.M.; Richter, K.E.; Schmidt, K.; Nelson, F.R.; Schachter, J.B. Peripheral elevation of IGF-1 fails to alter Abeta clearance in multiple in vivo models. *Biochem. Pharmacol.* **2008**, *75*, 1093–1103. [CrossRef]
239. Selles, M.C.; Fortuna, J.T.S.; Zappa-Villar, M.F.; de Faria, Y.P.R.; Souza, A.S.; Suemoto, C.K.; Leite, R.E.P.; Rodriguez, R.D.; Grinberg, L.T.; Reggiani, P.C.; et al. Adenovirus-Mediated Transduction of Insulin-Like Growth Factor 1 Protects Hippocampal Neurons from the Toxicity of Abeta Oligomers and Prevents Memory Loss in an Alzheimer Mouse Model. *Mol. Neurobiol.* **2020**, *57*, 1473–1483. [CrossRef]
240. Sevigny, J.J.; Ryan, J.M.; van Dyck, C.H.; Peng, Y.; Lines, C.R.; Nessly, M.L.; Group, M.K.P.S. Growth hormone secretagogue MK-677: No clinical effect on AD progression in a randomized trial. *Neurology* **2008**, *71*, 1702–1708. [CrossRef] [PubMed]
241. Tian, J.; Wang, T.; Wang, Q.; Guo, L.; Du, H. MK0677, a Ghrelin Mimetic, Improves Neurogenesis but Fails to Prevent Hippocampal Lesions in a Mouse Model of Alzheimer's Disease Pathology. *J. Alzheimers Dis.* **2019**, *72*, 467–478. [CrossRef] [PubMed]
242. Gonzalez, S.L. Progesterone for the treatment of central nervous system disorders: The many signaling roads for a single molecule. *Neural. Regen. Res.* **2020**, *15*, 1846–1847. [CrossRef] [PubMed]
243. Qin, Y.; Chen, Z.; Han, X.; Wu, H.; Yu, Y.; Wu, J.; Liu, S.; Hou, Y. Progesterone attenuates Abeta(25-35)-induced neuronal toxicity via JNK inactivation and progesterone receptor membrane component 1-dependent inhibition of mitochondrial apoptotic pathway. *J. Steroid. Biochem. Mol. Biol.* **2015**, *154*, 302–311. [CrossRef]
244. Jolivel, V.; Brun, S.; Biname, F.; Benyounes, J.; Taleb, O.; Bagnard, D.; De Seze, J.; Patte-Mensah, C.; Mensah-Nyagan, A.G. Microglial Cell Morphology and Phagocytic Activity Are Critically Regulated by the Neurosteroid Allopregnanolone: A Possible Role in Neuroprotection. *Cells* **2021**, *10*, 698. [CrossRef] [PubMed]
245. Ye, J.N.; Chen, X.S.; Su, L.; Liu, Y.L.; Cai, Q.Y.; Zhan, X.L.; Xu, Y.; Zhao, S.F.; Yao, Z.X. Progesterone alleviates neural behavioral deficits and demyelination with reduced degeneration of oligodendroglial cells in cuprizone-induced mice. *PLoS ONE* **2013**, *8*, e54590. [CrossRef]
246. El-Etr, M.; Rame, M.; Boucher, C.; Ghoumari, A.M.; Kumar, N.; Liere, P.; Pianos, A.; Schumacher, M.; Sitruk-Ware, R. Progesterone and nestorone promote myelin regeneration in chronic demyelinating lesions of corpus callosum and cerebral cortex. *Glia* **2015**, *63*, 104–117. [CrossRef]
247. El-Etr, M.; Akwa, Y.; Rame, M.; Schumacher, M.; Sitruk-Ware, R. Nestorone((R)), a 19nor-progesterone derivative boosts remyelination in an animal model of demyelination. *CNS Neurosci. Ther.* **2021**, *27*, 464–469. [CrossRef]
248. Bouhrara, M.; Reiter, D.A.; Bergeron, C.M.; Zukley, L.M.; Ferrucci, L.; Resnick, S.M.; Spencer, R.G. Evidence of demyelination in mild cognitive impairment and dementia using a direct and specific magnetic resonance imaging measure of myelin content. *Alzheimers Dement.* **2018**, *14*, 998–1004. [CrossRef]
249. Yilmaz, C.; Karali, K.; Fodelianaki, G.; Gravanis, A.; Chavakis, T.; Charalampopoulos, I.; Alexaki, V.I. Neurosteroids as regulators of neuroinflammation. *Front. Neuroendocrinol.* **2019**, *55*, 100788. [CrossRef] [PubMed]
250. Aryanpour, R.; Pasbakhsh, P.; Zibara, K.; Namjoo, Z.; Beigi Boroujeni, F.; Shahbeigi, S.; Kashani, I.R.; Beyer, C.; Zendehdel, A. Progesterone therapy induces an M1 to M2 switch in microglia phenotype and suppresses NLRP3 inflammasome in a cuprizone-induced demyelination mouse model. *Int. Immunopharmacol.* **2017**, *51*, 131–139. [CrossRef]
251. Gutzeit, O.; Segal, L.; Korin, B.; Iluz, R.; Khatib, N.; Dabbah-Assadi, F.; Ginsberg, Y.; Fainaru, O.; Ross, M.G.; Weiner, Z.; et al. Progesterone Attenuates Brain Inflammatory Response and Inflammation-Induced Increase in Immature Myeloid Cells in a Mouse Model. *Inflammation* **2021**, *44*, 956–964. [CrossRef]
252. Litim, N.; Morissette, M.; Di Paolo, T. Effects of progesterone administered after MPTP on dopaminergic neurons of male mice. *Neuropharmacology* **2017**, *117*, 209–218. [CrossRef]

253. Castelnovo, L.F.; Thomas, P. Membrane progesterone receptor alpha (mPRalpha/PAQR7) promotes migration, proliferation and BDNF release in human Schwann cell-like differentiated adipose stem cells. *Mol. Cell Endocrinol.* **2021**, *531*, 111298. [CrossRef] [PubMed]
254. Acosta, J.I.; Hiroi, R.; Camp, B.W.; Talboom, J.S.; Bimonte-Nelson, H.A. An update on the cognitive impact of clinically-used hormone therapies in the female rat: Models, mazes, and mechanisms. *Brain Res.* **2013**, *1514*, 18–39. [CrossRef]
255. Singh, M.; Su, C. Progesterone-induced neuroprotection: Factors that may predict therapeutic efficacy. *Brain Res.* **2013**, *1514*, 98–106. [CrossRef]
256. Yu, L.; Moore, A.B.; Castro, L.; Gao, X.; Huynh, H.L.; Klippel, M.; Flagler, N.D.; Lu, Y.; Kissling, G.E.; Dixon, D. Estrogen Regulates MAPK-Related Genes through Genomic and Nongenomic Interactions between IGF-I Receptor Tyrosine Kinase and Estrogen Receptor-Alpha Signaling Pathways in Human Uterine Leiomyoma Cells. *J. Signal. Transduct.* **2012**, *2012*, 204236. [CrossRef]
257. Xu, S.; Yu, S.; Dong, D.; Lee, L.T.O. G Protein-Coupled Estrogen Receptor: A Potential Therapeutic Target in Cancer. *Front. Endocrinol.* **2019**, *10*, 725. [CrossRef] [PubMed]
258. Park, S.; Nozaki, K.; Smith, J.A.; Krause, J.S.; Banik, N.L. Cross-talk between IGF-1 and estrogen receptors attenuates intracellular changes in ventral spinal cord 4.1 motoneuron cells because of interferon-gamma exposure. *J. Neurochem.* **2014**, *128*, 904–918. [CrossRef] [PubMed]
259. Tian, J.; Berton, T.R.; Shirley, S.H.; Lambertz, I.; Gimenez-Conti, I.B.; DiGiovanni, J.; Korach, K.S.; Conti, C.J.; Fuchs-Young, R. Developmental stage determines estrogen receptor alpha expression and non-genomic mechanisms that control IGF-1 signaling and mammary proliferation in mice. *J. Clin. Investig.* **2012**, *122*, 192–204. [CrossRef]
260. Ding, Q.; Vaynman, S.; Akhavan, M.; Ying, Z.; Gomez-Pinilla, F. Insulin-like growth factor I interfaces with brain-derived neurotrophic factor-mediated synaptic plasticity to modulate aspects of exercise-induced cognitive function. *Neuroscience* **2006**, *140*, 823–833. [CrossRef]
261. Aberg, M.A.; Aberg, N.D.; Palmer, T.D.; Alborn, A.M.; Carlsson-Skwirut, C.; Bang, P.; Rosengren, L.E.; Olsson, T.; Gage, F.H.; Eriksson, P.S. IGF-I has a direct proliferative effect in adult hippocampal progenitor cells. *Mol. Cell Neurosci.* **2003**, *24*, 23–40. [CrossRef]
262. Nieto-Estevez, V.; Oueslati-Morales, C.O.; Li, L.; Pickel, J.; Morales, A.V.; Vicario-Abejon, C. Brain Insulin-Like Growth Factor-I Directs the Transition from Stem Cells to Mature Neurons During Postnatal/Adult Hippocampal Neurogenesis. *Stem Cells* **2016**, *34*, 2194–2209. [CrossRef]
263. Cohen, E.; Paulsson, J.F.; Blinder, P.; Burstyn-Cohen, T.; Du, D.; Estepa, G.; Adame, A.; Pham, H.M.; Holzenberger, M.; Kelly, J.W.; et al. Reduced IGF-1 signaling delays age-associated proteotoxicity in mice. *Cell* **2009**, *139*, 1157–1169. [CrossRef]
264. Yuan, L.J.; Zhang, M.; Chen, S.; Chen, W.F. Anti-inflammatory effect of IGF-1 is mediated by IGF-1R cross talk with GPER in MPTP/MPP(+)-induced astrocyte activation. *Mol. Cell Endocrinol.* **2021**, *519*, 111053. [CrossRef] [PubMed]
265. Smith, C.C.; McMahon, L.L. Estrogen-induced increase in the magnitude of long-term potentiation occurs only when the ratio of NMDA transmission to AMPA transmission is increased. *J. Neurosci.* **2005**, *25*, 7780–7791. [CrossRef] [PubMed]
266. Clements, L.; Harvey, J. Activation of oestrogen receptor alpha induces a novel form of LTP at hippocampal temporoammonic-CA1 synapses. *Br. J. Pharmacol.* **2020**, *177*, 642–655. [CrossRef] [PubMed]
267. Lewis, M.C.; Kerr, K.M.; Orr, P.T.; Frick, K.M. Estradiol-induced enhancement of object memory consolidation involves NMDA receptors and protein kinase A in the dorsal hippocampus of female C57BL/6 mice. *Behav. Neurosci.* **2008**, *122*, 716–721. [CrossRef]
268. Vedder, L.C.; Smith, C.C.; Flannigan, A.E.; McMahon, L.L. Estradiol-induced increase in novel object recognition requires hippocampal NR2B-containing NMDA receptors. *Hippocampus* **2013**, *23*, 108–115. [CrossRef]
269. Snyder, M.A.; Cooke, B.M.; Woolley, C.S. Estradiol potentiation of NR2B-dependent EPSCs is not due to changes in NR2B protein expression or phosphorylation. *Hippocampus* **2011**, *21*, 398–408. [CrossRef] [PubMed]
270. Tanaka, M.; Sokabe, M. Bidirectional modulatory effect of 17beta-estradiol on NMDA receptors via ERalpha and ERbeta in the dentate gyrus of juvenile male rats. *Neuropharmacology* **2013**, *75*, 262–273. [CrossRef]
271. Boulware, M.I.; Weick, J.P.; Becklund, B.R.; Kuo, S.P.; Groth, R.D.; Mermelstein, P.G. Estradiol activates group I and II metabotropic glutamate receptor signaling, leading to opposing influences on cAMP response element-binding protein. *J. Neurosci.* **2005**, *25*, 5066–5078. [CrossRef]
272. Benquet, P.; Gee, C.E.; Gerber, U. Two distinct signaling pathways upregulate NMDA receptor responses via two distinct metabotropic glutamate receptor subtypes. *J. Neurosci.* **2002**, *22*, 9679–9686. [CrossRef]
273. Sahab-Negah, S.; Hajali, V.; Moradi, H.R.; Gorji, A. The Impact of Estradiol on Neurogenesis and Cognitive Functions in Alzheimer's Disease. *Cell Mol. Neurobiol.* **2020**, *40*, 283–299. [CrossRef] [PubMed]
274. Sha, S.; Hong, J.; Qu, W.J.; Lu, Z.H.; Li, L.; Yu, W.F.; Chen, L. Sex-related neurogenesis decrease in hippocampal dentate gyrus with depressive-like behaviors in sigma-1 receptor knockout mice. *Eur. Neuropsychopharmacol.* **2015**, *25*, 1275–1286. [CrossRef]
275. Liu, S.B.; Zhao, M.G. Neuroprotective effect of estrogen: Role of nonsynaptic NR2B-containing NMDA receptors. *Brain Res. Bull.* **2013**, *93*, 27–31. [CrossRef]
276. Androvicova, R.; Pfaus, J.G.; Ovsepian, S.V. Estrogen pendulum in schizophrenia and Alzheimer's disease: Review of therapeutic benefits and outstanding questions. *Neurosci. Lett.* **2021**, *759*, 136038. [CrossRef] [PubMed]
277. Watanabe, T.; Inoue, S.; Hiroi, H.; Orimo, A.; Muramatsu, M. NMDA receptor type 2D gene as target for estrogen receptor in the brain. *Brain Res. Mol. Brain Res.* **1999**, *63*, 375–379. [CrossRef]

278. Ikeda, K.; Fukushima, T.; Ogura, H.; Tsukui, T.; Mishina, M.; Muramatsu, M.; Inoue, S. Estrogen regulates the expression of N-methyl-D-aspartate (NMDA) receptor subunit epsilon 4 (Grin2d), that is essential for the normal sexual behavior in female mice. *FEBS Lett.* **2010**, *584*, 806–810. [CrossRef] [PubMed]
279. Hynd, M.R.; Scott, H.L.; Dodd, P.R. Differential expression of N-methyl-D-aspartate receptor NR2 isoforms in Alzheimer's disease. *J. Neurochem.* **2004**, *90*, 913–919. [CrossRef] [PubMed]
280. Clayton, D.A.; Browning, M.D. Deficits in the expression of the NR2B subunit in the hippocampus of aged Fisher 344 rats. *Neurobiol. Aging* **2001**, *22*, 165–168. [CrossRef]
281. Magnusson, K.R.; Kresge, D.; Supon, J. Differential effects of aging on NMDA receptors in the intermediate versus the dorsal hippocampus. *Neurobiol. Aging* **2006**, *27*, 324–333. [CrossRef]
282. Jullienne, A.; Montagne, A.; Orset, C.; Lesept, F.; Jane, D.E.; Monaghan, D.T.; Maubert, E.; Vivien, D.; Ali, C. Selective inhibition of GluN2D-containing N-methyl-D-aspartate receptors prevents tissue plasminogen activator-promoted neurotoxicity both in vitro and in vivo. *Mol. Neurodegener.* **2011**, *6*, 68. [CrossRef] [PubMed]
283. Lefort, R. Reversing synapse loss in Alzheimer's disease: Rho-guanosine triphosphatases and insights from other brain disorders. *Neurotherapeutics* **2015**, *12*, 19–28. [CrossRef] [PubMed]
284. Bai, N.; Hayashi, H.; Aida, T.; Namekata, K.; Harada, T.; Mishina, M.; Tanaka, K. Dock3 interaction with a glutamate-receptor NR2D subunit protects neurons from excitotoxicity. *Mol. Brain* **2013**, *6*, 22. [CrossRef] [PubMed]
285. Namekata, K.; Kimura, A.; Kawamura, K.; Guo, X.; Harada, C.; Tanaka, K.; Harada, T. Dock3 attenuates neural cell death due to NMDA neurotoxicity and oxidative stress in a mouse model of normal tension glaucoma. *Cell Death Differ.* **2013**, *20*, 1250–1256. [CrossRef]
286. Liu, S.B.; Zhang, N.; Guo, Y.Y.; Zhao, R.; Shi, T.Y.; Feng, S.F.; Wang, S.Q.; Yang, Q.; Li, X.Q.; Wu, Y.M.; et al. G-protein-coupled receptor 30 mediates rapid neuroprotective effects of estrogen via depression of NR2B-containing NMDA receptors. *J. Neurosci.* **2012**, *32*, 4887–4900. [CrossRef] [PubMed]
287. Roberts, H.; Lethaby, A. Phytoestrogens for menopausal vasomotor symptoms: A Cochrane review summary. *Maturitas* **2014**, *78*, 79–81. [CrossRef]
288. Zand, R.S.; Jenkins, D.J.; Diamandis, E.P. Steroid hormone activity of flavonoids and related compounds. *Breast Cancer Res. Treat.* **2000**, *62*, 35–49. [CrossRef]
289. Oseni, T.; Patel, R.; Pyle, J.; Jordan, V.C. Selective estrogen receptor modulators and phytoestrogens. *Planta Med.* **2008**, *74*, 1656–1665. [CrossRef]
290. Clement, Y.N.; Onakpoya, I.; Hung, S.K.; Ernst, E. Effects of herbal and dietary supplements on cognition in menopause: A systematic review. *Maturitas* **2011**, *68*, 256–263. [CrossRef]
291. Lim, D.W.; Lee, C.; Kim, I.H.; Kim, Y.T. Anti-inflammatory effects of total isoflavones from Pueraria lobata on cerebral ischemia in rats. *Molecules* **2013**, *18*, 10404–10412. [CrossRef] [PubMed]
292. Ganai, A.A.; Khan, A.A.; Malik, Z.A.; Farooqi, H. Genistein modulates the expression of NF-kappaB and MAPK (p-38 and ERK1/2), thereby attenuating d-Galactosamine induced fulminant hepatic failure in Wistar rats. *Toxicol. Appl. Pharmacol.* **2015**, *283*, 139–146. [CrossRef]
293. Yu, J.; Bi, X.; Yu, B.; Chen, D. Isoflavones: Anti-Inflammatory Benefit and Possible Caveats. *Nutrients* **2016**, *8*, 361. [CrossRef] [PubMed]
294. Yoon, G.A.; Park, S. Antioxidant action of soy isoflavones on oxidative stress and antioxidant enzyme activities in exercised rats. *Nutr Res. Pract.* **2014**, *8*, 618–624. [CrossRef] [PubMed]
295. Wei, J.; Yang, F.; Gong, C.; Shi, X.; Wang, G. Protective effect of daidzein against streptozotocin-induced Alzheimer's disease via improving cognitive dysfunction and oxidative stress in rat model. *J. Biochem. Mol. Toxicol.* **2019**, *33*, e22319. [CrossRef]
296. Chatterjee, G.; Roy, D.; Khemka, V.K.; Chattopadhyay, M.; Chakrabarti, S. Genistein, the Isoflavone in Soybean, Causes Amyloid Beta Peptide Accumulation in Human Neuroblastoma Cell Line: Implications in Alzheimer's Disease. *Aging Dis.* **2015**, *6*, 456–465. [CrossRef]
297. Bagheri, M.; Joghataei, M.T.; Mohseni, S.; Roghani, M. Genistein ameliorates learning and memory deficits in amyloid beta(1-40) rat model of Alzheimer's disease. *Neurobiol. Learn. Mem.* **2011**, *95*, 270–276. [CrossRef]
298. Xi, Y.D.; Li, X.Y.; Ding, J.; Yu, H.L.; Ma, W.W.; Yuan, L.H.; Wu, J.; Xiao, R. Soy isoflavone alleviates Abeta1-42-induced impairment of learning and memory ability through the regulation of RAGE/LRP-1 in neuronal and vascular tissue. *Curr. Neurovasc. Res.* **2013**, *10*, 144–156. [CrossRef] [PubMed]
299. Devi, K.P.; Shanmuganathan, B.; Manayi, A.; Nabavi, S.F.; Nabavi, S.M. Molecular and Therapeutic Targets of Genistein in Alzheimer's Disease. *Mol. Neurobiol.* **2017**, *54*, 7028–7041. [CrossRef] [PubMed]
300. Son, S.H.; Do, J.M.; Yoo, J.N.; Lee, H.W.; Kim, N.K.; Yoo, H.S.; Gee, M.S.; Kim, J.H.; Seong, J.H.; Inn, K.S.; et al. Identification of ortho catechol-containing isoflavone as a privileged scaffold that directly prevents the aggregation of both amyloid beta plaques and tau-mediated neurofibrillary tangles and its in vivo evaluation. *Bioorg. Chem.* **2021**, *113*, 105022. [CrossRef]
301. Lazennec, G.; Bresson, D.; Lucas, A.; Chauveau, C.; Vignon, F. ER beta inhibits proliferation and invasion of breast cancer cells. *Endocrinology* **2001**, *142*, 4120–4130. [CrossRef] [PubMed]
302. Uddin, M.S.; Hasana, S.; Hossain, M.F.; Islam, M.S.; Behl, T.; Perveen, A.; Hafeez, A.; Ashraf, G.M. Molecular Genetics of Early- and Late-Onset Alzheimer's Disease. *Curr. Gene Ther.* **2021**, *21*, 43–52. [CrossRef]
303. Selkoe, D.J. Alzheimer's disease: Genotypes, phenotypes, and treatments. *Science* **1997**, *275*, 630–631. [CrossRef] [PubMed]

304. Rebeck, G.W.; Reiter, J.S.; Strickland, D.K.; Hyman, B.T. Apolipoprotein E in sporadic Alzheimer's disease: Allelic variation and receptor interactions. *Neuron* **1993**, *11*, 575–580. [CrossRef]
305. Strittmatter, W.J.; Saunders, A.M.; Schmechel, D.; Pericak-Vance, M.; Enghild, J.; Salvesen, G.S.; Roses, A.D. Apolipoprotein E: High-avidity binding to beta-amyloid and increased frequency of type 4 allele in late-onset familial Alzheimer disease. *Proc. Natl. Acad. Sci. USA* **1993**, *90*, 1977–1981. [CrossRef]
306. Castellano, J.M.; Kim, J.; Stewart, F.R.; Jiang, H.; DeMattos, R.B.; Patterson, B.W.; Fagan, A.M.; Morris, J.C.; Mawuenyega, K.G.; Cruchaga, C.; et al. Human apoE isoforms differentially regulate brain amyloid-beta peptide clearance. *Sci. Transl. Med.* **2011**, *3*, 89ra57. [CrossRef] [PubMed]
307. Rosenberg, R.N.; Lambracht-Washington, D.; Yu, G.; Xia, W. Genomics of Alzheimer Disease: A Review. *JAMA Neurol.* **2016**, *73*, 867–874. [CrossRef]
308. Chen, L.H.; Fan, Y.H.; Kao, P.Y.; Ho, D.T.; Ha, J.C.; Chu, L.W.; Song, Y.Q. Genetic Polymorphisms in Estrogen Metabolic Pathway Associated with Risks of Alzheimer's Disease: Evidence from a Southern Chinese Population. *J. Am. Geriatr. Soc.* **2017**, *65*, 332–339. [CrossRef]
309. Song, Y.; Lu, Y.; Liang, Z.; Yang, Y.; Liu, X. Association between rs10046, rs1143704, rs767199, rs727479, rs1065778, rs1062033, rs1008805, and rs700519 polymorphisms in aromatase (CYP19A1) gene and Alzheimer's disease risk: A systematic review and meta-analysis involving 11,051 subjects. *Neurol. Sci.* **2019**, *40*, 2515–2527. [CrossRef]
310. Andrews, S.J.; Fulton-Howard, B.; Goate, A. Interpretation of risk loci from genome-wide association studies of Alzheimer's disease. *Lancet Neurol.* **2020**, *19*, 326–335. [CrossRef]
311. Lin, E.; Lin, C.H.; Lane, H.Y. Deep Learning with Neuroimaging and Genomics in Alzheimer's Disease. *Int. J. Mol. Sci.* **2021**, *22*. [CrossRef] [PubMed]
312. Hersi, M.; Irvine, B.; Gupta, P.; Gomes, J.; Birkett, N.; Krewski, D. Risk factors associated with the onset and progression of Alzheimer's disease: A systematic review of the evidence. *Neurotoxicology* **2017**, *61*, 143–187. [CrossRef] [PubMed]
313. Kim, C. Does menopause increase diabetes risk? Strategies for diabetes prevention in midlife women. *Womens Health* **2012**, *8*, 155–167. [CrossRef] [PubMed]
314. Tini, G.; Scagliola, R.; Monacelli, F.; La Malfa, G.; Porto, I.; Brunelli, C.; Rosa, G.M. Alzheimer's Disease and Cardiovascular Disease: A Particular Association. *Cardiol. Res. Pract.* **2020**, *2020*, 2617970. [CrossRef] [PubMed]

MDPI
St. Alban-Anlage 66
4052 Basel
Switzerland
Tel. +41 61 683 77 34
Fax +41 61 302 89 18
www.mdpi.com

International Journal of Molecular Sciences Editorial Office
E-mail: ijms@mdpi.com
www.mdpi.com/journal/ijms